DESIGN WITH MICROPROCESSORS
FOR MECHANICAL ENGINEERS

McGraw-Hill Series in Mechanical Engineering

Consulting Editors
Jack P. Holman, Southern Methodist University
John R. Lloyd, Michigan State University

Anderson: *Modern Compressible Flow: With Historical Perspective*
Arora: *Introduction to Optimum Design*
Bray and Stanley: *Nondestructive Evaluation: A Tool for Design, Manufacturing, and Service*
Culp: *Principles of Energy Conversion*
Dally: *Packaging of Electronic Systems: A Mechanical Engineering Approach*
Dieter: *Engineering Design: A Materials and Processing Approach*
Eckert and Drake: *Analysis of Heat and Mass Transfer*
Edwards and McKee: *Fundamentals of Mechanical Component Design*
Heywood: *Internal Combustion Engine Fundamentals*
Hinze: *Turbulence*
Howell and Buckius: *Fundamentals of Engineering Thermodynamics: English/SI Version*
Hutton: *Applied Mechanical Vibrations*
Juvinall: *Engineering Considerations of Stress, Strain, and Strength*
Kane and Levinson: *Dynamics: Theory and Applications*
Kays and Crawford: *Convective Heat and Mass Transfer*
Kimbrell: *Kinematics Analysis and Synthesis*
Martin: *Kinematics and Dynamics of Machines*
Norton: *Design of Machinery: An Introduction to the Synthesis and Analysis of Mechanisms and Machines*
Phelan: *Fundamentals of Mechanical Design*
Raven: *Automatic Control Engineering*
Rosenberg and Karnopp: *Introduction to Physics*
Schlichting: *Boundary-Layer Theory*
Shames: *Mechanics of Fluids*
Sherman: *Viscous Flow*
Shigley: *Kinematic Analysis of Mechanisms*
Shigley and Uicker: *Theory of Machines and Mechanisms*
Shigley and Mischke: *Mechanical Engineering Design*
Stiffler: *Design with Microprocessors for Mechanical Engineers*
Stoecker and Jones: *Refrigeration and Air Conditioning*
Vanderplaats: *Numerical Optimization: Techniques for Engineering Design, with Applications*
White: *Viscous Fluid Flow*
Zeid: *CAD/CAM Theory and Practice*

Also Available from McGraw-Hill

Schaum's Outline Series in Mechanical Engineering

Most outlines include basic theory, definitions, and hundreds of solved problems and supplementary problems with answers.

Titles on the Current List Include:

Acoustics
Basic Equations of Engineering
Continuum Mechanics
Engineering Economics
Engineering Mechanics, 4th edition
Fluid Dynamics, 2d edition
Fluid Mechanics & Hydraulics, 2d edition
Heat Transfer
Introduction to Engineering Calculations
Lagrangian Dynamics
Machine Design
Mathematical Handbook of Formulas & Tables
Mechanical Vibrations
Operations Research
Statics & Mechanics of Materials
Strength of Materials, 2d edition
Theoretical Mechanics
Thermodynamics, 2d edition

Schaum's Solved Problems Books

Each title in this series is a complete and expert source of solved problems containing thousands of problems with worked out solutions.

Related Titles on the Current List Include:

3000 Solved Problems in Calculus
2500 Solved Problems in Differential Equations
2500 Solved Problems in Fluid Mechanics and Hydraulics
1000 Solved Problems in Heat Transfer
3000 Solved Problems in Linear Algebra
2000 Solved Problems in Mechanical Engineering Thermodynamics
2000 Solved Problems in Numerical Analysis
700 Solved Problems in Vector Mechanics for Engineers: Dynamics
800 Solved Problems in Vector Mechanics for Engineers: Statics

Available at your College Bookstore. A complete list of Schaum titles may be obtained by writing to: Schaum Division
McGraw-Hill, Inc.
Princeton Road, S-1
Hightstown, NJ 08520

DESIGN WITH MICROPROCESSORS FOR MECHANICAL ENGINEERS

A. Kent Stiffler

Mechanical and Nuclear Engineering Department
Mississippi State University

McGraw-Hill, Inc.

New York St. Louis San Francisco Auckland Bogotá Caracas Lisbon London Madrid
Mexico Milan Montreal New Delhi Paris San Juan Singapore Sydney Tokyo Toronto

This book was set in Times Roman by Beacon Graphics Corporation.
The editors were John J. Corrigan and James W. Bradley;
the production supervisor was Louise Karam.
The cover was designed by John Hite.
R. R. Donnelley & Sons Company was printer and binder.

DESIGN WITH MICROPROCESSORS FOR MECHANICAL ENGINEERS

1 2 3 4 5 6 7 8 9 0 DOC DOC 9 0 9 8 7 6 5 4 3 2 1

ISBN 0-07-061374-5

Library of Congress Cataloging-in-Publication Data

Stiffler, A. Kent.
  Design with microprocessors for mechanical engineers/A. Kent Stiffler.
    p.   cm.—(McGraw-Hill series in mechanical engineering)
  Includes index.
  ISBN 0-07-061374-5
  1. Microprocessors.   2. Programmable controllers—Design and construction.
I. Title.   II. Series.
TJ223.M53S75   1992
621.39' 16—dc20   91-26487

ABOUT THE AUTHOR

A. Kent Stiffler is a Hearin-Hess professor and a member of the mechanical engineering faculty at Mississippi State University. Professor Stiffler obtained a bachelor of science degree from Lehigh University and both a master of science and doctorate from Pennsylvania State University, all in mechanical engineering. His teaching areas include system dynamics, automatic controls, and microprocessors in design. Professor Stiffler has conducted research for the National Science Foundation, the National Aeronautics and Space Administration, the Air Force Office of Scientific Research, and the Naval Research and Development Center. He won the 1986 award for outstanding contribution in research from the Southeastern Section of the American Society of Engineering Education. He is a member of the American Society of Mechanical Engineers.

CONTENTS

PREFACE

The microprocessor (MPU) has radically changed the concept of design. Consumers are now inundated with automobiles and appliances that have microprocessors as built-in controllers. *Mechatronics* is the Japanese word for this integration of mechanical and electronic components. This book is written so that mechanical engineers can become familiar with the microprocessor as a design tool at the chip/board level. For mechanical engineers who are using personal computers (PCs) in data acquisition and control, the text gives new understanding to the hardware interface between PCs and end components such as motors.

This textbook is intended for upper division undergraduates in nonelectrical engineering fields and would also be suitable for electrical technology programs. The material can be covered in a typical three-hour lecture or two-hour lecture, plus a three-hour laboratory per week. A laboratory concurrent with the lectures is highly recommended. It is assumed that the reader has been exposed to gates and gate logic; however, gate features are reviewed in Chapter 9. All pertinent electronics are explained when they become relevant in the text.

Most books on microprocessors cover the essentials of microcomputers: MPU architecture, assembly language, and memory/port communication with the MPU. They are written by electrical engineers who are PC oriented and think of interface hardware in terms of printers and disk drives. The hardware in mechatronics, such as switches, displays, sensors, analog converters, timers, motors, receive only cursory treatment, if that. The shortcomings are quickly apparent when an individual actually attempts to implement a design in the laboratory.

Design with Microprocessors for Mechanical Engineers presents a *thorough* explanation of end components and the integrated circuits (ICs) that are necessary to interface the MPU at the breadboard level. A description of representative IC pin assignments removes the mystery surrounding their function. Pin-by-pin wiring diagrams for common MPU-driven subsystems are presented. Many of the newer, smart ICs that combine logic and power drive to simplify the interfacing task are introduced. Finally, assembly language routines are given to bring these subsystems to life.

The book is built around the Motorola 6802 microprocessor. Personal computer users may view the 6802 as outdated when compared to the Motorola 68000/68020 and the Intel 8088/80386. However, our goal is control board design for consumer products, etc. The 8-bit 6802 (and its first cousin, the Apple 6502) is simple enough for the beginner to grasp quickly. Yet it is powerful enough for mechatronic applications where speed is usually not important, memory needs are few, and cost is the driving factor. The powerful MPUs have a more complicated architecture and language set which require a large investment in time to learn programming.

Design with Microprocessors for Mechanical Engineers places the emphasis on hardware and interfacing. In this respect the book is an excellent reference for the PC user. A logical niche for the mechanical engineer is microcontrollers. These microcomputers-on-a-chip, which combine simplified MPUs with memory, ports, timers, and analog-to-digital (A/D) converters onto a single chip, are the control unit in half the smart products. The 6802 MPU is a natural bridge to the highly popular Motorola 6805 and 6811 micro-controllers. The 6805 is discussed thoroughly in Chapter 17.

Chapters 3 through 8 constitute the digital side of the hardware and programming. The material can be covered in approximately five or six weeks, leaving the bulk of the semester to hardware interfacing. Chapter 3 covers the essentials of MPU architecture. Chapters 4 and 5 explain the details of assembly language, showing that the fear of assembly language is groundless. Students can be writing reasonably sophisticated programs in a short time.

Chapter 6 on memory is most important because it explains how the MPU communicates with a given chip on the bus. Ports are the gateway to the world outside the MPU/memory system. In Chapter 7 basic ports, including D flip-flops, and the versatile peripheral interface adapter (PIA) are presented. Interrupts in Chapter 8 can be postponed until later in the course. They are introduced at this point for continuity with MPU architecture and the PIA.

Chapter 9 is a review of diodes, general-purpose transistors, and gate properties. The main impetus of this chapter is to explain the drive capability of TTL and CMOS logic which underlies the output of most chips. The chapter is for reference and not part of the author's lecture sequence, although a short summary on gate drive is given during class sessions. In Chapters 10 through 15 the details of interfacing switches, displays, stepping motors, DC motors, analog timers, and power electronics are shown.

Each chapter after Chapter 7 stands alone. The instructor can select or omit any chapter regardless of order. The information on motors and power electronics cannot be found elsewhere in a single source. Interfacing DC motors in Chapter 12 necessarily contains elements of feedback control theory. Yet much can be gained by the student even though his or her control background may be limited. Chapter 13 on interfacing analog is equally divided among op amps, digital-to-analog converters, and analog-to-digital converters. Timers, examined in Chapter 14, play an important role in automobile engine control.

Chapter 15 on power electronics was written to pull together various sources on boosting TTL/CMOS power levels to drive motors, incandescent lights, heaters, etc., with both DC and AC power. Some aspects of power are introduced in the motor chapters, but Chapter 15 expands on the topic. The discussion of power transistors duplicates certain material in Chapter 9 on general-purpose transistors.

Chapter 16 brings the board design process together with a project to build a digital readout bathroom scale. Consideration is given to power supply, hardware design procedure, wiring tips, software design procedure, and troubleshooting. The project is repeated in Chapter 17 with the 6805 microcontroller, illustrating reduced wiring with microcontrollers.

Laboratory assignments can implement many of the wiring diagrams found throughout the book. The author is presently using the Heathkit 3400 trainer as the microprocessor system to interface external memory or PIA ports. Displays, motors, A/D, etc., are then interfaced to the permanent external port. The trainer memory holds the program for single-step software debugging. Hand assembly is no problem because the drive programs are relatively short. The user should be aware that a bus extender IC is placed between the Heathkit MPU data bus and its access pins. The extender is enabled through common sockets labeled RE. Thus, the external chip (PIA) enabled from the external decoder must be NANDed with R/W to enable the bus extender at the same time the external chip is enabled. During the last two or three weeks of the course students design and test their own microprocessor-based system on a breadboard.

An alternate approach to the laboratory teaches the microcontroller concurrently with the microprocessor in lecture. A PC-driven development system is introduced with the assembly language during the first five weeks. With a port already on-board the microcontroller, laboratory exercises could follow the book's interfacing chapters.

The author would also like to thank all parties that made a difference: C.T. Carley, department head, who offered constant encouragement; the Mississippi Chemical Corporation, who contributed funds at a critical juncture; my colleagues who also believed; and my wife Elizabeth and daughter Amy, who understood when I said, "I'm too busy."

McGraw-Hill and the author would like to thank the following reviewers for their many helpful comments and suggestions: Behnam Bahr, Wichita State University; Mark S. Darlow, Rensselaer Polytechnic Institute; Francis Donovan, University of South Alabama; Robert B. Keller, University of Michigan; John Lamancusa, Pennsylvania State University; Stephen McNeil, University of South Carolina; Glenn Masada, University of Texas at Austin; Andrzej Olas, Oregon State University; Rahmat Shoureshi, Purdue University; and Frederic L. Swern, Stevens Institute of Technology.

A. Kent Stiffler

Note Added in Proof

Motorola has recently added a new member to the 6805 family, which could be ranked as the smartest power IC yet. The 68HC05H2 CMOS microcontroller has eight DMOSFET ports rated at 300 mA and 6 V. Four of these power devices can be on at once to control up to 7.2 W. Applications include direct control of small motors, relays, and lamps in consumer, industrial, or automotive products. The ports can drive external power MOSFET transistors for high-power applications. EPROM versions are available. Future plans include the ability of each port to handle up to 1 A and/or 40 V continuously.

DESIGN WITH MICROPROCESSORS
FOR MECHANICAL ENGINEERS

CHAPTER
1

INTRODUCTION

1.1 MICROPROCESSORS IN DESIGN

Electronics is changing society and the way we think about engineering practice to the extent that it is called a second industrial revolution. The cause of this new revolution is the microprocessor—a nail head–size piece of silicon with thousands of transistors which can be programmed to control our environment. Mechanical engineers now use personal computers for routine calculations and data acquisition systems. Workstations with more powerful microprocessors can draft designs and optimize them for stresses and dynamics: computer-aided design (CAD). The finished design can then be realized through microprocessor-controlled machine tools: computer-aided manufacturing (CAM). Another aspect of CAM is the automated factory which relies on programmable controllers and robots. In all these applications the primary role of the mechanical engineer is software development or software implementation.

Mechatronics

Another path taken by microprocessor-based designs is mechatronics. *Mechatronics* is the science that integrates mechanical devices with electronic controls. The word was proposed in the early 1970s by Yaskawa Electric Corporation. Mechatronics is at the heart of precision consumer products in which the Japanese excel, such as cameras, videotape machines, and cassette players. Integrated circuit (IC) costs have dropped to the point where microprocessors and support chips can be purchased for a few dollars. Microprocessors

TABLE 1.1
Products in the home with embedded microprocessors

Baking timer	Lawn sprinkler
Bathroom scale	Microwave oven
Blood pressure recorder	Refrigerator
Camcorder	Sewing machine
Camera	Stereo
Carpenter's level	Studfinder
Clothes dryer	Tape measure
Clothes washer	Telephone
Dishwasher	Television
Exercise equipment	Thermostat
Fishfinder	Typewriter
Food processor	Video recorder
HVAC system	Walkman

are finding their way into most home appliances as an embedded controller. Table 1.1 gives a partial list of these marvels. Many of these items are old standards that have been upgraded in features and performance but remain cost-competitive. Bathroom scales with digital readout have driven their mechanical relatives from the marketplace. Other items, such as camcorders, would not be economically possible without microprocessors.

Microprocessors in mechatronics are widely used for distributed control of larger machines. Many robots are equipped with motor position controllers for each axis of rotation. Trajectory calculations are performed by a master computer, and coordinates are continually downloaded to the controllers for execution. The Boeing 747 is reported to have over 1000 microprocessor-based subsystems.

A decade ago automobiles were primarily mechanical products with electrical applications limited to ignition, lighting, and entertainment systems. Auto engineers used the first microprocessors as a way to help meet emerging federal fuel economy and automotive emission requirements. The electronic engine control system continually measures the oxygen content of the exhaust to establish air-fuel mixtures. Crankshaft position along with manifold pressure and coolant temperature measurements allow the microprocessor to calculate the ignition timing. Table 1.2 shows microprocessor applications that are now available for the car buyer.

TABLE 1.2
Automobile items with microprocessor-based controllers

Now	
Active suspension	Four-wheel drive
Antiskid brakes	Fuel/trip computer
Climate control	Level control
Cruise control	Odometer display
Engine control	Speedometer display
Entertainment system	Transmission
Near future	
Collision avoidance	Electronic muffler
Drive by wire	Navigation

In the near future, electronic subsystems will be integrated to permit power train control or "drive by wire." Engine parameters will be combined with wheel speed/slip data to determine transmission shift scheduling and torque control. Electronic mufflers will wipe out engine sound by generating audio waveforms in opposite phase. Road maps, which continually pinpoint the location of both the vehicle and its destination, will be displayed on an instrument panel screen.

Some high-priced cars already have 30 or more separate electronic modules. By the mid-1990s, some observers predict, the average electronic content per vehicle will be worth $2000.

Why MEs?

Before 1980 the idea of an automobile suspension system which could instantaneously change the ride characteristics was unthinkable, at least for a mechanical engineer (ME). Since their invention, weighing scales have consisted of levers, weights, and springs. Today the mechanical engineering graduate would be hard pressed to recognize any components under the cover. And tomorrow? Colt Industries has patented a gun with a display that indicates if the gun is loaded and how many shots remain in the chamber. Its microprocessor also fires the bullet. Such a gun could be coded to fire only for the owner. It will be safer, less expensive, and more reliable. The imagination knows no bounds. Virtually every new mechanical design should consider a microprocessor subsystem as a possible solution. Old designs should be reevaluated for an electronic option if the product is to remain competitive.

International competition is forcing economics on manufacturers. At one time the electronic section of a mechanical device would be subcontracted. Now the mechanical functions and electronics are often designed together in-house. Electrical engineers (EEs) generally have a better understanding of mechanical systems than MEs have of electronic systems. As a result, MEs are less likely to be selected for overall project design responsibility.

Yesterday's sewing machine consisted of a single motor driving a series of cams. Today the cam action is replaced by a microprocessor driving several stepping motors. Mechanical design should evolve from the ground level to the finished product under the direction of an ME. What is the nature of the design if he or she knows only cams and gears? At a minimum the ME should be familiar with the capabilities of a microprocessor and the support electronics for digital control in mechatronics.

Mechatronics System

Microprocessor applications in consumer products, appliances, or factory machines consist of a small board that contains the microprocessor and support chips for controlling the mechanism. A general configuration of the system hardware is shown in Fig. 1.1. The intelligence is composed of a microprocessor and a memory for the program. They are separated from the real world by an interface that permits input and output of both binary and analog signals. Input devices are generally sensors or switches. Sensor signals are conditioned (amplified) to match the voltage levels of the interface. Output devices are switches, displays, and actuators. Actuators are essentially motors or solenoids. Signals

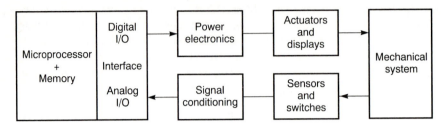

FIGURE 1.1
General configuration of a mechatronics system.

from the interface must be boosted in power to drive most output devices. In some cases, particularly robots, a computer sends data to or receives data from the controller board.

1.2   SOME EXAMPLES

CARPENTER'S LEVEL. Wedge Innovations of Sunnyvale, California, sells a computerized carpenter's level called *SmartLevel* shown in Fig. 1.2. Unlike the traditional level which is operated by positioning a bubble in an arched vial, SmartLevel has a liquid crystal display (LCD) for all 360° of inclination. A bubble indicates only level or plumb conditions and can be difficult to read. SmartLevel can be viewed up to 10 ft away. The readout flips when the tool is turned upside-down.

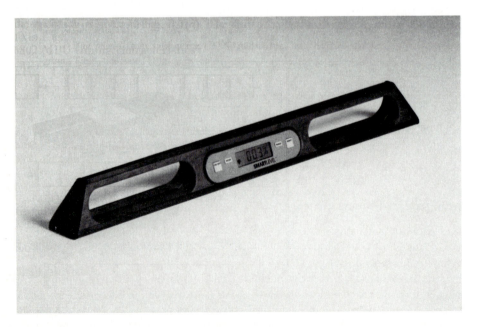

FIGURE 1.2
SmartLevel by Wedge Innovations. (*Courtesy of Wedge Innovations, copyright 1990.*)

SmartLevel contains a single patented solid-state sensor. The microprocessor converts the measurement into four different modes: digital angle (degrees of slope), percent of slope (grade), inches per foot of run (pitch), or an analog electronic bubble.

Bubble vial levels can get knocked out of balance. By pushing a button, SmartLevel can be recalibrated by the user to provide consistent correct measurements. It features three different precision settings, ranging to within one-tenth a degree or one-fifth a degree for exacting work and to within one-half a degree for approximate measurements.

BATHROOM SCALE. Today's bathroom scale features the basic components of many electronic consumer products: sensor, computer, and display. In this case a strain gauge generates a millivolt signal proportional to force (weight). The signal is amplified to voltage level for an analog-to-digital converter. A microprocessor performs various arithmetic and timing calculations before displaying the result. Two calculations are important. First, a "look-up table" routine determines the binary code for each display unit, lighting the proper segments to represent the number/letter. Second, only an average or several average weights are displayed near the end of the weighing cycle because the strain gauge output changes with slight shifting of body weight. Fluctuating readings can be disconcerting to the customer.

The Sunbeam Model 9360 Promptor is shown in Fig. 1.3 with the cover removed exposing the electronics. The microcontroller is under the display. Select the pound or kilogram mode with the position switch located on the right-hand side. Push the start button on the front. WAIT will appear on the display but do not step on; it is adjusting to zero setting. The display will change to STEP and a beep will sound when the scale is ready to weigh. After you step on the scale the display will change to WAIT (averaging?) for several seconds. A beep will sound again and your weight will be displayed. Shut off is automatic.

An interesting feature of the scale is object weighing. Step on the scale and have someone push the start button. WAIT will be displayed while it zeros for your weight. When STEP appears and a beep sounds, step off the scale. Pick up the object and step back on the scale within 8 s. The unit will beep again and display *only* the weight of the object.

CAMERA. Cameras are not just optics anymore. Since the early 1980s several companies have offered microprocessor-based cameras that take the difficulty out of creating great photographs. Ricoh's Mirai 35-mm camera represents the latest in technology. See Fig. 1.4.

A selector dial controls the autopower zoom lens. This specially created built-in lens system can zoom from a wide-angle 35-mm setting all the way to a 135-mm telephoto setting. The Mirai uses a variable focal lens that is controlled by the microprocessor and can produce all necessary focal lengths continuously throughout the range. An automatic light-metering system indicates when to use the built-in flash unit.

Mirai's programmed fully automatic exposure (AE) system sets the proper shutter speed (to 1/2000 s) and aperture combination according to the subject brightness. Four different AE programs are automatically selected according to the lens focal length. There are separate programs for wide, standard, telephoto, and macro-shooting modes which shift automatically as you zoom.

FIGURE 1.3
Bathroom scale with digital display (Sunbeam Model 9360 Promptor). (*Courtesy of Hanson Scales Co., a division of Sunbeam Corp.*)

Five motors are designed to perform specific tasks. The film advance motor advances and rewinds the film and powers the autoload system. The quick return motor cocks the shutter and lifts the mirror during exposure. The focusing motor operates the automatic focus (AF) mechanism, directed by an optical sensor built into the motor itself. The zoom motor drives the zoom mechanism. The aperture control motor adjusts the lens aperture using a stepper motor to minimize parts.

Mirai's viewfinder LED (light-emitting diode) display provides the essential information needed to take a picture: shutter speed, f/stop, automatic focus, exposure program and compensation, and flash charged. An LCD data panel gives other pertinent information: battery condition, film loaded correctly, focal length, frame number, subject distance, timer mode, and continuous shooting mode.

Two microprocessors control the camera. The AE processor has responsibility for the aperture, shutter, light measurement, and LCD circuits. The AF processor controls the lens, zoom, and focusing motors. It also handles the LED display and distance measurement circuit.

FIGURE 1.4
Ricoh's Mirai 35-mm camera uses two microprocessors and five motors to automate picture-taking fully.
(*Courtesy of Ricoh Corp.*)

GAS FURNACE. Super-high-efficiency gas furnaces utilize a condensing heat exchanger which reduces waste heat up the chimney to less than 5 percent. Water vapor is cooled to release the heat of vaporation, lowering annual heating costs by more than one-third. Smart super-high-efficiency furnaces by Carrier (58SXB) and Bryant (398B) further improve the overall performance using a microprocessor to adjust its operation constantly. Heating needs are controlled by a two-stage gas burner and variable speed DC motors. When the furnace first comes on, the high heat setting delivers hot air immediately. The microprocessor then switches to the low heat setting as demands are met. DC motors control the blower which circulates the warm air and the combustion air fan which draws air from the outside. Burner cycling is reduced to deliver a more even heat. Also, the motors require less power at slower speeds.

HEAT PUMP/AIR CONDITIONER. The Trane XV1500, shown in Fig. 1.5, is a variable speed heat pump/air conditioner that greatly improves comfort and lowers costs 20 to 50 percent over conventional units. A variable speed system adjusts its cooling or heating

FIGURE 1.5
Heat pump varies its capacity from approximately 12,000 to 36,000 Btu/h under microprocessor control. (*Courtesy of Trane Unitary Products Group, a division of American Standard Inc., Tyler, Tex.*)

capacity to match exactly the load (heat gain or loss in the conditioned space). The unit runs continually, cycling only for the mildest conditions.

Continual operation produces a more efficient system for two reasons. First, cycling losses are eliminated. These losses include the energy remaining in heat exchangers that is lost when a unit cycles off, plus the energy spent recooling or reheating heat exchangers when the unit cycles back on. Second, heat exchangers are sized for maximum capacity. At reduced operating speeds refrigerant flow is less, and the heat exchangers are oversized; heat transfer improves.

The heat pump/air conditioner's microprocessor controls three separate DC brushless motors to vary continually the speed of the compressor, indoor fan, and outdoor fan. Under conditions of high humidity lower airflow runs a colder evaporator, developing a higher latent capacity to remove moisture from the air.

The microprocessor determines load from the set-point temperature and indoor ambient temperature difference (error) and time. The error is modified by a proportional plus integral controller to integrate the difference time. Next it computes the capacity from the load after correcting for indoor dry bulb and wet bulb conditions. Then the microprocessor calculates the compressor speed and blower speed to meet the capacity.

An integral part of any heat pump system is a defrost cycle. In the winter the outside coil functions as an evaporator. Under certain conditions of outdoor temperature and humidity refrigerant temperature drops well below 32°F and frost forms on the coils. Frost buildup blocks the airflow and reduces efficiency. Conventional units use a timer to shunt hot refrigerant periodically through the coils whether or not they need defrosting. Trane improves the operation of their heat pumps by defrosting only when it is needed—demand defrost control. Figure 1.6 shows the controller board. A microprocessor learns (stores) the dry coil parameters after start-up. It then monitors the dry coil temperature and outdoor temperature. When certain conditions are met, the defrost cycle is activated.

SENSORS. Electronics is contributing greater intelligence to sensors. Digital outputs have made signal processing more accurate and versatile. Programmable temperature sensors use microprocessors at the sensing point for monitoring, signal conditioning, and control.

FIGURE 1.6
Electronic demand defrost control for residential heat pumps adjusts defrosting cycle to maximize system performance. (*Courtesy of Trane Unitary Products Group, a division of American Standard Inc., Tyler, Tex.*)

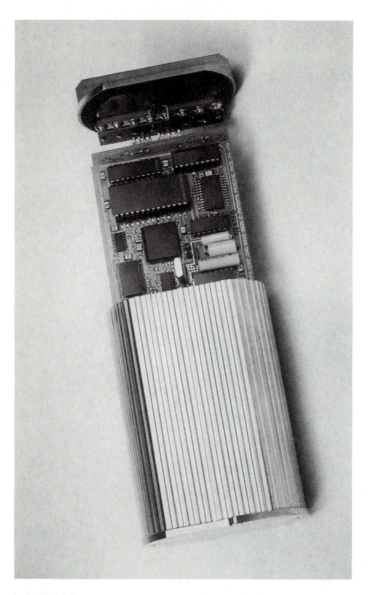

FIGURE 1.7
Accutech's Al-2000 smart temperature transmitter. (*Courtesy of Accutech, a division of Adaptive Instruments Corp.*)

Accutech makes the Al-2000 smart temperature transmitter shown in Fig. 1.7. The Al-2000 can function as a conventional analog two-wire transmitter. Or it can be accessed anywhere along the same pair of wires for digital communication with a computer. The device accepts and individually linearizes most popular thermocouples and RTD sensor types.

The microprocessor on-board makes several measurements before reporting temperature. First, it checks memory for sensor set-up information and then calls up the linearization table for that sensor. Second, the processor reads the internal calibration thermometer for ambient temperature and compares it with factory calibration characteristics. Third, it checks the input temperature, converts it to a digital value, and then linearizes the signal. The output is fully temperature compensated. Finally, the microprocessor converts the signal back to analog, matching the zero to full-scale setting stored in its memory.

There are several housekeeping chores performed by the microprocessor. It checks to see if the main computer wants to reconfigure it to handle another type of input. Also, it runs a series of internal diagnostics to make sure it is working properly.

ROBOTS. Unimation's PUMA industrial robots are a six-axis revolute design with on-board controllers for each axis of revolution. See Fig. 1.8. It has a high-level robot programming language which incorporates user-friendly instructions for the inexperienced operator. Both teach pendant and computer terminal inputs are possible. There are communication links for supervisory computers to give full factory automation integration.

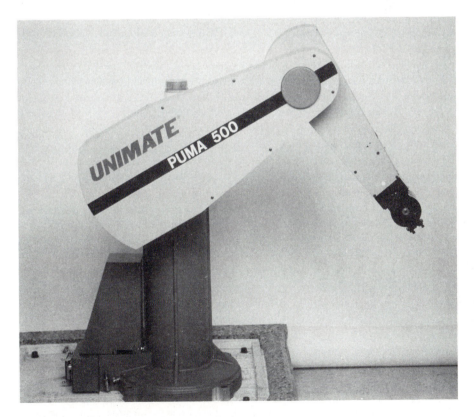

FIGURE 1.8
Unimate's PUMA robot with microprocessor-based controllers for each of six axes of revolution. (*Courtesy of AEG Westinghouse Industrial Automation Corp.*)

The robot computer calculates the trajectory and transforms world coordinates to individual joint angle coordinates. The trajectory algorithm determines the straightline path between successive tool locations and establishes the optimum accelerations and decelerations between endpoints. Intermediate tool locations are found by applying inverse kinematic mathematics.

The trajectory is a series of points (interpolations) separated in time by the algorithm execution time. Joint positions are references (set point) for the on-board axes controllers. Each closed-loop controller is microprocessor-based. The controller compares reference and feedback signals to generate instantaneous voltages for the DC motor drive unit. The main feedback device is an encoder which senses the actual rotational position. The microprocessor control algorithm finds the motor voltage necessary for quick nonoscillatory responses between interpolations.

ENGINE CONTROL. The automobile market was a major factor in the early development of microcontrollers. The motivation for electronic engine control came from two government requirements in the 1970s: (1) Environmental Protection Agency (EPA) rules for exhaust emissions and (2) regulation of the average fuel economy.

The basic unit controls the air-fuel mixture, ignition timing, and idle speed. Sensor inputs for the controller include engine crankshaft position, manifold absolute pressure (MAP), throttle position, and exhaust gas oxygen (EGO) content. Figure 1.9 shows a control unit's electronics.

The primary purpose of fuel control is to maintain the air/fuel ratio at or near stoichiometry (a ratio of 14.7 for gasoline). The microprocessor accomplishes this task by controlling the fuel injector ON time or the fuel pulse cycle. There are several different control modes. When the engine is turned ON, a start mode keeps the air/fuel ratio low (rich) to prevent stalling in cold weather. After the coolant temperature rises the microprocessor switches to an open-loop mode. For a constant air/fuel ratio, the quantity of fuel per shot (injector ON time) is directly proportional to the volumetric efficiency of the engine. Volumetric efficiency is a function of engine revolutions per minute (rpm) and intake manifold pressure—a fixed relationship for a given engine. Therefore, a look-up table in memory will give a suitable fuel pulse time for every speed/pressure measurement.

The EGO sensor is continually monitored for a minimum voltage which indicates the sensor has warmed up. Then the microprocessor switches to a closed-loop mode. The amount of oxygen in the exhaust depends upon the air/fuel ratio. The air/fuel ratio is controlled by measuring the EGO, comparing it to the desired value, and altering the fuel injector pulse time to correct for a rich or lean mixture. The control strategy includes input from engine temperature and throttle position sensors.

Under hard acceleration the microprocessor chooses a richer mixture to give maximum torque. The air/fuel ratio is made leaner during deceleration periods to reduce emissions. Afterward control is returned to the closed-loop mode.

The engine ignition system supplies a spark to ignite the fuel mixture. Timing of the spark relative to piston top dead center (TDC) is critical because it affects engine performance and exhaust emissions. All distributorless ignition systems employ some type of toothed wheel attached to the crankshaft or camshaft. A Hall-effect sensor with some signal conditioning generates a square wave output. By counting the number of pulses, the

FIGURE 1.9
Automobile engine control module. (*Courtesy of Ford Motor Co.*)

microprocessor can calculate engine speed or simply determine engine position relative to TDC.

The proper spark advance depends upon engine temperature, rpm, MAP, and barometric pressure. Look-up tables store the contribution of each measurement to the overall spark advance. The microprocessor loads a counter with a number representing the computed spark advance. After the count down counter is decremented to zero, the microprocessor opens the coil primary to fire the spark plug. Spark plug selection is determined by the microprocessor, each cylinder having its own coil.

1.3 WHICH MICROPROCESSOR?

In the 1960s the basic components of a microprocessor (registers, arithmetic units, control, instruction decoders, etc.) existed as separate entities in mainframe computers. By 1970 these components were being manufactured as individual integrated circuits (ICs). In 1971 all components were integrated into a single IC, called a microprocessor, by Intel:

TABLE 1.3
Evolution of microprocessors

Motorola	Intel	Year	Architecture	Clock (MHz) Intel	Motorola
	4004	1971	4-bit	0.7	
6800	8080	1974	8-bit	0.8	1
6802/6808	8085	1977	8-bit	6.2	2
6809	8086/8088	1978	16-bit	8	2
68000/68020	80286/80386	1985	32-bit	16+	10+

the 4004 with 4-bit architecture. The first widely used microprocessors were the 8-bit Intel 8080 and Motorola 6800, both introduced in 1974.

Other manufacturers have joined the fray; however, Intel and Motorola have dominated the market. Table 1.3 gives the evolution of each company's microprocessor. The evolution was driven by the need for more powerful personal computers (PCs) and workstations. Power means more addressable memory and faster execution time. It is surprising how each company produced an improved competitive microprocessor within a similar time frame. Although Intel clock speeds are higher than Motorola's for a comparable machine, Motorola microprocessors use fewer clock cycles per instruction.

Some microprocessors are better recognized by their PC application:

Apple II	6502
Radio Shack Color	6809
IBM PC	8086/8088
Apple MacIntosh	68000/68020

The 6502 was built in 1975 by MOS Technology Inc. (now a division of Commodore International Ltd.). Its design philosophy was derived from the 6800 so that its internal structure and language set are very similar to the 6800. Also, the Zilog Z-80 has all the hardware and software features of the Intel 8080 but with several additions. Intel's 8085 a year later was a response to the Z-80. Both machines replaced the 8080's external clock IC with an on-board clock.

Motorola microprocessors use different registers, instruction types, and addressing features from Intel microprocessors. Therefore, it is always easier to grasp the basics of the next improvement than to cross company lines. In the past, schools have adopted the 8085 or 8086/8088 as a teaching tool because of the popularity of the IBM PC. Presently the 68000 is beginning to gain favor. Electrical engineers are inherently interested in building PCs along with support hardware and software. Mechanical engineers have been interested in using the PC for data acquisition and control. In going to greater computing power, the microprocessor has become increasingly complex. There are more registers, longer bit words and addresses, more pins, and a greatly expanded instruction set. A measure of this fact is the 1464 instructions for the 6809 compared to 197 instructions for the 6802. A first course in microprocessors has traditionally been a software endeavor.

The 6802/6808

In mechatronics we are not designing "number crunching" computers; we are building a mechanism control board. Large addressable memory is not necessary, and moderate clock speeds are more than acceptable. The more powerful machines would be out of place and certainly not cost-effective. Most mechatronic systems can be handled by 8-bit microprocessors.

The Motorola 6802/6808 are 8-bit data and 16-bit address microprocessors, which are really 6800s with the clock moved on-board. The 6802 is a 6808 with a few added bytes of random access memory (RAM). Both the 6802 and the 8085 are widely used in mechatronics. The 6802 is preferable because it has a simpler control strategy with non-multiplexed buses. Interfacing tasks are easier. In addition, it has a smaller yet equally strong instruction set. With the 6802, the beginning student can be writing programs in a few weeks. This allows time for learning the hardware and electronics that are necessary for interfacing real-world components to the microprocessor.

Microcontrollers

Microprocessors which have combined memory *and* various input/output (I/O) features on a single chip are called *microcontrollers*. At times they are referred to as a microcomputer-on-a-chip. I/O features may include parallel and serial ports, timers and/or analog-to-digital converters. Generally, they have a simplified design for the microprocessor, which makes them easier to manufacture but limits their instruction set.

Microcontrollers now make up over 50 percent of the microprocessors used in mechatronics. The market for microcontrollers is estimated to be nearly $2 billion per year and growing, twice as large as the market for microprocessors in PCs. There are a wide variety of microcontrollers available to the buyer. Only the more recent powerful micro-controllers, e.g., the Motorola 6811 or the Intel 8096, have external data and address pins which allow them to function as a microcontroller or as a stand-alone microprocessor. However, the cheapest microcontroller is the one with just enough features to do the task.

Chapter 17 will review the Motorola 6805 family of microcontrollers. It is a simpli-fied version of the 6802 microprocessor with a similar instruction set. The 6805 was in-troduced in 1979, and additions to the family continue to be made. Over the years it has been the most popular microcontroller on the market.

Development Systems

A host of hardware and software tools are available to the engineer for system develop-ment. Most manufacturers supply "educational boards" to introduce the engineering com-munity to their products. A basic inexpensive laboratory system is called an evaluation kit or trainer. They are typically single-board microcomputers with little memory and a lim-ited operating system.

Figure 1.10 shows the Heathkit 3400A Microprocessor Trainer for the 6802. The operator enters the program in machine language by way of a keypad (hand assembly). A set of six 7-segment displays give addresses and their content. The program can be run in

FIGURE 1.10
Heathkit 3400A Microprocessor Trainer with a 6802 microcomputer system. (*Courtesy of Heath Co., Benton Harbor, Mich.*)

real time or single step mode for debugging. At any point in the program microprocessor registers or memory addresses can be examined for content. Every pin on the microprocessor is accessible on the trainer through individual wire sockets. ICs and device drivers can be mounted on the breadboard strip in front and interfaced to the trainer system. Trainers offer a convenient method for learning to interface basic components in a mechatronic system before a separate board with the microprocessor is introduced.

For a more sophisticated approach to system development, software assistance can be purchased for an IBM PC. "Assembler" software allows the user to write the system program in either assembly or a high-level language. It checks the program for syntax errors and then compiles the program into machine code. Simulation software accepts the machine code and mimics the microprocessor's execution of the program. The simulator uses memory locations in the PC to represent microprocessor internal registers, memory, and I/O ports. Thus, the program can be debugged one line at a time. An emulator is an option that includes both hardware and software to debug system hardware. The microprocessor in the prototype is removed from the system, and the emulator board output connector is inserted into the empty slot. The emulator receives the machine code from the computer. Since the program is in the PC memory under control of the emulator software commands, the actual system I/O circuits and devices can be tested against the emulator.

Simulation and emulation extract a cost/time penalty that may not be warranted by the beginner. But they are indispensable to the practitioner.

1.4 PROLOGUE TO THE TEXT

Mechanical engineering students will discover that the study of microprocessors is different from their usual world of equation solving. Only an occasional equation will be found, mostly of the algebraic type. But attention to detail is very important. Software skills must be developed to program the microprocessor. Hardware knowledge must be gained to interface the microprocessor to real-world devices.

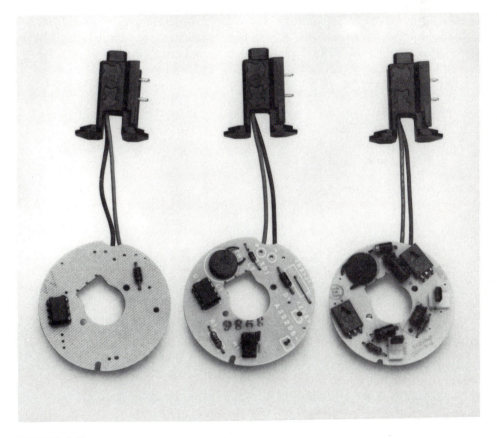

FIGURE 1.11
Evolution of chips to control a brushless motor. From 11 components to 6 to 2: the Sprague Hall power IC and one diode for cut-off protection. (*Courtesy of Sprague Semiconductor Group.*)

The only background necessary to begin our study of microprocessor systems is an introduction to logic gates and flip-flops. Exposure to transistors and op amps would be helpful for interfacing tasks. Most ME curriculums include an EE core group which covers these topics. It is doubtful if any ME student feels comfortable with his or her EE preparation. Therefore, a review of pertinent material is given throughout the textbook as needed.

The best source of information on interfacing ICs or chips as they are called is manufacturer's data sheets. They give the novice a full description of pin assignments (pin-outs) and show wiring diagrams on how the IC should be connected to microprocessors and other chips. We have tried to maintain this approach by giving the reader wiring diagrams and software for all basic interfacing circuits.

One important trend today bodes well for the ME in electronics—function integration into a single chip. We have stated how microcontrollers have eliminated the bus structure of microprocessors by combining microprocessor, memory, and I/O ports onto one IC. Outputs then require large amounts of power gain and additional control logic to operate actuators and displays. Protection against voltage and current overloads are often necessary. Smart power ICs are merging these functions to reduce the number of packages and connections. Sprague Electric Company reduced the components of a brushless fan motor control from 11 to 6 to 2. See Fig. 1.11 on the previous page. Similar integration is commonplace; the driving force is lower costs in the auto industry.

There are exciting opportunities for MEs to design with microprocessors today.

CHAPTER

2

NUMBER SYSTEMS AND ARITHMETIC

Everyone is familiar with the *decimal* number system. From early childhood base 10 numbers have been the foundation of everyday arithmetic and monetary exchange. Therefore, it is difficult to imagine counting and calculating by another number system. Yet this is precisely what digital computers do. Virtually all digital computers (or microprocessors) operate with a numbering system called *binary* (from the Latin *bis* = two ways). This is due to the ease of fabricating two state cells which are the basis of all logic/memory operations. Other important numbering systems include *octal* and *hexadecimal*. Both are closely related to binary since they are notations representing binary for our convenience. Many older computers used octal numbers for addresses and machine code instructions. Today all computers (or microprocessors) use hexadecimal numbers. However, it must be made clear that computers read only binary. An understanding of binary arithmetic is essential for programming computers at the machine code level.

2.1 DECIMAL SYSTEM (BASE 10)

The decimal system was invented by Hindu mathematicians around A.D. 800, borrowed by the Arabs, and introduced into Europe about A.D. 1200, where it was known as the Arabic number system. The system is based on 10 symbols or digits: 0, 1, 2, 3, 4, 5, 6, 7, 8, 9. Any number N can be represented by a set of these digits, e.g., $(3186)_{10}$. The subscript indicates the base. Digit order is important because each digit position carries a particular weight which determines the magnitude:

$$\begin{array}{cccc} \ldots & 10^3 & 10^2 & 10^1 & 10^0 & \text{weight} \\ \ldots & 1000 & 100 & 10 & 1 & \text{value} \end{array}$$

$$\begin{array}{rl} 6 \times 10^0 & 6 \\ + \ 8 \times 10^1 & 80 \\ + \ 1 \times 10^2 & 100 \\ + \ 3 \times 10^3 & + \ 3000 \\ \hline & (3186)_{10} \end{array}$$

We see that the weight of each digit increases by a factor of 10 as we proceed from left to right; the digit itself is a multiplier. The sum of all multiplier-weight products is the number. The subscript is omitted in everyday arithmetic, but we will find it helpful when mixing bases.

All number systems considered are based on this multiplier-weight format. A general rule can be written as

$$N = (b_n \cdots b_1 b_0)_B = \sum_{i=0}^{n} b_i B^i$$

where B = base and n = number of digits less one.

2.2   BINARY SYSTEM (BASE 2)

Binary numbers have two states, which are assigned the Arabic symbols 0 and 1. They are referred to as *bi*nary digi*ts* or bits. In digital circuits a certain voltage range defines logic 1 and another voltage range defines logic 0. Circuits called TTL (transistor-transistor logic), e.g., having any voltages between 0 and 0.8 V represent logic 0 and any voltages between 2 and 5 V represent logic 1. Because the range is so broad, signal attenuation, noise, and drift are not a problem unlike analog systems. Accuracy is limited only by the number of bits employed.

A typical number in the base 2 system is written as $(11010010)_2$. The set of 8 bits is called a *byte*. Each bit position carries a particular weight similar to the decimal system. The multiplier can be only 0 or 1. Following the preceding formula, the number is interpreted as

8th bit or bit 7	7th bit or bit 6	6th bit or bit 5	5th bit or bit 4	4th bit or bit 3	3rd bit or bit 2	2nd bit or bit 1	1st bit or bit 0	
2^7	2^6	2^5	2^4	2^3	2^2	2^1	2^0	weight
128	64	32	16	8	4	2	1	value

$$\begin{array}{rl} 0 \times 2^0 & 0 \\ + \ 1 \times 2^1 & 2 \\ + \ 0 \times 2^2 & 0 \\ + \ 0 \times 2^3 & 0 \\ + \ 1 \times 2^4 & 16 \\ + \ 0 \times 2^5 & 0 \\ + \ 1 \times 2^6 & 64 \\ + \ 1 \times 2^7 & + \ 128 \\ \hline & (210)_{10} \end{array}$$

The numerical weight of each digit increases by a constant factor of 2 as the number is traversed from the *least significant bit* (LSB) with weight 2^0 to the *most significant bit* (MSB) with weight 2^7. Therefore, the number $(11010010)_2$ is equivalent to $(210)_{10}$.

2.3 OCTAL SYSTEM (BASE 8)

The octal number system is based on eight digits: 0, 1, 2, 3, 4, 5, 6, 7. All numbers are represented by these eight digits, e.g., $(753)_8$. We can interpret this number as before:

$$
\begin{array}{cccc|l}
\ldots & 8^3 & 8^2 & 8^1 & 8^0 & \text{weight} \\
\ldots & 512 & 64 & 8 & 1 & \text{value}
\end{array}
$$

$$
\begin{array}{lr}
 & 3 \times 8^0 & 3 \\
+ & 5 \times 8^1 & 40 \\
+ & 7 \times 8^2 & \underline{+\ 448} \\
 & & (491)_{10}
\end{array}
$$

The octal form is shorthand for binary numbers with the bits grouped in threes:

$$(491)_{10} = (753)_8 = (111101011)_2$$
$$7 \quad 5 \quad 3$$

2.4 HEXADECIMAL SYSTEM (BASE 16)

The hexadecimal system is based on 16 digits and/or symbols: 0, 1, 2, 3, 4, 5, 6, 7, 8, 9, A, B, C, D, E, F. The letters have no significance other than the need for single symbols to represent the higher digits: 10 = A, 11 = B, 12 = C, 13 = D, 14 = E, and 15 = F. It may seem strange because we are using letters. Just forget that there is a base 10 system. Counting will be awkward at first until practice brings familiarity.

Hexadecimal numbers will have the following appearance: $(B3A)_{16}$. We can interpret this number in the same manner:

$$
\begin{array}{ccccl}
\ldots & 16^3 & 16^2 & 16^1 & 16^0 & \text{weight} \\
\ldots & 4096 & 256 & 16 & 1 & \text{value}
\end{array}
$$

$$
\begin{array}{lll}
\ \ A \times 16^0 & \quad 10 \times 16^0 & 10 \\
+\ 3 \times 16^1 \quad \text{or} & +\ 3 \times 16^1 & 48 \\
+\ B \times 16^2 & +\ 11 \times 16^2 & \underline{+\ 2816} \\
 & & (2874)_{10}
\end{array}
$$

Hexadecimal numbers are similar in structure to octal numbers, with each symbol representing 4 bits in the binary number. Four bits have 16 different combinations. Thus,

$$(2876)_{10} = (B3A)_{16} = (101100111010)_2$$
$$\text{B} \quad 3 \quad \text{A}$$

A relationship between the four systems can be seen more clearly in Table 2.1. Hexadecimal is a shorter, more convenient method for writing binary numbers. It simplifies

TABLE 2.1
Comparing number systems

Decimal	Binary	Octal	Hexadecimal
0	0000	0	0
1	0001	1	1
2	0010	2	2
3	0011	3	3
4	0100	4	4
5	0101	5	5
6	0110	6	6
7	0111	7	7
8	1000	10	8
9	1001	11	9
10	1010	12	A
11	1011	13	B
12	1100	14	C
13	1101	15	D
14	1110	16	E
15	1111	17	F
16	10000	20	10

data entry and display to a greater degree than octal. Thus, hexadecimal numbers are widely employed to represent data and addresses in microprocessors. For example,

$$\text{8-bit data:} \qquad (10110001)_2 = (B1)_{16}$$
$$\text{16-bit address:} \qquad (0011111101101010)_2 = (3F6A)_{16}$$

2.5   BINARY CODED DECIMAL (BCD) SYSTEM

Most digital instruments and equipment have visual outputs in the form of decimal digits 0 through 9. Each digit can be generated by decoding a 4-bit binary number. The binary coded decimal (BCD) system is a binary system based on a 8421 weighting of each group of 4 bits. Thus, the decimal number 873 would be represented in BCD as

8421	8421	8421	weight
1000	0111	0011	BCD
8	7	3	decimal equivalent

The binary number $(001101101001)_2 = (873)_{10}$ is difficult to interpret immediately although it could be decoded for the decimal number. There is a distinct advantage in hardware decoding and in data recognition by the engineer when working with BCD. Since microprocessors operate in binary, it will be necessary to convert binary numbers to their BCD equivalent before BCD is sent to the output port.

2.6   DECIMAL CONVERSION

The preceding number systems have demonstrated the process for converting from base B to base 10. In working with microprocessors, we will need to convert a specific decimal number to binary or hexadecimal. Conversions can be done by successive divisions with the desired base. With each division, the remainder will be a base digit. The remainders form the equivalent binary number.

Decimal to Binary

For decimal conversion to binary, the divisor is 2 and the remainder is 1 or 0. For example, the decimal number $(38)_{10}$ is converted into its equivalent binary number as follows:

$$
\begin{array}{lll}
38 \div 2 = 19 & \text{remainder } 0 & \text{(LSB)} \\
19 \div 2 = 9 & \text{remainder } 1 & \\
9 \div 2 = 4 & \text{remainder } 1 & \\
4 \div 2 = 2 & \text{remainder } 0 & \\
2 \div 2 = 1 & \text{remainder } 0 & \\
1 \div 2 = 0 & \text{remainder } 1 & \text{(MSB)}
\end{array}
$$

The original number is divided by 2, and the remainder is recorded as the LSB. The quotient is divided by 2, and again the remainder is recorded. The process is repeated until the result is zero, the MSB always being 1. Then collect the remainders in reverse order beginning with the last remainder or MSB. The number is $(100110)_2 = (38)_{10}$.

Decimal to Hexadecimal

Decimal conversion to hexadecimal is carried out in the same manner as decimal to binary. In this case the divisor is 16, and the remainders can be any hexadecimal digit/symbol from 0 to F. For example, convert the decimal number $(41,976)_{10}$ to hexadecimal.

$$
\begin{array}{lll}
41,976 \div 16 = 2623 & \text{remainder } 8 & \text{(LSD)} \\
2623 \div 16 = 163 & \text{remainder } 15 = F & \\
163 \div 16 = 10 & \text{remainder } 3 & \\
10 \div 16 = 0 & \text{remainder } 10 = A & \text{(MSD)}
\end{array}
$$

The number is $(A3F8)_{16} = (41,976)_{10}$. The reader should find that it is easier to convert to hexadecimal as an intermediate step when converting decimal to binary. Thus,

$$(38)_{10} = (26)_{16} = (0010\ 0110)_2$$

2.7 BINARY ADDITION AND SUBTRACTION

Mathematical operations such as addition, subtraction, multiplication, and division can be performed on binary numbers as well as on decimal numbers. The most common operations are addition and subtraction. Binary, because of its few symbols, has simple rules of manipulation.

Addition

Addition is simply counting up one unit at a time. When a base sum is reached in one column, the count is replaced by zero and a carry of one is added to the digit on the left. There are four rules for adding binary numbers:

1. $0 + 0 = 0$

2. $\left.\begin{array}{c} 0 + 1 \\ 1 + 0 \end{array}\right\} = 1$

3. $1 + 1 = 10$ (0 + carry 1)

4. $1 + 1 + 1 = 11$ (1 + carry 1)

Note: $1 + 1 + 1 = (3)_{10} = (11)_2$

Example 2.1

```
        [1] [1]     [1] [1]
      0   0   1   1   0   1   1   0      (54)
    + 1   0   1   1   0   0   1   1      (179)    [  ] denotes carry.
 [0]  1   1   1   0   1   0   0   1      (233)
```

Example 2.2

```
                        [1] [1]
      1   1   0   0   1   0   0   1      (201)
    + 1   0   1   1   0   0   1   1      (179)
 [1]  0   1   1   1   1   1   0   0      (380)
```

Eight-bit (unsigned) binary numbers are limited to $(255)_{10}$. In Example 2.1 the answer is within the allowable range and no carry (to the ninth bit) occurs. In Example 2.2 the answer exceeds $(255)_{10}$. The answer is correct as a 9-bit number, the carry is the ninth bit, but it is incorrect within the bounds of an 8-bit number. Data registers for the Motorola 6802 are 8 bits wide. A carry, resulting from the addition of binary numbers, is placed in a separate register. *When (unsigned) binary numbers are added, the answer is incorrect if the carry is set.* The programmer must check the carry after every addition if he or she believes any results will fall out of the allowable range.

Subtraction

Subtraction is similar to addition but corresponds to a down count where units are removed from the total. Binary subtraction is carried out exactly like decimal subtraction. In place of a carry the operation extracts a borrow from the next digit when the subtrahend (bottom) is larger than the minuend (top). If the answer contains a net borrow, the microprocessor sets the *carry*. The subtraction rules follow addition rules:

$$0 - 0 = 0$$
$$1 - 0 = 1$$
$$1 - 1 = 0$$
$$0 - 1 = 10 - 1 + \text{borrow}$$
$$= 1 + \text{borrow}$$

Note: $(10)_2 = (2)_{10}$.

Example 2.3

$$
\begin{array}{r}
1\ 1\ 0\ 0\ 1\ 0\ 0\ 1 \quad (201) \\
-\ 1\ 0\ 1\ 1\ 0\ 0\ 1\ 1 \quad (179) \\
\hline
[0]\ \ 0\ 0\ 0\ 1\ 0\ 1\ 1\ 0 \quad (22)
\end{array}
$$

Example 2.4

$$
\begin{array}{r}
0\ 0\ 1\ 1\ 0\ 1\ 1\ 0 \quad (54) \\
-\ 1\ 0\ 1\ 1\ 0\ 0\ 1\ 1 \quad (179) \\
\hline
[-1]\ \ 1\ 0\ 0\ 0\ 0\ 0\ 1\ 1 \quad (131)
\end{array}
$$

In Example 2.3 we are subtracting a smaller number from a larger number. The answer has no net borrow and is therefore correct. In Example 2.4 we are subtracting a larger number. A net borrow of one (carry set) has occurred and the answer is incorrect. *When (unsigned) binary numbers are subtracted, the answer is incorrect if the carry is set.*

The problem is handled differently if the calculation is with pencil and paper. In cases where the subtrahend is greater than the minuend the rows can be interchanged and the answer is labeled negative. This is the normal decimal procedure.

2.8   BINARY MULTIPLICATION AND DIVISION

Multiplication

Multiplication is the process where a number is to be added to itself a given number of times. This is a slow process, but an algorithm for computers is very straightforward. However, the usual method for multiplying numbers is demonstrated with the base 10:

$$
\begin{array}{rl}
621 & \text{multiplicand} \\
\times\ 248 & \text{multiplier} \\
\hline
4968 & \text{first partial product} \\
2484 & \text{second partial product} \\
1242 & \text{third partial product} \\
\hline
154008 & \text{product}
\end{array}
$$

This short method of multiplication uses each digit of the multiplier to multiply the multiplicand. Each partial product is shifted and added for the answer. The procedure depends upon memorization of the familiar multiplication table. Binary multiplication follows the same procedure except the table (rules) are much shorter.

$$0 \times 0 = 0$$
$$\left.\begin{array}{l} 0 \times 1 \\ 1 \times 0 \end{array}\right\} = 0$$
$$1 \times 1 = 1$$

Example 2.5

$$
\begin{array}{r}
1101 \qquad (13) \\
\times\ 1001 \qquad (9) \\
\hline
1101 \\
0000 \\
0000 \\
1101 \\
\hline
1110101 \qquad (117)
\end{array}
$$

Comments

1. The product needs as many (digit) places as the sum of the multiplicand and multiplier places. Thus, the product of two 8-bit numbers takes 16 places unless the product range is known to be less.
2. The partial products are always the multiplicand or zeros shifted by the number of places in the multiplier digit.

Shifting and adding are easily implemented with hardware or software. The more powerful microprocessors in today's PCs have multiplication circuits on-board and a MUL(tiply) instruction. The 6802 microprocessor must have a subroutine which multiplies two numbers. A possible algorithm may be constructed as follows:

Store multiplicand at location A.
Store multiplier at location B; place a marker on LSB.
Clear location C for product.
Store number of multiplier bits at location D.

1. Is the marker bit in B equal to one?
   a. Yes, add A to C.
   b. No, continue.
2. Move the marker to the next bit on the left.
3. Shift all bits in A one place to the left.
4. Decrement D.
5. Is D equal to one?
   a. Yes, stop.
   b. No, branch to step 1.

Division

Division is the process where the number of times a divisor may be subtracted from a dividend is calculated (counted). In place of repetitive subtraction a faster technique is the familiar decimal long division. Binary division is done the same way as decimal division:

$$0 \div 0 \qquad \text{meaningless as in decimal system}$$
$$1 \div 0 \qquad \text{meaningless as in decimal system}$$
$$0 \div 1 = 0$$
$$1 \div 1 = 1$$

Example 2.6

```
                           1010      quotient (10)
divisor (5)   101)110010             dividend (50)
                 -101
                 ─────
                  0010
                  0000
                  ────
                   101
                   101
                   ────
                   0000
                   0000
                   ────
```

In this case the remainder is zero. With control applications, any remainder is ignored because only whole numbers are used. Large digital computers work with decimal point arithmetic. The procedure for binary follows decimal calculations.

Example 2.7

```
                        1010.11
            100)101011.00
                100
                ───
                010
                000
                ───
                101
                100
                ───
                011
                000
                ───
                110
                100
                ───
                100
                100
                ───
```

To convert $(1010.11)_2$ to the decimal system, each successive place to the right of the decimal is assigned a base weight: $2^{-1}, 2^{-2}, 2^{-3}, \ldots$. Thus,

$$(1010.11)_2 = (1 \times 2^{-1}) + (1 \times 2^{-2}) + (0 \times 2^0) + (1 \times 2^1)$$
$$+ (0 \times 2^2) + (1 \times 2^3) = (10.75)_{10}$$

Binary division is carried out with an algorithm and subsequent subroutine. Subroutines for both multiplication and division will be given in Chap. 4.

2.9 BINARY CODED DECIMAL ADDITION

The programmer may find it necessary to add BCD numbers. There are two alternatives. (1) Convert BCD to binary, perform the addition, and then convert back to BCD. (2) Build circuitry within the microprocessor to follow BCD rules. Remember, the six numbers 1010 to 1111 are illegal. These rules are as follows:

1. Add the two BCD numbers as though they are binary numbers.
2. If the least significant 4 bits (nibble) of the result is a number greater than 9, add 6 to these 4 bits. Or, if there is a carry from the fourth bit to the fifth bit, add 6 to the 4 bits. This carry is called the auxiliary carry (AC). Otherwise no change is made.
3. After step 2 is completed, if the most significant 4 bits of the result is a number greater than 9, add 6 to these 4 bits. Or, if the normal carry (C) is set, add 6 to the most significant 4 bits. Otherwise no change is made.

Example 2.8. Add 59 and 41 in BCD.

$$
\begin{array}{llll}
 & 0101\ 1001 & (59) \\
 & \underline{0100\ 0001} & (41) \\
\text{AC} = 0 \quad 0] & 1001\ 1010 & \text{sum } 1010 > 9 \\
 & \underline{0000\ 0110} & \text{add } 6 \\
\text{AC} = 1 \quad 0] & 1010\ 0000 & \text{step 2 complete} \\
 & & \\
 & 1010\ 0000 & \text{sum } 1010 > 9 \\
 & \underline{0110\ \quad \underline{}} & \\
\text{AC} = 0 \quad 1] & 0000\ 0000 & (100)\ \text{step 3 complete}
\end{array}
$$

All addition routines for (unsigned) binary numbers can be modified to handle (unsigned) BCD addition by inserting an instruction digital adjust accumulator (DAA) after the addition instruction.

2.10 SIGNED NUMBERS (TWOS COMPLEMENT NUMBERS)

Up to now we have been addressing unsigned binary numbers, i.e., numbers that carry a sign symbol separate from themselves. They have separate sets of rules governing addition and subtraction. Signed numbers have the sign inherent within the number, important for computers since they cannot carry a sign.

The simplest approach to the problem of representing negative numbers is to use the most significant bit (MSB) to denote the sign of the number. For example, suppose we drive a car both forward and backward with a three-digit odometer showing zero initially. Representative numbers would be

$$
\begin{array}{ll}
998 & (-2) \\
999 & (-1) \\
000 & (0) \\
001 & (+1) \\
002 & (+2)
\end{array}
$$

It is easy to see that 999 appears when the car backs 1 mi although it may seem strange to accept 999 as a signed number equivalent to $(-1)_{10}$. Yet the arithmetic shows

$$
\begin{array}{rl}
999 & (-1) \\
+\ \ 001 & (+1) \\
\hline
1]\ \ \ 000 & (0)
\end{array}
$$

ignore carry

Furthermore, the range can be divided in half with positive numbers 000 to 499 and with their negative equivalents 999 to 500. The negative number N_C is said to be the base complement of positive number N. It is given by the definition: $N_C = B^{n+1} - N$. Thus, for the base 10 with our three-digit numbers, the 10s complement of 325 is

$$
\begin{array}{rl}
1000 & 10^3 \\
-\ \ 325 & N \\
\hline
675 & N_C
\end{array}
$$

$$
\begin{array}{rll}
325 & (+325) \\
+\ \ 675 & (-325) \\
\hline
1]\ \ \ 000 & (0) & \text{check}
\end{array}
$$

ignore carry

Comments

1. The carry is always ignored unlike unsigned arithmetic.
2. The odometer can be used for forward travel where the range of unsigned numbers is 000 to 999. It is only our interpretation of the reading that differs.

Twos complement numbers represent both positive and negative binary numbers. Consider the negative or 2s (twos) complement of $(011)_2$.

$$
\begin{array}{rl}
1000 & 2^3 \\
-\ \ 011 & 3 \\
\hline
101 & -3
\end{array}
$$

$$
\begin{array}{rll}
011 & (3) \\
+\ \ 101 & (-3) \\
\hline
1]\ \ \ 000 & (0) & \text{check}
\end{array}
$$

ignore carry

The range of a 3-bit binary number is 0 to 7 as an unsigned number. But as a signed number, 0, 1, 2, 3 are positive and 7, 6, 5, 4 are negative. In general, 2s complement positive numbers range from 0 to $(2^{n-1} - 1)$. The 2s complement negative numbers range from (-1) to -2^{n-1}. This range of values is not symmetrical since there is one more negative number than there is positive. Several 8-bit numbers are listed in 2s complement form along with their decimal equivalents. See Table 2.2.

TABLE 2.2
Eight-bit signed binary numbers

Binary	Sign bit -2^7 -128	2^6 64	2^5 32	2^4 16	2^3 8	2^2 4	2^1 2	2^0 1	Weight value	Decimal equivalent
	0	1	1	1	1	1	1	1		+127
	0	1	1	1	1	1	1	0		+126
					⋮					
	0	0	0	0	0	0	0	1		+1
	0	0	0	0	0	0	0	0		0
	1	1	1	1	1	1	1	1		−1
	1	1	1	1	1	1	1	0		−2
					⋮					
	1	0	0	0	0	0	0	1		−127
	1	0	0	0	0	0	0	0		−128

Comments

1. All positive numbers have *zero* in the MSB. All negative numbers have *one* in the MSB. This bit is called the sign bit.

2. The most negative number (-128) has no positive counterpart ($+128$).

3. To determine the magnitude of any unknown negative number, take its 2s complement. The result is a positive number whose magnitude equals that of the original number.

If subtraction rules were necessary to find a 2s complement, some advantage would be lost. There are two alternative methods for taking the "2s complement" of a 2s complement number without subtraction.

Method I

Take the 1s complement and add one. The 1s complement is formed by changing all 1s to 0s and all 0s to 1s, i.e., complementing the number. The 1s complement was a popular method of representing negative numbers in the early days of computers. As an example, find the 2s complement of $(00000110)_2 = (6)_{10}$.

$$
\begin{array}{lll}
 & 00000110 & (6) \\
\text{complement} & 11111001 & \\
\text{add 1} & 1 & \\
\hline
 & 11111010 & (-6)
\end{array}
$$

Method II

Start with the least significant bit (LSB) of the number. Proceed toward the remaining bits until a one is found. It may be the LSB. Complement all bits *after* the first "one" bit.

$$
\begin{array}{lll}
 & 11111010 & (-6) \\
\text{complement} & \longleftarrow\!\!\lrcorner & \\
 & 00000110 & (+6)
\end{array}
$$

Example 2.9. Find the binary equivalent of $(-43)_{10}$.

$$(+43)_{10} = (00101011)_2$$

Twos complement the binary number:

$$(-43)_{10} = (11010101)_2$$

The design of digital circuitry to implement subtraction is more complex than addition circuitry. It makes economic sense to use addition circuits for both. Digital computers perform subtraction by taking the 2s complement of the subtrahend and adding it to the minuend.

2.11   TWOS COMPLEMENT ARITHMETIC

When an addition operation is performed on signed (2s complement) numbers, the numbers are added regardless of their signs. The answer is a 2s complement number with the correct sign. We have seen that *any carries out of the MSB are meaningless and should be ignored.* The sum is correct only if it is within the allowed range. Unsigned numbers signal this fact by setting the carry. Signed numbers exceed the range under two conditions: (1) Two positive numbers are added to produce a negative sum or (2) two negative numbers are added to produce a positive sum. These results are not possible but arise with 2s complement numbers because of arithmetic overflow. Hardware circuits within the microprocessor simply check the signs (MSB) of the two numbers to be added. When they are the same, the MSB of the result must agree with the two MSBs. Otherwise the *overflow V is set, signaling an incorrect answer for signed numbers.* When a positive 2s complement and a negative 2s complement number are added, the sum is always correct.

Example 2.10. Add positive and negative.

$$
\begin{array}{ll}
\ 00001000 & (+8) \\
+\ 11110100 & (-12) \\
\hline
\ 11111100 & (-4) \qquad V = 0 \qquad C = 0 \quad \text{(ignore)} \\
& \text{CORRECT} \\
\end{array}
$$

$$\ 00000100 \qquad (+4) \qquad \text{complement to check}$$

Example 2.11. Add positive and positive (no overflow).

$$
\begin{array}{ll}
\ 00110011 & (+51) \\
+\ 01000001 & (+65) \\
\hline
\ 01110100 & (+116) \qquad V = 0 \qquad C = 0 \quad \text{(ignore)} \\
& \text{CORRECT} \\
\end{array}
$$

Example 2.12. Add positive and positive (overflow).

$$
\begin{array}{ll}
\ 01111111 & (+127) \\
+\ 00000001 & (+1) \\
\hline
\ 10000000 & (+128) \qquad V = 1 \qquad C = 0 \quad \text{(ignore)} \\
& \text{ERROR} \\
\end{array}
$$

Example 2.13. Add negative and negative (no overflow).

$$
\begin{array}{ll}
11111011 & (-5) \\
\underline{11111001} & (-7) \\
11110100 & (-12) \quad V = 0 \quad C = 1 \quad \text{(ignore)} \\
& \hspace{2.2cm} \text{CORRECT}
\end{array}
$$

Example 2.14. Add negative and negative (overflow).

$$
\begin{array}{ll}
10000011 & (-125) \\
\underline{11111010} & (-6) \\
01111101 & (-131) \quad V = 1 \quad C = 1 \quad \text{(ignore)} \\
& \hspace{2.2cm} \text{ERROR}
\end{array}
$$

Signed numbers overflow because they have a finite number of digits. The result in Example 2.14 would be correct if the numbers were 9 bits long.

$$
\begin{array}{ll}
110000011 & (-125) \\
\underline{111111010} & (-6) \\
101111101 & (-131) \quad V = 0 \quad C = 1 \\
& \hspace{2.2cm} \text{CORRECT}
\end{array}
$$

Digital computer software must use addition algorithms to check for overflow error since they are number crunching machines which sooner or later exceed their range. Microprocessor control applications often need only 8-bit arithmetic but rarely (if ever) exceed 16 bit arithmetic. Therefore, these "error" algorithms are not necessary. Sixteen-bit arithmetic with 8-bit registers will be demonstrated in a later chapter.

The adder in the microprocessor accepts both numbers to be added as inputs, with the sum appearing on the output. How does the adder know that the numbers are signed 2s complement and not unsigned binary numbers? The answer is it doesn't! It adds the numbers as unsigned binary numbers, but the answer is correct whether the numbers are unsigned or signed.

Unsigned numbers

$$
\begin{array}{ll}
\ \ 10001101 & (+141) \\
\underline{+\ \ 00010110} & (+22) \\
0]\ \ \ 10100011 & (+163) \quad C = 0 \\
& \hspace{1.8cm} \text{CORRECT}
\end{array}
$$

Signed numbers

$$
\begin{array}{ll}
\ \ 10001101 & (-115) \\
\underline{+\ \ 00010110} & (+22) \\
0]\ \ \ 10100011 & (-93) \quad V = 0 \\
& \hspace{1.8cm} \text{CORRECT}
\end{array}
$$

Notice that (1) the bit pattern is the same and (2) the answer is correct as both unsigned number addition and signed number addition. It is only our interpretation of these bit patterns which decides the arithmetic. Twos complement arithmetic allows us to work with either signed or unsigned numbers using the same internal adding circuit.

2.12 HEXADECIMAL ARITHMETIC

Programmers often find it necessary to add or subtract hexadecimal numbers, particularly when writing branch instructions. The operation can be performed in decimal and converted to hexadecimal by referring to tables. However, hexadecimal numbers can be added quickly with the following procedure:

1. Add the hex digits in a given column, substituting their decimal equivalent.
2. If the sum is 15 or less, write down the hexadecimal number.
3. If the sum is 16 or more, subtract 16 and carry one to the next column of digits on the left.

Example 2.15. Add $(3D4B)_{16}$ and $(C6A9)_{16}$.

$$
\begin{array}{ccccc}
(15+1) & 19 & (14+1) & 20 \\
3 & D & 4 & B \\
\underline{C} & \underline{6} & \underline{A} & \underline{9} \\
1]\quad 0 & 3 & F & 4 \\
\end{array}
$$

Hexadecimal subtraction is similar to rules for decimal subtraction. But in keeping with the signed number's concept, we suggest using the 16s complement (or forming the negative equivalent of the hex number) and adding. Negation of hex numbers follows these rules:

1. Add the hex digit to the least significant hex digit which makes the sum equal to 16.
2. Add digits to all other hex digits which make the sum equal to 15.
3. The added digits form negative equivalent.

Example 2.16. Find the negative equivalent of the hex number 4C0A.

$$
\begin{array}{ll}
A = 10 + 6 & 6 \\
0 = 0 + 15 & F \\
C = 12 + 3 & 3 \\
4 = 4 + 11 & B \\
\end{array}
$$

$$
\begin{array}{lc}
-(4C0A) = & B3F6 \\
& +\ 4C0A \\
\hline
1] & 0000 \quad \text{check} \\
\end{array}
$$

PROBLEMS

2.1. What is the minimum number of bits to represent the number of days in a year?
2.2. If an address bus has 16 lines, what is the number of bytes that can be addressed?
2.3. Why do some states use seven characters in license plates? Give a quantitative argument.
2.4. Think about a number system to the base 3. What are the digits? Demonstrate it by adding $(31)_{10}$ to $(64)_{10}$.
2.5. Convert the following decimal numbers to binary: (a) 48, (b) 631, (c) 1074, (d) −2067.
2.6. Convert the following binary numbers to decimal: (a) 11010, (b) 11010010, (c) 01111001, (d) 100011101, (e) 011110011. Assume (1) unsigned, (2) signed numbers.

2.7. Convert $(536)_{10}$ to BCD.

2.8. Convert the binary numbers in Prob. 2.6 to hexadecimal.

2.9. Convert the following hex numbers to decimal numbers: (*a*) 432, (*b*) C4, (*c*) A2F, (*d*) F4C5.

2.10. Convert the decimal numbers in Prob. 2.5 to hex.

2.11. Multiply 1110×1010. Express the answer in hex.

2.12. Solve for the quotient: $11001011 \div 101$.

2.13. Find the 1s complement of 11110000.

2.14. Find the 2s complement of 10101111.

2.15. What is the range of a 12-bit number if (1) unsigned, (2) signed?

2.16. Add $(112)_{10}$ and $(63)_{10}$ in BCD.

2.17. Add the following numbers and determine if the sum is correct if (1) unsigned, (2) signed.

(*a*) 10011101   (*b*) 01011100   (*c*) 01110010
    11000110   10010111   01011001

(*d*) 11001011   (*e*) 00010110   (*f*) 11011101
    11101101   01001101   01100011

2.18. Subtract the numbers in Prob. 2.17 if they are in 2s complement form.

2.19. Subtract the hexadecimal number 0042 from the number 0010.

CHAPTER
3

MICROPROCESSOR ARCHITECTURE/ ASSEMBLY LANGUAGE

All digital computers contain the following subsystems:

1. Arithmetic and logic unit
2. Control unit
3. Memory
4. Input/output (I/O)

The microprocessor is a sequential machine which contains the arithmetic/logic and control units on a single integrated circuit (IC). ICs have anywhere from a dozen to thousands of gates on a single chip. This was made possible in the late 1960s when MOS (metal oxide semiconductor) technology enabled the transistor to be reduced to "micro" cell size. The first microprocessor was manufactured in 1971 by Intel Corporation: the 4004. It contained a little over 2000 transistors on a small piece of semiconductor less than $\frac{1}{4}$ inch square. Today's 32-bit word size microprocessors in personal computers (PCs) contain hundreds of thousands of transistors and rival the speed and power of many mainframe computers.

While we may be anxious to learn how to interface the microprocessor for real-world control solutions, hardware (components and wiring) must first be put aside for software (programs). Microprocessor programming begins with the internal organization or *microprocessor architecture*.

3.1 MICROPROCESSOR/MICROCOMPUTER

The term "microprocessor" is often abbreviated as MPU (MicroProcessor Unit) or CPU (Central Processing Unit). The MPU is usually placed on a dual-in-line (DIP) plastic or ceramic package ($\sim$2 in $\times$ 0.5 in). Figure 3.1 shows the 40-pin package and pin assignment for the 6802. Computer systems which use microprocessors are called *microcomputers*. A block diagram of a microcomputer is given in Fig. 3.2. It includes the MPU, memory, and an input/output port called I/O. Microcomputers can be the desktop PCs familiar to everyone or a single board with three or four chips having a total cost under $10. Everything in the outside world connects to the system through the I/O port. The I/O data enters or leaves the port as binary. When the data is analog, suitable converter chips must be employed. Many manufacturers incorporate some memory and I/O capability with the MPU onto a single IC, called a microcomputer-on-a-chip or *microcontroller*. We will study microcontrollers in Chap. 17.

Information (binary 1s or 0s) is exchanged between the MPU and the other two elements of the system: memory and I/O. The details on how the system is organized to transfer this information is called the *system architecture*. Binary transfers travel on a collection of wires called *buses*.

The actual data appears on the *data bus*. A data word in the 6802 is 8 bits. We can see in Fig. 3.1 that the 6802 has eight data pins (D0 to D7). All eight data wires run in common to other chips. Data can be transferred in both directions, to or from the MPU.

The location of data in memory or I/O is called the *address*. The *address bus* consists of 16 wires from the corresponding MPU pins (A0 to A15). That represents 2^{16} or 65,536 addresses. All control applications need only a fraction of these addresses; thus, some address lines will be absent from a given memory or I/O. Address bits travel *from* the microprocessor only.

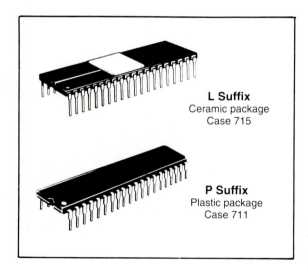

L Suffix
Ceramic package
Case 715

P Suffix
Plastic package
Case 711

(a)

FIGURE 3.1
The Motorola 6802/6808 8-bit microprocessor. (*a*) Package; (*b*) pin assignment. (*Reprinted with permission of Motorola, Inc.*)

Pin Assignment

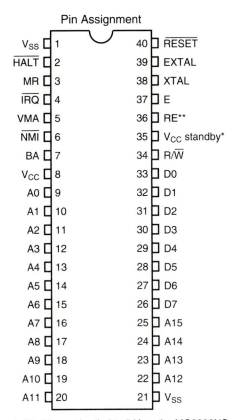

V_{SS}	1	40	$\overline{RESET}$
$\overline{HALT}$	2	39	EXTAL
MR	3	38	XTAL
$\overline{IRQ}$	4	37	E
VMA	5	36	RE**
$\overline{NMI}$	6	35	V_{CC} standby*
BA	7	34	$R/\overline{W}$
V_{CC}	8	33	D0
A0	9	32	D1
A1	10	31	D2
A2	11	30	D3
A3	12	29	D4
A4	13	28	D5
A5	14	27	D6
A6	15	26	D7
A7	16	25	A15
A8	17	24	A14
A9	18	23	A13
A10	19	22	A12
A11	20	21	V_{SS}

* Pin 35 must be tied to 5 V on the MC6802NS
** Pin 36 must be tied to ground for the MC6808

(*b*)   **FIGURE 3.1 (continued)**

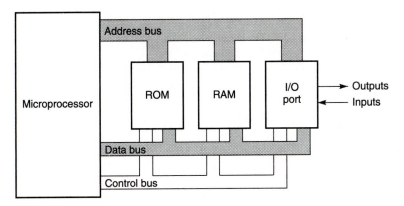

FIGURE 3.2
Microcomputer system.

The microprocessor controls and synchronizes the data transfer to or from one chip at a time by means of the *control bus.* All control signals are *from* the microprocessor only. One control line is the read/write ($R/\overline{W}$) line. The signal is HIGH when the microprocessor is ready to read (receive data), and the signal is LOW when the microprocessor is ready to write (send data). Control lines will be discussed under system architecture in Chap. 6.

3.2 PROGRAM/MEMORY

Computers receive data from external devices such as sensors, switches, or keyboards. They perform arithmetic and logic operations based on this data. The results are sent to displays, disks, or printers. In control applications commands are issued to open a valve, step a motor, or display a warning light. Although the MPU has remarkable capabilities, it will do only what it is told to do. Operations that the MPU is told to perform are called *instructions.* Typical instructions are ADD, AND, SHIFT, and STORE. Instructions are executed in the exact order they are presented. A group of sequential instructions for the MPU is called a *program.*

A comparison can be made between hand calculators and computers. Calculators have a similar, although limited, instruction set and the I/O is restricted to a keyboard and display. Instructions, along with numbers (data) to be operated on, are executed one step at a time. The result of an instruction is displayed after each entry. Programmers are rarely concerned with the speed of operation.

A given instruction can be executed in several microseconds. Since speed is paramount with both number crunching and control applications, it makes sense to store the program and execute it at once. (In fact, some calculators do just this.) *Memory* is an indispensable part of any microcomputer system because it stores the program. The MPU reads the first instruction stored in memory and acts upon this instruction. When it is completed, the MPU goes back to memory for the next instruction. The process is repeated indefinitely unless the MPU reaches a WAIT or HALT state.

Memory

Memory stores the program as well as other data, tables, calculation results, etc. A PC can have a memory size of over 256K words (bytes), while an appliance control board memory may only have a few hundred bytes of instruction. Unlike standard engineering notation where K is an abbreviation for 1000, K is 1024 when it is applied to memories.

Each byte of data in memory must have a location, i.e., address. As shown in Fig. 3.3, a 4K memory has 4096 ($= 2^{12}$) bytes of data. The address requires 12 bits to identify 4096 locations. However, the data is only 8 bits long. The 6802 uses a 16-bit address bus to identify up to 64K data bytes. Because of language features presented later, it is convenient to partition a hex address, e.g., 91FA, into 2 bytes. The most significant byte (MSByte) 91 is called the *page.* Page data location FA is the least significant byte (LSByte). There are 256 pages (FF) with 256 locations on each page. When the MPU signals memory, data is transferred between the data bus and memory. Figure 3.4 illustrates the page concept for a 64K memory. Only address lines A0 to A11 would be needed for the 4K memory.

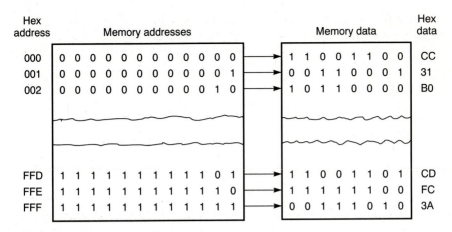

FIGURE 3.3
A typical 4K memory stores 4096 words of 8-bit length.

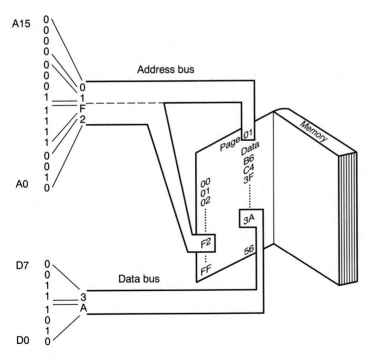

FIGURE 3.4
Addressing memory using page concept.

MEMORY READ. Memories operate in two basic ways: READ or WRITE. An MPU *reads* memory when it requests information (data) stored at a particular address. The steps are:

1. Location requested is placed on the address bus.
2. A READ command is sent to memory.
3. Data is transferred from memory to data bus.

A word at a particular address may be read many times. Each READ merely copies the data onto the data bus without altering the memory contents.

MEMORY WRITE. An MPU *writes* to memory when it wants to store data at a particular address. The steps are:

1. Location sought is placed on the address bus.
2. Data is transferred from the MPU to the data bus.
3. A WRITE command is sent to memory.

Written data is saved for future use. Therefore, the writing process destroys the previous data in memory at the sought address.

Most microprocessor systems use two types of memory: ROM and RAM. *ROM* is an acronym for *read-only memory;* the MPU cannot write to it. ROMs consist of hard-wired logic gates or programmable logic arrays (PLA) which output a given data byte when its address appears on the input pins. ROM programs are permanent and do not disappear after the system power is turned OFF. PCs use ROM to store a program called the *monitor.* It serves several functions:

1. Contains basic instructions for storing data in the RAM
2. Contains instructions for disk drive, keyboard, and cathode ray tube (CRT)
3. Gives programmer access to memory locations

Other ROMs serve as math coprocessors (floating-point arithmetic operations) or BASIC/FORTRAN interpreters. Control units for automobile subsystems and appliances, which run the same instructions over and over, are ROMs. *EPROMs* (erasable programmable ROMs) can be programmed by the user with special equipment, the procedure being quite simple. Contents can be erased with ultraviolet light in a matter of minutes. Figure 3.5 shows an EPROM. EPROMs are ideal for laboratory program development. They are cheaper for commercial applications when production runs are short.

RAM (random access memory) is memory when data in any given location (address) is directly accessible without going through other locations. But ROM locations are randomly accessible too! RAM is really READ/WRITE memory (RWM) for which RAM has become the accepted term. PC users WRITE programs into RAM from external

FIGURE 3.5
EPROM IC with quartz window to erase memory.

devices (keyboard or disk drive). There, it can be easily changed. Control systems in mechatronics will have very little RAM.

RAMs are designated as static or dynamic. Static RAMs (SRAMs) use flip-flops to store each binary bit. They are the most common and simplest to use. Dynamic RAMs (DRAMs) use capacitors to store the bit. They require special circuitry to pulse (refresh) the contents every few milliseconds or the contents are lost. However, dynamic RAM contains more memory for a given size, important for mainframe computers. Both RAMs are said to be volatile because the content is lost after the system power is turned OFF.

3.3 MICROPROCESSOR ARCHITECTURE

The word "architecture" is used to describe the microprocessor internal components, function, bus arrangement, and interrelationships. This structure is so complex that it would take a team several years to design one. We only want to use the microprocessor. But a general understanding of these components will make you a better programmer. Figure 3.6 gives an internal block diagram of the 6802. Components can be placed into four categories: (1) control unit (CU), (2) arithmetic and logic unit (ALU), (3) clock, and (4) programming registers. The CU and ALU are not accessible by the programmer, and they process instructions automatically. Register is the name given to any RAM-like storage cells that are found in ICs outside of memory. The 6802 has six programming registers that are accessible by the programmer. Their function is essential to learning the instruction set.

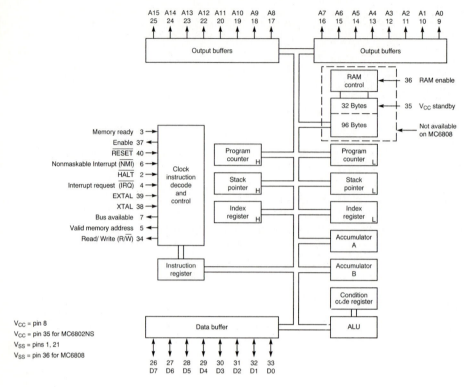

FIGURE 3.6
Block diagram of the Motorola 6802/6808 microprocessor. (*Reprinted with permission of Motorola, Inc.*)

Control Unit

The control unit is made up of the following:

1. *Data register.* This register acts as a buffer between the external data bus pins and the internal 8-bit bus which connects all components. After a READ it passes the instruction code to the instruction register prior to decoding. It temporarily holds data from a programming register, called the *accumulator*, just before a WRITE.

2. *Address register.* This 2-byte register acts as a buffer between the address bus pins and the internal 8-bit bus. It temporarily holds each instruction address prior to a READ/WRITE.

3. *Instruction decoder.* This circuit examines the data in the instruction register and decodes it into a set of signals for the logic control.

4. *Logic control.* This unit uses inputs from the master clock and the instruction decoder to derive timing signals. Some timing signals go to external pins and become the control bus. A READ/WRITE signal is one of them. Other external pins are inputs to the logic control, which can halt the microprocessor or interrupt the program. In addition, internal timing signals regulate the transfer of information among data/address registers, programming registers, and the ALU.

Arithmetic/Logic Unit

All ALUs perform the following arithmetic and logic operations:

1. Binary addition and subtraction
2. Logical AND, OR, XOR
3. Shift register contents left or right
4. Complement

Later more powerful processors have multiplication and division circuits. All circuits are hard-wired gates.

The ALU has two inputs and one output. See Fig. 3.7. One input is from the accumulator; another input is from memory by way of the internal data register. This forms the two math operands. The result of the operation is copied into the accumulator.

The ALU also sets a number of flip-flops, called *flags,* which store information about the result. Among the flags are the carry and overflow bits.

Clock

When the microprocessor executes an instruction, it must do so in a particular order and at a specific time. External control signals must be given to memory or I/O. Data is latched into the data register and passed to the instruction register for decoding. Bytes are shunted from register to register. New addresses are formed for the address bus. These critical timing requirements are controlled with an internal *two-phase clock.* Figure 3.8 shows the waveform. A clock is simply a periodic rectangular (almost) pulse. A two-phase clock generates two signals, called $\phi1$ and $\phi2$, which are 180° out of phase for the 6802. Various changes within the MPU are triggered by the sharp leading and trailing edges of each pulse. Signal HIGHs have a smaller time base than LOWs so that edges are nonoverlapping. The $\phi2$ signal also appears at pin 37 labled E. It is one of the control bus lines which synchronizes external ICs sharing common buses.

The clock period is determined by an external *quartz crystal* attached to pins 38 (XTAL) and 39 (EXTAL). Most 6802 systems operate at 1 MHz. Because there is an internal ÷4 circuit, a 4-MHz crystal is required.

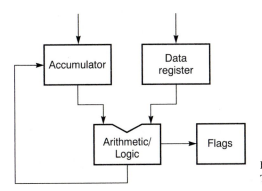

FIGURE 3.7
The arithmetic/logic unit operation.

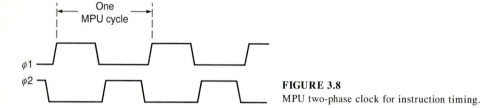

FIGURE 3.8
MPU two-phase clock for instruction timing.

Programming Registers

On the one hand, we have ICs which are external to the MPU. Memory contains the program. I/O is supplying and delivering information to the external world. On the other hand, we have the inaccessible intricate machinery of the control unit carrying out instructions ultimately for the I/O. The bridge spanning these two worlds is the programming registers. They are so closely bound to the instructional set that it is difficult to appreciate all of their utilities until later. Figure 3.9 gives the individual registers.

ACCUMULATORS. The 6802 has two accumulators called ACCumulator A (ACCA) and ACCumulator B (ACCB). Other microprocessors such as the 6502 or the 8085 have a single accumulator although they have other registers that function similarly. Both ACCA and ACCB are 8-bit registers which are indistinguishable in their behavior. An accumulator is the most important register in the MPU. All key operations funnel through it (them). First, the ACC is a temporary holding register before the ALU performs an operation. After the operation the ACC holds the answer: sum, difference, or logic. Second, the ACC is the source of all data sent to the I/O and is the destination of all data received from the I/O. Thus, it "accumulates" results.

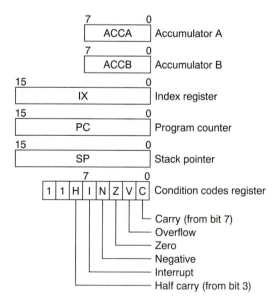

FIGURE 3.9
Programming registers for the 6802 microprocessor. (*Reprinted with permission of Motorola, Inc.*)

INDEX REGISTER. The index register (IX) is a 16-bit register that will allow us to write more compact programs. When the IX contains an address, it can be used to point at memory data. This will permit quick access to memory for table searches. In addition, a 16-bit register can be incremented or decremented 65,536 times compared to 256 times for an 8-bit accumulator. The range for counting is greatly extended.

PROGRAM COUNTER. The program counter (PC) is a 16-bit register that contains the address of the next byte in memory. It is a very powerful tool that functions as part of the control unit. Recall that a list of instructions in memory must be executed in exact numerical order by address. Each address must be placed on the address bus to locate the data byte. The program starting address is loaded into the PC. After each byte of instruction is transferred to the MPU for decoding the PC is incremented. After the MPU finishes each operation it places the PC address into the address register (then onto the address bus) in anticipation of receiving the next program byte from memory. Thus, the PC continually points to the next instruction to be executed.

STACK POINTER. The stack pointer (SP) is another special 16-bit register. It allows the MPU to use part of the RAM for temporary storage. The programmer can use instructions specifically for this purpose as a programming aid. Furthermore, the MPU will automatically store part of the programming registers during interrupts or branch to subroutines. The SP contains the address where bytes are to be stored. It is decremented after each byte is stored.

CONDITIONAL CODE REGISTER. The conditional code (CC) register is an 8-bit register in which each bit is called a flag. Flags are labeled H, I, N, Z, V, and C as shown in Fig. 3.9. Set bits in the two MSBs of the CC are mere fillers because these bits are not used. The function of flags is to retain information about the result of the last arithmetic or memory operation. The half-carry flag (H) is used by the MPU for the DAA instruction only. The interrupt mask (I) can be set by the programmer to prevent an interrupt from some outside event. Negative (N) and zero (Z) flags are set according to the arithmetic result. Carry (C) and overflow (V) flags have been discussed in Chap. 2. Branch instructions use the CC to determine if the MPU should execute the next instruction or go (branch) to an instruction elsewhere in the program.

3.4 ASSEMBLY LANGUAGE

Now that we have some understanding of the registers and internal circuits found in the microprocessor, we will see how a simple program is written. As an example, suppose we want to add 3 to 4 and place the result in memory at address 9. Recall that one of the accumulators must be involved because the accumulator "accumulates" the result of arithmetic operations. A program may look like this:

```
LOAD 04            first instruction
ADD 03             second instruction
STORE (at) 0009    third instruction
```

The first instruction loads the accumulator with data 04, the accumulator being implied. Data words are 8 bits long. With hexadecimal notation, 4 becomes 04. The second instruction takes data 03 into the ALU and adds the accumulator content to this number. The result is copied back into the accumulator. The third instruction stores the accumulator content at address 0009. Again, the accumulator as a source for the store operation is implied. Hexadecimal address 0009 is two data words long. Since all programs in memory are a string of data words, the address (as data) must be split into 2 bytes with the HIGH-byte given first. Thus,

> LOAD
>   04
> ADD
>   03
> STORE
>   00
>   09

Operation Code

The words LOAD, ADD, STORE cannot be understood by the microprocessor. Everything must be given to the MPU in binary. Therefore, operations LOAD, ADD, STORE are given as an 8-bit binary code expressed in hex: 86, 8B, B7. These numbers are called operation codes or *OP codes* for short. An MPU with 8-bit data words can execute 256 different operations. Our program is now listed in memory, starting at address 0000. See Table 3.1. Binary numbers that the MPU will actually use are called *machine language*. Sometimes the term is also applied to the hexadecimal equivalent.

Instructions consist of OP codes plus up to two more data bytes (numbers, addresses, etc.) following each OP code. There is no distinction between OP codes and data bytes except in their order of presentation. MPUs always interpret the first memory word as an OP code to be decoded. The decoder determines how many of the following bytes are part of the instruction. When the instruction is complete, it knows the next byte is an OP code, and the process is repeated.

TABLE 3.1
Memory content of "add program"

Address	Memory content	Hex equivalent
0000	10000110	86
0001	00000100	04
0002	10001011	8B
0003	00000011	03
0004	10110111	B7
0005	00000000	00
0006	00001001	09
...		
...		
0009	00000111	07

Mnemonics

It would soon become maddening if programmers worked entirely with machine language. LOAD, ADD, STORE are much easier concepts for people, if not for machines. Therefore, every manufacturer of MPUs defines a three- or four-letter code that describes the function of each instruction. The code is known as a *mnemonic* (ni′-man-ik), which means an aid to programming. Mnemonics are "concocted" from a description of the operation itself and distinguish which accumulator is being addressed. Thus,

Mnemonic	Description
LDA A	LoaD Accumulator A
ADD A	ADD to accumulator A
STA A	STore Accumulator A
LDX	LoaD indeX register
WAI	WAIt
CLR A	CLeaR accumulator A
BRA	BRanch Always

Every programmer must become familiar with a given MPU's mnemonics. Figure 3.10 lists the instruction set for the 6802. Each MPU has its own instruction set although the 6800, 6802, and 6808 share the same set. There are 72 mnemonics. Although the list may seem formidable, an individual can quickly learn about two dozen mnemonics that will carry him or her through most programming tasks. The analogy is similar to everyday speech, which is a small fraction of an individual's total vocabulary. A final format for the preceding "add program" is given in Table 3.2.

A set of instructions written in mnemonics is referred to as *assembly language*. Programs written in assembly language are called *source programs*. Higher level language

ABA	Add Accumulators	CLR	Clear	PUL	Pull Data
ADC	Add with Carry	CLV	Clear Overflow	ROL	Rotate Left
ADD	Add	CMP	Compare	ROR	Rotate Right
AND	Logical And	COM	Complement	RTI	Return from Interrupt
ASL	Arithmetic Shift Left	CPX	Compare Index Register	RTS	Return from Subroutine
ASR	Arithmetic Shift Right	DAA	Decimal Adjust		
BCC	Branch if Carry Clear	DEC	Decrement	SBA	Subtract Accumulators
BCS	Branch if Carry Set	DES	Decrement Stack Pointer	SBC	Subtract with Carry
BEQ	Branch if Equal to Zero	DEX	Decrement Index Register	SEC	Set Carry
BGE	Branch if Greater or Equal Zero			SEI	Set Interrupt Mask
BGT	Branch if Greater than Zero	EOR	Exclusive OR	SEV	Set Overflow
BHI	Branch if Higher	INC	Increment	STA	Store Accumulator
BIT	Bit Test	INS	Increment Stack Pointer	STS	Store Stack Register
BLE	Branch if Less or Equal	INX	Increment Index Register	STX	Store Index Register
BLS	Branch if Lower or Same			SUB	Subtract
BLT	Branch if Less than Zero	JMP	Jump	SWI	Software Interrupt
BMI	Branch if Minus	JSR	Jump to Subroutine	TAB	Transfer Accumulators
BNE	Branch if Not Equal to Zero	LDA	Load Accumulator	TAP	Transfer Accumulators to Condition Code Reg
BPL	Branch if Plus	LDS	Load Stack Pointer	TBA	Transfer Accumulators
BRA	Branch Always	LDX	Load Index Register	TPA	Transfer Condition Code Reg to Accumulator
BSR	Branch to Subroutine	LSR	Logical Shift Right	TST	Test
BVC	Branch if Overflow Clear			TSX	Transfer Stack Pointer to Index Register
BVS	Branch if Overflow Set	NEG	Negate	TXS	Transfer Index Register to Stack Pointer
		NOP	No Operation		
CBA	Compare Accumulators	ORA	Inclusive OR Accumulator	WAI	Wait for Interrupt
CLC	Clear Carry				
CLI	Clear Interrupt Mask	PSH	Push Data		

FIGURE 3.10
6802 instruction set illustrating mnemonics. (*Reprinted with permission of Motorola, Inc.*)

TABLE 3.2
Final format for the "add program"

Address	Memory content	Mnemonics	Comments
0000	86	LDA A 04	Load accumulator A
0001	04		with 04.
0002	8B	ADD A 03	Add 03 to accumulator A.
0003	03		
0004	B7	STA A (0009)	Store accumulator A
0005	00		at address 0009.
0006	09		
0007	3E	WAI	Wait.
...			
0009	—	—	Result.

statements consist of a series of source programs. Once source programs have been written to perform a given task, they must be converted to machine language in one of two ways:

1. *Hand assembly.* Write the source program. Look up OP codes in the MPU manufacturer's manual and load the string of hex data bytes (as binary) into successive RAM addresses. This can be done directly with a personal computer using the monitor. Trainers, such as the Heath 3401, and EPROM programmers have ROM programs which perform the same task. Hand assembly is suitable for short programs which include all the programs in this text.

2. *Assembler.* For longer programs hand assembly is tedious and requires accurate formatting. An assembler is an independent program designed to convert the assembly language into machine code. Each microcomputer must have a resident or *self assembler* to convert its own assembly language into machine language. The program may be on-board the ROM or loaded into RAM via a floppy disk. Most microprocessors from a given company have different architecture and slightly different mnemonic codes. The diversity is even greater between companies. An assembler designed to work on one computer to produce machine language code for another computer or MPU is called a *cross-assembler.* The advantage of assemblers is that memory allotment and content as well as "debugging" is handled automatically.

To help the assembler identify the various types of numbers which may appear in a program, the following symbols are used as prefixes:

1. $ sign indicates the number is a hex number ($13 is hex, equivalent to 19_{10}).
2. @ sign indicates an octal number.
3. % sign indicates a binary number.
4. No symbol indicates a decimal number.

Since all programs in this text are intended for hand assembly, we will write hex numbers only and drop the $.

3.5 INSTRUCTION INFORMATION SHEET

Appendix A gives an information sheet on each mnemonic in the 6802 instruction set. The sheets appear in alphabetical order according to the mnemonic placed in an upper corner. Figure 3.11 shows the sheet for LDA.

Most instructions ask the MPU to transfer data between registers or between a register and memory. The line

$$\text{Operation:} \qquad (\text{ACCX}) \leftarrow (\text{M})$$

explains the transfer process in shorthand notation. The notation is read from right to left:

(M)	the contents of memory location M
←	are transferred to
(ACCX)	accumulator X (substitute A or B)

All branching operations are determined by flag settings in the condition code (CC) register. Therefore, the instruction preceding the branch instruction is important. Information sheets tell how each CC register bit is affected by the instruction. Usually, only select bits are affected. Here, the N bit is set if the loaded number is negative, and the Z bit is set if the loaded number is zero. The V bit is cleared.

The table at the bottom of Fig. 3.11 suggests that there are four addressing methods to load accumulators A and B. These methods are called addressing modes. Each has its

---

LDA	Load Accumulator	LDA

Operation: ACCX ← (M)

Description: Loads the contents of memory into the accumulator. The condition codes are set according to the data.

Condition Codes: H: Not affected.
I: Not affected.
N: Set if most significant bit of the result is set; cleared otherwise.
Z: Set if all bits of the result are cleared; cleared otherwise.
V: Cleared.
C: Not affected.

Addressing Modes	Execution Time (No. of cycles)	Number of bytes of machine code	Coding of First (or only) byte of machine code		
			HEX.	OCT.	DEC.
A IMM	2	2	86	206	134
A DIR	3	2	96	226	150
A EXT	4	3	B6	266	182
A IND	5	2	A6	246	166
B IMM	2	2	C6	306	198
B DIR	3	2	D6	326	214
B EXT	4	3	F6	366	246
B IND	5	2	E6	346	230

FIGURE 3.11
A sample instruction information sheet. (*Reprinted with permission of Motorola, Inc.*)

own OP code. OP code 86, which was the LDA instruction in the previous "add program," is A IMM (immediate) mode. It takes 2 bytes of machine code: OP code + data. Execution time to complete the instruction is two clock cycles. Addressing modes are discussed in the following section.

3.6 ADDRESSING MODES

Every instruction starts with an OP code to tell the MPU what operation to perform. Either none, one, or 2 bytes, called the *operand,* follow the code to complete the instruction. The OP code itself determines where the data is found (addressed) in memory, what registers to use, and the number of bytes needed. To a large extent, the programmer must know this information to select the proper OP code. The selection procedure is organized into groups called addressing modes. They include the (1) immediate (IMM), (2) extended (EXT), (3) direct (DIR), (4) inherent (INH), (5) index (IND), and (6) relative (REL).

Immediate Mode (IMM)

Immediate instructions are used if the data is permanent and known to the programmer when the program is written. Instructions generally consist of 2 or 3 bytes (OP code + data). The byte which follows immediately after the OP code is the data for one of the MPU programming registers. See Fig. 3.12. Two data bytes will be necessary when addressing the index register or stack pointer. A pound sign (#) is the assembler symbol for an immediate instruction. Examples follow.

Memory content	Mnemonics	Comments
86	LDA A #04	Put next byte 04
04		into accumulator A.
8B	ADD A #03	Add next byte 03
03		to accumulator A.

The "add program" in Table 3.2 was written in the IMM mode.

Extended Mode (EXT)

Extended instructions are used when a program address for the data is elsewhere rather than the next address following the OP code. Thus, data can be loaded or stored into specific locations after the program is written. Instructions consist of 3 bytes: OP code plus the address of the data in memory. The second byte contains the MSByte or page of the address, while the third byte contains the LSByte or line in the page. See Fig. 3.13. As an example of extended addressing, the "add program" in Table 3.2 is now written in the EXT mode. See Table 3.3.

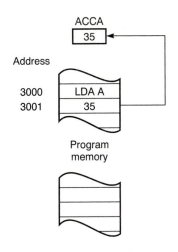

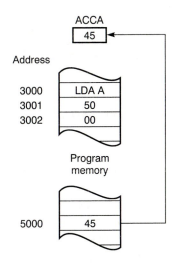

FIGURE 3.12
Immediate mode transfer of data to
accumulator A.

FIGURE 3.13
Extended mode transfer of data to accumulator A.

TABLE 3.3
"Add program" written in extended mode

Address	Memory content	Mnemonics	Comments
0000	04		
0001	03		
0002	—		Result.
0003	B6	LDA A 0000	Get byte 04 at 0000
0004	00		for accumulator A.
0005	00		
0006	BB	ADD A 0001	Add byte 03 at 0001
0007	00		to accumulator A.
0008	01		
0009	B7	STA A 0002	Store A at 0002.
000A	00		
000B	02		
000C	3E	WAI	Wait.

Direct Mode (DIR)

The direct mode is very similar to extended mode for addressing: OP code followed by the address of the data. However, the address must start on page 0. Instructions consist of 2 bytes: Op code plus the LSByte of the address. Since the MPU supplies a 16-bit address to the bus, the MSByte of the address is set to 00 automatically by the MPU. Memory locations which can be addressed by a direct instruction are limited to 0000 to 00FF. Extended addressing has the advantage of being able to select any memory location 0000 to FFFF. Direct addressing has the advantage of one less byte in the instruction. Fewer

TABLE 3.4
"Add program" written in direct mode

Address	Memory content	Mnemonics	Comments
0000	04		
0001	03		
0002	—		Result.
0003	96	LDA A 00	Get byte 04 at 0000
0004	00		for accumulator A.
0005	9B	ADD A 01	Add byte 03 at 0001
0006	01		to accumulator A.
0007	97	STA A 02	Store A at 0002.
0008	02		
0009	3E	WAI	Wait.

memory spaces are needed, and the execution time is less. As an example of direct addressing, the "add program" in Table 3.2 is now written in the DIR mode. See Table 3.4.

Inherent Mode (INH)

Inherent instructions, often called implied instructions, consist of only one byte: the OP code. Generally, the information required by the OP code is within an MPU register, i.e., the MPU does not have to WRITE memory to complete the instruction. The WAI instruction is an example. Other mnemonics which have multiple OP codes often include a single-byte accumulator instruction. For example, the mnemonic CLR has the following modes:

> A(INH)
> B(INH)
> EXT
> IND

The acronym INH is "implied" but never written.

Indexed Mode (IND)

As the name suggests, the indexed mode makes use of the index register (IX). Most manufacturers refer to the index register with an X. Indexed instructions consist of 2 bytes: OP code plus OFFSET. The address of the data is the offset byte added to the contents of the index register as an unsigned number. See Fig. 3.14. There is no change in the index register. It is only used to calculate the address. A distinction must be made between indexed instructions and index register instructions, such as LDX, which change the index register content. The "add program" in Table 3.2 is now written in the IND mode. See Table 3.5.

A base address (0000) is loaded into the index register. An INDEXED LOAD finds data 04 at base address plus 00. Data 03 is INDEXED ADDED to 04 in accumulator A by finding 03 at base address plus 01. Finally, an INDEXED STORE places the result in accumulator A at base address plus 02.

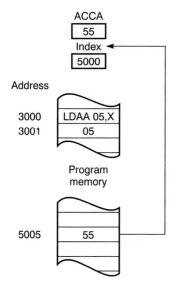

FIGURE 3.14
Indexed mode transfer of data to accumulator A.

TABLE 3.5
"Add program" written in indexed mode

Address	Memory content	Mnemonics	Comments
0000	04		
0001	03		
0002	—		Result.
0003	CE	LDX #0000	Load 0000 into IX.
0004	00		
0005	00		
0006	A6	LDA A 00,X	Get byte 04 at 0000 + 00
0007	00		for accumulator A.
0008	AB	ADD A 01,X	Add byte 03 at 0000 + 01
0009	01		to accumulator A.
000A	A7	STA A 02,X	Store A at 0000 + 02.
000B	02		
000C	3E	WAI	Wait.

If this program does not appear efficient, it isn't. Indexed instructions are most useful, if not indispensable, when it is necessary to move or change blocks of data in memory and when it is desirable to access tables in memory.

Relative Mode (REL)

The relative mode is restricted to branch instructions and is the instruction's only mode. Relative instructions have 2 bytes: OP code plus displacement. The displacement is relative to the program counter (PC) and is added to the PC as a signed number. Thus, program control can be transferred forward or backward. Consider the following instruction:

Address	Memory content	Mnemonics
6000	20	BRA 10
6001	10	
6002	XX	NEXT INST

The PC summary is

$$PC = 6000 \quad \text{before BRA 10 executed}$$
$$PC = 6002 \quad \text{while MPU interpreting displacement 10}$$
$$PC = 6002 + 10 \quad \text{after BRA 10 executed}$$
$$= 6012$$

3.7 INSTRUCTION TIMING

When the microprocessor executes a program, it proceeds from instruction to instruction in sequential order. The process never ends, branching back to an earlier instruction(s), unless a rare WAI instruction is encountered. Two basic steps are repeated over and over as shown in Fig. 3.15: *fetch* the instruction OP code, *execute* the OP code. During the fetch phase the OP code is taken from memory and placed in the MPU data register. This operation is the same for every instruction and takes one clock cycle. Executing the OP code can take anywhere from 1 to 11 clock cycles. IMM and INH modes are shorter than EXT or IND modes.

During each clock cycle several events occur within the MPU. Events are timed by the rising and falling edges of the two-phase clock. We will illustrate these events for the execution of the following instruction:

Address	Memory content	Mnemonics
0007	B7	STA A (1234)
0008	12	
0009	34	

Figure 3.16 shows a timing diagram for the instruction. The $\phi 1$ clock is principally responsible for placing PC content on the address bus and incrementing the PC for the next program byte. The $\phi 2$ clock is chiefly responsible for signaling memory to place the byte on the data bus and then latching the byte into the MPU data register.

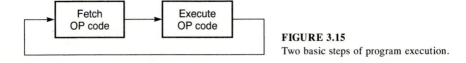

FIGURE 3.15
Two basic steps of program execution.

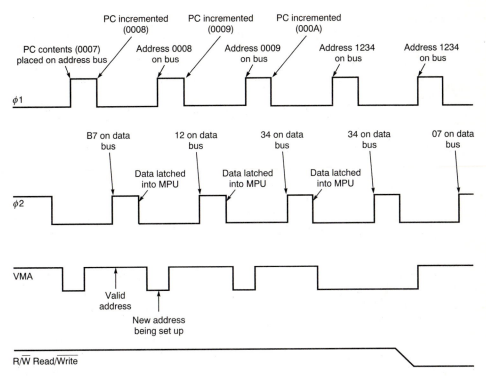

FIGURE 3.16
Timing diagram for instruction STA A (1234).

Cycle one

1. Contents (0007) of PC are placed on the address bus from the MPU address register.
2. PC is incremented by one to 0008. There is no change on the address bus.
3. Contents (B7) of memory at address 0007 are placed on the data bus.
4. Data (B7) is latched into the MPU data register and decoded as a store A extended.

Cycle two

5. Contents (0008) of PC are placed on the address bus from the MPU address register.
6. PC is incremented by one to 0009.
7. Contents (12) of memory at address 0008 are placed on the data bus.
8. Data (12) is latched into the MPU data register. This data is the MSByte of a store A address.

Cycle three

9. Contents (0009) of PC are placed on the address bus from the MPU address register.
10. PC is incremented by one to 000A.

11. Contents (34) of memory at address 0009 are placed on the data bus.
12. Data (34) is latched into the MPU data register—LSByte and MSByte of address 1234 to the MPU address register.

Cycle four

13. Contents (1234) of the MPU address register are placed on the address bus.
14. Contents of the accumulator are prepared for transfer to the MPU data register.

Cycle five

15. Contents of the accumulator is gated to the data bus.
16. Read/Write line is put in a LOW state (write), signaling memory to store data byte at address 1234.

Control lines. There are three control lines (signals) from the MPU to the memory chip: $\phi2$ clock, valid memory address (VMA), and Read/$\overline{\text{Write}}$ (R/$\overline{\text{W}}$). They communicate with memory for each stage of the instruction execution. Memory chip control will be discussed in Chap. 6.

PROBLEMS

3.1. What is a program?
3.2. When the MPU adds two numbers, where are the numbers held? Where is the sum held?
3.3. When the MPU accepts data from the bus, where does the data go first?
3.4. What is an OP code?
3.5. How can a program be executed sequentially?
3.6. Give three steps taken by the MPU to write the number B6 into memory at location 6000.
3.7. What is the first byte of any instruction?
3.8. How many bytes make up an instruction using the (*a*) extended mode, (*b*) implied mode, (*c*) relative mode, (*d*) direct mode, (*e*) indexed mode?
3.9. How much time does it take to execute the following instructions? (*a*) ADD A #03 (*b*) CLR 06,X (*c*) STA B B6 (*d*) WAI.
3.10. The program counter (PC) contains address 2005. What is the address of the present memory byte being processed by the MPU?
3.11. The index register contains the hex number 06B4. What register or memory location is cleared by the instruction CLR 06,X?
3.12. Give the mnemonics for each general register transfer statement.
(*a*) (ACCB) ← (ACCB) + 01
(*b*) (3B1F) ← (ACCA)
(*c*) (ACCA) ← 00
(*d*) (IXH) ← (07DE) (IXL) ← (07DF)
3.13. Accumulator A contains 7C. What flags are affected by each instruction? (*a*) DEC A, (*b*) ADD A #9A, (*c*) SUB A #B3, (*d*) ADD A #04.

3.14. Show the memory content for the following program starting at address 0000:

LDA A #4E
LDA B #2A
INC A
SBA
STA A DF36
WAI

3.15. Memory contains the list of numbers 26, 36, 46, 56, 66, 76 starting at address 0000. Write a program to add the list and store the sum at address 0100.

3.16. Show a timing diagram for the instruction LDA A #01.

CHAPTER
4

INSTRUCTION SET/ PROGRAMMING

Each main subsystem—microprocessor, memory, and input/output (I/O) port—can be thought of as a collection of registers (bit storage devices). We will discover in Chap. 6 that I/O is seen by the microprocessor as addresses just like memory. There are two basic operations performed by the microprocessor as it executes the program: (1) data transfer among the subsystem registers and (2) arithmetic and logic transformations within the ALU. The individual kinds of transfers and transformations make up the *instruction set*.

A particular microprocessor executes its own instruction set. Sets are quite similar among microprocessors made by the same company. The 6802/6808 have an instruction set which includes 72 separate instructions. However, each instruction can have more than one addressing mode telling the MPU where to find the data. Thus, there are 197 valid OP codes. An information sheet for each instruction is given in App. A.

It is common practice to divide an instruction set into groups of functionally similar instructions for ease in learning. One grouping is:

1. *Data transfer group.* Instructions which move data between registers.
2. *Arithmetic group.* Instructions which add, subtract, increment, or decrement the registers.
3. *Logic group.* Instructions which AND, OR, complement, etc., data in the accumulator and memory.

4. *Shift group.* Instructions which move all bits in a given location one place to the right or left.

5. *Branch group.* Instructions which transfer the program execution sequence to a specified address. This enables the programmer to make decisions.

Instructions involving the stack, subroutines, and interrupts will be covered in separate chapters.

4.1 DATA TRANSFER INSTRUCTIONS

LDA A	Load accumulator A		
LDA B	Load accumulator B		
Immediate	86 F1	LDA A #F1	Load A with data F1.
Direct	D6 A3	LDA B A3	Load B with data at address 00A3.
Extended	B6 03 CF	LDA A 03CF	Load A with data at address 03CF.
Indexed	E6 FF	LDA B FF,X	Load B with data at address IX + FF.

STA A	Store accumulator A		
STA B	Store accumulator B		
Direct	97 36	STA A 36	Store A at address 0036.
Extended	F7 F0 00	STA B F000	Store B at address F000.
Indexed	A7 00	STA A 00,X	Store A at address IX + 00.

LDX	Load index register (IX)		
Immediate	CE C8 40	LDX #C840	Load data C840 into IX.
Direct	DE A1	LDX A1	Load data at address 00A1 into IX HIGH-byte and data at address 00A2 into IX LOW-byte.
Extended	FE 43 16	LDX 4316	Load data at address 4316 into IX HIGH-byte and data at address 4317 into IX LOW-byte.
Indexed	EE 26	LDX 26,X	Load data at address IX + 26 into IX HIGH-byte and data at address IX + 27 into IX LOW-byte.

STX	Store index register (IX)		
Direct	DF 2A	STX 2A	Store IX HIGH-byte at address 002A and LOW-byte at address 002B.
Extended	FF 1C 7E	STX 1C7E	Store IX HIGH-byte at address 1C7E and LOW-byte at address 1C7F.
Indexed	EF 06	STX 06,X	Store IX HIGH-byte at address IX + 06 and LOW-byte at address IX + 07.

| TAB | Transfer accumulator A to accumulator B |
| TBA | Transfer accumulator B to accumulator A |

| Inherent | 16 | TAB | A is copied into B. |
| Inherent | 17 | TBA | B is copied into A. |

CLR A	Clear accumulator A
CLR B	Clear accumulator B
CLR	Clear operand address

Inherent	4F	CLR A	
Inherent	5F	CLR B	
Extended	7F 60 00	CLR 6000	Data 00 stored at address 6000.
Indexed	6F 15	CLR 15,X	Data 00 stored at address IX + 15.

CLR A or CLR B instruction does the same thing as LDA A #00 or LDA B #00, but it takes one less byte of program memory. There is no instruction such as CLX to clear the index register. Use LDX #0000 instead.

Conditional Code Register

The conditional code (CC) register has certain bits which are set or cleared according to the results of instruction execution. The 8 bits of the CC register are as follows:

bit 0:   C (carry or borrow flag)
bit 1:   V (2s complement underflow or overflow flag)
bit 2:   Z (zero result flag)
bit 3:   N (negative result flag)
bit 4:   I (interrupt mask)
bit 5:   H (half carry, bit 3 to bit 4)
bit 6 and 7:   (always one)

Bits can be read by the programmer with the CC register transfer to accumulator A. Similarly, bits can be modified with an accumulator A transfer to the CC register. Also, these operations would be performed if the program had to preserve the present contents of the CC register to use the CC register in a subroutine.

| TPA | Transfer CC register to accumulator A |
| TAP | Transfer accumulator A to CC register |

| Inherent | 07 | TPA | CC is copied into A. |
| Inherent | 06 | TAP | A is copied into CC. |

Specific instructions exist to set or clear the C, V, and I bits.

CLC	Clear carry flag
SEC	Set carry flag
CLV	Clear overflow flag
SEV	Set overflow flag
CLI	Clear interrupt flag
SEI	Set interrupt flag

Inherent	0C	CLC	CC register bit 0 = 0.
Inherent	0D	SEC	CC register bit 0 = 1.
Inherent	0A	CLV	CC register bit 1 = 0.
Inherent	0B	SEV	CC register bit 1 = 1.
Inherent	0E	CLI	CC register bit 4 = 0.
Inherent	0F	SEI	CC register bit 4 = 1.

Miscellaneous instruction

NOP	No operation

Inherent	01	NOP	Does nothing.

NOP instruction causes the program counter (PC) to be incremented. No other internal or external registers are affected. Blocks of NOPs can be used as placeholders for program sections removed or inserted later. They can be placed in timing loops to slow execution time. Each NOP takes two cycles of machine time.

4.2 ARITHMETIC INSTRUCTIONS

ADD A	Add operand to accumulator A
ADD B	Add operand to accumulator B

Immediate	8B FF	ADD A #FF	Add FF to A, sum in A.
Direct	DB FF	ADD B FF	Add to B the data stored at address 00FF.
Extended	BB 00 FF	ADD A 00FF	Add to A the data stored at address 00FF.
Indexed	EB 07	ADD B 07,X	Add to B the data stored at address IX + 07.

If the sum exceeds decimal 255, the carry flag is set. The flag effectively is the ninth bit. Multiple-precision addition can be done by adding the carry to the next higher byte. Add with carry instructions facilitate the process.

ADC A	Add operand plus carry to accumulator A		
ADC B	Add operand plus carry to accumulator B		

Immediate	89 0B	ADC A #0B	Add 0B plus carry to A.
Direct	D9 A0	ADC B A0	Add to B the carry plus the data stored at address 00A0.
Extended	BB 02 36	ADC A 0236	Add to A the carry plus the data stored at address 0236.
Indexed	EB 02	ADC B 02,X	Add to B the carry plus the data stored at address IX + 02.

Example 4.1. Write a program to add 16-bit binary numbers (double-precision addition). One addend is found at addresses 0000 and 0001, LOW-byte first. The second addend is found at addresses 0002 and 0003. Results are stored at addresses 0004 and 0005.

Address	Comments
0000	LOW-byte of addend 1
0001	HIGH-byte of addend 1
0002	LOW-byte of addend 2
0003	HIGH-byte of addend 2
0004	LOW-byte of sum
0005	HIGH-byte of sum

Program

Address	Content	Mnemonics	Comments
0010	D6 00	LDA B 00	Load LOW-byte direct.
0012	DB 02	ADD B 02	Add both LOW-bytes.
0014	D7 04	STA B 04	Store LOW-byte sum.
0016	D6 01	LDA B 01	Load HIGH-byte direct.
0018	D9 03	ADC B 03	Add both HIGH-bytes and carry.
001A	D7 05	STA B 05	Store HIGH-byte sum.
001C	3E	WAI	Wait.

SUB A	Subtract operand from accumulator A
SUB B	Subtract operand from accumulator B
SBC A	Subtract operand plus borrow from accumulator A
SBC B	Subtract operand plus borrow from accumulator B

Subtract instructions take the operand (subtrahend) from the accumulator (minuend), storing the difference in the accumulator. If the operand is larger than the accumulator, a borrow occurs, setting the carry (borrow) flag. Multiple-byte subtraction routines can be written with the SBC instruction. The addressing format for the operand is identical to the ADD and ADC instructions.

| ABA | Add accumulator B to accumulator A |
| SBA | Subtract accumulator B from accumulator A |

| Inherent | 1B | ABA | B is added to A; sum is stored in A. B is unchanged. |
| Inherent | 10 | SBA | B is subtracted from A; difference is stored in A. B is unchanged. |

INC A	Increment accumulator A
INC B	Increment accumulator B
INC	Increment contents of operand address

Inherent	4C	INC A	Add one to A.
Inherent	5C	INC B	Add one to B.
Extended	7C 50 A2	INC 50A2	Add one to address 50A2.
Indexed	6C 03	INC 03,X	Add one to address IX + 03.

DEC A   Decrement accumulator A

DEC B   Decrement accumulator B

DEC   Decrement contents of operand address

DEC instructions subtract one from the indicated register. Addressing format is the same as the INC instruction. DEC is often used to decrement a counter (register) to zero, branching to another address when the count equals zero.

| INX | Increment the index register |
| DEX | Decrement the index register |

| Inherent | 08 | INX | Add one to IX. |
| Inherent | 09 | DEX | Subtract one from IX. |

INX advances the index register, pointing to a new address, when tables are searched. DEX decrements the index register as a 16-bit counter.

CMP A	Compare operand to accumulator A
CMP B	Compare operand to accumulator B
CBA	Compare accumulator A to accumulator B

Immediate	81 C5	CMP A #C5	Compare C5 to A.
Direct	D1 C5	CMP B C5	Compare data at address 00C5 to B.
Extended	B1 26 A4	CMP A 26A4	Compare data at address 26A4 to A.
Indexed	E1 08	CMP B 08,X	Compare data at address IX + 08 to B.
Inherent	11	CBA	Compare B to A.

CMP instructions subtract memory or "immediate" data from an accumulator. Unlike a SUB instruction, the actual results of the subtraction do not appear in the accumulator or anywhere. Both memory and accumulator contents are unchanged. The purpose of a CMP is to set the conditional code flags prior to a branch instruction.

Example 4.2. Determine if the contents of address 26A4 are less than, equal to, or greater than $(36)_{16}$.

The following program will set the CC flags to make our decision.

C6 36	LDA B #36	Load B with number.
F1 26 A4	CMP B 26A4	Compare data at address 26A4 to B.

1. N = 1, Z = 0: Memory content is greater than 36.
2. Z = 1: Memory content and 36 are equal.
3. N = Z = 0: Memory content is less than 36.

CPX	Compare operand to index register		
Immediate	8C 60 00	CPX #6000	Compare 6000 to IX.
Direct	9C 60	CPX 60	Compare contents of addresses 0060 and 0061 to IX.
Extended	BC 60 00	CPX 6000	Compare contents of addresses 6000 and 6001 to IX.
Indexed	AC 60	CPX 60,X	Compare contents of addresses IX + 60 and IX + 61 to IX.

The CPX instruction subtracts memory or "immediate" data from the 16-bit index register. Contents of the indicated address are compared to the index register HIGH-byte, and contents of the next address are compared to the index register LOW-byte. The register contents do not change. CPX is used to determine when the end of a list is reached. For example, a list of numbers resides between addresses 6000–6020. A list search is terminated by a CPX 6020, branching on Z = 1. Contrary to Motorola literature, the N and V flags set improperly, and the reader should not attempt to branch from these bits.

TST A	Test accumulator A contents		
TST B	Test accumulator B contents		
TST	Test contents of operand address		
Inherent	4D	TST A	Subtract 00 from A.
Extended	7D 40 00	TST 4000	Subtract 00 from the contents of address 4000.
Indexed	6D 0A	TST 0A,X	Subtract 00 from the contents of address IX + 0A.

The TST instruction subtracts zero from memory or accumulator but does not disturb them. TST A and TST B are equivalent to CMP A #00 and CMP B #00. Its advantage is the saving of byte 00 in the program.

DAA   Decimal adjust accumulator A

Inherent   19   DAA   Corrects binary sum to provide BCD result.

The DAA instruction allows direct addition of BCD numbers. Otherwise BCD numbers would need to be converted to binary, added, and converted back to BCD. Suppose we want to add $(37)_{10}$ and $(24)_{10}$, stored as BCD numbers:

$$
\begin{array}{rclcc}
(37)_{10} & = & & 0011 & 0111 \\
+\ (24)_{10} & = & + & 0010 & 0100 \\
\hline
(61)_{10}\ \text{ans.} & & 0] & 0101 & 1011 \\
& & & 5 & B
\end{array}
$$

The result 5B is not a valid BCD number. A DAA instruction following the binary addition instruction corrects the result by these rules:

1. Add 06 to accumulator A if the lower nibble (bits 0 to 3) exceeds 9 or if the half carry flag $H = 1$ (a carry from bit 3 to bit 4 occurred).
2. After step 1 add 60 to accumulator A if the upper nibble exceeds 9 or if the carry flag $C = 1$. Thus,

$$
\begin{array}{rcc}
& 0101 & 1011 \\
+ & 0000 & 0110 \\
\hline
0] & 0110 & 0001 \\
& 6 & 1
\end{array}
$$

DAA does not work with subtract instructions or with accumulator B. BCD subtraction is accomplished by complementation. The subtrahend is replaced by its 10s complement, and the numbers are added as in Example 4.3. BCD addition is useful in programs for sales registers, calculators, and games—less useful in control applications.

Example 4.3. The decimal numbers 4688 and 3736 are stored as BCD numbers, upper byte first, in addresses 0000–0003. Write a program to add the numbers and store a BCD result in addresses 0004–0005.

Address	Content	Mnemonic	Comments
0006	96 01	LDA A 01	Load LOW-byte of augend.
0008	9B 03	ADD A 03	Add LOW-byte of addend.
000A	19	DAA	Adjust LOW-byte result.
000B	97 05	STA A 05	Store LOW-byte result.
000D	96 00	LDA A 00	Load HIGH-byte of augend.
000F	99 02	ADC A 02	Add HIGH-byte of addend.
0011	19	DAA	Adjust result.
0012	97 04	STA A 04	Store HIGH-byte result.

4.3 LOGIC INSTRUCTIONS

| AND A | AND accumulator A |
| AND B | AND accumulator B |

Immediate	84 01	AND A #01	AND A with 01.
Direct	D4 0F	AND B 0F	AND B with data at address 000F.
Extended	B4 C6 01	AND A C601	AND A with data at address C601.
Indexed	E4 B0	AND B B0,X	AND B with data at address IX + B0.

Each bit of the operand is ANDed with the corresponding bit of the accumulator. The result is left in the accumulator. Thus, the instructions

    LDA A #AF
    AND A #01

are implemented as

Old accumulator A	10101111	AF
Operand data	00000001	01
New accumulator A	00000001	01

Because we have ANDed the higher bits with zero, only the status of bit 0 is left in the accumulator. The procedure of "zeroing out" selected bits is called *masking*. It is used to determine the status of individual devices connected to a port.

Example 4.4. Eight devices are connected to an 8-bit port located at address 6000; device N is connected to port bit N. Write a program to check if device 3 is ON (port bit 3 HIGH).

    LDA A 6000
    AND A #08

Logic operations set the N and Z flags according to the result. In this case $Z = 1$ if the device is OFF, and $Z = 0$ if the device is ON. A decision is made with a branch instruction.

| BIT A | Bit test accumulator A |
| BIT B | Bit test accumulator B |

BIT instructions serve the same function as AND except that the accumulator data is not changed. In effect, it sets the flags without storing the result in the accumulator. BIT is analogous to CMP in that one byte of memory is saved.

| ORA A | OR accumulator A |
| ORA B | OR accumulator B |

Immediate	8A F0	ORA A #F0	OR A with F0.
Direct	DA 60	ORA B 60	OR B with data at address 0060.
Extended	BA 60 00	ORA A 6000	OR A with data at address 6000.
Indexed	EA 03	ORA B 03,X	OR B with data at address IX + 03.

ORA instructions OR the accumulator with memory and immediate data bit by bit also. The result is stored in the specified accumulator. Thus, the instructions

LDA A #B6
ORA A #1D

are implemented as

Old accumulator A	10110110	B6
Operand data	00011101	1D
New accumulator A	10111111	BF

EOR A   Exclusive-OR accumulator A
EOR B   Exclusive-OR accumulator B

Each bit of operand is exclusively ORed with each corresponding bit of an accumulator. If bit pairs are 0 and 1, the result is 1. If bit pairs are both 0 or both 1, the result is 0. Thus, the instructions

LDA A #B6
EOR A #1D

are implemented as

Old accumulator A	10110110	B6
Operand data	00011101	1D
New accumulator A	10101011	AB

Example 4.5. Eight devices are monitored to determine if their status has changed. Port address is 6000. Write the monitoring program.

Mnemonics	Comments
LDA A 6000	Load port old data.
STA A 0010	Save it.
LDA A 6000	Load port new data.
EOR A 0010	Exclusive-OR new and old data.

The result in accumulator A will have 1s in the bits whose port device switched either OFF-to-ON or ON-to-OFF.

COM A	Complement accumulator A
COM B	Complement accumulator B
COM	Complement contents of operand address

Inherent	43	COM A	Complement A.
Inherent	53	COM B	Complement B.
Extended	73 60 00	COM 6000	Complement data at address 6000.
Indexed	63 00	COM 00,X	Complement data at address IX + 00.

COM instruction inverts an accumulator or memory by changing 1s to 0s and 0s to 1s. The operation is also known as the 1s complement.

NEG A   Negate accumulator A
NEG B   Negate accumulator B
NEG   Negate contents of operand address

NEG instruction complements an accumulator or memory and adds one. In effect, a negate of data is the same as the 2s complement of data.

4.4 SHIFT AND ROTATE INSTRUCTIONS

ASL A   Arithmetic shift left accumulator A
ASL B   Arithmetic shift left accumulator B
ASL   Arithmetic shift left contents of operand address

Inherent	48	ASL A	Shift A one place to left.	
Inherent	58	ASL B	Shift B one place to left.	
Extended	78 40 01	ASL 4001	Shift data at address 4001 one place to left.	MSB to C
Indexed	68 10	ASL 10,X	Shift data at address IX + 10 one place to left.	0 to LSB

ASL instructions move each bit in an accumulator or memory to the next higher bit position. The MSB moves into carry C while a zero moves into the LSB. A diagram of the operation is given in Fig. 4.1(a). The shift is called arithmetic because it effectively

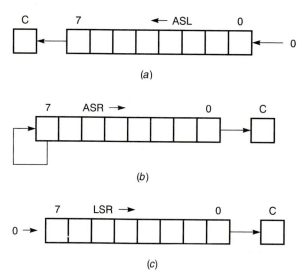

(a)

(b)

(c)

FIGURE 4.1
Bit progression for shift instructions. (a) Arithmetic shift left; (b) arithmetic shift right; (c) logic shift right.

multiplies the content value by 2 for both unsigned and signed numbers. Results cannot exceed the capacity of an 8-bit register: $(-128)_{10}$ to $(+127)_{10}$. Examples for $44 \times 2 = 88$ and $-44 \times 2 = -88$ follow:

$$00101100 = \quad (44)_{10}$$
$$\text{ASL} \leftarrow 01011000 = \quad (88)_{10}$$
$$11010100 = (-44)_{10}$$
$$\text{ASL} \leftarrow 10101000 = (-88)_{10}$$

The V flag is set if an ASL causes the number to exceed its range: $V = N \oplus C$. After the shift C is the sign of the old number; N is the sign of the new number. If the two flags differ, the number has changed signs, i.e., exceeded the allowable range.

ASR A   Arithmetic shift right accumulator A
ASR B   Arithmetic shift right accumulator B
ASR   Arithmetic shift right contents of operand address

ASR instructions move each bit in an accumulator or memory to the next lower bit position. The MSB does not change while the LSB moves into the carry. Figure 4.1(*b*) illustrates the bit changes. An ASR divides all content numbers by 2 within the signed arithmetic range. For example,

$$(-45)_{10} = 11010011$$
$$(-23)_{10} = 11101001 \longrightarrow \text{ASR}$$

If the number is odd, the odd bit (bit 0) is moved to the carry. Thus, it is lost and the number rounds down to the next lower digit.

LSR A   Logic shift right accumulator A
LSR B   Logic shift right accumulator B
LSR   Logic shift right contents of operand address

LSR instructions move each bit in an accumulator or memory to the next lower bit position. Unlike ASR, a zero is moved to the MSB. The LSB also moves into the carry. Figure 4.1(*c*) shows the bit changes. LSR can be used to divide a binary number by 2; however, it differs from ASR in that the MSB no longer is a sign bit. For unsigned numbers, the range is extended by a factor of 2. There is no LSL instruction because it would be identical to ASL.

ROR A		Rotate right accumulator A	
ROR B		Rotate right accumulator B	
ROR		Rotate right contents of operand address	
ROL A, ROL B, ROL		Rotate left	
Inherent	46	ROR A	Rotate A one place to right.
Inherent	56	ROR B	Rotate B one place to right.
Extended	76 20 00	ROR 2000	Rotate contents of address 2000 one place to right.
Indexed	66 10	ROR 10,X	Rotate contents of address IX + 10 one place to the right.

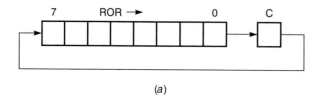

(a)

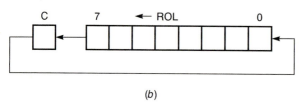

(b)

FIGURE 4.2
Bit progression for rotate instructions. (*a*) Rotate right; (*b*) rotate left.

ROR (rotate right) and ROL (rotate left) instructions are shown in Fig. 4.2. They represent a circular bit shifting in which the carry acts as a ninth bit. Bits coming off one end are placed into the carry while the carry is transferred into the vacancy at the opposite end. After nine rotates the bit pattern is the same again. ROTATE instructions are very useful for polling the status of port devices. The port byte is read and successive bits are rotated into the carry. If the carry is set (e.g., a pressure transducer voltage exceeds safe operating conditions), a branch instruction diverts the program to a service routine (turn off apparatus).

Example 4.6. We can extend the multiplication and division by 2 range by using double-precision numbers. Write a program to divide a double-precision signed number by 2 (or multiples of 2 by repeating the program).

Address	Mnemonics	Comments
0000		HIGH-byte of number.
0001		LOW-byte of number.
0002		HIGH-byte result.
0003		LOW-byte result.
0004	CLC	Clear carry.
0005	LDA A 00	Load HIGH-byte direct.
0007	ASR A	Shift HIGH-byte to right.
0008	LDA B 01	Load LOW-byte direct.
000A	ROR B	Rotate LOW-byte; bit 8 to bit 7 via carry.
000B	STA A 02	Store HIGH-byte result.
000D	STA B 03	Store LOW-byte result.
000F	WAI	

4.5 BRANCH AND JUMP INSTRUCTIONS

Past programs have been very simple because they lacked instructions to make decisions. For the task of monitoring a peripheral switch, the MPU must be able to decide if the switch is ON or OFF. Instructions are executed sequentially unless the microprocessor can transfer control to another instruction other than the next one in the sequence. The decision to transfer control is based on the conditional code register. The programmer must set up a condition which will change a flag(s) setting when the decision should be made. Such transfers are called *conditional transfers*. At other times the transfer always takes place; no decision is necessary. They are called *unconditional transfers*. BRANCH and JUMP are the unconditional transfer instructions.

Unconditional Transfers

JMP	Jump program to operand address		
Extended	7E 18 00	JMP 1800	Jump to address 1800.
Indexed	6E 06	JMP 06,X	Jump to address IX + 06.

JMP instructions load the program counter (PC) with the specified address; the program immediately executes sequentially at the new address. Since any 16-bit address is available, the program can jump to any location in memory.

BRA	Branch always with relative address		
Relative	20 F8	BRA F8	Branch always to address PC + F8.

All branch instructions follow the OP code with a byte called the *relative address*, and the relative addressing mode is reserved for branch instructions only. *The relative address is a displacement that is added to the PC as an 8-bit signed number.* If the MSB is one, the negative number will cause the program to branch backward toward lower memory. Otherwise it branches forward. The maximum positive number represented by 8 bits is $(127)_{10}$, and the most negative number is $(-128)_{10}$. BRA and JMP instructions are equivalent. BRA takes one less byte to execute, but JMP has much more range.

Perhaps the single most common programming mistake is the erroneous calculation of the relative address. First, the programmer must be aware of the PC contents. While the MPU is processing the relative address, the PC has already been advanced to point at the next instruction. *Therefore, the relative address is added to the address of the next instruction.* Second, the programmer often counts the bytes to the new location and converts the count to hexadecimal. It is easier simply to subtract the two hexadecimal addresses.

Figure 4.3 illustrates forward and backward branching. It is assumed that the program is written, and we wish to check the branch instruction. Care must be taken when adding 8-bit numbers to 16-bit addresses. If the branch is forward, the 8-bit relative address is between 00 and 7F. Fill in the remaining bits with 00. If the branch is backward,

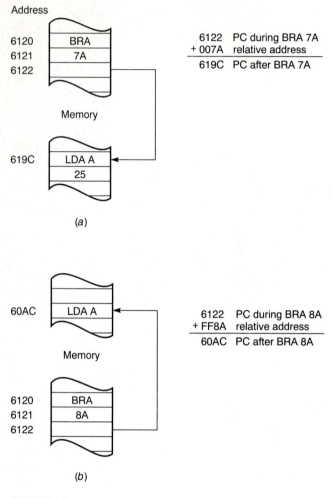

FIGURE 4.3
Branching with the relative mode. (*a*) Forward branching; (*b*) backward branching.

the 8-bit relative address is between 80 and FF. Fill in the remaining bits with FF. Likewise, when calculating a relative address, throw away the MSByte of the 16-bit number:

$$
\begin{array}{ll}
60AC & \text{new PC} \\
-\ \underline{6122} & \text{old PC} \\
FF8A & \text{relative address (use 8A)}
\end{array}
$$

Conditional Transfers

Conditional transfers take place only if specific conditions are satisfied. The microprocessor makes these decisions by using the status bits: negative (N), zero (Z), overflow (V), and carry (C). The MPU tests one or more of these flags each time it executes a conditional branch. Flags may be reset (cleared) or set. There is a branch instruction for each. It is generally the instruction preceding the branch instruction which determines the conditional code bit status.

BCC	Branch if carry clear
BCS	Branch if carry set
BNE	Branch if result not zero
BEQ	Branch if result zero
BVC	Branch if overflow clear
BVS	Branch if overflow set
BPL	Branch if result plus
BMI	Branch if result minus

Relative	24	BCC	Branch if C = 0.
Relative	25	BCS	Branch if C = 1.
Relative	26	BNE	Branch if Z = 0.
Relative	27	BEQ	Branch if Z = 1.
Relative	28	BVC	Branch if V = 0.
Relative	29	BVS	Branch if V = 1.
Relative	2A	BPL	Branch if N = 0.
Relative	2B	BMI	Branch if N = 1.

Conditional branch instructions can be used to terminate unsigned (C = 1) and signed (V = 1) arithmetic when the solution exceeds the valid range. Branching on the N bit is rare. In fact, over 90 percent of the branch instructions ever used will be BNE or BEQ because the Z bit is the most commonly tested bit among all instructions.

Example 4.7. A pressure switch is connected to bit 0 of an 8-bit port at address 3000. If the switch signal goes HIGH, the program must jump to a shutdown routine beginning at address 6000. Otherwise the program continues with the remaining tasks. Write this section of the overall program.

Address	Content	Mnemonics	Comments
0010	B6 30 00	LDA A 3000	Read the port.
0013	84 01	AND A #01	Mask off all bits but bit 0.
0015	27 03	BEQ 03	If switch LOW, branch over JMP.
0017	7E 60 00	JMP 6000	If switch HIGH, no branch and execute JMP.
001A		CONTINUE	

In many situations the programmer must use a branch instruction to set up the decision test. Yet he or she may want to branch beyond the relative address range. The preceding program shows how this can be accomplished with the addition of a JMP instruction.

Other Conditional Transfers

There are 14 conditional branches in the instruction set. Eight of them test the C, Z, V, and N flags directly. Six additional branch instructions test logical combinations of conditional code flags. All six instructions follow immediately after any of the compare or subtract instructions: CBA, CMP, SBA, SUB. It is helpful to remember that their form is accumulator A minus accumulator B (A − B) or accumulator minus operand (A − M and B − M). Compare and subtract instructions are used to test data to determine if it is positive, negative, or zero. Once the instruction is executed, N and Z flags will be affected accordingly. Then a branch occurs depending upon the flag status. Two branch instructions are reserved for unsigned arithmetic. Four are reserved for signed arithmetic.

Unsigned numbers

BHI	Branch if accumulator higher than operand	
BLS	Branch if accumulator lower or the same as operand	

Relative	22	BHI	Branch if $C + Z = 0$.
Relative	23	BLS	Branch if $C + Z = 1$.

The instruction previous to BHI subtracts the operand from an accumulator. If the accumulator is smaller, a borrow occurs ($C = 1$). The Z flag is set ($Z = 1$) if the result is zero. Thus, the condition is C (OR) $Z = 0$ for the branch to take place. BLS complements the BH1 instruction.

Signed numbers

BGE	Branch if accumulator greater than or equal to operand
BLT	Branch if accumulator less than operand

Relative	2C	BGE	Branch if $N \oplus V = 0$.
Relative	2D	BLT	Branch if $N \oplus V = 1$.

BGE instructions are intended for signed numbers. Note that −2 is greater than −3. If the accumulator and operand are both positive or both negative, a subtraction will produce $N = 0$ and $V = 0$ when the accumulator is greater. However, if the subtrahend is a large negative number, both the N and V flag are set. The criteria for a branch is met by N (EOR) $V = 0$. BLT complements the BGE instruction.

BGT	Branch if accumulator greater than operand
BLE	Branch if accumulator less than or equal to operand

Relative	2E	BGT	Branch if $Z + (N \oplus V) = 0$.
Relative	2F	BLE	Branch if $Z + (N \oplus V) = 1$.

BGT instructions are the same as BGE instructions except that we must exclude a zero result, i.e., the Z flag being set. BLE complements the BGT instruction.

Example 4.8. Accumulator A contains 20 while accumulator B contains A0. As unsigned numbers, B > A but as signed numbers A > B. Demonstrate the BHI and BGT instructions when paired with CBA (A − B).

Binary subtraction		Twos complement subtraction	
00100000	20	00100000	
− 10100000	A0	+ 01100000	
1] 10000000	80	0] 10000000	

     With straight binary subtraction the result is 80 and the borrow is set. The MPU uses 2s complement arithmetic for the same result but the borrow is lost. For this reason all subtract or compare instructions set the carry when the absolute value of the subtrahend is larger than the absolute value of the minuend. Thus,

$$N = 1$$
$$V = 1$$
$$Z = 0$$
$$C = 1$$

Unsigned numbers

$$\left.\begin{array}{l} \text{CBA} \\ \text{BH1} \end{array}\right] \quad C + Z = 1 \quad \text{or it will } not \text{ branch}$$

Signed numbers

$$\left.\begin{array}{l} \text{CBA} \\ \text{BGT} \end{array}\right] \quad Z + (N \oplus V) = 0 \quad \text{or it will branch}$$

4.6 PROGRAMMING CONCEPTS

Loops

The most common attribute of any strong program is the ability to repeat a sequence by looping back to an earlier instruction. A decision to branch back is accomplished by placing a conditional branch instruction in the program.

     Looping can be easily demonstrated with a multiplication routine. Many microprocessors, 6802 included, do not have an instruction to multiply two numbers. One method to multiply numbers with software is a repeated addition program. Multiplication is nothing more than adding the multiplicand to itself. The number of repeated additions is given by the multiplier.

     There are many ways to configure a program. Any solution that works is acceptable. Experience will enable the programmer to write more compact codes which will run faster. Each assembly language listing should be preceded by a short paragraph explaining the procedure. A generous sprinkling of comments through the listing will further assist the reader. Good comments refer to groups of assembly lines and explain what is happening. Avoid at all cost a line-by-line explanation of the source code. This makes sense only when the reader does not know the instruction set.

Example 4.9. Write a program to multiply two 8-bit numbers with an 8-bit product.

This program multiplies by repeated addition. The multiplier, multiplicand, and product are stored at addresses 0000–0002, respectively. The product, initially zero, is repeatedly added to the multiplicand in ACC A. The count is kept in ACC B. No allowance is made for the product exceeding $(255)_{10}$.

Program

Address	Mnemonics	Comments
0000	—	Multiplier
0001	—	Multiplicand
0002	—	Product
0003	CLR 0002	Clear product.
0006	LDA B 0000	Get count.
0009	LDA A 0001	Get multiplicand.
000C	ADD A 0002	Add multiplicand to product.
000F	STA A 0002	
0012	DEC B	Are we done adding?
0013	BNE F4	
0015	WAI	

Labels

Labels are mnemonics that represent addresses. The labeled addresses can be instruction operands or they can refer to locations that are the objects of program branches and jumps. When labels are operands, their equivalent addresses are defined at the beginning of the program. Location labels are listed in a separate column to the left of the mnemonics column.

Labels are an integral part of any assembler software to produce a program hexadecimal listing. The use of labels eliminates the problem of calculating offsets. The assembler calculates the offset automatically and inserts it into the listing. Each assembler has different rules for what can or cannot be a label. For the 6802 assembler, valid labels should:

1. Consist of one to six alphanumeric characters
2. Not be A, B, or X alone
3. Begin with an alphabetic character
4. Use directive EQU to assign label values
5. Use directive ORG to denote the program beginning (origin) address

Example 4.10. The readability of programs can be greatly improved with labels. Rewrite the multiplication problem in Example 4.9 using labels:

```
          ORG      0003
MLTPR     EQU      0000
MLTPD     EQU      0001
PROD      EQU      0002
          CLR      PROD
          LDA B    MLTPR
SUM       LDA A    MLTPD
          ADD A    PROD
          STA A    PROD
          DEC B
          BNE      SUM
          WAI
```

Algorithms and Flowcharts

Once a software function has been defined, there are few times that the program can be written directly into assembly language. First, the solution must be expressed by means of an algorithm. An *algorithm* is a step-by-step specification of the solution to a given problem. The algorithm may be expressed in any language or symbolism but generally is a verbal outline of the solution.

Suppose we want to write a program to find the largest number in a list of numbers. The list is located in a specified block of memory. An algorithm for the solution would look like this:

1. Clear answer address.
2. Get list starting address.
3. Load number.
4. Compare to answer.
5. Store answer if bigger.
6. Otherwise branch over store.
7. Are we at the end of the list?
8. Yes, stop.
9. No, get next number.
10. Loop to step 3.

In many cases the programmer may find it convenient to transfer the algorithm to a flowchart. A *flowchart* is a graphical representation of the algorithm expressed in function blocks with lines showing the program progression. Flowcharts are used also as a part of programming documentation to explain the program to others who are not familiar with it. Flowchart symbols are given in Fig. 4.4. Ovals are used to denote the beginning and end of a program. Rectangles are the executable statements. Parallelograms are inputs

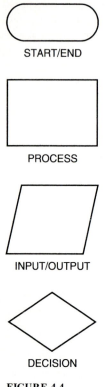

START/END

PROCESS

INPUT/OUTPUT

DECISION

FIGURE 4.4
Flowcharting symbols.

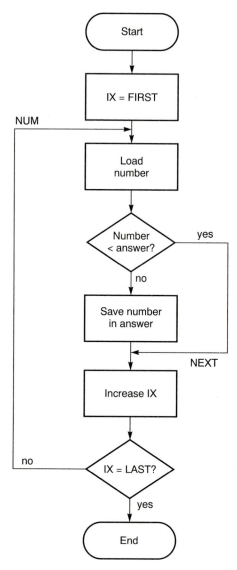

FIGURE 4.5
Flowchart for biggest number routine.

and outputs. They include READ and WRITE commands especially from external devices. Decisions are represented by diamonds. There are typically two output paths: the yes/no answers from a branch instruction. Sometimes a compare instruction will generate three exit paths corresponding to less than, equal to, or greater than possibilities.

Figure 4.5 shows a flowchart for the "biggest number" algorithm. You will notice that flowcharts display the solution after very little study, while algorithms must be followed line by line to be understood. This ability to visualize relationships between program sections is an advantage over algorithms. Although they are easy to read, flowcharts are time-consuming to draw. Furthermore, flowcharts cannot be drawn on computer word processors when using assemblers to write the code. More people advocate flowcharts than use them to develop assembly code. Documentation is another matter.

Example 4.11. Write the "biggest number" program by following the flowchart.

```
            ORG     0000
FIRST       EQU     0020
LAST        EQU     0030
ANSWR       EQU     0031

            CLR     ANSWR
            LDX     #FIRST
NUM         LDA A   0,X
            CMP A   ANSWR
            BLS     NEXT
            STA A   ANSWR
NEXT        INX
            CPX     LAST
            BNE     NUM
            WAI
```

Top-Down Programming

Top-down programming is a widely accepted approach to planning large programs. The technique is to make a steady progression from a general problem statement down to the detail of a program step. The purpose is to break the problem down into a series of smaller and smaller problems. It is essential that at an intermediate stage the problem statements are self-contained, i.e., do not loop back to other "general" statements. In this way the smaller problems can be worked on independently and a team of programmers can attack the solution at once. At this point each programmer further subdivides the smaller problem into several other units.

A microprocessor-based automobile engine control system is a good candidate for the team approach. The first stages in the solution would proceed as follows:

Stage I.   Control of a gasoline engine

Stage II.   1. Input sensors
            2. Control fuel
            3. Control ignition

Stage III. 1. (a) Input exhaust gas O_2
               (b) Input throttle position
               (c) Input crankshaft position
               (d) Input manifold pressure
               (e) Input coolant temperature
               (f) Input airflow
               (g) Input barometric pressure

2. (a) Calculate fuel injection
   (b) Output pulse length to injectors
3. (a) Calculate engine speed
   (b) Calculate dwell time
   (c) Calculate ignition point
   (d) Output ignition pulse
4. (a) Calculate digital display
   (b) Output to dash

A natural result of top-down programming is *linear or straightline* programs. Straightline programs require all programming elements or building blocks to have a single entry point and a single exit point. There are three basic flow structures that together can solve any programming problem: sequence element, branch element, and loop element. Figure 4.6 shows the structures. All three are demonstrated in the "biggest number" flowchart. When they are combined similar to Fig. 4.7, the code is easy to follow and sections can be solved independently for assembly later.

4.7 LOOK-UP TABLES

Tables are a very convenient method to organize engineering data into rows and columns for easy reference. A *look-up table* is a one-dimensional table with two equivalent sets of numbers side-by-side. One set is the input; the other set is the output. Outputs are stored as data in memory. Inputs are used to access the data when they function as offsets in the indexed addressing mode. There are two interesting applications: (1) codes and (2) equations.

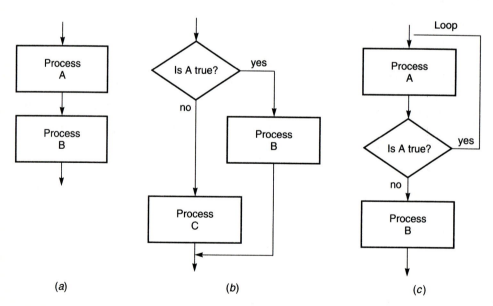

FIGURE 4.6
Basic flowchart structures for straightline programming. (*a*) Sequence; (*b*) branch; (*c*) loop.

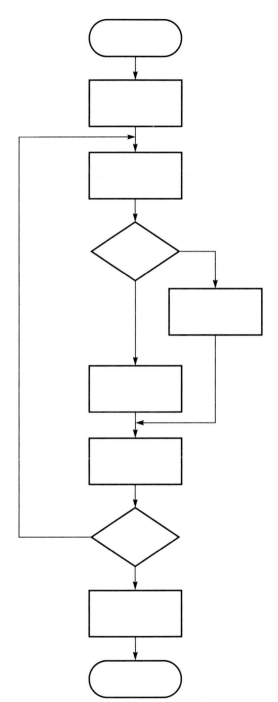

FIGURE 4.7
Straightline coding simplifies top-down
programming.

Codes

Probably the most common peripheral in mechatronics is the seven-segment display. Each display has a corresponding I/O port and address. Each segment of a display is controlled by a port bit. To make a given decimal number appear on the display, only one-bit pattern should be applied to the display port. The bit pattern or code is stored in a table at an arbitrary address in memory similar to Table 4.1. A standard look-up table routine retrieves the code.

Example 4.12. A seven-segment code is loaded into memory starting at address 3046. A display is connected to a port at address 5002. The number to be displayed is presented to an input port at address 5000. Write a routine to display the number via a look-up table.

INPUT	EQU	5000
OUTPUT	EQU	5002
TABLE	EQU	3046

Address	Content	Label	Mnemonics
1000	B6	LK-UP	LDA A INPUT
1001	50		
1002	00		
1003	B7		STA A OFFSET
1004	10		
1005	0A		
1006	CE		LDX # TABLE
1007	30		
1008	46		
1009	A6		LDA A 00,X
100A	00	OFFSET	
100B	B7		STA A OUTPUT
100C	50		
100D	02		

The input could just as well be a memory location holding the result of an arithmetic operation. It is placed (stored) at address 100A, in effect, writing over the spacer offset 00. When

TABLE 4.1
Look-up table for seven-segment code

Number	Address	Code
0	3046	C0
1	3047	F9
2	3048	A4
3	3049	B0
4	304A	99
5	304B	92
6	304C	82
7	304D	F8
8	304E	80
9	304F	98

the index register is loaded with the table beginning address, the stored offset or input counts down to point to the correct code address for the accumulator load instruction. The routine is adaptable to any input, output, or table location. It should be placed in RAM (random access memory). An alternative look-up table for EPROMs (erasable programmable read-only memory) is found in Chap. 16.

Equations

A more powerful application of look-up tables is the "solution" of equations. Good candidates are functions such as sin θ. Large computers use an nth-order polynomial to approximate sin θ.

The equation is accurate and suitable for floating-point arithmetic. However, most 8-bit microprocessors need algorithms to multiply two numbers. The preceding calculation would be a prohibitive time-consuming task. Furthermore, 8-bit accuracy with whole numbers is sufficient for mechatronics. A fast algorithm uses a look-up table of precalculated functions.

Example 4.13. Write a program to obtain the sine of an angle. The angle can be 0° to 360°, requiring 9 bits or double precision. It will be stored at addresses 0000–0001. The result, sin θ, will be stored at address 0002. The look-up table begins at address 2000.

To solve this problem, we must construct a table of sine data. First, the sine of an angle falls between $+1.00$ and -1.00. Since we cannot store decimal fractions, the result will be scaled between $+100$ and -100. Thus, sin 30° = 0.5, and our table will contain $(50)_{10}$ or $(32)_{16}$. Likewise, sin 210° = -0.5 and the stored result will be $-(50)_{10}$ or $(CE)_{16}$. A look-up table is given in Table 4.2.

TABLE 4.2
Look-up table for $y = \sin \theta$

Label	Address	Data	Function
TABLE	2000	00	sin 0°
	2001	01	sin 1°
	2002	02	sin 2°
	$\vdots$	$\vdots$	$\vdots$
	205A	65	sin 90°
	$\vdots$	$\vdots$	$\vdots$
	20B4	00	sin 180°
	$\vdots$	$\vdots$	$\vdots$
	20FF	A0	sin 255°
NTABLE	2100		sin 256°
	$\vdots$	$\vdots$	$\vdots$
	210E	9C	sin 270°
	$\vdots$	$\vdots$	$\vdots$
	2168	00	sin 360°

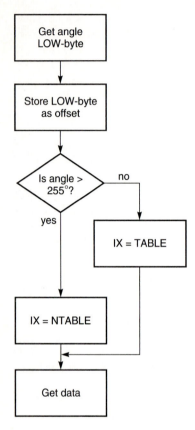

FIGURE 4.8
Flowchart to find sin θ via look-up table.

Second, the full range of unsigned angles is 360, which requires more than an 8-bit data representation. Therefore, the input angle must reside in two memory locations: 0000-HIGH-byte and 0001-LOW-byte. What happens when the input exceeds 255 = FF? The HIGH-byte changes from 00 to 01, and the LOW-byte starts its count over again. A standard look-up routine can accommodate this transition by simply reloading the index register with a new table address: NTABLE = TABLE + 0100. A flowchart is given in Fig. 4.8.

Sine Program

The following program finds the sine of any integer angle between 0° and +360°. A standard look-up routine is modified for angles greater than 255° ≡ FF. The double-byte angle is placed in memory at addresses 0000–0001. The result is found at address 0002. Transition from 255° to 256° is handled by reloading the index register with a new address offset by 0100 from the old address.

```
ANGLHI    EQU    0000
ANGLLO    EQU    0001
RESULT    EQU    0002
TABLE     EQU    2000
NTABLE    EQU    2100
```

```
SINE        LDA A   ANGLLO
            STA A   OFFSET
            TST     ANGLHI
            BNE     NEW
            LDX     #TABLE
            BRA     GET
NEW         LDX     #NTABLE
GET         LDA A   00,X

OFFSET      STA A   RESULT
```

Messages

Seven-segment displays will show characters including most, but not all, letters of the alphabet (see Table 10.5). Upon power up, the Heath ET-3400-A Microprocessor Trainer shows the message: CPU UP. Each character code is stored in memory before the program is executed. Suppose there is a corresponding display for each character and each display has its own I/O port, which is an address. Thus, the process of displaying a message is one of *transferring a string of characters from a given memory location to another location.*

Message Program

This program transfers a string of characters from address 5000 to address 6000. If a letter code table is stored starting at address 5000, and seven-segment LEDs are assigned addresses starting at 6000, a message is displayed:

```
0000    EQU     50          First character address
0001    EQU     00
0002    EQU     60          First LED address
0003    EQU     00

LOOP    LDX     #0000       Get letter.
        LDA A   00,X
        LDX     #0002       Display letter.
        STA A   00,X
        INC     0001        Point to next letter and next LED.
        INC     0003
        LDA B   0001        Do we have last letter?
        CMP B   #LAST
        BNE     LOOP
```

PROBLEMS

4.1. After a TPA instruction has been executed the contents of accumulator A read D3. Which conditional code flags are set?

4.2. Address 0000 has been assigned the number 88. The contents are to be shifted eight times. Give the instruction(s) (*a*) to shift right and return the contents intact, (*b*) to shift right and leave 00.

4.3. What is the difference between CLR A and LDA A #00?

4.4. A 16-bit number resides in addresses 0000–0001. A second 16-bit number is placed in addresses 0002–0003, HIGH-byte first. It must be determined if the second number is larger than the first number by setting the carry (borrow) flag. Will the following instructions accomplish the task? Can you suggest another way to do the problem?

      LDX   #0000
      CPX   0002

4.5. Write a routine to set bits 5, 3, and 1, leaving all other bits unchanged. The byte is (a) in accumulator A, (b) in memory location 0010.

4.6. Memory locations 0000 to 0100 contain as contents the LOW-byte of the address. For example, address 00C4 has data C4. Accumulator A contains 06, and index register IX contains 0010. What is the content of A after executing each of the following instructions?

(a) B4,00,40   ADD A 0040
(b) AB,06   ADD A 06,X
(c) B7,00,10   STA A 0010
(d) B6,01   LDA A 01,X
(e) EE,01   LDX 01,X

4.7. Obtain the machine code for the "biggest number" routine in Sec. 4.6.

4.8. Write the instruction sequence to branch elsewhere if the contents of memory location 005A is less than the contents of memory location 005B. Data is (1) signed, (2) unsigned.

4.9. Consider the following branch instructions. Where does the program counter branch to?

	Address	Mnemonics
(a)	4010	BCS E9
(b)	26F6	BHI 6A

4.10. What are the compare and branch instructions to take a branch if accumulator A data is less negative than $(-100)_{10}$?

4.11. Write a program to do triple-precision addition. Specify memory locations for the input data and the result.

4.12. A double-precision number is stored at addresses 2000–2001. Write a routine to multiply this number by 2. Use shift instructions.

4.13. Write a routine to multiply two 16-bit numbers by repeated addition.

4.14. Write a routine to divide a 16-bit number by an 8-bit number. Use repeated subtraction.

4.15. A list of 10 numbers in memory starts at address 1000. Develop a program to move the list to addresses 10A0–10A9.

4.16. Write a program to display the message: GO BULLDOGS. The message code is stored at starting address 00C0 and the output seven-segment LEDs start at address 00D0.

4.17. Change the routine in Prob. 4.16 so that the message blinks ON and OFF at one-second intervals. Use the instruction JSR DELAY to represent a one-second delay.

4.18. Modify the message program in Sec. 4.7 so that characters scroll across the displays, advancing one character every second. The displays are initially blank, and the characters start their appearance from the right. Use instruction JSR DELAY to represent a one-second delay.

4.19. Program development work is being done on a new MPU-based carpenter's level. You must write a short program to find the correct display code and to send this code to a display port at address 4002. The sensed angle is found at port address 4000. The code is stored in a table from address 1000 (angle 0°) to address 105A (angle 90°).

4.20. Suppose you want to calculate the equation $Y = \sqrt{X} + 10$ as part of a program. An appropriate algorithm could be used, but you choose a look-up table routine. The table of solutions starts at address 0100. Input X is placed at address 0000 and output Y is found at address 0001. Write a look-up table program to find Y. Show the first three and last three numbers in the table. Use integers only.

CHAPTER

5

STACK/
SUBROUTINES

The *stack* is a successive block of RAM (random access memory) locations set aside by the programmer for temporary storage of data. The stack length is limited only by the total RAM available; however, it usually amounts to only a few tens of bytes. Stack area is reserved when the microprocessor system is planned. It should be used only for this purpose, well removed from the main program. Special stack instructions allow the programmer to store information in the stack. Later the data can be retrieved when it is needed. The advantage over conventional LDA, STA instructions is efficiency—fewer bytes of instruction.

The stack is also used by the MPU to save automatically several of its internal registers when other functions are performed outside the main program. For example, when a *subroutine* call instruction is encountered, the MPU leaves the main program and executes the subroutine program stored at another address. After the separate program is completed the microprocessor returns to the main program. A return can occur only if the previous program counter (PC) has been saved (on the stack). Interrupts use the stack similarly to subroutines except that they are initiated by external signals. Interrupts are discussed in Chap. 8.

5.1 STACK POINTER

The stack is analogous to a "stack" of dishes in a cupboard. Initially, the cupboard is empty. This represents the *highest* available address in RAM, starting the stack as far away from the program as possible. Dishes (data) are placed one at a time onto the stack. If dishes are needed for a meal, they are pulled from the top of the stack. The stack works on the principle that the last data placed on the stack is the first data pulled from the stack—LIFO: last in, first out.

With dishes we know where the next one is placed. Microprocessors must be told where the top of the stack can be found. The *stack pointer* (SP) is a 16-bit register in the MPU that contains the address of the next available memory location in the stack. Data is stored on the stack at the address "pointed to" by the SP, *after* which the SP is automatically *decremented* by one. Thus, the SP always points to the next lower available address relative to stack data. Conversely, when data is removed from the stack, the SP is *incremented* by one *before* the data is retrieved.

We will now review instructions pertinent to stack manipulation.

LDS Load stack pointer register

Immediate	8E 02 00	LDS #0200	Load SP with 0200.
Direct	9E F0	LDS F0	Load data at address 00F0 into SP HIGH-byte and data at address 00F1 into SP LOW-byte.
Extended	BE F0 00	LDS F000	Load data at address F000 into SP HIGH-byte and data at address F001 into SP LOW-byte.
Indexed	AE 10	LDS 10,X	Load data at address IX + 10 into SP HIGH-byte and data at address IX + 11 into SP LOW-byte.

STS Store stack pointer register

Direct	9F 40	STS 40	Store SP HIGH-byte at address 0040 and LOW-byte at address 0041.
Extended	BF 01 00	STS 0100	Store SP HIGH-byte at address 0100 and LOW-byte at address 0101.
Indexed	AF 01	STS 01,X	Store SP HIGH-byte at address IX + 01 and LOW-byte at address IX + 02.

INS Increment stack pointer
DES Decrement stack pointer

| Inherent | 31 | INS | Add one to SP contents. |
| Inherent | 34 | DES | Subtract one from SP contents. |

TSX Transfer stack pointer to index register
TXS Transfer index register to stack pointer

| Inherent | 30 | TSX | Load IX with contents of SP + 1. |
| Inherent | 35 | TXS | Load SP with contents of IX − 1. |

Remember, the SP contains the address of an empty (irrelevant) data location, one less than the address of stored data. Therefore, the retrieved data is at one higher address than the SP. Conversely, the SP should be loaded with one lower address than the index register address of relevant data.

All the preceeding stack transactions are available for the user to write more compact code. However, only the LDS instruction has any importance for the beginning programmer. RAM should be reserved for the stack if the program needs the following:

1. Temporary data storage
2. Subroutines
3. Interrupts

The SP is loaded with an address at the very beginning of the program. This address represents the top of an empty stack. It is generally, but not always, located at the highest available RAM address.

> **Example 5.1.** A microcomputer system consists of the 6802 microprocessor with on-board RAM, ROM, EPROM, and I/O. Memory space is allotted as follows:
>
> | | |
> |---|---|
> | RAM: | addresses 0000–007F |
> | ROM: | addresses 1000–3FFF |
> | EPROM: | addresses F000–FFFF |
> | I/O: | addresses 6000–6003 |
>
> What instructions are required at the beginning of the program to initialize the SP?
>
> The SP is loaded with the highest available RAM location. Therefore, the first program instruction is
>
> LDS   #007F

5.2 PUSH/PULL INSTRUCTIONS

Direct data transfer to and from the stack is accomplished with two instructions: "push" and "pull" (called "pop" in some assembly language sets). The push instruction deposits one data byte onto the stack top. The pull instruction removes one data byte from the stack top. Data transfer is through the accumulators.

PSH A	Push accumulator A onto stack		
PSH B	Push accumulator B onto stack		
Inherent	36	PSH A	A onto stack; SP decremented.
Inherent	37	PSH B	B onto stack; SP decremented.

PUL A	Pull accumulator A from stack		
PUL B	Pull accumulator B from stack		
Inherent	32	PUL A	SP incremented; A loaded from stack.
Inherent	33	PUL B	SP incremented; B loaded from stack.

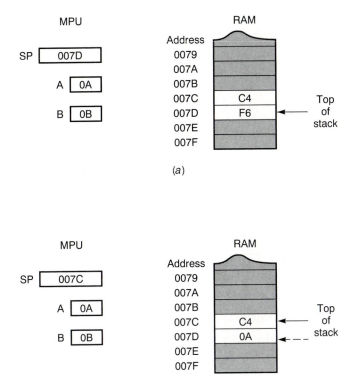

FIGURE 5.1
Execution of PSH instruction. (*a*) Before PSH A; (*b*) after PSH A.

Figure 5.1 illustrates a push onto the stack. Before the instruction PSH A is executed, the SP contains the address 007D, and accumulator A contains data 0A. The SP points to the top of the stack and irrelevant data F6. If the instruction PSH A is executed, the contents of accumulator A are pushed into RAM at address 007D. Then the SP is automatically decremented to 007C. Address 007C is now at the top of the stack.

We want to follow PSH A instruction with a PUL B instruction. Remember, the top of the stack 007C is the next available location to place data, and it points to irrelevant content. If the stack was initialized with address 007F, stored relevant data occupies addresses 007D–007F. Therefore, the SP must be incremented first to locate the last byte placed on the stack, Fig. 5.2. Then data 0A at address 007D is transferred to accumulator B. Actually, data 0A is copied into accumulator B. The value remains in the stack, but it is considered irrelevant data to be written over with the next PSH instruction.

Example 5.2. Data is to be transferred from memory location 1000 to an I/O port at address 6000. The content of both accumulators must be saved for further operations. Write instructions to save temporarily accumulator A while it is being used for the transfer.

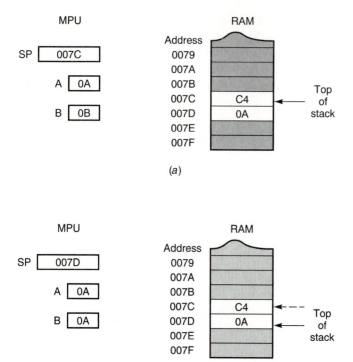

FIGURE 5.2
Execution of PUL instruction. (*a*) Before PUL B; (*b*) after PUL B.

Accumulator A can be saved anywhere in memory wi ... (extended) instruction, and the content can be retrieved with a subsequent LDA ins ... However, a routine in-volving the stack will take fewer bytes of instruction:

```
PSH A
LDA A   1000
STA A   6000
PUL A
```

Example 5.3. The 6802 microprocessor has no instruction which will swap accumulators. Write a swap routine using the stack

```
PSH A
PSH B
PUL A
PUL B
```

The sequence follows the LIFO procedure for accessing the stack. Since the content of B was the last data onto the stack, it is the first data pulled from the stack and placed in A.

There are several observations that can be made in regard to operating a stack:

1. Pull data from the stack in the reverse order that it was pushed onto the stack.
2. All pushes eventually should have an equal number of pulls. Otherwise each time the program loops through the stack instructions, the stack grows. Eventually, the stack creeps into RAM occupied by the main program and writes over it.
3. Data is copied from an accumulator and written over previous RAM data during a push. Conversely, data is copied from the stack and written over previous accumulator data during a pull.

5.3 MULTIPLE STACKS

Normally, the stack consists of a single block of successive RAM locations. However, instructions in the main program can change the SP address without storing or retrieving information onto or from the stack. In such a case there exists two or more stacks, each consisting of a block of successive locations in memory. One application of multiple stacks is using the SP as a second index register.

> **Example 5.4.** In Sec. 4.7 a message program transferred a string of characters at addresses 5000–5006 to a set of displays at addresses 6000–6006. The routine used dual index registers to point to each starting address. It solves the problem by setting up dual memory locations to be incremented while alternately loading these locations into the index register for indexed addressing instructions. Rewrite the program using the SP for the second index register. Assume the SP has already been initiated with address FFF0.

Revised Message Program

	LDX #5006	Point to each ending
	LDS #6006	address.
NEXT	LDA A 0,X	Get character.
	PSH A	Display character.
	DEX	Get next character.
	CPX $4FFF	Gone passed first character?
	BNE NEXT	
	LDS #FFF0	

The new routine is more efficient than the old routine. In normal stack operations data is pushed onto the stack and retrieved within a specified number of instructions. Thus, the SP does not need to be preserved prior to entering the message routine.

5.4 SUBROUTINES

Up to this point all programs have followed the principle of straightline programming with one entry and one exit point. If a particular group of instructions are needed at several different locations in the main program, they must be rewritten for each location. Microcomputer programs often use the same routine repeatedly. For example, time delays to slow the microprocessor are essential for operating number displays and stepping motors. To repeat the delay loop programming steps each time the loop is required would significantly increase the program size. Subroutine programming avoids this repetition. A *subroutine* is a small group of instructions that is used more than once by the main program. Multiplication and division programs are typical subroutines. Special function routines such as square root and sine are good candidates also. The subroutine is written once and stored in a specific memory location where it can be called up any number of times by the main program. The result is a more streamlined main program which allows the programmer to develop a library of insertable subroutines.

Subroutines are called from the main program by one of two special instructions: JUMP TO SUBROUTINE (JSR) and BRANCH TO SUBROUTINE (BSR). These instructions are similar to their JMP and BRA counterparts with one important exception. *JSR and BSR cause the microprocessor automatically to push the current program counter onto the stack* before transferring control to the subroutine. When the subroutine is finished, we must transfer control back to the main program (calling program). Every subroutine ends with a RETURN FROM SUBROUTINE (RTS) instruction to accomplish this result.

Figure 5.3 illustrates a subroutine call from several locations in the main program and subsequent return. The subroutine MULT(iply) is placed at address 2010. At address 0102 in the main program, JSR instruction transfers control to the subroutine by saving present program counter (PC) contents 0105 in the stack and loading 2010 into the PC. After the microprocessor marches to the end of MULT an RTS instruction restores the old PC contents 0105. There is no need to specify a return address. RTS automatically pulls it from the stack. This allows repeated calls from several different points in the main program.

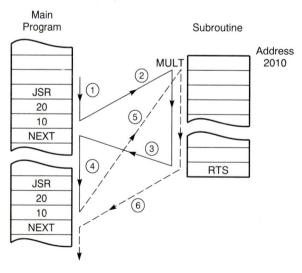

FIGURE 5.3
Transfer of control to a subroutine.

Subroutine Instructions

JSR	Jump to subroutine		
BSR	Branch to subroutine		
RTS	Return from subroutine		

Extended	BD 20 10	JSR 2010	Address 2010 to PC; next instruction address to stack.
Indexed	AD 4A	JSR 4A,X	Address IX + 4A to PC; next instruction address to stack.
Relative	8D 4A	BSR 4A	Address PC + 4A to PC; next instruction address to stack.
Inherent	39	RTS	Restore PC from stack.

The JSR instruction can load the subroutine address into the PC with extended or indexed addressing. A BSR can be used instead if the starting address of the subroutine is within +127 to −128 locations of the next instruction. The advantage of BSR is that it uses one less byte in the main program. The disadvantage is that the offset can be easily miscalculated. Changes within the MPU during a JSR instruction are:

1. LOW-byte of the PC is pushed onto the stack.

2. SP is decremented.

3. HIGH-byte of the PC is pushed onto the stack.

4. SP is decremented.

5. Operand address of JSR is loaded into the PC.

The RTS instruction reverses this sequence of events:

1. SP is incremented.

2. Top byte in the stack is loaded into the HIGH-byte of the PC.

3. SP is incremented.

4. Top byte in the stack is loaded into the LOW-byte of the PC.

With the old PC now restored, the MPU continues the main program at the point where it left off before entering the subroutine.

Disadvantage of Subroutines

Subroutines do not always save memory and programming steps because the JSR instruction takes up 3 bytes of memory and the RTS instruction takes up one byte. Thus, four memory locations are needed for each subroutine call. Before considering a subroutine, the programmer should weigh the size of the subroutine and how many times it will be executed. Certainly, a single execution would be better embedded in the main program. Let L be the subroutine length without RTS. If the subroutine is executed twice, we will save L steps over embedding it twice. However, the cost is two times four subroutine instruction steps. A similar argument for three or more executions will demonstrate that a subroutine is worthwhile, providing subroutine length

$$L > \frac{4N}{N-1}$$

where N = the number of executions.

Nested Subroutines

In a few situations the main program calls a subroutine, which then calls a second subroutine. The second subroutine is named a *nested subroutine* because it returns control to the first subroutine. Figure 5.4 shows a nested subroutine sequence and the status of the stack.

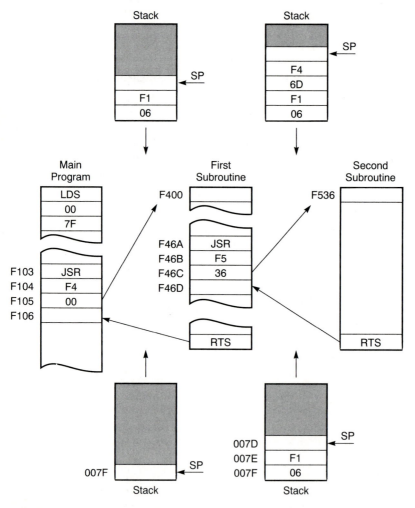

FIGURE 5.4
Nested subroutines and stack content.

At the start of the main program the stack is loaded with 007F. When the JSR F400 instruction at addresses F103–F105 is executed, the next instruction address F106 is pushed onto the stack. With F400 on the PC, the first subroutine runs until encountering JSR F536 at addresses F46A–F46C. This pushes address F46D onto the stack also before any RTS instruction is reached. The second subroutine begins at address F536 but has no nested subroutines of its own. As the RTS instruction is executed, the MPU pulls the return address F46D from the top of the stack and places it in the PC. This causes the MPU to jump (back) to the departure point in the first subroutine. When the subroutine finishes, an RTS pulls return address F106 from the stack and places it in the PC. The MPU is now back to the main program. As long as enough RAM is reserved for the stack, there can be any number of nested subroutines.

Preserving Registers

One of the more common programming mistakes occurs when the main program and the subroutine share MPU registers. The main program uses all the registers eventually. During the execution of a subroutine one or more registers will have their contents replaced by subroutine values. If the main program simply loads data via accumulator A and stores it to an I/O port, the accumulator data is no longer needed. A subroutine uses an accumulator A with irrelevant data. The problem arises during loops when several registers are involved with loop count data and the conditional code (CC) register is critical to a branching operation. Register contents often must be the same after the jump to subroutine as they were before the jump.

The most common method for preserving MPU register contents during a subroutine is to push the needed registers onto the stack at the beginning of the subroutine. At the end of the subroutine the registers are pulled from the stack to restore their contents before returning to the main program. In many cases only a single register must be preserved. The following sequence of instructions within subroutine SUBR saves, then restores, all the registers:

```
IXHI   EQU   0000    Temporary RAM for IX;
IXLO   EQU   0001    other registers to stack.

SUBR   PSH A          Save A.
       PSH B          Save B.
       TPA            Save CCR.
       PSH A
       STX IXHI       Save IX.

        —             Subroutine primary
        —             instructions.

        —
       LDX IXHI       Restore IX.
       PUL A          Restore CCR.
       TAP
       PUL B          Restore B.
       PUL A          Restore A.
       RTS
```

Passing Parameters

Sometimes a subroutine operates independently from the main program. However, many subroutines must process data generated by the main program or return data for subsequent calculations by the main program. Transferring data to and from the subroutine is called *passing parameters*. There are several ways to pass data. The method depends upon the amount of data passed. Programmers can use one of the following register/memories:

1. Internal registers
2. RAM locations
3. Stack

When the number of data bytes to be passed is fewer than five, it can be placed in the A, B, and IX internal registers. For example, time delays are generated with timing loops (see next section), often by counting down the index register. If two different time delays are necessary in the main program, the count can be passed via IX to the delay subroutine before the subroutine is executed. Thus, no separate subroutines are necessary.

When a subroutine requires a large number of data bytes or internal register contents should be saved for later execution steps, parameters can be passed through memory. Data is stored in a specified RAM location by the main program. The subroutine is programmed to accept data from these locations. For example, multiplication subroutine in Sec. 5.6 has 16-bit arithmetic. The multiplicand and multiplier are placed in RAM addresses 0000–0002 prior to a subroutine call. The product is returned to the main program in addresses 0003–0004.

The stack is an alternative method for passing parameters. Data is pushed onto the stack by the main program and pulled from the stack by the subroutine. Some difficulty is presented by the method since return addresses will be pushed on top of the data by the JSR instruction. The subroutine must remove and save the old PC before stack data can be accessed. Assuming we do not save the internal registers for main program use, the following instructions can circumvent the problem:

```
PCHI   EQU   0000
PCLO   EQU   0001

SUBR   PUL A
       STA A   PCHI
       PUL A
       STA A   PCLO
       —
       —                    Subroutine primary
       —                    instructions.
       —
       LDA A   PCLO
       PSH A
       LDA A   PCHI
       PSH A
       RTS
```

5.5 TIMING LOOPS

Microprocessors are very fast, much faster than we would like in some applications. For example, a pulse to a stepper motor integrated circuit (IC) will advance the motor one step. A motor step rate is limited and cannot possibly step as fast as the MPU can send pulses. Therefore, the MPU must stall between pulses for a specified length of time. A *timing loop* is a program routine to idle the microprocessor for a given time period. These delay loops are commonly used for blinking signs, switch-bounce elimination, and signal generation among other applications.

A delay is generated by branching around a loop many times. The amount of elapsed time is determined as follows:

$$\text{Total delay} \quad T_d = [(N \times I_L) + I_S]P$$

where N = number of times through the loop
I_L = instruction cycles in the loop
I_S = instruction cycles out of the loop
P = clock period

There is no preferred way to write the loop although most timing loops decrement the index register. The following program is for a one-second delay:

```
DELAY   LDX   #F424
LOOP1   DEX              one-
        INX              second
        DEX              delay
        BNE   LOOP1      loop
        RTS
```

The index register count is found by constructing a table.

Instruction	Cycles	Times executed	Total cycles
LDX #	3	1	3
DEX	4	N	$4N$
INX	4	N	$4N$
DEX	4	N	$4N$
BNE	4	N	$4N$
RTS	5	1	5
		Total	$16N + 8$

The clock period depends upon the external crystal driving the MPU. In most cases the period is 1 μs. Thus,

$$1 \text{ s} = (16N + 8) \text{ cycles} \times 10^{-6} \text{ s/cycle}$$
$$N = 62499.5 \cong (62500)_{10} \quad \text{or} \quad (F424)_{16}$$

If the number of times through the loop must be rounded to the nearest whole number, the delay is limited in accuracy to $\pm 0.5 \, I_L P$ or 8 μs in the one-second loop. More precise timing can be achieved with a programmable timer IC. See Chap. 14.

Nested Loops

We now have a timing loop to create a one-second delay. Longer delays can be produced by executing the one-second loop more than once. The number of times we can repeat a given loop can be controlled with another counting loop outside the delay routine. Loops within loops are called *nested loops*. An M-second delay routine would look like the following:

```
DELAY    LDA A    #(M)16
LOOP2    LDX      #F424       M-
LOOP1    DEX                  second
         INX                  delay
         DEX                  nested
         BNE      LOOP1       loop
         DEC A
         BNE      LOOP2
         RTS
```

The program is a two-level system of nested loops. There is no limit to the number of levels.

There is one drawback to timing loops. The microprocessor does nothing but "twiddle its thumbs" during the delay routine. It would be more productive if the MPU was able to perform other tasks during the delay rather than dedicate itself to the loop. This approach is possible if we employ interrupts and programmable timers, discussed in Chap. 14.

> **Example 5.5.** The previous nested delay routine is programmed for 60 s. What is the timing error?
>
> The return from subroutine (RTS) instruction has been moved to the end of the program. Thus, the inner loop delay is
>
> $$[16(62500) + 3] \text{ cycles}$$
>
> The inner loop is executed 60 times while the additional DEC A and BNE LOOP2 instructions are also executed 60 times. LDA A and RTS instructions are executed once. So,
>
> $$T_d = [60[16(62500) + 3 + 2 + 4] + 2 + 4]10^{-6}$$
> $$= 60.000546 \text{ s}$$

5.6 USEFUL SUBROUTINES

Multiplication

Earlier we learned a routine to multiply two numbers by repeated addition; i.e., the multiplicand was added to itself N times, where N is equal to the multiplier. This method is undesirable because it is very slow when the multiplier is large. A much faster procedure uses the shift and add process of everyday arithmetic done by hand. A procedural review of multiplication is now given:

$$
\begin{array}{rcl}
(14)_{10} & \longleftarrow \text{multiplicand} \longrightarrow & 1110 \\
\underline{(12)_{10}} & \longleftarrow \text{multiplier} \quad \longrightarrow & \underline{1100} \\
28 & & 0000 \\
\underline{14} & & 0000 \\
(168)_{10} & & 1110 \\
& & \underline{1110} \\
& \text{product} \longrightarrow & 10101000
\end{array}
$$

For the binary example, each multiplier bit multiplies the multiplicand. The result is zero or the multiplicand. After each multiplication the result is shifted to the left. The product is formed by adding the shifted multiplicands. Notice that the shifted multiplicands could have been added after each shift rather than at the end. This procedure suggests the following algorithm:

1. Clear the product location.
2. Check the LSB of multiplier. If it is one, continue. If it is zero, branch to step 4.
3. Add the multiplicand to the product.
4. Shift the multiplicand to the left.
5. Shift the multiplier to the right for a new LSB.
6. If the shifted multiplier is zero, stop.
7. Branch to step 2.

Multiplication Subroutine

The following program multiplies a 16-bit multiplicand by an 8-bit multiplier and gives a 16-bit product. Enter the program by having the multiplicand at addresses 0000–0001 (MSByte first) and the multiplier at address 0002. The product is left at addresses 0003–0004.

```
MCNDH   EQU     0000
MCNDL   EQU     0001
PLIER   EQU     0002
PRODH   EQU     0003
PRODL   EQU     0004

        CLR     PRODH       Clear product.
        CLR     PRODL
LOOP    LSR     PLIER       Is LSB one?
        BCC     SHIFT       No, don't add.
        LDA A   PRODL
        ADD A   MCNDL       Add multiplicand
        STA A   PRODL       to product.
```

```
          LDA A   PRODH
          ADC A   MCNDH
          STA A   PRODH
          TST     PLIER      Is multiplier zero?
          BEQ     END        Yes, stop.
SHIFT     ASL     MCNDL      Shift multiplicand
          ROL     MCNDH      to left.
          BRA     LOOP
END       RTS
```

Division

Many microprocessors do not have division instructions as well as multiplication instructions. Software routines can be written which implement division of two numbers by repeatedly subtracting the divisor from the dividend. The number of times the divisor is subtracted from the dividend without a borrow is the quotient. These routines are simple but slow like repeated addition routines for multiplication.

A much faster program uses successive subtraction but starts with the most significant bit (MSB) of the quotient. The number of steps is equal to the bit size of the quotient. Alternatively, repeated subtraction increments the quotient one at a time, and the number of steps is equal to the quotient. Suppose we set up the following division problem.

$$\frac{10}{5\overline{)50}} \quad \text{or} \quad \frac{00001010}{0101\overline{)00110010}} \quad \begin{array}{l} \text{quotient} \\ \text{dividend} \end{array}$$

divisor

		dividend:	50		
Test 1.	5(divisor) $\times$	128(bit 7):	640 > 50	bit 7 = 0
Test 2.	5(divisor) $\times$	64(bit 6):	320 > 50	bit 6 = 0
Test 3.	5(divisor) $\times$	32(bit 5):	160 > 50	bit 5 = 0
Test 4.	5(divisor) $\times$	16(bit 4):	80 > 50	bit 4 = 0
Test 5.	5(divisor) $\times$	8(bit 3):	40 $\leq$ 50	bit 3 = 1
			-40	
Test 6.	5(divisor) $\times$	4(bit 2):	20 > 10	bit 2 = 0
Test 7.	5(divisor) $\times$	2(bit 1):	10 $\leq$ 10	bit 1 = 1
			-10	
Test 8.	5(divisor) $\times$	1(bit 0):	5 > 0	bit 0 = 0

$$40 + 10 = 50$$

The procedure tests each bit of the quotient from the MSB (weight 128) to the least significant bit LSB (weight 1) and keeps those bits which make up the answer. Multiply the MSB of the quotient times the divisor. If the product is less than the dividend, the bit should be set. Subtract the product from the dividend and continue with the next bit. Otherwise clear the MSB bit and test the next bit.

To make the procedure work in binary, the divisor is first shifted seven places to the left, i.e., multiplied by 128.

Test 1

$$
\begin{array}{r}
00110010 \quad \text{dividend} \\
-\ 01010000000 \quad \text{divisor} \times 128 = \text{product} \\
\hline
1]\quad 10110110010 \quad\quad\quad
\end{array}
$$

Notice two memory locations must be saved for the divisor. A subtraction causes a borrow; therefore, the product is larger than the dividend. Bit 7 is cleared. After four more tests, we have

Test 5

$$
\begin{array}{r}
00110010 \quad \text{dividend} \\
-\ 00101000 \quad \text{divisor} \times 8 \\
\hline
0]\quad 00001010 \quad\quad\quad
\end{array}
$$

There is no borrow, so bit 3 is set. Each test is similar to the preceding cases. The divisor is successively shifted to the right for each test. The quotient bit is set if the subtraction produces no borrow. We now have the basis for a division routine.

Division Subroutine

The division routine should be entered with 16-bit dividends and 8-bit divisors. Quotients are 8 bits. Two addresses are saved for the product HIGH-byte and LOW-byte. The divisor is entered into the product HIGH-byte; thus, it is automatically shifted eight times to the left (divisor $\times$ 256). A subsequent shift to the right will start the test with the correct product. To set the proper quotient bits, 80 is loaded into an address called MARK. It is shifted to the right after each test. MARK is ORed with the quotient when a quotient bit is to be set. At the program end the divisor will be found in the product LOW-byte, but the dividend will be destroyed. The dividend LOW-byte will contain any remainder:

```
QUOT      EQU    0000
DVNDHI    EQU    0001
DVNDLO    EQU    0002
PRODHI    EQU    0003        (divisor)
PRODLO    EQU    0004
MARK      EQU    0005

          CLR    QUOT
          CLR    PRODLO
          LDA A  #80         Set mark at bit 7
          STA A  MARK        of quotient.
TEST      LSR    PRODHI      Start product
          ROR    PRODLO      at divisor × 128.
          LDA A  DVNDHI      Get dividend.
          LDA B  DVNDLO
```

	SUB B PRODLO	Subtract product.
	SBC A PRODHI	
	BCS BITO	Set quotient bit?
	STA A DVNDHI	Yes, same result.
	STA B DVNDLO	
	LDA A QUOT	
	OR A MARK	Set quotient bit.
	STA A QUOT	
BITO	LSR MARK	
	TST MARK	Is MARK bit shifted through bit 0?
	BNE TEST	No, do next test.
	RTS	

PROBLEMS

5.1. Assume all memory locations have contents equal to the LOW-byte of each address; i.e., address 6040 has contents equal to 40. If the following program is executed, what addresses or registers have changed and give their contents.

```
LDS     #F7FF
LDA A   6000
PUL B
PSH A
```

5.2. The following program is executed.
(*a*) What are the addresses of data 0A, 0B, and 0C?
(*b*) Where is the stack pointer (SP) after PSH B?
(*c*) Give a program to recover 0A and store it at address 0001.

```
LDS     #001F
LDA A   #0A
PSH A
LDA B   #0B
PSH B
LDA A   #0C
PSH A
WAI
```

5.3. Why is it necessary to initialize the SP? Where is this done in the program?

5.4. Determine the number of machine cycles to store and retrieve accumulators A and B in memory assuming there is no stack. How does this compare with using the stack?

5.5. Can you think of any reasons to save MPU registers at the beginning of the subroutine rather than in the main program before the subroutine is executed?

5.6. Write a subroutine to read port A at 6000 and to store the byte at address 0100. Accumulator A is to be used and must be preserved for the main program when the subroutine is done. The subroutine starts at address 0040.

5.7. Given the instruction JSR 1000 and the following register content: IX 0050, A 01, B 02, SP 8000. Show the registers which change when the instruction is executed and the new content.

5.8. Two equal length blocks of data are stored at addresses 5000–5016 and 6000–6016, respectively. Write a subroutine which compares the two blocks to see if the data is identical. Return with the carry set when the blocks are equal. Assume accumulators A and B contents are to be restored if they are used.

5.9. Each bit of port B activates a device when the bit is logic HIGH. Write a program to turn on each device successively for 1 min. Assume port B is address 6000.

5.10. Construct a timing loop to give a 10-ms delay.

5.11. What is the maximum delay possible by counting down accumulator A given a 1-MHz clock?

5.12. Can you improve the timing loop accuracy in Example 5.5?

5.13. The multiplication routine in Sec. 5.6 allows for a 16-bit product. In some cases this range can be exceeded. Suggest a programming change which would branch or jump to an error message display starting at address ERROR if the range is exceeded.

5.14. For the reader who wants a challenge, modify the division routine in Sec. 5.6 to allow for 16-bit quotients.

CHAPTER
6

INTERFACING MEMORY

Memory is the essential part of any computer system both inside and outside the MPU. Memory cost is related to its speed, the time required to perform a read or write operation. In particular, fast memories tend to be expensive while inexpensive memories tend to be slow. Therefore, a general-purpose computer system can be expected to have three types of memories:

1. Fast elements composed of discrete flip-flops which make up the registers within the MPU. These units have an operating speed in the range of nanoseconds (ns) (1 ns = 1×10^{-9} s).

2. A much larger *main memory* made up of RAM and ROM integrated circuits (ICs). Capacity is usually measured in kilobytes. Individual bit storage is composed of flip-flops or capacitors, but IC speed is on the order of microseconds (μs) (1 μs = 1×10^{-6} s). Memory ICs are interfaced directly to the bus structure.

3. An even larger, slower, and cheaper *secondary memory* is composed of magnetic disk or tape units. Common floppy disks have a storage capacity approximately equal to 1,000,000 bytes. Hard disks provide a much larger capacity than floppy disks. Such disks can store over one hundred times more information than floppy disks and store or retrieve information nearly one hundred times faster. Secondary memories are attached directly to I/O structures. They typically store programs and data to be downloaded into the main memory when needed. Loading time per instruction is on the order of milliseconds.

In this chapter we review the construction of bit storage cells for both RAM and ROM. Cell organization into memory ICs are examined along with pin assignments of several memory ICs. Our main goal is the interfacing of memory ICs to the MPU.

Microprocessor systems use a common shared data bus. *Only one memory chip may share data with the MPU at a given time.* If two ICs use the bus at the same time, bus contention results. This condition produces invalid data on the bus. In the extreme case where one IC drives the connection with logic 1 while a second IC drives the same connection to logic 0, excessive current may destroy a device immediately. We have seen that data is transferred to and from the MPU in less than 1 μs. Thus, the data transfer must be carefully timed. Microprocessor control lines and address lines make up a decoding scheme that establishes which device is activated for bus use. Control lines regulate the timing while address lines select the proper device. An understanding of these processes will lay the foundation for interfacing *all* ICs to the data bus.

6.1 RAM CONSTRUCTION

Semiconductor memories are built from bipolar or MOS (metal oxide semiconductor) transistors. Bipolar technology is older but produces very fast memories ($\sim$50 ns). MOS technology is approximately 10 times slower than bipolar, but it has these important advantages: (1) it has a higher packing density, (2) it consumes less energy, and (3) it is cheaper. Consequently, all larger memories currently use MOS technology. Individual transistors are made for general-purpose switching (Chap. 9) or power switching (Chap. 15). However, memory contains miniaturized versions with little power and bit cells are buffered to drive the memory output.

Memory Cell

A basic memory cell consists of a bistable latch or flip-flop. Figure 6.1 shows a flip-flop constructed from a pair of bipolar transistors. (Flip-flops also can be constructed from NAND or NOR gates. See Chap. 10 on switch debounce.) The transistors act as a simple switch. If a voltage is applied to base resistance R1 or R2, the transistor is switched ON.

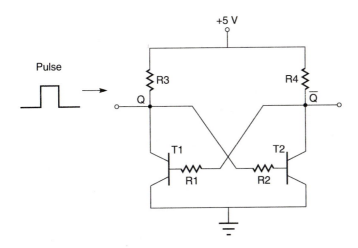

FIGURE 6.1
Basic bipolar flip-flop memory cell.

ON-state resistance of a transistor is very low. The voltage drop across resistors R3 and R4 assures that point Q is a logic LOW (near ground). If the transistors are OFF, point Q is a logic HIGH (near 5 V). Suppose we apply a positive voltage pulse to the input. Transistor T2 is switched ON, causing current to flow through the transistor. Point $\overline{Q}$ goes LOW. A low voltage to base resistance R1 switches transistor T1 OFF. Point Q stays HIGH after the pulse is removed. This means that the ON-state of transistor T2 keeps transistor T1 in the OFF-state, and vice versa. A similar argument applies when the input pulse is a logic LOW. In this case point Q is LOW and point $\overline{Q}$ is HIGH. The state of a memory bit is given by Q.

Because of the dual feedback from transistor collectors to opposite bases, flip-flops are very insensitive to external disturbances, such as power fluctuations or induced noise. Flip-flop memories will hold their state indefinitely until a new input changes their state or the power is turned OFF. Such memory is called *static RAM* or SRAM.

Storing bits into RAM is somewhat more complicated than the preceding. A group of bits (cells) must be selected at one time to receive or dispense data. An actual static MOS RAM cell looks like Fig. 6.2. The storage cell consists of six MOS transistors.

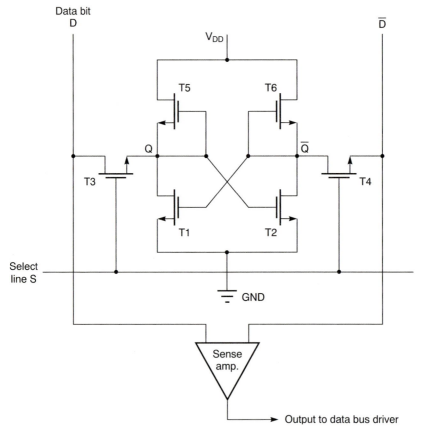

FIGURE 6.2
Static (MOS type) RAM storage cell with sense amplifier.

Transistors T5 and T6 serve as resistors. A single select line S (also called a word line) is used to select a group of bits for both reading and writing. The remaining two transistors T3 and T4 isolate a given cell from the input and are switched ON by the select line.

During a read operation the cell is selected by setting S equal to logic 1, which switches T3 and T4 ON and connects stored signals Q and $\overline{Q}$ to the data bit lines D and $\overline{D}$, respectively. If Q = 1, D = 1 and $\overline{D}$ = 0. The resulting bit pairs are processed by an internal amplifier called a *sense amplifier*. A sense amplifier outputs logic 1 if Q = 1 and logic 0 if Q = 0. It serves two purposes: (1) amplifies the power of a very weak storage cell charge and drives the data bus and (2) rejects stray noise interference by operating as a differential amplifier, processing both D and $\overline{D}$ to stabilize output D.

Complete RAM

The six transistor cells and sense amplifier represent a single-bit storage. There are thousands of these cells in a memory IC. Each byte (word) in memory is associated with a unique address. The address bus must be interfaced to the byte select lines by means of an internal decoder.

DECODER. A (simple) decoder converts an *N*-bit input, representing the binary number *j*, into a 2^N-bit word in which the *j*th bit is 1 while the remaining bits are 0. The decoder is called a 1-of-2^N decoder. For example, a 128-byte memory is accessed with seven address lines: $2^7 = 128$. The memory IC has seven address pins to an internal decoder. A 1-of-128 decoder diagram is shown in Fig. 6.3. The address decoder is made up of 128 7-input AND gates and seven inverters. Each AND gate output is connected to one select line. Notice that gate inputs are wired in such a way as to produce a logic 1 only at the AND gate output corresponding to the address. Thus, byte select line one is HIGH and all remaining select lines are LOW when the binary address is 0000001. Byte select line 64 is HIGH only when the binary address is 1000000.

Memory ICs do not have separate lines for input data and output data. Instead, they use the same data pins for both transfers. This is possible because the MPU cannot read and write simultaneously. The MPU read/write control line (R/$\overline{W}$) switches the memory IC to the desired data direction. Internal control is accomplished with a tristate logic device.

TRISTATE LOGIC DEVICE. Tristate logic devices can be gates, buffers, or amplifiers. They are distinguished from their more common relatives by an extra input called the *enable*. See Fig. 6.4. When enable E is HIGH, the tristate merely passes (and boosts the power of) the input to the output. However, when the enable is LOW, the device has a very high impedance, effectively disconnecting the input. More information on tristate gates is given in Chap. 9.

Figure 6.5 shows a simplified arrangement of storage cells and tristates to construct a complete 2 × 2 bit RAM. A single address line A0 goes to a 1-of-2 decoder. When A0 is LOW, select line S_0 is HIGH, connecting the first row of cells to the data bus. The second row of cells is connected when A0 is HIGH. Chip select line CS controls memory access to the data bus. If CS is HIGH, the memory chip is connected to the data bus for

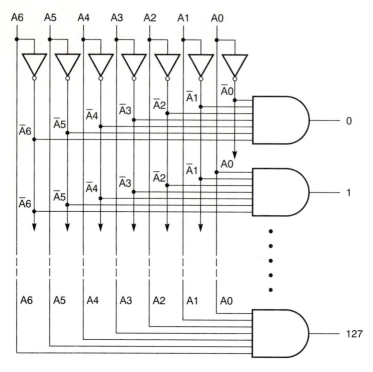

FIGURE 6.3
1-of-128 address decoder.

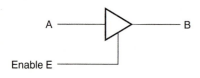

A E	B
0 1	0
1 1	1
0 0	Z
1 0	Z

(High impedance)

FIGURE 6.4
Tristate logic device.

a read or write. A LOW on CS disconnects the chip from the bus by disabling all tristates. When the R/$\overline{\text{W}}$ line is HIGH, the tristate sense amplifiers are active, placing cell data on the bus for a read. Conversely, a LOW on the R/$\overline{\text{W}}$ line disables the sense amplifiers and enables the two pairs of write amplifiers. Data is placed into the cells by circumventing the sense amplifiers.

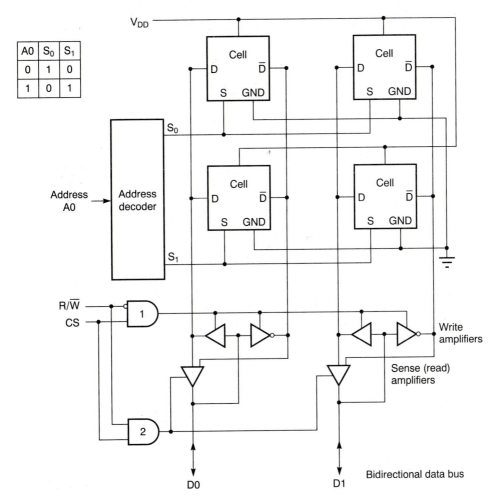

A0	S_0	S_1
0	1	0
1	0	1

FIGURE 6.5
A 2 × 2 bit RAM showing enable for bidirectional data bus.

Dynamic RAM

Another important class of RAMs is called dynamic RAMs or DRAMs. They use a memory cell with a simple capacitor as a storage element rather than flip-flops. Figure 6.6 shows a dynamic RAM cell constructed of a capacitor and single transistor. To write to the cell, data is applied to line D. The select line is set to 1. Voltage on the line charges the capacitor to write a logic 1 and discharges to write a logic 0. For reading the contents, line D is converted to an output line and the cell is again selected by setting S = 1. If capacitor C is charged, it discharges into line D, producing a pulse which is converted by the sense amplifier to logic 1. A discharged capacitor (storing logic 0) will be converted to logic 0 by the amplifier.

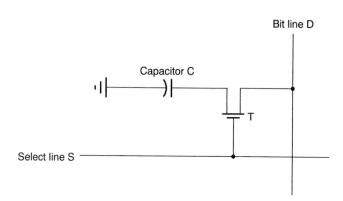

FIGURE 6.6
Dynamic RAM cell.

Unfortunately, DRAM cell storage is not ideal. First, a read operation is destructive since reading a cell always leaves it in a discharged state of logic 0. If the cell originally stored logic 1, a read must be immediately followed by a write operation to restore the orginal state. Second, the capacitor insulation is less than perfect. Even if the cell is never read, a stored charge leaks away within a few milliseconds. To prevent RAM contents from being lost, they must be restored at regular intervals, a process called refreshing. DRAMs require external logic to provide refreshing, cycle timing, and address multiplexing. Logic is provided by a DRAM controller IC. This controller simplifies the interface of DRAMs to the microprocessor, in some cases making the DRAM appear as a SRAM to the MPU.

DRAM disadvantages are more than outweighed by its cell simplicity. A single capacitor and transistor require much less cell area than the six transistors of a static memory cell. As a result, DRAMs have at least four times the packing density, which means DRAMs can store four times as much data as SRAMs on a chip of the same size. Memory is proportionally cheaper. DRAMs also consume less power than SRAMs because flip-flops have at least four transistors ON regardless of whether or not they are accessed.

DRAMs and their external logic controller are justified in large scale memory systems down to the personal computer level. They are not practical in mechatronics where a few thousand bytes of memory are more than sufficient.

6.2   ROM CONSTRUCTION

A read-only memory (ROM) contains permanently stored data which can be read but cannot be changed (written to). The program is not lost when the power is turned OFF. Stored data may include mathematical functions, tables, display or service routines. Large systems place high-level language interpreters and compilers for BASIC, FORTRAN, etc. into ROMs. Most microprocessor-based systems for automobiles, appliances, and other consumer products use ROM exclusively.

The basic storage cell of ROMs is a simple semiconductor switch that is permanently turned ON or OFF, thus generating a constant 1 or 0 data bit. Otherwise ROM structure is similar to RAM structure except that the data flow is one direction only. Figure 6.7

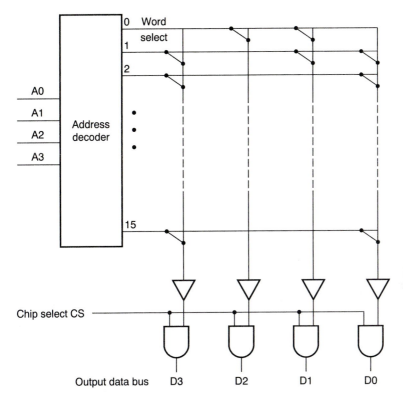

FIGURE 6.7
Basic ROM structure.

shows a hypothetical 16 × 4 bit ROM. Four address lines are decoded into 16 separate word select lines. Only one word select line is HIGH for each address. Select lines are connected to the vertical bit lines at each junction only by diagonal switches. If a switch is closed anywhere along the HIGH select line, the decoder drives that particular bit line HIGH. Disconnected switches will leave the bit line LOW. For example, address 1111 causes select line 15 to go HIGH—all other select lines are LOW. Only select line 15 can drive the data bus. Switches for bit lines D3 and D0 are permanently closed. Thus, the word 1001 appears on the data bus. There is no read/write line. A single CS enables the output buffers.

ROMs are manufactured in several varieties. The type of application will determine which ROM is used.

Masked ROM

Masked ROMs are always programmed by the manufacturer. The customer develops the ROM programming and forwards it to the manufacturer according to specifications. The fabrication process involves a number of steps: diffusion, etching, and masking. A mask of the customer's program defines the 1s and 0s. The transistor gate is an oxide layer. If the oxide layer is left thin, a MOS transistor exists across the select-line/bit-line diagonal.

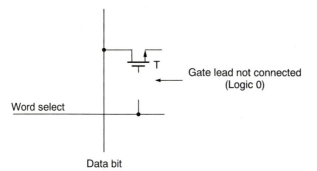

FIGURE 6.8
Disconnected MOS cell in ROM.

However, a thick oxide layer insulates the gate from the select line and no functioning connection exists. See Fig. 6.8. The term "ROM" is often restricted to memories manufactured by the mask process.

Manufacturers charge several thousand dollars for a mask. The customer should be certain of his or her program because the ROM is useless if a single bit is wrong. The ROM itself is very cheap. Thus, it is most economical to order several thousand ROMs to spread the mask cost.

Programmable Read-Only Memory

Programmable read-only memories or *PROMs* are programmed by the customer rather than the manufacturer. There are no mask charges but individual PROMs are more expensive than ROMs. They are more economical for low production applications. The customer must have a programming unit to "burn in" the desired bit patterns. Two cells of a PROM are shown in Fig. 6.9. Most cell transistors are bipolar. A fuse link is placed be-

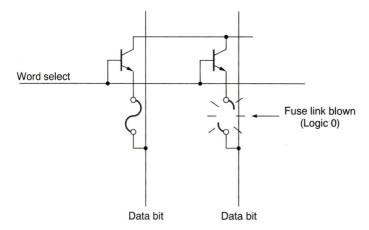

FIGURE 6.9
Two cells in a bipolar PROM with fuse links for programming logic 1 or logic 0.

tween the transistor output and the bit line. If the fuse is intact, a select line HIGH turns ON the transistor and the data bit reads logic 1. However, the user can remove the link to produce a logic 0. The programming unit applies an excessive voltage pulse to the transistor, blowing the fuse wherever a logic 0 is required. PROMs are useful during the development phase of a new microprocessor-based system.

Erasable Programmable Read-Only Memory

Erasable PROMs or *EPROMs* are even more advantageous for system development. The memory's IC is visible through a quartz window. See Fig. 3.5. When the circuit is exposed to high-intensity ultraviolet (UV) light for ~30 min, the stored program is erased. Chips are programmed with an instrument called an EPROM programmer. The EPROM is removed and tested in the microprocessor-based system. If modifications are necessary, the EPROM is removed, erased, and reprogrammed as often as necessary. After the user is satisfied the window is taped over (daylight will erase memory in about two months), and the EPROM functions as the system ROM.

An EPROM memory cell resembles the masked ROM MOS transistor but has an added feature: a floating gate. The floating gate is completely surrounded by a silicon dioxide (SiO_2) insulator. See Fig. 6.10. With masked ROM transistor action, a select gate voltage causes the region between drain (input) and source (output) to conduct. However, EPROMs are controlled by a floating gate. The cell is programmed by applying a relatively high voltage, ~25 V, to the select gate and drain. Electrons penetrate the insulating SiO_2 layer, and a negative charge builds up on the gate. The charge remains until it is erased. In this programmed state the charged gate blocks the positive voltage on the select gate from inducing a conductive channel. The cell is said to be storing a logic 0.

The cell is returned to its unprogrammed state (erased) by illuminating the cell with UV light. Electrons on the floating gate receive enough energy from the light to overcome the insulating barrier and leak to the substrate. Now the select gate voltage causes the cell to conduct. The cell is said to store a logic 1.

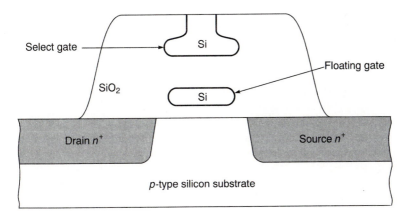

FIGURE 6.10
EPROM memory cell with charged floating gate.

EPROM Programmers

EPROM programmers are usually one of two types: stand-alone or personal-computer (PC)-driven. In each case the cost varies from $500 to $1000. Stand-alone varieties have a keyboard to load hexadecimal programs into an on-board RAM. The device to be programmed is placed in a programming socket, and the RAM-stored program is downloaded into the EPROM in one step. In order to handle a wide variety of EPROMs, hardware personality modules are used. The module adapts the programmer to individual voltages, currents, and pin configurations.

PC-driven systems rely on software to control all operations including personality module functions. Figure 6.11 shows a typical PC-driven programmer by BP Microsystems. It requires an IBM PC, XT, AT, PS/2, or 386 compatible computer with 512K RAM and a standard parallel printer port. The unit will program or read virtually every EPROM or EEPROM (below) and can be updated with new floppy disks.

Electrically Erasable PROMs

Electrically erasable PROMs or *EEPROMs* are similar to EPROMs, having a floating gate structure in the MOS cell. A very thin oxide layer between the gate and substrate is used to charge or discharge the floating gate. The EEPROM memory chip is placed in the MPU system much like a RAM. A read/write line is activated to write or erase data. A 20-V

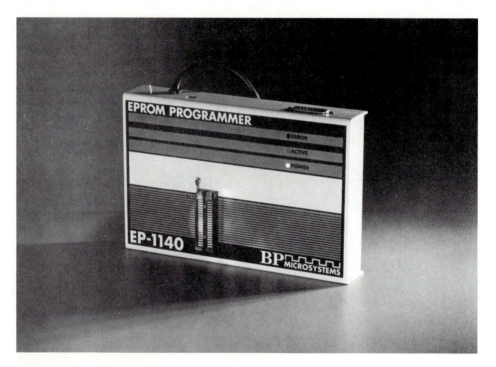

FIGURE 6.11
PC-driven EPROM/EEPROM programmer. (*Courtesy of BP Microsystems.*)

supply voltage is required. Write cycle times for each byte are on the order of 10 ms, so a write process is only practical during the loading phase. It functions in the system only as a ROM. Read cycle times are approximately several hundred nanoseconds.

There is an advantage to using ROMs that do not have to be removed from the system when they are programmed or erased. However, EEPROMs are expensive compared to other types of ROMs. Also, they do not retain their gate charge as long as EPROMs. EEPROM use will grow as technology improves its performance.

6.3 THE MOTOROLA 6810 RAM

The Motorola 6810 static RAM is a 1024-bit (128-word $\times$ 8-bit) memory in a 24-pin package. A memory of 128 bytes may seem ridiculously small by today's personal computer standards. But many consumer product control boards need only several hundred bytes of RAM. The 6810 is designed for use with the 6800/6802/6808 MPU family, but it can be integrated with other microprocessor systems. However, slow memories cannot be used with fast microprocessors. Memory timing is discussed in Sec. 6.7. Table 6.1 lists three 6810 types.

Figure 6.12 shows the pin assignment for the 6810 RAM. All pins are TTL compatible. They include:

V_{CC}	The chip requires a 5-V power supply (V_{CC}) and ground (GND).
$R/\overline{W}$	The read/write signal is obtained directly from the MPU. When $R/\overline{W}$ is HIGH, data is read from the data bus. In effect, the RAM tristate buffers drive the MPU. A LOW on $R/\overline{W}$ enables buffers to write data into the memory cells.
D0–D7	Eight data lines connected between the 6810 and MPU constitute the data bus. Other ICs are connected to the data bus also.
A0–A6	Only seven address lines are necessary to locate each 2^7 or 128 bytes of stored data. They are connected to the lower equivalent address pins of the MPU. Only part of the 16-line address bus is connected directly to a single memory IC (unless a 64K DRAM is used).
CS0 CS3 CS1 $\overline{CS2}$ CS4 $\overline{CS5}$	A signal must be generated to enable tristate buffers before the data bus is connected internally to memory cells. On many ICs a single pin labeled CS (chip select), CE (chip enable), or simply E (enable) turns ON the chip. But here, as many as six CSs are used. Two CSs should have HIGH signals

TABLE 6.1
6810 RAM types

Type	Access time (ns)	Maximum MPU clock frequency (μs)
MCM6810P*	450	1.0
MCM68A10P	360	0.667
MCM68B10P	250	0.5

*P suffix is a plastic package; L suffix is a ceramic package.

Pin Assignment

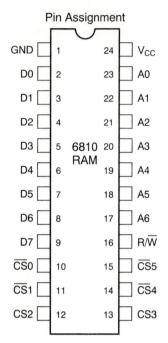

FIGURE 6.12
Pin assignment for the Motorola 6810 static RAM (128 × 8-bit). (*Reprinted with permission of Motorola, Inc.*)

and four should have LOW signals. A gating scheme is shown in Fig. 6.13. An ANDing of the CSs enables the 6810 RAM:

$$\text{ENABLE} = \text{CS0} \cdot \overline{\text{CS1}} \cdot \overline{\text{CS2}} \cdot \text{CS3} \cdot \overline{\text{CS4}} \cdot \overline{\text{CS5}}$$

The origin of these CS signals will be explained in Sec. 6.6 on address decoding.

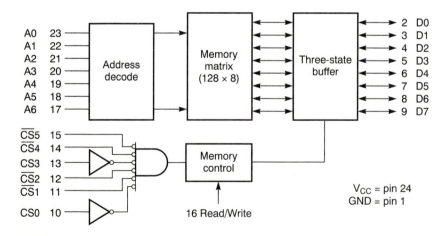

FIGURE 6.13
Block diagram of the 6810 RAM with multiple chip selects. (*Reprinted with permission of Motorola, Inc.*)

Example 6.1. In anticipation of the section on address decoding, suppose we were to connect address lines A15–A12 to the first four CSs and ground to the last two CSs. What addresses placed on the address bus would enable the 6810 RAM?

$$
\begin{array}{cccc}
\text{A15} & \text{A14} & \text{A13} & \text{A12} \quad \dots \\
\downarrow & \downarrow & \downarrow & \downarrow \\
\text{CS0} & \overline{\text{CS1}} & \overline{\text{CS2}} & \text{CS3} \\
\downarrow & \downarrow & \downarrow & \downarrow \\
1 & 0 & 0 & 1 \\
& & 9
\end{array}
$$

The answer is any hexadecimal address with 9 as the first digit: 9 _ _ _. The example is illustrative only because we have not accounted for the role played by MPU control lines.

6.4 THE INTEL 2114A RAM

The Intel 2114A static RAM is a 4096-bit (1K $\times$ 4-bit) memory in an 18-pin package. It has a larger capacity than the 6810 RAM but the data lines are only 4 bits wide. Fewer pins make a cheaper package. Figure 6.14 shows the 2114A data sheet as it appears in the *Intel Memory Data Book*. It is TTL compatible, and with access times from 100 to 250 ns it can be interfaced to the Motorola 6802 MPU. The 2114A address, data, and control pins differ from the 6810:

I/O_1 to I/O_4	Labeling differs among manufacturers. Data input/output (I/O) notation is equivalent to data notation D0–D3. Since our data bus is 8 bits wide, the 2114A must be paired to form byte size storage. Another unit would be used for D4–D7.
A0–A9	Ten address lines are necessary to read or write 1 of 1024 storage locations: $2^{10} \equiv 1024$.
$\overline{\text{WE}}$	The write enable (WE) pin is brought LOW when writing to memory and raised HIGH when memory is read. Thus, the $\overline{\text{WE}}$ line is equivalent to a R/$\overline{\text{W}}$ line and connects directly to the MPU R/$\overline{\text{W}}$ pin.
$\overline{\text{CS}}$	A single CS replaces the six CSs on the 6810 RAM. The 2114A RAM is enabled by a LOW on the $\overline{\text{CS}}$ pin.

Example 6.2. How are two 2114A RAMs connected to form a 1K $\times$ 8-bit memory for the 6802 MPU?

1. Connect I/O_1 to I/O_4 pins on RAM #1 to data lines D0 to D3.
2. Connect I/O_1 to I/O_4 pins on RAM #2 to data lines D4 to D7.
3. Tie both R/$\overline{\text{W}}$ pins together to read (write) both at the same time.
4. Tie both CS pins together, forming a common CS. An enable signal will read (write) the HIGH data nibble from (to) RAM #2 and the LOW data nibble from (to) RAM #1.

A wiring diagram is given in Fig. 6.15.

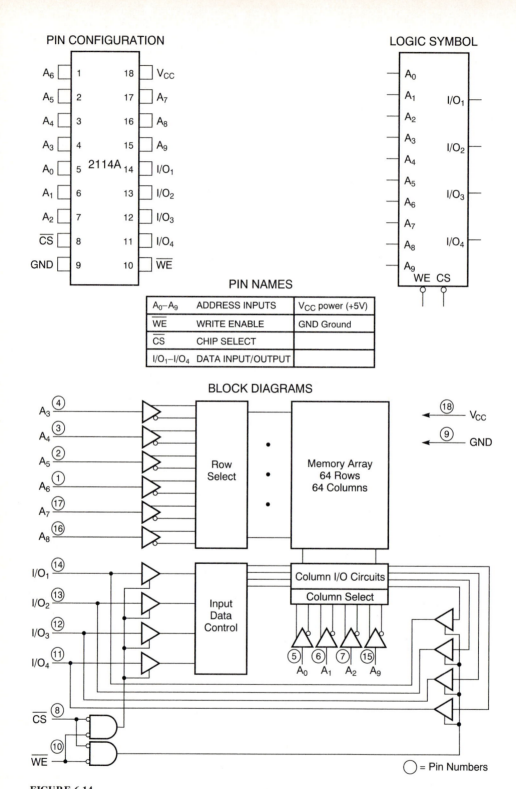

FIGURE 6.14
The Intel 2114A static RAM (1024 × 4-bit). (*Reprinted with permission of Intel Corp.*)

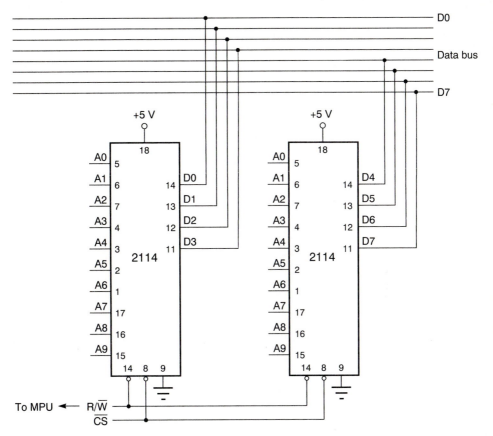

FIGURE 6.15
Two 2114A RAMs arranged as a 1024 × 8-bit memory.

6.5 THE INTEL EPROMS

The Intel 2732A EPROM is a 4K × 8-bit UV erasable memory, which is also TTL compatible. Most EPROMs have 8-bit data lines. Standard access time is 250 ns. Other versions include the 2732A-2 (200 ns), 2732A-3 (300 ns), and the 2732A-4 (450 ns). Figure 6.16 gives the pin assignment for the 2732A as well as other Intel EPROMs. Memory capacity is determined by the number of address pins.

Type	Address Pins	Capacity
2716	11	2K
2732	12	4K
2764	13	8K
27128	14	16K

Most EPROMs follow a universal 24- or 28-pin standard configuration adopted by the Joint Electronic Devices Engineering Council (JEDEC). These standard pin functions

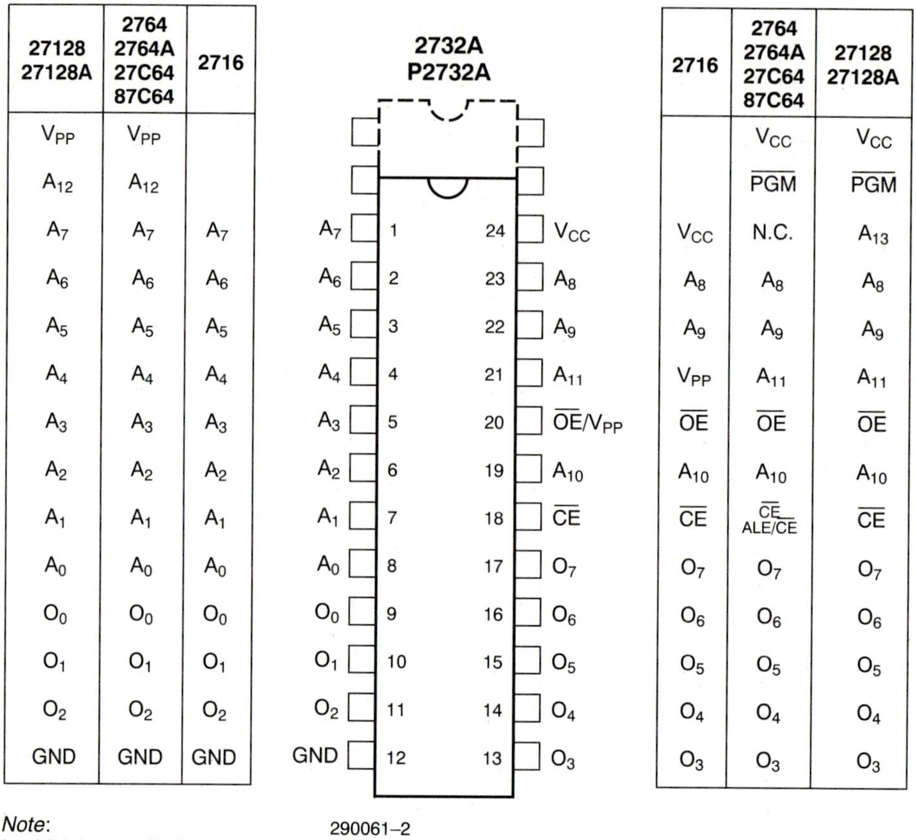

Left side (address/data pins):

27128 / 27128A	2764 / 2764A / 27C64 / 87C64	2716
V_{PP}	V_{PP}	
A_{12}	A_{12}	
A_7	A_7	A_7
A_6	A_6	A_6
A_5	A_5	A_5
A_4	A_4	A_4
A_3	A_3	A_3
A_2	A_2	A_2
A_1	A_1	A_1
A_0	A_0	A_0
O_0	O_0	O_0
O_1	O_1	O_1
O_2	O_2	O_2
GND	GND	GND

2732A / P2732A pinout:

Left pin	#	#	Right pin
A_7	1	24	V_{CC}
A_6	2	23	A_8
A_5	3	22	A_9
A_4	4	21	A_{11}
A_3	5	20	$\overline{OE}/V_{PP}$
A_2	6	19	A_{10}
A_1	7	18	$\overline{CE}$
A_0	8	17	O_7
O_0	9	16	O_6
O_1	10	15	O_5
O_2	11	14	O_4
GND	12	13	O_3

Right side:

2716	2764 / 2764A / 27C64 / 87C64	27128 / 27128A
	V_{CC}	V_{CC}
	$\overline{PGM}$	$\overline{PGM}$
V_{CC}	N.C.	A_{13}
A_8	A_8	A_8
A_9	A_9	A_9
V_{PP}	A_{11}	A_{11}
$\overline{OE}$	$\overline{OE}$	$\overline{OE}$
A_{10}	A_{10}	A_{10}
$\overline{CE}$	$\overline{CE}$ ALE/$\overline{CE}$	$\overline{CE}$
O_7	O_7	O_7
O_6	O_6	O_6
O_5	O_5	O_5
O_4	O_4	O_4
O_3	O_3	O_3

Note: 290061–2
Intel "Universal Site" compatible EPROM configurations are shown in the blocks adjacent to the 2732A pins.

FIGURE 6.16
Pin assignments for Intel EPROMs. (*Reprinted with permission of Intel Corp.*)

are given in the figure. EPROMs with 24 pins are positioned at the bottom of the 28-pin universal socket on an EPROM programmer.

Pin groupings for an EPROM are almost identical to RAMs. There are address lines, data lines, power, and a chip enable (select). Because EPROMs are read only in circuit, there is no read/write line ($R/\overline{W}$ or $\overline{WE}$). It is replaced by a programming pin ($\overline{OE}/V_{PP}$).

Programming Mode

1. Erase the EPROM leaving all bits in the "1" state. Only "0s" are selectively programmed.
2. $\overline{OE}/V_{PP}$ is held at 21 V.
3. Data to be programmed is applied to pins O_0 to O_7.

4. When the address and data are stable, a 20-ms minimum (50-ms typical) active LOW TTL pulse is applied to $\overline{CE}$. A pulse is applied for each address/data input.

5. The 2716 EPROM has separate $\overline{OE}$ and V_{PP} pins. During its programming the $\overline{OE}$ pin is held HIGH (5 V).

Read mode

1. $\overline{OE}/V_{PP}$ is tied to ground (held LOW), enabling the output buffers.

2. For the 2716 EPROM, $\overline{OE}$ is tied to ground and V_{PP} is tied to 5 V (held HIGH).

3. The address is sent to the address bus via an LDA instruction.

4. The chip enable ($\overline{CE}$) pin is brought LOW. This signal comes from the address decoder.

Normally, a commercial EPROM programmer is used for the programming mode. So, the circuit designer is chiefly concerned with the read mode. Figure 6.17 shows a wiring diagram that implements the read mode for the 2716 EPROM. Pin 21 is wired to address line A11 for the 2732A.

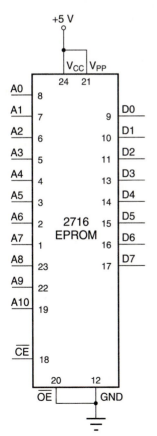

FIGURE 6.17
Pin wiring for a 2716 EPROM after it is programmed to use.

6.6 ADDRESS DECODING

Memory requirements of a typical microprocessor system often cannot be met with a single memory IC. In addition, many I/O devices have internal registers that either contain data or control how the device operates. Other I/O devices are simply tristate buffered with the data bus. With Motorola microprocessors, each I/O register or buffer is given a unique address and the I/O device is handled just as though it was memory; i.e., use the LDA and STA instructions to transfer the data. The method is called *memory-mapped I/O.* (Intel microprocessors use special instructions, IN and OUT, to communicate with I/O. This method is called isolated I/O.)

Microprocessors can communicate with only one chip at a time. *The process of selecting one chip among many on the data bus is called address decoding.* Each chip selected can have one or many addresses. Therefore, memory space must be allocated for each chip. The total addressable space with 16-bit address lines is 65,536 bytes. A chart giving the addressable chip type and its assignment space is called a *memory map.* Figure 6.18 shows a memory map for the Heathkit Microprocessor Trainer. It includes space for memory (RAM and ROM) and I/O (keyboard and displays). Unused space is available for the designer to add chips to the system.

An instruction consisting of several data bytes is executed in microseconds. *An added task of address decoding is to transfer data only when valid data is on the bus.* This duty is assigned to the control bus. There are three MPU control lines of interest:

$R/\overline{W}$ The read/write line gates writable memory or I/O registers for data direction. It connects directly to an IC pin, designated $R/\overline{W}$.

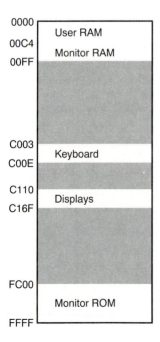

FIGURE 6.18
Memory map for the Heathkit Microprocessor Trainer.

VMA   The valid memory address line goes HIGH when the address on the bus is stable. It is used as part of address decoding logic to select (enable) the chip only when the address is valid.

E (or $\phi2$)   The enable (E) pin on the 6802 is really the phase 2 ($\phi2$) clock. It is connected to similarly named pins on some Motorola family members such as the 6820/21 PIA (Chap. 7). The $\phi2$ clock is HIGH when data on the bus is valid. Unless the $\phi2$ signal is already connected to the IC, it is included as part of address decoding logic also. For writable devices (RAMs and latches), it keeps the device from being enabled during the first half of the $\phi2$ cycle when the data is invalid. Since the MPU accepts data on the falling edge of the $\phi2$ clock, *read-only devices do not need $\phi2$ in the CS logic*. In fact, we will see in Sec. 6.7 that its inclusion is detrimental to reading slow EPROMs.

Decoding for the Intel 2716 EPROM

The 2716 EPROM has 11 address pins, labeled A0 through A10 for 2K data bytes. It has a single chip enable ($\overline{CE}$), active LOW. *Generally, the low-order address lines are connected to memory and used to address individual memory bytes or registers within the chip.* Memory space for the EPROM can be any one of 32 spaces, 2K each:

    0000 to 07FF
    0800 to 0FFF
    1000 to 17FF
    1800 to 1FFF
        .
        .
        .
    F800 to FFFF

For example, if all 11 EPROM address pins have LOWs, the memory's first data byte is placed on the bus for any address in the first column above once the chip is enabled. *It is the remaining higher address lines, in addition to VMA and $\phi2$, that are available to select the EPROM* for a particular space.

FULL DECODING. Suppose we choose the space F000 to F7FF. The resulting bit pattern on the MPU address lines from a STA instruction is

A	15	14	13	12	11	10	9	8	7	6	5	4	3	2	1	0
	1	1	1	1	0	X	X	X	X	X	X	X	X	X	X	X

where X denotes "don't care." X is then 1 or 0, depending upon which address is chosen. An *address decoder* must be built which will enable the EPROM for the preceding bit pattern of higher addresses: A15 to A11. The decoder input should include the VMA line, which is logic 1 when the address is valid. The $\phi2$ clock is logic 1 when the data is valid. Its input to the decoder is necessary for RAMs but is omitted for EPROMs.

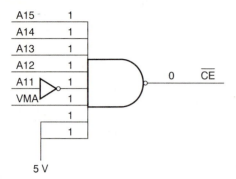

FIGURE 6.19
Full decoding for 2716 EPROM with a 74LS30
8-input NAND gate. Address space F000 to F7FF.

The address decoder can be any type of logic circuit that meets the previous requirements. Perhaps the easiest circuit to use is the 74LS30 8-input NAND gate. See Fig. 6.19. NAND gates are active LOW only when *all* inputs are logic 1. Address lines, which must be LOW for proper decoding, have inverters placed between the address line and the gate input. If memory is enabled by an active HIGH, the NAND gate output is simply inverted. When all the remaining high address lines are connected to the decoder as demonstrated here, each memory location in the selected chip has only one address. The addresses are said to be *fully decoded*.

PARTIAL DECODING. With *partial decoding,* all higher address bits are not decoded. What happens if we omit one or more of the high MPU addresses from the address decoder logic? In particular, let us omit address line A11:

A	15	14	13	12	11	10	9	8	7	6	5	4	3	2	1	0
	1	1	1	1	·	X	X	X	X	X	X	X	X	X	X	X

where · denotes an unconnected address line. The MPU will write A11 = logic 0 for instruction STA (F000) but will write A11 = logic 1 for instruction STA (F800). In either case the EPROM is enabled and the first data byte is transmitted to the MPU. Two 2K-memory spaces are enabled. The lower space F000 to F7FF is assigned to the EPROM. The higher space F800 to FFFF is said to be *redundant.* The EPROM responds to these addresses, so the space cannot be occupied by other ICs. A LDA F800 will fetch the first data byte as well as LDA F000. Figure 6.20 shows the partially decoded 2716 EPROM interfaced to the MPU.

While omitting address lines from the address decoder destroys potential memory space for other chips, the technique poses no threat to systems which use only a small portion of the available space. On the contrary, partial decoding schemes are frequently used because they save decoding logic. Here an inverter IC is no longer necessary.

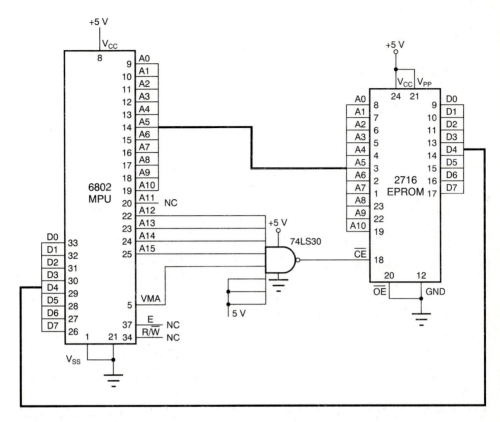

FIGURE 6.20
2716 EPROM interfaced to 6802 MPU. Decoded for memory space F000 to F7FF (F800 to FFFF redundant).

Example 6.3. Suppose address line A15 is not connected to the address decoder but address line A11 is connected for the partially decoded EPROM. What is the effect on memory space?

A	15	14	13	12	11	10	9	8	7	6	5	4	3	2	1	0
	·	1	1	1	1	X	X	X	X	X	X	X	X	X	X	X

Now either a logic 1 or logic 0 on A15 will enable the EPROM for the given bit pattern. The decoder address space is

7800 to 7FFF
F800 to FFFF

As more of the highest available address lines for decoding are *not* connected, the whole address space takes on the appearance of swiss cheese. The highest available address lines should be connected to the decoder for a more ordered approach to assigning other chips to the memory space. See Fig. 6.21.

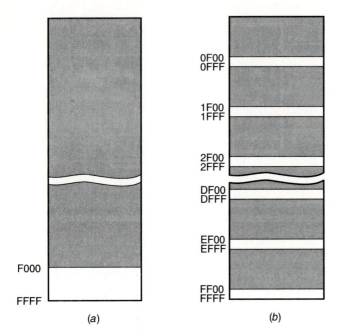

FIGURE 6.21
Decoded space for 256-byte memory using two different groups of four address lines:
(a) 1111 · · · · XXXXXXXX; (b) · · · · 1111XXXXXXXX. Dot denotes not connected.

Interfacing the Motorola 6810 RAM

The 6810 RAM has seven address pins for 128 data bytes. It also comes with six CSs, four active LOW and two active HIGH, which must be active simultaneously to enable the RAM. The advantage of multiple CSs is easily recognized. Partial decoding is done on the IC with a minimum number of gates external to RAM. Connect VMA and the $\phi2$ clock to the active HIGH CSs. This leaves active LOW CSs for the four highest available address lines. Inverters will be needed for HIGH address lines. If any selects are not used, they should be tied to 5 V or ground as the logic dictates.

> **Example 6.4.** Suppose we have allocated the entire 4K of memory space 5000 to 5FFF for a 6810 RAM. Determine an interface with the 6802 MPU.
>
> Most of the space is sacrificed to redundant addresses but 60K remain for other chips. Addresses are assigned as follows:

A	15	14	13	12	11	10	9	8	7	6	5	4	3	2	1	0
	0	1	0	1	·	·	·	·	·	X	X	X	X	X	X	X

Address lines A14 and A12 should be inverted before they are tied to the active LOW CSs. Data pins and the R/$\overline{\text{W}}$ pin on the 6810 are interfaced directly to the MPU. A wiring diagram is given in Fig. 6.22.

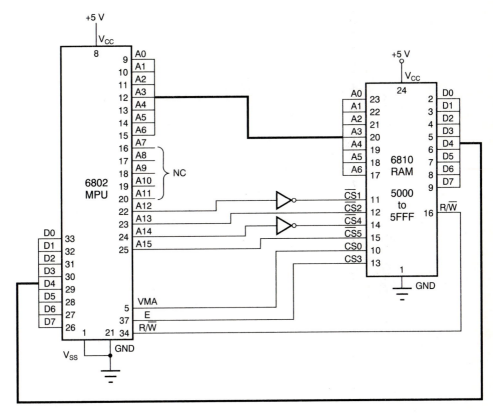

FIGURE 6.22
6810 RAM interfaced to 6802 MPU. Decoded for memory space 5000 to 5FFF.

Decoder ICs

An alternate method to provide address decoding is through decoder ICs. They are efficient when more than three or four chips are to be interfaced to the data bus. Figure 6.23
shows the pin assignment and truth table for a 74LS138 1-of-8 decoder, which converts a

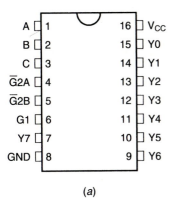

(a)

FIGURE 6.23
74LS138 decoder IC. (a) Pin assignment. (*Reprinted by permission
of Texas Instruments.*)

'LS138, 'S138
Function Table

Inputs					Outputs							
Enable		Select										
G1	$\overline{G2}$*	C	B	A	Y0	Y1	Y2	Y3	Y4	Y5	Y6	Y7
X	H	X	X	X	H	H	H	H	H	H	H	H
L	X	X	X	X	H	H	H	H	H	H	H	H
H	L	L	L	L	L	H	H	H	H	H	H	H
H	L	L	L	H	H	L	H	H	H	H	H	H
H	L	L	H	L	H	H	L	H	H	H	H	H
H	L	L	H	H	H	H	H	L	H	H	H	H
H	L	H	L	L	H	H	H	H	L	H	H	H
H	L	H	L	H	H	H	H	H	H	L	H	H
H	L	H	H	L	H	H	H	H	H	H	L	H
H	L	H	H	H	H	H	H	H	H	H	H	L

*$\overline{G2} = \overline{G2A} + \overline{G2B}$
H = high level, L = low level, X = irrelevant

(*b*)

FIGURE 6.23 (continued)
74LS138 decoder IC. (*b*) Truth table. (*Reprinted by permission of Texas Instruments.*)

3-bit binary input to one of eight outputs. Unlike decoders for memory cells, decoder IC outputs are active LOW. Thus, a binary input of 101, denoting the digit 5, produces a LOW on output number 5 (pin 10). All remaining outputs are HIGH. Each output connects to an active LOW chip enable ($\overline{CE}$) on separate devices. There is virtually no chance to enable two chips on the data bus at the same time.

The 74LS138 has three enable pins (G1, $\overline{G2A}$, $\overline{G2B}$) which enable the decoder only if G1 = 1, $\overline{G2A}$ = 0, and $\overline{G2B}$ = 0. Normally, G1 is connected to VMA; $\overline{G2A}$ and $\overline{G2B}$ are tied permanently to ground. However, more complete decoding can be done if the LOW enables are used for two additional address lines.

A basic wiring diagram for address decoding with a 74LS138 decoder is shown in Fig. 6.24. Inputting the top three address lines effectively splits up the memory map into eight spaces.

A15	A14	A13	Output	Space
0	0	0	Y0	0000 to 1FFF
0	0	1	Y1	2000 to 3FFF
0	1	0	Y2	4000 to 5FFF
0	1	1	Y3	6000 to 7FFF
1	0	0	Y4	8000 to 9FFF
1	0	1	Y5	A000 to BFFF
1	1	0	Y6	C000 to DFFF
1	1	1	Y7	E000 to FFFF

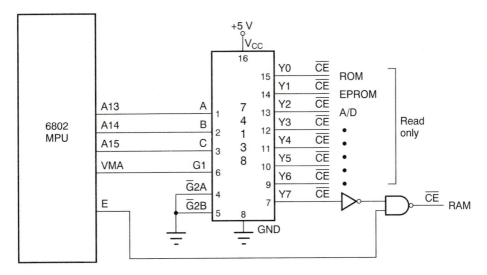

FIGURE 6.24
Using a 74138 decoder to divide the 64K-memory space into eight distinct regions.

Outputs can directly drive CSs on any read-only devices. Since the $\phi2$ clock should bypass the decoder IC, RAMs need further decoding with $\phi2$ and its 74LS138 outputs. Other decoder ICs include the 7445 BCD-to-decimal decoder and the 74154 1-of-16 decoder.

6.7   TIMING REQUIREMENTS

When the MPU reads data from the bus or writes data to the bus, the data and address are valid within a given time period. Similarly, memory ICs need a certain amount of time to transfer data to and from the bus after the address is valid. Timing computations should be made to ensure that the microprocessor is compatible with memory and I/O systems. Compatibility is assured between members of a Motorola family (6802 MPU, 6810 RAM, and 6821 PIA).

   If timing requirements are not met, there are two alternatives: (1) Lengthen the MPU clock cycle time by driving the internal clock with a lower frequency external crystal. At higher frequencies the 6802 is limited to a clock speed of 1 μs. We have assumed in this text that the MPU clock is operating at 1 μs. High clock speed is desirable because program runtime is proportional to the clock period. (2) Purchase faster memories although memory costs increase with speed. Bus timing calculations determine if a given memory unit is fast enough.

   Figure 6.25 shows the timing diagram for a 6802 MPU. The first four time lines are relative to the microprocessor. Parameters on the two data time lines are specific to a read or write. Table 6.2 gives specifications for the 6802 MPU series. Notice that later versions, the 68A02 and 68B02, have faster minimum cycle times and will require correspondingly faster memories. This information represents "worst case" values from a random selection of chips. The CS time line has timing parameters important to memory. Values are found in the manufacturer's data sheets. Table 6.3 gives the specifications for the 6810 RAM series.

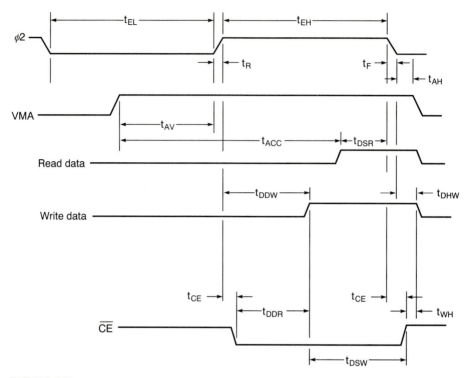

FIGURE 6.25
Timing diagram between the 6802 MPU and memory with chip enable $\overline{CE}$.

TABLE 6.2
6802 MPU timing characteristics

Time	6802		68A02		68B02		Units
	Min	Max	Min	Max	Min	Max	
Cycle t_{CYC}	1.0	10	0.67	10	0.5	10	μs
$\phi2$ low t_{EL}	450	5000	280	5000	210	5000	ns
$\phi2$ high t_{EH}	450	9500	280	9700	210	9700	ns
Clock rise and fall t_R, t_F	—	25	—	25	—	25	ns
Address hold t_{AH}	20	—	20	—	20	—	ns
Address valid t_{AV}	160	270	100	—	50	—	ns
Read data set-up t_{DSR}	100	—	70	—	60	—	ns
Read data hold t_{DHR}	10	—	10	—	10	—	ns
Write data delay t_{DDW}	—	225	—	170	—	160	ns
Write data hold t_{DHW}	30	—	20	—	20	—	ns

Timing is controlled by the $\phi1$ and $\phi2$ clocks, which are nonoverlapping for a cycle time $t_{CYC} = 1 \ \mu$s. The clock HIGH t_{EH} and LOW t_{EL} have a guaranteed minimum time of 450 ns with a clock rise time t_R or fall time t_F not exceeding 25 ns. When the $\phi1$ clock goes HIGH (and the $\phi2$ clock goes LOW), the MPU places the address on the bus. Because of capacitive bus loading, the address lines, as well as the VMA and R/$\overline{W}$, are stable

TABLE 6.3
6810 RAM timing characteristics

Time	6810 Min	6810 Max	68A10 Min	68A10 Max	68B10 Min	68B10 Max	Units
Cycle t_{CYC}	450	—	360	—	250	—	ns
Access time (read) t_{ACC}	—	450	—	360	—	250	ns
Data delay (read) t_{DDR}	—	230	—	220	—	180	ns
Data set-up (write) t_{DSW}	190	—	80	—	60	—	ns
Write hold (write) t_{WH}	0	—	0	—	0	—	ns

~300 ns later. The 6802 data sheet references an address valid time t_{AV} from the end of the LOW $\phi2$-clock signal. (*Note:* $t_{EL} - t_{AV} \cong 300$ ns.)

Read Cycle Timing

MPU PERSPECTIVE. The MPU latches data into its internal data register upon a HIGH-to-LOW transition of $\phi2$. Data must be valid on the bus for t_{DSR}, the read data set-up time, before the trailing edge of $\phi2$. After the fall of $\phi2$ the address, R/$\overline{W}$, and VMA (valid memory address) will hold for time t_{AH}.

MEMORY PERSPECTIVE. When the address changes, internal memory cell circuits begin processing the changes. Memory access time $(t_{ACC})_{MEM}$ is the time it takes memory to process the address change and place data on the bus. Independent of the memory access time is the time to enable the memory data buffers once the CS (or CE) is activated. Normally, this time is faster than the access time and the buffers are already enabled when the data change arrives from the memory cells. If the decoder uses address and VMA lines only, the equivalent CS access time $(t_{ACC})_{CS}$ is the decoder propagation delay t_{CE} (15 ns per gate, LS-TTL) plus the read data delay time t_{DDR} found in the memory specifications. If an added $\phi2$-clock line is wired to the decoder, CS access time is lengthened by the time between the rise of VMA and the rise of $\phi2$. This presents no problem for most RAMs; however, EPROMs are another matter.

READ CRITERIA. The microprocessor should not try to read data faster than memory can place data on the bus. A basic measure of the operating speed for both the MPU and memory is access time. It is measured from the rise of a valid memory address to the appearance of data on the bus. The MPU's access time is defined by

$$(t_{ACC})_{MPU} = t_{AV} + t_R + t_{EH} - t_{DSR}$$

Two read timing criteria must be met before a microprocessor is interfaced with memory:

1. $(t_{ACC})_{MPU} > (t_{ACC})_{MEM}$
2. $(t_{ACC})_{MPU} > (t_{ACC})_{CS}$ or $(t_{ACC})_{MPU} > t_{CE} + t_{DDR}$ (without $\phi2$)
   $(t_{ACC})_{MPU} > t_{CE} + t_{DDR} + t_{AV} + t_R$ (with $\phi2$)

Example 6.5. Examine the read timing compatibility between the 6802 microprocessor and both the 6810 RAM and the 2716 EPROM.

6810 RAM (with $\phi2$)

$$(t_{ACC})_{MPU} = t_{AV} + t_R + t_{EH} - t_{DSR}$$
$$= 160 + 25 + 450 - 100 \text{ ns}$$
$$= 535 \text{ ns}$$

$$(t_{ACC})_{MEM} = 450 \text{ ns}$$
$$535 > 450 \text{ ns} \quad \textit{check}$$

$$(t_{ACC})_{CS} = t_{CE} + t_{DDR} + t_{AV} + t_R$$
$$= 30 \text{ (two-gate decoder)} + 230 + 160 + 25 \text{ ns}$$
$$= 475 \text{ ns}$$
$$535 > 445 \text{ ns} \quad \textit{check}$$

2716 EPROM

$$(t_{ACC})_{MEM} = 450 \text{ ns} \quad \text{(faster versions available)}$$
$$535 > 450 \text{ ns} \quad \textit{check}$$

$$(t_{ACC})_{CS} = t_{CE} + t_{DDR} + t_{AV} + t_R \quad \text{(with } \phi2)$$
$$= 60 + 450 + 160 + 25$$
$$= 695 \text{ ns}$$
$$535 < 695 \text{ ns} \quad \textit{fails}$$

$$(t_{ACC})_{CS} = t_{CE} + t_{DDR} \quad \text{(without } \phi2)$$
$$= 60 + 450$$
$$535 > 510 \text{ ns} \quad \textit{check}$$

Write Cycle Timing

MPU PERSPECTIVE. During a write cycle the microprocessor has the responsibility of placing data on the bus. The MPU applies data to the bus after a delay t_{DDW}, the write data delay time, following the rise of the $\phi2$ clock. After the fall of $\phi2$ data remains valid for a short time t_{DHW}, the write data hold time.

MEMORY PERSPECTIVE. Memory terminates the data write cycle when the CS goes HIGH. Data must be on the bus at least t_{DSW}, the write data set-up time, ahead of the CS change so that memory has time for a read. A falling $\phi2$ line changes the CS after a decoder delay t_{CE}. Some RAMs require data to be held at time t_{WH}, the write hold time, after the CS goes HIGH. However, t_{WH} is zero for the 6810 RAM.

WRITE CRITERIA. The write cycle begins and ends on the rise and fall of the $\phi2$ clock. Over this period the time for the MPU to place data on the bus plus the time for memory to read the data cannot exceed the period:

$$t_{DDW} + t_{DSW} < t_{EH} + t_{CE}$$

A second criteria examines whether data is on the bus when the CS changes at the end of a write. Decoder propagation delays and write hold time can exceed the MPU data hold time after the fall of $\phi 2$:

$$t_F + t_{DHW} > t_{WH} + t_{CE}$$

Example 6.6. Determine write timing compatibility between the 6802 microprocessor and the 6810 RAM.

6810 RAM (with $\phi 2$)

$t_{DDW} + t_{DSW} < t_{EH} + t_{CE}$
$225 + 190 < 450 + 30$ (two gates)
$415 < 480$ ns *check*
$t_F + t_{DHW} > t_{WH} + t_{CE}$
$25 + 30 > 0 + 30$ (two gates)
$55 > 30$ ns *check*

PROBLEMS

6.1. List four basic sections of a RAM IC.

6.2. List four types of read-only memory (ROM).

6.3. What is the difference between an EPROM and EEPROM?

6.4. What MPU lines are inputs to address decoders for (*a*) RAMs, (*b*) ROMs?

6.5. What are the advantages and disadvantages of partial decoding?

6.6. Give the data in each address of an unerased EPROM.

6.7. What characteristics of the 7421 4-input AND gate and the 7423 4-input NOR gate make them ideal for address decoding?

6.8. A RAM is enabled under the following conditions: CE = A15 · A14 · $\overline{\text{A13}}$ · $\overline{\text{A12}}$ · A11. Find the enable address space. What address will access the first byte location?

6.9. A 4K RAM uses a NAND gate decoder. Address lines A15 and A14 only are inputs. Each line is inverted before inputting the gate. What address space is selected? Give the redundant addresses. How many addresses access the first byte in memory?

6.10. A 6810 should use how many address lines to the decoder for full decoding?

6.11. A 6810 RAM is to be fully decoded with access to base address 3600. Show the decoding logic.

6.12. Figure 6.22 shows a 6810 RAM decoded for memory space 5000 to 5FFF. Show a decoding arrangement for space 5000 to 50FF.

6.13. A 6810 RAM is designated address space 0000 to 00FF. Determine the decoder configuration. Show a wiring diagram interface with the 6802 MPU.

6.14. Construct a 4K × 8-bit memory using 2114A RAMs. Decode for address space 8xxx.

6.15. A 2716 EPROM shares the address space with only one other device, an I/O port having four addressable bytes located in space 5000 to 5FFF. Determine the minimum number of address lines needed for partial decoding of the EPROM. Show an interface with the 6802 MPU. The decoder can be a 4-input NAND gate or a 4-input NOR gate.

6.16. The 74138 decoder in Fig. 6.24 grounds the enable pins: $\overline{\text{G2A}}$ and $\overline{\text{G2B}}$. Suppose these pins are tied directly to A12 and A11. What memory space is enabled for each decoder output?

6.17. A 6802 system has a 2K ROM, a 1K RAM, and a 4-byte I/O. You have the entire 64K-memory space in which to locate these ICs. Each has a single chip enable LOW. Determine a memory map for the system. Show the partial decoding for each IC.

6.18. Investigate the timing relationships between a 2716 EPROM and the 68A02 MPU operating with a minimum clock cycle.

6.19. A 68B10 RAM should interface with a 68B02 MPU with no timing problems because they are in the same family. Investigate the timing criteria for both read and write.

6.20. A certain RAM has the characteristics listed below. Analyze its probability of successfully interfacing a 68B02 MPU operating at minimum clock speed.

$$t_{ACC} = 200 \text{ ns}$$
$$t_{DDR} = 200 \text{ ns}$$
$$t_{DSW} = 100 \text{ ns}$$

CHAPTER
7

INTERFACING PORTS: THE PIA

A microprocessor/memory system would be useless if it could not interact with external devices or peripherals, collectively called input/output or I/O for short. Controllers receive binary switch status or analog transducer signals which are converted to binary form for the MPU. After the information is processed the controller sends binary commands to solid-state relays, displays, or analog converters for motor control. The 8-bit parallel interface between data coming from an external device to the MPU is called an *input port*. The 8-bit parallel interface between data going from the MPU to an external device is called an *output port*.

We have seen that bus data is constantly changing as the program is executed. Microprocessors transmit or receive data only for a fraction of a microsecond. An output port must grab and hold bus data at the instant it is sent while an input port must pass data to the bus only when it is requested. Ports function as though they are memory within a memory-mapped I/O structure:

1. Each port is assigned one, sometimes several addresses. This contrasts with memory ICs which can have thousands of assigned addresses.
2. STA (address) instruction sends data to an output port. LDA (address) instruction loads port data into the MPU.
3. Each port has one or more chip select lines. An address decoder enables the port when data is transmitted to the bus by the instruction.

Although single hard-wired ports will be discussed first, the main topic of this chapter is a programmable I/O port, the most commonly used chip today. Motorola was the first manufacturer to develop a programmable port, called the 6820 Peripheral Interface Adapter (PIA). A later version is the 6821. Similar programmable port ICs include the Intel 8255 Programmable Peripheral Interface (PPI) and the Zilog Programmable Input/Output (PIO). Programmable ports offer several advantages over hard-wired ports: (1) One PIA functions as several ports with on-board decoding. It simplifies a design by employing fewer chips. (2) Software control of port configuration is very versatile. It can be changed while the program is being executed.

7.1   A SIMPLE OUTPUT PORT

Output ports hold data in the same manner as static RAMs: flip-flops. The most common type is the D-type flip-flop, which can function as a single bit port among its varied uses in digital electronics. Figure 7.1 illustrates the 74LS75 D-type transparent latch, four to a package. It has a single data input D and two outputs Q and $\overline{Q}$. $\overline{Q}$ is the complement of Q. Only one output is connected, generally Q. Outputs are controlled by the enable C. As long as the enable is LOW, output Q holds its last value. When the enable goes HIGH, the output follows the data presented at input D and the flip-flop is referred to as *transparent*. Upon a HIGH-to-LOW transition of the enable, output Q is held at the data level just prior to transition. The data at transition is said to be *latched* (saved) on the trailing edge of the positive enable pulse.

An alternate flip-flop is the 74LS74 edge-triggered D-type flip-flop. See Fig. 7.2. There are two to a package. The 74LS74 is quite similar to the 74LS75. However, data is latched into the flip-flop upon the LOW-to-HIGH transition of the enable. Also, the 74LS74 is not transparent. Only the data level at the leading edge of the positive enable pulse appears on the output. The value is held until the next positive (LOW-to-HIGH) transition. Edge-triggered enables are referred to as clock inputs. They are distinguished from transparent latch enables by a sideways triangle at the input.

The 74LS74 D-type flip-flop has two additional inputs called *preset* and *clear*. A small bubble on the flip-flop symbol indicates that these inputs are active LOW. Active LOW on a pin assignment is demonstrated with a bar over the word. A LOW on preset ($\overline{PRE}$) sets the output (Q = 1) while a LOW on clear ($\overline{CLR}$) "clears" the output (Q = 0).

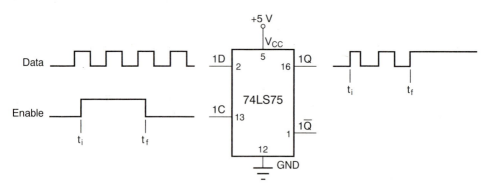

FIGURE 7.1
Operation of a 74LS75 transparent latch. Passes data on t_i but latches data on t_f.

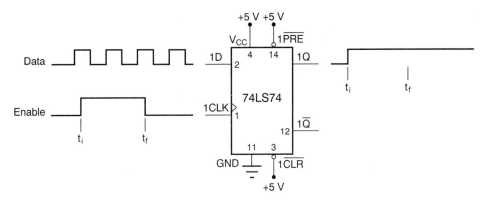

FIGURE 7.2
Operation of a 74LS74 positive edge-triggered flip-flop. Latches data only on t_i.

Both output states take precedent over data or clock input causality. Normally, the preset and clear are tied permanently HIGH (5 V).

A practical output port should have eight data bits. Two 8-bit registers, called octal registers, are available as ports for microprocessor systems: the 74LS373 octal D-type transparent latch and the 74LS374 octal D-type positive edge-triggered flip-flop. Both are shown in Fig. 7.3. Each bit has a common enable (clock). The enable is inverted twice, so it functions the same as single flip-flops discussed earlier. The complement $\overline{Q}$ of flip-flop outputs is inverted by three-state buffers to pass the latched data. Buffers are controlled by input $\overline{OC}$. When $\overline{OC}$ is LOW, the flip-flop states are connected to the output. When $\overline{OC}$ is HIGH, the outputs are in their high impedance state (disconnected).

Pin assignments for the 74LS373 latch and 74LS374 flip-flop are identical. To be used as an output port, pin $\overline{OC}$ is simply grounded, connecting internal flip-flops permanently to output pins. The enable C is driven by an address decoder designed specifically for the port address. During a write operation, STA (port address), data should be latched on the trailing edge of the enable pulse because the data may not be valid on the leading edge. For the 74LS373 latch, the address decoder must generate a HIGH pulse since the IC latches on a negative transition of the enable. For the 74LS374 flip-flop, the address decoder must generate a LOW pulse for the enable. Decoder inputs include address, VMA, ϕ2, and R/$\overline{W}$ lines. The latter is LOW for a write. Figure 7.4 gives a port wiring diagram for a 74LS373. Partial decoding is for space 7800 to 7FFF. The port is enabled for any address within this range.

7.2 A SIMPLE INPUT PORT

The function of an input port is simpler than that of an output port. No data storage is necessary since data from an external device is waiting for the microprocessor. Data should be placed onto the bus only at the instant the MPU performs a read operation, LDA (port address). A simple input port consists of three-state buffers. The buffers connect the input data to the bus only when the port address decoder enables the buffers. Decoder inputs include address, VMA, and R/$\overline{W}$ lines.

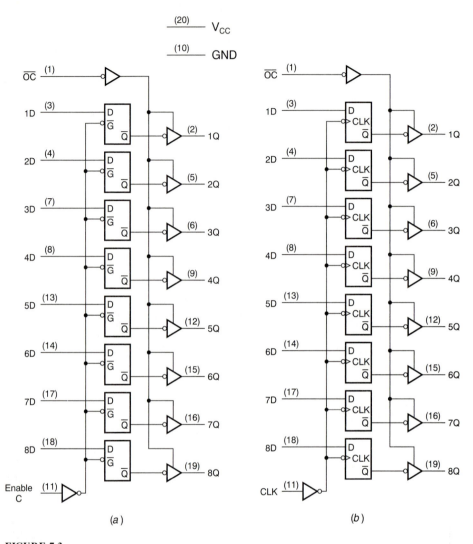

FIGURE 7.3
Pin assignments for 74LS373 octal D-type latch and 74LS374 octal D-type flip-flop. (*a*) 74LS373 transparent latch; (*b*) 74LS374 edge-triggered flip-flop. (*Reprinted by permission of Texas Instruments.*)

A standard input port is the 74LS244 octal buffer. Each input line connects to a three-state buffer which drives the corresponding data line. The first four buffers are enabled together by $1\overline{G}$ while the remaining four buffers are enabled by $2\overline{G}$. Figure 7.5 gives a port wiring diagram. The address space is the same as the previous 74LS373 output port. Both could use the same address in a system because one is write only while the other is read only. Decoding with the $R/\overline{W}$ line guarantees both ICs will not use the bus at the same time.

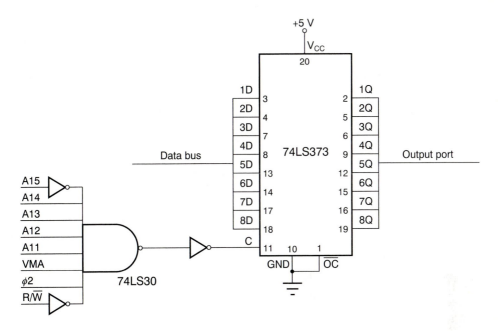

FIGURE 7.4
Output port using a 74LS373 octal latch. Inverter following NAND gate decoder is omitted when the port is a 74LS374 octal flip-flop. Decoded for address space 7800 to 7FFF.

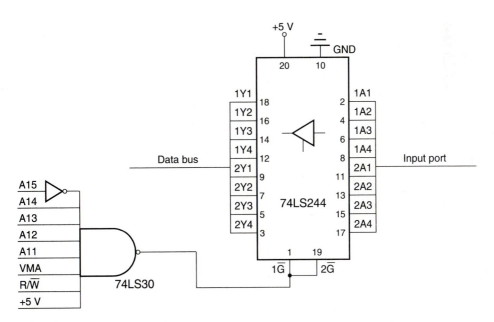

FIGURE 7.5
Input port using a 74LS244 octal buffer. Decoded for address space 7800 to 7FFF.

7.3   THE PERIPHERAL INTERFACE ADAPTER (PIA)

The Motorola 6820/6821 PIA was developed to support the 6800/6802/6808 microprocessor series although it can be used in many designs built with other MPUs if proper precaution is taken with control signals. The 6820 is an earlier version which is no longer listed in the catalog. Differences are slight, but the 6821 is preferred for its port drive characteristics given later. The PIA is constructed with two ports to connect peripherals in the outside world. Each port has eight data lines which can be programmed individually to act as input or output bits. In most instances a port is programmed as a group to form a parallel 8-bit input port or output port.

Another feature of the PIA is four separate interrupt lines which can be programmed in several modes. Two additonal handshake lines enhance peripheral interrupts control. These features are independent of port operation and will be discussed in Chap. 8.

Pin Assignment

Figure 7.6 gives the pin assignment for the 6821 PIA. Since the IC is positioned between the outside world and the microprocessor, one group of pins connects to peripheral devices and another group connects to the MPU.

PERIPHERAL SIDE. The PIA has two 8-bit parallel ports to the outside world labeled port A and port B. Pins on either port can be programmed to act as an input bit or output bit. All the lines on the peripheral side of the PIA are compatible with standard TTL logic. Specifics of TTL logic and CMOS logic will be explained in Chap. 9. There are several differences between ports outlined here.

Port PA0–PA7. Port A bits have an internal pull-up resistor which raises a typical TTL HIGH voltage from 3.4 to 5 V. Thus, port A can drive both TTL and CMOS logic. As a result, port A requires more drive current than port B for the input mode, a maximum of 1.5 standard TTL loads. For the output mode, port A can drive a maximum of two TTL loads.

Port PB0–PB7. In constrast, port B uses internal three-state buffers. Lines cannot drive CMOS without external pull-up resistors. But three-state capacity allows them to enter a high impedance state when the peripheral data lines are not enabled. A major advantage of the buffers is the output drive current. Port B pins in the output mode will source at least 1.0 mA (typically 2.5 mA) at 1.5 V. This power is sufficient to switch ON general-purpose transistors and some power transistors (Darlingtons). For this reason port B is usually configured as an output port.

CA1,CA2, CB1,CB2. These pins are assigned to peripheral interrupt inputs or to interrupt handshake control. They are compatible with standard TTL as well. CA1 and CA2 have internal pull-up resistors with load characteristics of port A. CB1 and CB2 buffers are similar to port B. In a handshake output mode they can source 1.0 mA at 1.5 V. *Pins CA1, CA2, CB1, and CB2 are not connected when the PIA is used only for ports.*

Pin Assignment

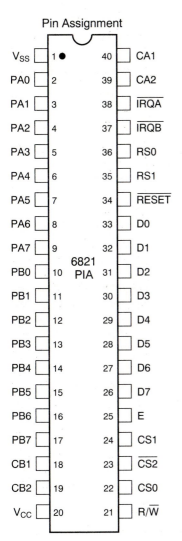

V$_{SS}$	1 ●	40	CA1

FIGURE 7.6
Pin assignment for the 6821 Peripheral Interface Adapter (PIA).
(*Reprinted by permission of Motorola, Inc.*)

MPU SIDE. We are already familiar with most RAM-like PIA lines which interface the microprocessor: an 8-bit bidirectional data bus, chip select lines, address lines, and a read/write line. New lines include reset and enable.

Data lines D0–D7. The 8-bit data bus allows data transfer between the MPU and PIA.

Address lines RS0 and RS1. Two lines, called register selects (RSs), are used to select the various programmable registers inside the PIA. They are analogous to address lines on a RAM. RS0 and RS1 are connected to address line bits A0 and A1, respectively. Four register addresses can be selected. The register select lines should be stable for the duration of the enable (E) HIGH pulse on pin 25.

Chip select lines CS0, CS1, and $\overline{\text{CS2}}$. Three CSs enable the address space for the PIA registers and allow the MPU to communicate only with the PIA for its assigned addresses. CS0 and CS1 must be HIGH, and $\overline{\text{CS2}}$ must be LOW for the PIA selection. The higher address lines are tied directly to the CSs for partial decoding. One HIGH CS should be used for the valid memory address VMA. An external decoder gate(s) may be necessary if more complete decoding is desired. CS lines should be stable for the duration of the enable HIGH pulse.

Enable (E). Data is transferred between the PIA and microprocessor during the enable (E) signal HIGH if the PIA has been selected. Enable is always connected to the $\phi2$-clock line on the MPU, also labeled E. There is no need to use the $\phi2$-control signal as an external decoder input when selecting a PIA.

The E line presents some confusion with the 6802 MPU and 6821 PIA since the term "enable" is generally used to denote chip selection or chip enable (CE). We will maintain this designation and refer to the E line as the $\phi2$ clock.

Read/write (R/$\overline{\text{W}}$). The read/write line controls the direction of data transfers on the data bus. It connects to the MPU read/write line. A LOW on R/$\overline{\text{W}}$ allows data transfers from the MPU to the PIA on the falling edge of the $\phi2$ signal. A HIGH on R/$\overline{\text{W}}$ enables the PIA data buffers for data transfers to the MPU if the $\phi2$-clock line is HIGH.

$\overline{\text{Reset}}$. The active LOW reset line is used to reset *all* register bits in the PIA to logic 0 (LOW). It is connected to the $\overline{\text{reset}}$ pin of the 6802 MPU. When this pin is momentarily brought LOW by an external switch, the MPU starts the execution of the program. More will be said on this subject in Chap. 16.

Interrupt request lines $\overline{\text{IRQA}}$ and $\overline{\text{IRQB}}$. Interrupt request lines $\overline{\text{IRQA}}$ and $\overline{\text{IRQB}}$ are tied together and connected to the MPU interrupt request pin $\overline{\text{IRQ}}$. Any one of four devices wired to PIA peripheral side lines CA1, CA2, CB1, and CB2 can interrupt the MPU. Interrupt request lines are not connected (NC) if the PIA is used only as ports.

> **Example 7.1.** Each pin on the 6821 PIA has been explained earlier. Show an interface to the 6802 MPU that will enable PIA register addresses 4000–4003.
>
> A wiring diagram for the interface is given in Fig. 7.7. Two address lines and VMA are tied to the three CSs:
>
> CS0 to VMA
> CS1 to A14
> $\overline{\text{CS2}}$ to A15
>
> Thus, the partially decoded address space is 4000–7FFF, which includes base addresses 4000–4003. A more limited address space can be achieved with an external decoder, but this generally is not necessary. Six interrupt related pins are not connected (NC). All remaining lines except ports (PA0 to PA7 and PB0 to PB7), power (V_{CC}), and ground (V_{SS}) are tied to their counterparts on the microprocessor.

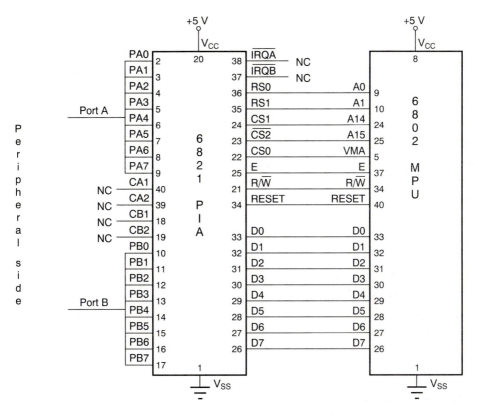

FIGURE 7.7
6821 PIA interfaced to the 6802 MPU. Decoded for address space 4000 to 7FFF.

7.4 PIA REGISTERS

PIA registers behave like memory locations in RAM. The MPU reads and writes to registers exactly the same. There are six registers; ports A and B each have three. Sides A and B have identical registers so that statements about one side apply to the other.

	Side A	Side B
1. Data Register	DRA	DRB
2. Data Direction Register	DDRA	DDRB
3. Control Register	CRA	CRB

Data Register

Data registers represent the ports. *During an MPU write operation* DRA (DRB) stores the data sent on the data bus. If the port lines have been programmed as output lines, the data will also appear on the port lines. A logic 1 written to the register will cause a HIGH on the corresponding port line while a logic 0 results in a LOW. Data can be written to lines that are programmed as inputs, but the register content will be controlled by the peripheral data on the port.

During an MPU read operation data placed on port lines by peripherals is transferred directly to the MPU data bus, provided the lines are programmed as inputs. If individual lines are programmed as outputs, a read operation will differ slightly between side A and side B. On side A peripheral data is transferred to the data bus regardless of whether the lines are programmed as input or output. On side B three-state buffers are between the port and data bus. So, the MPU will read the DRA (DRB) registers. Register content will likely differ from peripheral data. All peripheral data will be read correctly for binary 1 or 0 only if the data is compatible with TTL voltages.

Data Direction Register

The data direction registers set up the individual port lines as inputs or outputs. Each bit in the DDRA (DDRB) register controls the corresponding port data line, bit 7 controls PA7 (PB7), etc. If the bit is logic 1, the corresponding port line acts as an output, whereas a logic 0 in the DDRA (DDRB) bit causes the port line to act as an input.

The different port lines can be set up in a combination of inputs and outputs. In most cases a port will be configured as all inputs or all outputs. To set up all eight lines of port A as inputs, a STA instruction sends byte 00 to the address of DDRA. Similarly, port B can be configured as an output port by storing byte FF to the address of DDRB.

Control Register

We have six addressable registers on a PIA, yet two address lines RS0 and RS1 permit the addressing of only four different memory locations (registers). Two registers in each port, DRA (DRB) and DDRA (DDRB), must share the same address. Bit 2 in each control register (CRA or CRB) determines the selection of either the data register or the data direction register. See Fig. 7.8. A logic 1 in bit 2 allows access to the data register DRA (DRB) while a logic 0 causes the data direction register to be addressed.

All remaining bits in the control register control the operation of the four interrupt lines CA1, CA2, CB1, and CB2. These bits will be discussed in the following chapter. Interrupt lines are not connected when the PIA is used as ports only. Therefore, the contents of the remaining control register bits are immaterial.

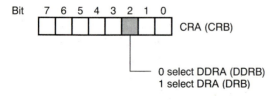

FIGURE 7.8
Control register bit 2 selects data direction register or data register.

Example 7.2. Port B is to be set up with PB0 to PB3 as inputs and PB4 to PB7 as outputs. What data byte is sent to the control register to address the data direction register? What data byte is sent to the data direction register?

> CRB: 00
> DDRB: F0

7.5   ADDRESSING THE PIA

Two MPU address lines connect the PIA directly through the two register select lines: A0 to RS0, A1 to RS1. This gives the PIA four addresses for six registers. RS1 determines which side of the PIA is addressed. When RS1 is LOW, side A is addressed. When RS1 is HIGH, side B is addressed. RS0 addresses registers on a particular side. When RS0 is HIGH, the control register is addressed. When RS0 is LOW, the data register or the data direction register is addressed. Even though they share the same address, DRA (DRB) and DDRA (DDRB) are selected individually depending upon bit 2 of the control register. A LOW bit 2 selects the data direction register while a HIGH selects the data register. A summary is given in Table 7.1.

Only the lowest two address bits select one of the on-board registers. Register addresses are as follows:

Address	Register
XXX0	DDRA or DRA
XXX1	CRA
XXX2	DDRB or DRB
XXX3	CRB

The complete address is determined by the CSs. If all the remaining address lines A2 to A15 (plus VMA) were connected to an external decoder which interfaced the CSs, the four fully decoded addresses would be unique. Since partial decoding is the rule, each register can have many addresses.

TABLE 7.1
PIA register addressing

RS1	RS0	CR2	Register
0	0	0	DDRA
0	0	1	DRA
0	1	x	CRA
1	0	0	DDRB
1	0	1	DRB
1	1	x	CRB

Example 7.3. Figure 7.7 shows a partially decoded PIA interfaced to the MPU. Address bits A15 and A14 tie to PIA chip selects $\overline{CS2}$ and CS1, respectively. VMA ties to CS0. What are the register addresses? Does port A (DRA), e.g., have other addresses?

The register base addresses are

4000	DDRA or DRA
4001	CRA
4002	DDRB or DRB
4003	CRB

The PIA is enabled for address space 4000–7FFF. Port A is addressed when both address lines A0 and A1 are LOW. Thus, all the following addresses are DRA:

4000, 4004, 4008, 400C
4010, 4014, 4018, 401C
⋮

4FF0, 4FF4, 4FF8, 4FFC
⋮

7000, 7004, 7008, 700C
7010, 7014, 7108, 710C
⋮

7FF0, 7FF4, 7FF8, 7FFC

7.6   INITIALIZING THE PIA

Before we can begin using the PIA, port I/O lines should be programmed for the desired direction of peripheral data flow. Furthermore, we should be able to address the ports themselves. Suppose side A is to be set up as an input port and side B is to be set up as an output port. Figure 7.9 shows the *final state* of registers on each side. The data direc-

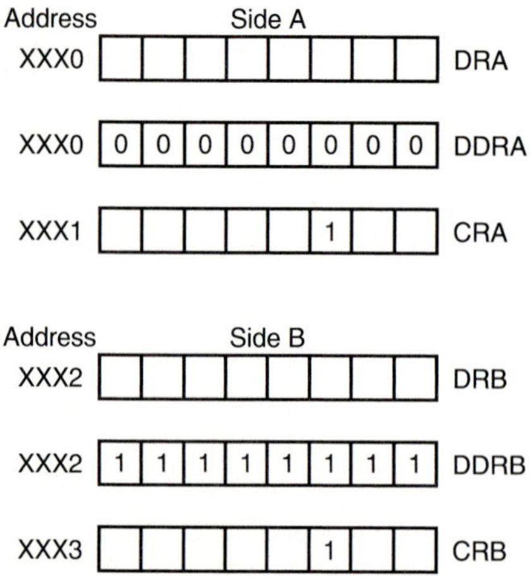

FIGURE 7.9
PIA register content when side A is an input port and side B is an output port.

tion register (DDRA) of side A must be left with all logic 0s for input lines. Side B data direction register (DDRB) must be left with all logic 1s for output lines. In each case control register bit 2 must be logic 1 so that data registers (ports) and not the data direction registers are addressed.

The *beginning state* of each register is controlled by the reset line, which receives its signal from the MPU. When the reset line goes LOW for a short period of time and then goes HIGH, the microprocessor begins execution of the program. This reset action also automatically *clears all registers in the PIA*. Microprocessor reset hardware will be given in Chap. 16.

A program is written for the PIA to transform register beginning states to final states. The execution of this program is called *initializing the PIA*. The initialization procedure involves the following steps:

1. Control register bit 2 is cleared by a reset. DDRA and DDRB are first addressed at XXX0 and XXX2, respectively.
2. Load all 0s into DDRA for an input port.
3. Load all 1s into DDRB for an output port.
4. Load 1 into bit 2 of both control registers. DRA and DRB are now addressed at XXX0 and XXX2, respectively.

Example 7.4. Write a program to initialize the PIA. Side A is an input port while side B is an output port. The PIA is enabled for base addresses 4000–4003.

```
INIT PIA   LDA A #00      side A input port
           STA A 4000
           LDA A #FF      side B output port
           STA A 4002
           LDA A #04      port A at 4000
           STA A 4001
           STA A 4003     port B at 4002
```

The PIA initialization program is placed at the beginning of the main program. It is executed only once. Thereafter the MPU can read peripheral data from input port A with instruction LDA A 4000. Also, the MPU can write peripheral data to output port B with instruction STA A 4002.

PROBLEMS

7.1. A 74LS75 D-type latch is used as a single-bit port to switch ON a transistor-controlled DC motor. The instructions below activate the motor. Discuss the items that would be connected to pins 16, 13, and 12 (see Fig. 7.1) for a functioning port.

```
LDA A #01
STA A 2000
```

7.2. The 74LS373 octal latch was proposed as a simple output port. However, it has three-state buffers which are permanently enabled when it is used as an output port. The three-state buffers allow the latch to be used as a simple input port also. Show an input port diagram similar to Fig. 7.5.

7.3. Can the 74LS374 edge-triggered flip-flops be used to replace the 74LS373 latches as input ports in Prob. 7.2?

7.4. Figure 7.7 shows an interface between the 6821 PIA and the 6802 MPU. Devise a complete address decoding scheme to replace the partial decoding.

7.5. Port B is configured with lines PB0 to PB6 as inputs and line PB7 as an output. If port B is read, what kind of data do you expect to find on each line?

7.6. Suppose the PIA initialization is not started from a hardware reset but from software. What additional instructions are needed?

7.7. It is proposed to use index register instructions to initialize the PIA. The method will work if register select lines are interchanged: A0 to RS1, A1 to RS0. Write the initialization routine for port A as an input port and port B as an output port. How many memory bytes are saved when compared to Example 7.4?

7.8. PIA port B alone is used to interface peripherals. Four sensors are monitored on PB0 to PB3 while four control signals are sent through PB4 to PB7. Write an initialization program.

7.9. Six temperature switches from a device are tied to a PIA. If any switch reads HIGH, the MPU sends a LOW through the PIA to shut down the device. Write the program including the PIA initialization routine.

CHAPTER
8

INTERRUPTS

In the preceding chapter all input/output transfers are controlled by the program. Input ports are read, arithmetic operations are performed, and data are sent to the output ports. With feedback systems, arithmetic calculations include real-time controller strategies. In other cases peripherals are only monitored. When peripherals need attention, they signal the MPU by changing the voltage level of an input line. The device is said to raise a *flag*. The MPU responds by jumping to a service routine for the given device. Suppose eight peripherals are connected to PIA port A. Each one signals HIGH only if it needs attention. A program to monitor the devices could be the following:

```
LOOP:   LDA A PORT      Is there a flag?
        BEQ LOOP
        LSR A           If device 0, service it.
        BCS ADDR 0
        LSR A           If device 1, service it.
        BCS ADDR 1
          .
          .
          .
        BRA LOOP        Continually monitor.
```

Service routines are located at addresses ADDR 0, etc. If long routines tend to exceed the branch range, JSR instructions can be placed at ADDR 0, etc.

All program control of I/O must be in a loop to read continually inputs and update outputs. The process of repeatedly checking the status of input ports is called *polling*. Polling is satisfactory for simple I/O operations, but it has disadvantages for more complex programs. As the number of instructions increases, the loop becomes longer and the inputs are checked less frequently. The possibility that peripheral data is missed should be carefully weighed. For example, switch closings (keyboards) are often momentary. Also, systems that use feedback control algorithms should minimize instruction execution time to improve system performance. The addition of device status checks lengthens loop time.

An alternative to program control is interrupt control. An *interrupt* is a direct means of control in which a peripheral activates a separate interrupt request line. An active line demands immediate attention from the microprocessor. To respond to an interrupt, the MPU must suspend execution of the main program and jump to a service routine. This temporary exchange of program control is exactly what occurs when a JSR instruction is executed. In the case of interrupts the action is initiated by an I/O device. A disadvantage of interrupts is that they occur randomly. Thus, it is difficult to test and debug the system. In addition, extra hardware in the form of latches or flip-flops may be required to ensure proper recognition of the interrupt signal. Interrupts are less useful if the rate of I/O data transfer is high.

This chapter explains interrupt control with the 6802 microprocessor. A major function of the 6821 PIA is enhanced interrupt capabilities for the MPU. PIA interrupt operations will also be discussed.

8.1 INTERRUPT USES

Important Events

Interrupts have priority over the main program being executed. Their fast response is usually touted as a major advantage over program polling particularly when polling is combined with other duties. However, most microprocessor-based controller boards run programs in milliseconds. There are few events that need attention faster than polling. One such event is a power failure. Impending loss of power is sensed with an *RC* circuit which alerts the MPU through an interrupt line. Before power is lost, many instructions can be executed. A service routine takes what steps are necessary to prepare peripherals for shutdown. One option includes switching to a backup power supply.

Infrequent Events

Interrupts are principally used to improve programming efficiency. For example, switch position changes infrequently. Suppose an auto engine speed controller must check the water temperature switch and display a warning when the engine is too hot. The controller's main function is real-time feedback control of auto speed. To poll a switch that may never trigger places an extra burden on the software.

Figure 8.1 shows a general hardware configuration for the system. A microprocessor is continually exchanging data with peripherals through an I/O port. The temperature switch uses the interrupt line. Once the switch is set, an interrupt service routine will display a warning. However, the switch will remain ON for an extended time. Some means must be found to disarm the signal or the system will be continuously interrupted to the detriment of other software duties. A D flip-flop latches the transition signal from the switch; the flip-flop output drives the interrupt line. Any subsequent LDA or STA instruction for the port address will enable the port and at the same time clear the flip-flop and interrupt line. Another transition of the switch setting is necessary to cause a subsequent interrupt.

External flip-flops are not necessary with PIAs. Their interrupt input lines have edge-triggered registers that are cleared by software.

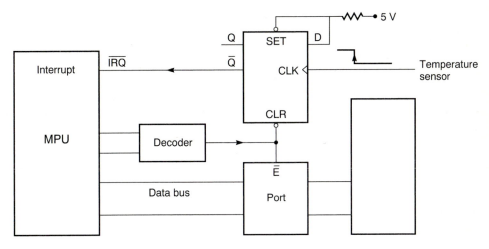

FIGURE 8.1
Generation of an interrupt which is cleared by a port read or write.

Slow Events

Some peripheral data should be read (or written) as soon as the peripheral is ready for data transfer. Yet the peripheral is sufficiently slow that the microprocessor could execute many instructions between transfers. Interrupts to signal ready data are preferable over dedicated polling of the port. One of the slower devices in mechatronics is the analog-to-digital converter. Every converter has a start of conversion pin which intitiates the process and an end of conversion pin which signals the process is finished. In Fig. 8.2 we demonstrate a start signal with the port enable. The MPU starts conversion by reading the port. An end of conversion line is connected to the MPU interrupt line. When the MPU reads the port, conversion of the analog voltage begins anew.

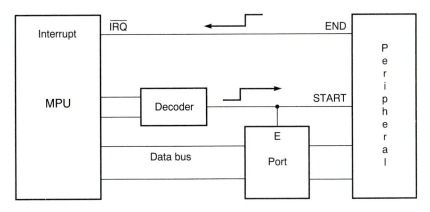

FIGURE 8.2
Interrupt handshake with analog-to-digital converter. A port read starts a new conversion.

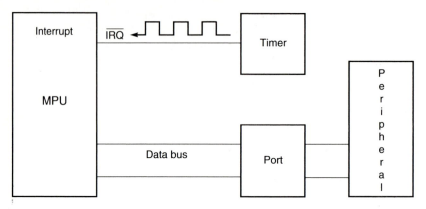

FIGURE 8.3
Periodic interrupts with an external timer.

Timed Events

A microprocessor is often called upon to transfer data on a timed basis. For example, data may be collected at set intervals. Peripherals may be updated or simply pulsed on a timely schedule, e.g., clock advances or auto fuel injection. To free the MPU, an external timer can be used for interval timing. In Fig. 8.3 a timer generates the periodic signal which interrupts the MPU on schedule. An interrupt service routine updates the peripheral port. Chapter 14 covers timers.

8.2 INTERRUPT INPUTS

Three pins on the 6802 MPU are interrupt inputs:

1. Reset ($\overline{\text{RST}}$)
2. Interrupt request ($\overline{\text{IRQ}}$)
3. Nonmaskable interrupt ($\overline{\text{NMI}}$)

Each input serves a slightly different purpose but all three have a common procedure for accessing their individual service routines. Normally, an interrupt pin on the 6802 MPU is held HIGH. When the line is brought LOW, the MPU jumps to the address of the interrupt routine. It does so by fetching 2 bytes for the address and placing them onto the program counter (PC). The procedure is similar to a jump to subroutine. If a service routine was always started at the same address, the address could be hard-wired into the MPU. However, the designer would be severely restricted in how he or she lays out the memory map. In practice, the designer selects the interrupt routine starting address by loading the MSB and LSB of each address into memory locations reserved for it. Starting addresses in the reserved locations are called *interrupt vectors* because they point to the interrupt service routine.

In any 6802 MPU system the upper eight memory locations (FFF8–FFFF) are reserved for vector addresses. In addition to three hardware interrupts, there is a software interrupt initiated by an instruction (SWI). Interrupt vector assignments are given in

TABLE 8.1
Interrupt vector addresses

	Vector	
Type	MSB	LSB
RST	FFFE	FFFF
NMI	FFFC	FFFD
SWI	FFFA	FFFB
IRQ	FFF8	FFF9

Table 8.1. Remember, these locations are in a particular RAM or ROM. It is not necessary to have a complete 64K memory; only enabling at least one memory IC for the vector addresses is required.

> **Example 8.1.** An external device is tied to $\overline{\text{IRQ}}$. The interrupt service routine starts at address 0140. What content should be loaded into the IRQ reserved spaces?

Address	Content
FFF8	01
FFF9	40

8.3   RESET ($\overline{\text{RST}}$)

By now you may have wondered how the microprocessor begins execution of the main program. This function is handled by the reset interrupt since it has no service routine per se. Instead, the reset vector address is the starting address of the main program. Once the starting address is loaded at locations FFFE and FFFF, the microprocessor begins operation when a reset switch activates the $\overline{\text{RST}}$ pin as shown in Fig. 8.4 (additional comments about reset hardware are given in Chap. 16). A 10-K pull-up resistor keeps the pin at logic 1. The microprocessor should not be activated until supply voltage reaches 4.75 V after power up. When the reset switch is pushed, the pin is grounded for logic 0. It must be held LOW at least three clock cycles because bus and control signals can be indeterminate. Grounding the $\overline{\text{RST}}$ line also clears all registers in the PIA. Unlike most active LOW

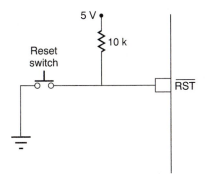

FIGURE 8.4
Microprocessor start-up hardware.

pins, the start sequence does not begin on the HIGH-to-LOW transition but, rather, on the LOW-to-HIGH transition when the switch is released.

Start Sequence

When the $\overline{\text{RST}}$ line is allowed to go HIGH, the following sequence of events occurs:

1. The I flag in the conditional code (CC) register is set. This flag, called the interrupt mask bit, prevents any of the remaining interrupts from interfering with the start sequence.
2. Contents of location FFFE are loaded into the HIGH-byte (MSB) of the program counter (PC).
3. Contents of location FFFF are loaded into the LOW-byte (LSB) of the PC.
4. PC contents go onto the address bus during $\phi 1$.

Every time the reset switch is pushed the system is interrupted and the MPU jumps to the beginning of the program. Thus, an operator can "reset" the system after a power failure or after an anomaly causes a system to lock up.

8.4   INTERRUPT REQUEST ($\overline{\text{IRQ}}$)

The line most commonly used by peripherals to interrupt the microprocessor is the interrupt request ($\overline{\text{IRQ}}$) line. A LOW signal on the line will initiate the interrupt sequence. However, the interrupt will be ignored if the I bit in the CC register is set (I = 1). The interrupt is said to be *masked*. The I bit is masked by one of three methods:

1. An MPU start (or reset) procedure which automatically sets the I flag.
2. A peripheral interrupt on the $\overline{\text{IRQ}}$ or $\overline{\text{NMI}}$ line which automatically sets the I flag.
3. An SEI (set interrupt mask) instruction.

The mask can be removed by a CLI (clear interrupt mask) instruction. If the I flag is clear, the MPU jumps to the $\overline{\text{IRQ}}$ service routine. Some provision for the MPU to return to the main program where it left off must be made. Like a subroutine call, the PC should be saved onto the stack. Unlike a subroutine call, the interrupt can occur virtually anywhere in the program. Microprocessor register content is unpredictable. Therefore, all MPU registers should be saved so that they can be restored to the same values held when the interrupt intruded.

$\overline{\text{IRQ}}$ Sequence

If the I bit in the CC register is clear and the $\overline{\text{IRQ}}$ line goes LOW for at least one clock cycle, the $\overline{\text{IRQ}}$ sequence begins:

1. Current instruction is completed.
2. MPU registers are placed onto the stack starting with the program counter LOW-byte (PCL) and ending with the CC register. See Fig. 8.5.

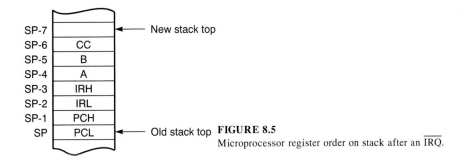

FIGURE 8.5
Microprocessor register order on stack after an $\overline{\text{IRQ}}$.

3. CC register I bit is set. MPU is masked from further interrupts.

4. Contents of location FFF8 are loaded into program counter HIGH-byte (PCH).

5. Contents of location FFF9 are loaded into PC LOW-byte.

6. PC contents go to the address bus during $\phi1$.

8.5 NONMASKABLE INTERRUPT ($\overline{\text{NMI}}$)

The other interrupt line available for peripherals is the nonmaskable interrupt ($\overline{\text{NMI}}$). As the name implies, the $\overline{\text{NMI}}$ line cannot be masked whereas the $\overline{\text{IRQ}}$ line can be masked. An $\overline{\text{IRQ}}$ interrupt proceeds only if the I bit is cleared by the programmer. An NMI interrupt takes place regardless of the I bit setting. There is no method to prevent the interrupt service routine from being executed. A nonmaskable interrupt is usually reserved for emergency routines such as a power failure shutdown. In the absence of "interrupt at any cost" the $\overline{\text{NMI}}$ line is used as a second interrupt line where it does function as a higher priority interrupt.

The nonmaskable interrupt $\overline{\text{NMI}}$ is active LOW, but it is edge-triggered, i.e., begins on the HIGH-to-LOW signal transition. It cannot interrupt again until the transition is repeated. In contrast, the $\overline{\text{IRQ}}$ interrupt is level sensitive and continually recognized as long as the line is LOW. An NMI interrupt sequence is similar to the $\overline{\text{IRQ}}$ sequence: MPU registers are loaded onto the stack as before, interrupt mask bit set. It differs only with the PC contents taken from locations FFFC and FFFD.

8.6 INTERRUPT INSTRUCTIONS

CLI	Clear interrupt mask		
SEI	Set interrupt mask		

Inherent	OE	CLI	Clear CC register bit 1.
Inherent	OF	SEI	Set CC register bit 1.

Every microprocessor starts running the program in memory upon reset. The reset sequence always masks the $\overline{\text{IRQ}}$ interrupt. Therefore, *an IRQ interrupt is allowed only if a CLI instruction is written near the beginning of the main program.*

In a few cases we may want to set the interrupt flag with software: the SEI instruction. Generally, the CLI instruction has been written and the programmer wants to protect a small segment of the main program from an interrupt interference. He or she then sandwiches that segment between SEI and another CLI instruction.

RTI   Return from interrupt

Inherent   3B   RTI   Pulls MPU registers from stack.

All interrupt service routines must end with the RTI instruction. It is similar to an RTS for a subroutine. RTI restores all old MPU registers from the stack in addition to the PC. Since there are 7 bytes to be stored, RTI takes 10 clock cycles, making it the longest 6802 instruction.

An interrupt service routine itself can be interrupted by an $\overline{\text{NMI}}$ interrupt before the routine is complete. Furthermore, another $\overline{\text{IRQ}}$ interrupt can be recognized before the first routine is complete if a CLI instruction appears at the beginning of the current routine. An interrupt of an interrupt is called a *nested interrupt.* Microprocessor register groups are stored onto the stack with each interrupt and are removed in proper order similar to nested subroutines.

Example 8.2. $\overline{\text{IRQ}}$ interrupts only are recognized. A CLI instruction is placed near the beginning of the main program. If the $\overline{\text{IRQ}}$ sequence automatically sets the interrupt flag, how can the $\overline{\text{IRQ}}$ line remain active after the service routine is completed?

The process is handled by the RTI instruction, i.e., the old CC register is restored. Its interrupt flag has been cleared by the CLI instruction.

SWI   Software interrupt

Inherent   3F   SWI   Jump to vector address stored at FFFA and FFFB.

The SWI instruction functions as a programmed interrupt. Steps in the interrupt sequence are similar to hardware interrupts except that the procedure is initiated by software, and the service routine vector address is located at FFFA and FFFB. Its primary purpose is a keyboard system debugging aid. Pauses can be inserted into programming, and the MPU registers are conveniently stored in the stack for readout. For example, the Heath 3400 Microprocessor Trainer employs the SWI instruction to single-step the program for examining registers and for tracking program errors.

WAI   Wait for interrupt

Inherent   3E   WAI   Save MPU registers and wait for hardware interrupt.

The WAI instruction was used in earlier chapters to halt short demonstrative programs. Its purpose is more meaningful than a simple program halt. When the WAI instruction is executed, the following sequence of events occurs:

1. All microprocessor registers are stored onto the stack in the same order produced by a hardware or software interrupt. Execution time is nine clock cycles.

2. The MPU goes into a dormant state with the data address, and R/W lines are in their high impedance state with the VMA held LOW.

3. The dormant state exists until an IRQ or NMI interrupt signal is received by the MPU.

4. The appropriate service routine vector address is fetched four clock cycles later.

Why not let the interrupt happen without the wait state? A WAI instruction is advantageous because it places the MPU registers onto the stack *before* the interrupt. The result is a faster interrupt. The programmer is cautioned to clear the interrupt mask bit at the beginning of a program if an IRQ interrupt line is recognized. Otherwise the system can hang up indefinitely. We can escape a wait state with the reset interrupt.

Example 8.3. A prototype distance finder is being developed. Figure 8.6 shows a schematic for the system. A pushbutton on the IRQ interrupt line starts the process. A sound pulse is emitted toward the object, and at the same instant a software timing loop begins counting. The returning pulse is detected by the NMI interrupt line before the MPU has exited the IRQ service routine. Its NMI service routine reads the now frozen counter, computes the distance, and displays the result. Outline the software.

Address	Content		Comments
FFF8	10		IRQ interrupt vector address
FFF9	00		
FFFC	20		NMI interrupt vector address
FFFD	00		
0000		PIA INIT:	PORT A OUTPUT
			PORT B OUTPUT
		MAIN PROG:	LDS #XXXX
			CLI
			WAI
			BRA FD
1000		IRQ:	SEND SOUND PULSE
			START COUNTER
			RTI
2000		NMI:	READ COUNTER
			COMPUTE DISTANCE
			DISPLAY IT
			RTI

Vector addresses (1000 and 2000) for IRQ and NMI service routines are loaded into the proper reserved addresses FFF8–FFF9 and FFFC–FFFD, respectively. Each routine ends

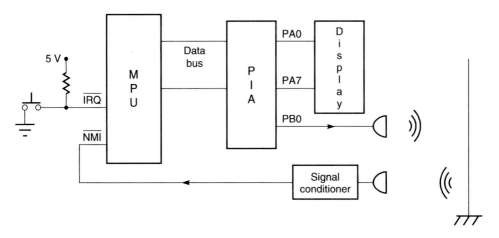

FIGURE 8.6
General schematic of a prototype distance finder.

with RTI. After $\overline{\text{NMI}}$ interrupts the counter and displays the distance the MPU returns to the $\overline{\text{IRQ}}$ routine. It stays in the count loop until the counter overflows (clears) then returns to the main program. Every program begins with the PIA initialization, stack pointer loading, and CLI instruction (only if $\overline{\text{IRQ}}$ is to be recognized). After returning from the $\overline{\text{IRQ}}$ service routine the MPU branches back to await the pushbutton request for a new reading.

8.7 INTERRUPTS AND THE PIA

Up to this point the PIA has been viewed as two 8-bit ports controlled by bit 2 of the control registers. It also has four additional lines, called control lines, that can be connected to peripherals: CA1, CB1, CA2, CB2. These lines are controlled by the remaining bits of the control registers. Control lines CA1 and CB1 are *input only* lines. But control lines CA2 and CB2 can be programmed as *input or output* lines. Thus, two-way communication or handshaking can be established between the MPU and peripheral(s).

Figure 8.7 shows the control register bit function in general. Bits 0 and 1 control CA1 (CB1) while bits 3 to 5 control CA2 (CB2). Bit 5 establishes line direction. Bits 6 and 7 are status bits. When CA1 (CB1) is active, a flag is raised in bit 7 (bit 7 is set). Bit 6 denotes a similar status for line CA2 (CB2).

Figure 8.8 illustrates two types of I/O operations performed with control lines. Their primary purpose is interrupt control. Four peripherals interrupting the MPU can be handled through the single $\overline{\text{IRQ}}$ line. An active peripheral on any control line will set its corresponding status bit. If either status bit is set in control register A (control register B), interrupt line $\overline{\text{IRQA}}$ ($\overline{\text{IRQB}}$) will go LOW. Each interrupt line is an open-collector output (see Chap. 9). This means that the lines can be tied (wire-ANDed) together to form a common line to the MPU $\overline{\text{IRQ}}$ pin. Any status flag will cause an interrupt. The interrupt routine must then read each control register to determine which peripheral needs service.

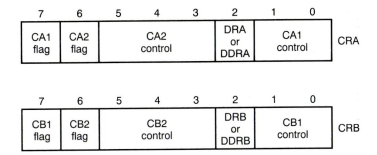

FIGURE 8.7
Control register bit functions.

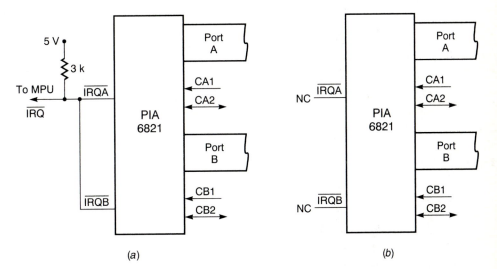

(a)   (b)

FIGURE 8.8
Two types of I/O operations with control lines. (a) Interrupt control; (b) program control.

The PIA manufacturer assumed that every activated control line will be followed by a read of its port. Thus, the PIA automatically resets the control register status bits when the MPU *reads the corresponding port (data register)*. The programmer must include the LDA (port) instruction in the service routine to clear the status bits, preventing a continuous interrupt.

A second I/O operation performed with control lines is program control. IRQA and IRQB are not connected for interrupts. Control lines are treated as a 4-bit "third" port. Input data is determined by a read of the control register status bits. Output data on CA2 or CB2 is configured by a write to control register bits 3 to 5.

A complete description of each control register bit now follows. Keep in mind that all PIA registers are cleared by hardware resets (RST).

CA1 (CB1) Control

Bits 1 0

0	0	interrupt masked; active CA1 (CB1) on negative edge
1	0	interrupt masked; active CA1 (CB1) on positive edge
0	1	interrupt nonmasked; active CA1 (CB1) on negative edge
1	1	interrupt nonmasked; active CA1 (CB1) on positive edge

If bit 0 is cleared, the interrupt line $\overline{\text{IRQA}}$ ($\overline{\text{IRQB}}$) is disabled. But an active input CA1 (CB1) sets status flag bit 7 for program I/O operation. If bit 0 is set, the interrupt line IRQA (IRQB) is enabled and pulled LOW when input CA1 (CB1) becomes active.

Bit 1 determines the conditions under which input CA1 (CB1) is active. If bit 1 is cleared, the input is active only when the peripheral signals with a HIGH-to-LOW (negative) transition on CA1 (CB1). If bit 1 is set, the input is active only when the peripheral signals with a LOW-to-HIGH (positive) transition. Either active designation will set status bit 7. A choice allows designers to adjust different peripherals to an active LOW only $\overline{\text{IRQ}}$ line on the MPU without resorting to extra hardware.

Register Control

Bit 2

0	MPU addresses data direction register DDRA (DDRB)
1	MPU addresses data register DRA (DRB)

CA2 (CB2) Control

Bits 5 4 3

0	X	X	CA2 (CB2) designated an input line
0	0	0	interrupt masked; active CA2 (CB2) on negative edge
0	1	0	interrupt masked; active CA2 (CB2) on positive edge
0	0	1	interrupt nonmasked; active CA2 (CB2) on negative edge
0	1	1	interrupt nonmasked; active CA2 (CB2) on positive edge

If bit 5 is cleared, CA2 (CB2) becomes an input line. Bits 3 and 4 determine the mask and active state of input CA2 (CB2) in the same manner as bits 0 and 1 for input CA1 (CB1). An active CA2 (CB2) sets the bit 6 status flag, and the $\overline{\text{IRQA}}$ ($\overline{\text{IRQB}}$) pin is pulled LOW if bit 3 is set by the programmer.

Bits 5 4 3

1	X	X	CA2 (CB2) designated an output line
1	0	0	complete handshake
1	0	1	partial handshake
1	1	0	program control: CA2 (CB2) = 0
1	1	1	program control: CA2 (CB2) = 1

If bit 5 is set, CA2 (CB2) becomes an output line. Status bit 6 clears automatically and is ignored. With an input line CA1 (CB1) and an output line CA2 (CB2), the MPU and peripheral can talk back and forth in a handshake mode. A *complete handshake* occurs when one device signals the other, and an acknowledgment signal is returned. A *partial handshake* occurs when the acknowledgment signal is omitted. Either program I/O or interrupt I/O can be used with handshaking.

Perhaps the single most important feature of the output line mode is the difference between port A and port B. The PIA handshake operation is structured so that peripherals should send data to port A for an MPU read and should receive data from port B with an MPU write. A detailed description of each follows.

Complete Handshake: Port A

The complete *input* handshake is initiated by configuring PIA control register A with 100 in bits 5, 4, and 3. Peripheral data lines are tied to port A for data transfer to the MPU. The peripheral should have a "data is ready" line which is tied to CA1. In addition, the peripheral should have a "data was read" line which is tied to CA2.

Figure 8.9 illustrates the protocol. When the peripheral has data ready to send to the MPU, it places data on the port lines and notifies the MPU via interrupt input line CA1. The "data is ready" signal can be a positive (shown here) or negative transition, but CRA bit 1 is initialized accordingly. At the instant the PIA receives an allowable transition it changes CA2 from LOW-to-HIGH, thereby informing the peripheral that the MPU will be taking action. *When port A is read,* CA2 automatically drops LOW upon the falling edge of $\phi 2$ during the LDA PORTA instruction. This transition informs the peripheral that the data was read. CA2 stays LOW until the peripheral sends another data ready signal.

With complete handshaking, the MPU essentially says: "Peripheral, I want to transfer data. But you are too slow. Ring me when you are ready. In the meantime I have other things to do. I'll notify you when I receive (send) it."

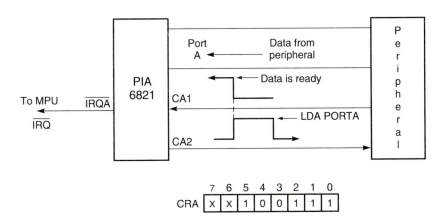

FIGURE 8.9
Complete input handshake with port A.

Example 8.4. An experimenter wants to record a time-varying displacement. The sensor generates a voltage proportional to position and interfaces to an analog-to-digital converter at port A. Successive readings are made and stored at memory locations 1000 to 1FFF. The converter, our peripheral in Fig. 8.9, is a slow device. Since we want new positions as fast as they are converted, a handshake arrangement would be prudent. The converter begins conversion upon a negative transition signal on its "START" pin. START is tied to CA2. The converter announces conversion is done with a positive transition signal on its "END CONVERSION" (EOC) pin. EOC is tied to CA1. Determine the PIA initialization and program outline.

CRB	Bit	Content	Comments
	0	1	Unmask interrupt on CA1
	1	1	active positive transition.
	2	1	Select CRB over DDRB.
	3	0	
	4	0	Complete handshake.
	5	1	
	6	0	Clear for interrupt.
	7	0	Clear for interrupt.

The following program first stores all the data. Following each interrupt by the converter, a byte is stored and programming waits for the next interrupt. After the data is collected the interrupt is masked and the data is analyzed.

Program

FFF8	FCB	01	Load IRQ
FFF9	FCB	00	vector address at 0100.
	ORG	0000	
0000	INIT:	CLR A	Initialize port A
		STA A 6000	for complete handshake.
		LDA A #27	
		STA A 6001	
		LDS #XXXX	
		CLI	
		LDX #1000	Point to first storage cell.
	WAIT:	WAI	
		INX	
		CPX #2000	Are data slots filled?
		BEQ PLOT	Yes, analyze data.
		BRA WAIT	

PLOT:	LDA B #26	Mask interrupt and
	STA B 6001	continue with data
	CONTINUE	analysis.
	⋮	
0100	LDA A 6000	Get data interrupt
	STA A 0,X	routine.
	RTI	

The last two examples have used the WAI instruction because a short program was dedicated to the interrupting peripheral. In practice, a program will have a number of tasks to perform with the interrupt occurring irregularly. WAI is unlikely to appear in most interrupt programs.

Complete Handshake: Port B

The complete *output* handshake is initiated by configuring PIA control register B with 100 in bits 5, 4, and 3. Peripheral data lines are tied to port B for data transfer from the MPU. The peripheral should have a "data is requested" line which is tied to CB1. In addition, the peripheral should have a "data was written" line which is tied to CB2.

Figure 8.10 illustrates the protocol. When the peripheral wants the MPU to send data, it notifies the MPU via interrupt line CB1. The "data is requested" signal can be a positive (shown here) or negative transition, but CRB bit 1 is initialized accordingly. At the instant the PIA receives an allowable transition it changes CB2 from LOW-to-HIGH, thereby informing the peripheral that the MPU is aware of the request. When data *is written* to port B, CB2 automatically drops LOW upon the falling edge of $\phi 2$ during the STA PORTB instruction. This transition informs the peripheral that data is available. The procedure is repeated until the MPU transfers all the necessary data.

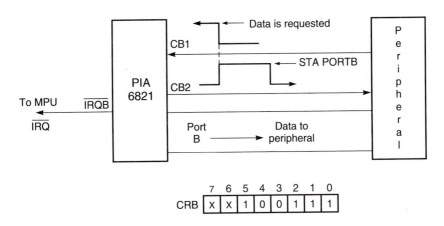

FIGURE 8.10
Complete output handshake with port B.

Notice that an MPU read causes the transition of CA2 while an MPU write causes the transition of CB2. This is why port A is used for input handshakes while port B is used for output handshakes.

Partial Handshake: Port A

The partial *input* handshake is initiated by configuring PIA control register A with 101 in bits 5, 4, and 3. Peripheral data lines are tied to port A for data transfer to the MPU. The peripheral should have a single control line tied to CA2. CA2 is normally HIGH in partial handshake whereas CA2 is normally LOW in the complete handshake mode.

Figure 8.11 illustrates the protocol. The peripheral sends data to port A but CA1 is not connected to notify the MPU, and there is no interrupt. The MPU will read port A on its own schedule. Each time the MPU reads port A, a LOW pulse is sent to the peripheral via CA2. Line CA2 goes LOW upon the falling edge of $\phi2$ during the LDA PORTA instruction but goes HIGH after *one* $\phi2$-clock cycle.

The pulse informs the peripheral that new data can be placed on the data lines. For example, position control with a DC motor requires a lengthy control program which can take longer to execute than the data conversion time of the analog-to-digital converter. The CA2 pulse starts a conversion which is finished before the MPU is ready for new data. Complete handshaking would not be practical.

Partial Handshake: Port B

The partial *output* handshake is initiated by configuring PIA control register B with 101 in bits 5, 4, and 3. Peripheral data lines are tied to port B for data transfer from the MPU. The peripheral should have a single control line tied to CB2. CB2 is normally HIGH.

Figure 8.12 illustrates the protocol. The MPU places data on port B for the peripheral with a STA PORTB instruction. At the same time line CA2 goes HIGH-to-LOW upon the falling edge of $\phi2$. After *one* $\phi2$ clock cycle CB2 returns HIGH. The LOW pulse informs the peripheral that data is available. The peripheral reads the data when it is

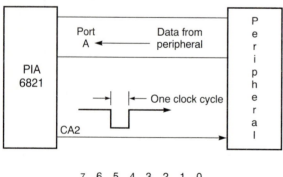

FIGURE 8.11
Partial input handshake with port A.

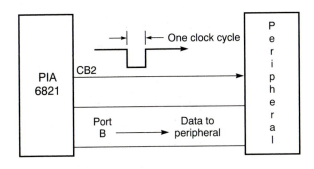

	7	6	5	4	3	2	1	0
CRB	X	X	1	0	1	1	X	0

FIGURE 8.12
Partial output handshake with port B.

ready, but the MPU is unaware of this action. CB1 is disconnected, and there is no interrupt of the MPU.

Program Control: CA2 (CB2)

Handshake control uses PIA lines CA2 and CB2 (output initialized) to pulse the peripheral. Program control allows the programmer to treat these lines as extra output bits. Respective control registers are initialized by setting bits 4 and 5. Bit 3 determines the output binary state. In the main program CA2 (CB2) can be changed by simply writing the control register: STA (CRA) or STA (CRB).

Bits	5	4	3	Output CA2 (CB2)
	1	1	0	LOW
	1	1	1	HIGH

Example 8.5. PIA port B drives a display. Line CB2 is interfaced to a stepper motor driver IC. Upon a LOW-to-HIGH transition signal the motor will step once. Show the PIA initialization and the subroutine instructions to step the motor. Assume the PIA occupies address space 7000–7003 and ignore port A.

Control register B is initialized with content 34 as follows:

CRB	Bit	Content	Comments
	0	0	Mask CA1 interrupt.
	1	0	Don't care, CA1 not connected.
	2	1	Address DRB.
	3	0	CB2 output LOW.
	4	1	Program control mode.
	5	1	Program control mode.
	6	0	Don't care, clear status.
	7	0	Don't care, clear status.

Program

PIA INIT:	CLR A	Port B all outputs.
	STA A 7002	
	LDA A #34	CB2 under
	STA A 7003	program control.
STEP:	LDA A #34	CB2 LOW
	STA A 7003	
	LDA A #3C	CB2 HIGH
	STA A 7003	
	RTS	

Status Bits

Status bits 7 and 6 in each control register are *read-only* bits. They are set when active transitions occur on *inputs* CA1 (CB1) *or* CA2 (CB2), respectively. This process is independent of the interrupt mask setting in bits 0 and 3. If CA2 (CB2) is initialized as an output line, bit 6 defaults to logic 0. Both bits 7 and 6 can be cleared only by *reading the port* DRA (DRB).

Under interrupt control, $\overline{\text{IRQA}}$ ($\overline{\text{IRQB}}$) is brought LOW if either bit 7 or bit 6 is set and the interrupt mask is clear. Under program I/O with the interrupt masked or $\overline{\text{IRQA}}$ ($\overline{\text{IRQB}}$) not connected, CA1 (CB1) and CA2 (CB2) can be treated as two (each) input ports. Their status can be determined by a read of the control register. This read must be followed by a read of the port to clear the status register. Input bits are activated by a transition (edge-triggered) signal. They will remain in a set state unless cleared by a port read.

> **Example 8.6.** A switch input to the PIA is through CA1. The PIA is at address space 4000–4003. What instructions are necessary to read the switch status?
>
> | | | |
> |---|---|---|
> | LDA A | 4001 | Get old CRA. |
> | LDA B | 4000 | Clear new CRA bit 7. |
> | AND A | #80 | Isolate switch status, old bit 7. |
> | BNE | ON | |

PROBLEMS

8.1. A temperature switch is interfaced to the MPU under interrupt control. A 7474 flip-flop is used for the port. After the switch goes HIGH to alert the system it can stay HIGH for an extended period. The flip-flop should be cleared immediately by the interrupt service routine to prevent multiple interrupts. Develop a wiring diagram.

8.2. How can an interrupt service routine be interrupted if the interrupt sequence sets the I bit?

8.3. Are there any interrupt types that do not use the stack?

8.4. A system's software includes an interrupt service routine. A short section of the main program should not be interrupted. How do you use the CLI and SEI instructions in the main program?

8.5. Write a short program to read accumulators A and B. Use the software interrupt instruction.

8.6. A timer pulses the $\overline{\text{IRQ}}$ line every second. Write the service routine for a 12-hour clock. What additional steps must be undertaken to prepare for interrupts? Store seconds, minutes, and hours at addresses 0000–0002. The service routine starts at address 1000.

8.7. In Example 8.4 the analog-to-digital converter starts conversion with a negative transition signal from CA2. What changes are necessary to handle a positive transition signal?

8.8. In Example 8.4 the system stores a block of data. The time between stored bytes is the analog-to-digital conversion time. Suppose we want a time graph of the data with 0.01 s between points. How would you change the program?

8.9. Initialize the PIA control registers for port A as an input port and port B as an output port. All four control lines CA1, CA2, CB1, and CB2 are inputs. Lines are activated by negative transition signals. The PIA occupies address space 6000–6003. Assume (*a*) program control, (*b*) interrupt control.

8.10. Initialize PIA control register B for control lines only. CB1 is an input line and CB2 is an output line. Input signals are normally LOW but are HIGH when active. Interrupts are used where possible. The PIA is at 4000–4003.

8.11. In Prob. 8.10 give the instructions for (*a*) reading CB1, (*b*) writing logic 0 to CB2.

8.12. Control register B content is 1D. How is port B controlled? If the content is 32?

8.13. How is the PIA $\overline{\text{IRQA}}$ line cleared after an interrupt? The $\overline{\text{IRQB}}$ line?

8.14. The prototype distance finder in Fig. 8.6 uses PB0 to send a sound pulse and $\overline{\text{NMI}}$ to receive a sound pulse. Reconsider the programming with CA2 replacing PB0 and CA1 replacing $\overline{\text{NMI}}$.

8.15. A system uses both PIA ports to drive displays. Two control lines are tied to separate indicator lights. Two control lines under interrupt control are tied to active LOW switches. Make an outline of all pertinent initialization, main program, and service routine considerations before the software is written.

CHAPTER
9

TRANSISTORS
AND THE
TTL/CMOS
LOGIC FAMILIES

The principal active devices in electrical circuits are the transistors. Two important types are the (1) bipolar, simply called a "transistor," and the (2) field-effect transistor or FET. When input currents are kept relatively small, the transistor serves as an amplifier. When the transistor is overdriven, it acts as a switch. It is this mode which is most useful in digital electronics. Bipolar transistors are used both in integrated circuits (TTL family) and as interfacing devices in general circuits. FET is used primarily in integrated circuits (CMOS family) and power switching devices.

9.1 DIODES

Diodes and transistors are made from silicon. Each silicon (Si) atom has four outer electrons which join with neighboring atoms to form a complete bond (insulator). If one of the Si atoms is replaced (doped) by arsenic, which has five outer electrons, one unpaired electron is present. The extra electron is easily freed by an electric field to form electricity readily. The carrier is referred to as an n-type, n standing for negative charge carriers. If the Si is doped by gallium, which has three outer electrons, an electron vacancy called a hole exists. Any free electron can wander in and fill this vacancy; in the process it also conducts. This positive charged carrier is referred to as a p-type.

If a p slice and n slice are joined together, the device is called a *diode*. The structure and symbols are shown in Fig. 9.1(a). At ~0.7 V the diode conducts in the forward direc-

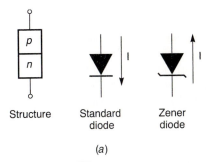

Structure   Standard   Zener
                 diode   diode

(a)

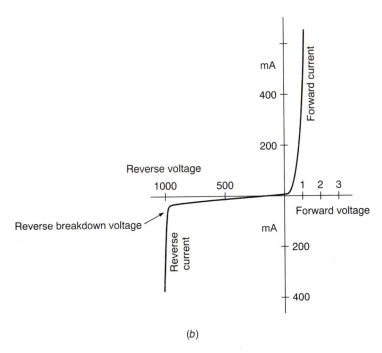

(b)

FIGURE 9.1
(a) Diode symbols; (b) electrical characteristics.

tion. Note that there is very little resistance to current in this direction. Typically, currents of several hundred amperes can flow in the forward direction with a voltage drop of ~1 V. Diode size is based on its power rating. On the other hand, resistance to current flow in the reverse direction is very high. Generally, current flow in this direction will be a few milliamperes when the reverse breakdown voltage is as high as ~1000 V. Thus, the diode acts as a one-way switch permitting current to flow in one direction and not the other.

Diodes, called zener diodes, can be made with low reverse breakdown voltages. Once the reverse breakdown voltage is reached, the zener has *very* little increase in voltage with large reverse currents. The rather constant voltage makes the diodes useful for precise voltage regulation.

9.2 BIPOLAR TRANSISTOR

The bipolar transistor consists of three slices of *p*-type and *n*-type material. The base (B), a very thin slice of one of them, is sandwiched by the complementary pair, thus the name "bipolar." Other segments are called the collector (C) and the emitter (E). Figure 9.2 shows the two variations in structure: *npn*-type and *pnp*-type.

Each type has three different ways the load can be wired across (C) and (E), in regard to position and current direction. The connections are common emitter, common collector, and common base. Observe that all current directions are reversed from the *npn*-type to the *pnp*-type. Also, the *npn* transistors are turned ON by applying a voltage HIGH to the base (B) while the *pnp* transistors are turned ON by applying a voltage LOW to the base.

The *npn* transistor is the most popular type available today because it costs less to produce. The *npn* transistor is also faster than the *pnp* transistor. It operates in the same manner for any configuration. Power gain is highest for the common emitter. This is the reason that the connection is used most frequently. The common base connection has the lowest input resistance and the highest output resistance. The common collector connection shows the highest input resistance and the lowest output resistance.

A typical transistor circuit is shown in Fig. 9.3(*a*). A small base current I_B is used to control a much larger collector current I_C or

$$I_E = I_C + I_B \approx I_C \tag{9.1}$$

The transistor parameter β, called the DC current gain, is defined as

$$\beta = \frac{\Delta I_C}{\Delta I_B} \quad \text{with } V_{CE} \text{ constant} \tag{9.2}$$

In most company literature, data sheets on transistors refer to the DC current gain symbol as h_{FE}. Gain usually is within the range 20 to 300 but is not constant for a given transistor because the device is nonlinear.

Transistor characteristics are shown in Fig. 9.3(*b*) and (*c*). Three states are possible with a transistor: cutoff, active, saturation.

Cutoff

The base characteristics illustrate the base current I_B versus the base-to-emitter voltage V_{BE}. We see that a base voltage below 0.6 V will not permit any current to flow through the transistor, including the base current and the load or collector current. The circuit is open and the transistor is said to be cutoff. When voltage V_{BE} is greater than 0.6 V, the transistor conducts, but the base current has little effect on the base-to-emitter voltage. The behavior is similar to a diode, with V_{BE} never exceeding 0.8 V.

Active

When the transistor is conducting, the transistor is said to be active and acts as a current gain device:

$$I_C = \beta I_B \tag{9.3}$$

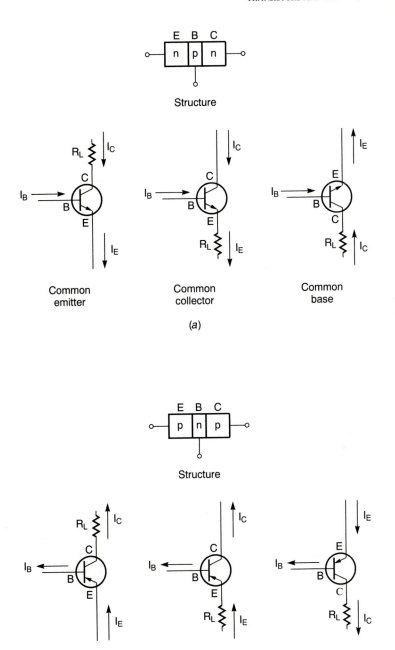

FIGURE 9.2
Transistor types and connections. (*a*) *npn* transistor; (*b*) *pnp* transistor.

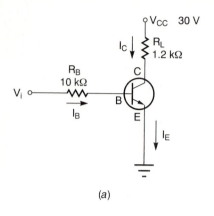

(a)

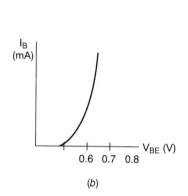

(b)

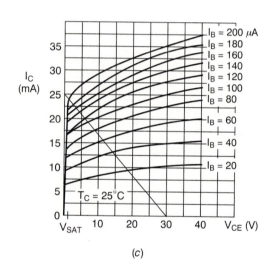

(c)

FIGURE 9.3

The *npn* common emitter transistor and characteristics. (*a*) Basic circuit; (*b*) base characteristics; (*c*) collector characteristics.

V_{BE} is generally taken to be 0.7 V when the circuit is conducting. In the controlled loop a supply voltage V_{CC} is divided between a calculated voltage drop across the load

$$V_L = I_C R_L \tag{9.4}$$

and the remaining drop across the transistor

$$V_{CE} = V_{CC} - V_L \tag{9.5}$$

Since the output voltage V_{CE} can vary from $V_{CE} = 0$ ($I_C = V_{CC}/R_L$) to V_{CC}, a resultant "load line" can be drawn across the collector characteristic curves. The interest in this line is its intersection with the "saturation line." Thus, the region of active DC current gain is $0 < I_B < 190$ μA for the particular transistor and load given. Once an operating point P is established, the current gain can be found from the graph by introducing a small change in the base current.

Saturation

The base current may be increased to any desired value, but the collector current I_C is limited to the saturation value $I_{CS} = 24$ mA. If $I_B \geq I_{CS}/\beta$, the transistor is said to be saturated. Notice that V_{CE} is not quite zero at saturation. Generally, the saturation voltage V_{SAT} for V_{CE} is about 0.2 V for most transistors. The high slope of the saturation line precludes any variation in V_{SAT} with the load R_L. The base-to-emitter voltage V_{BE} remains ~0.7 V once the transistor is turned ON.

Example 9.1. For the basic circuit and transistor characteristics shown in Fig. 9.3, find the pertinent voltage and currents if V_i is (a) 0.1 V, (b) 1.0 V, (c) 5 V.

(a) Since $V_{BE} < 0.6$ V the transistor is cutoff. $I_B = I_C = 0$. Therefore, $V_{CE} = 30$ V.

(b) Since $V_i > 0.6$ V, $V_{BE} \cong 0.7$ V (any transistor):

$$I_B = \frac{V_i - V_{BE}}{R_B} = \frac{1 - 0.7}{10 \text{ k}\Omega} = 30 \ \mu A$$

On the load line at $I_B = 30 \ \mu A$, $V_{CE} \cong 16$ V. Furthermore, $I_C \cong 11$ mA. Usually, characteristic curves are not readily available in graphical form and the key device specifications must be read from manufacturer's tables. Using a current gain of 250 yields

$$I_C = I_B\beta = 7.5 \text{ mA} \quad \text{compared to 11 mA previously}$$

$$V_{CE} = V_{CC} - I_C R_L = 30 \text{ V} - 7.5 \text{ mA} (1.2 \text{ k}\Omega) = 21 \text{ V} \quad \text{compared to 16 V}$$

(c) $I_B = \dfrac{5 - 0.7}{10 \text{ k}\Omega} = 430 \ \mu A$

Since $I_B > I_{BS}$, the transistor is saturated. Then $I_{CS} = 24$ mA. Alternatively to the graph, we have

$$I_{CS} = \frac{V_{CC} - V_{CE}}{R_L} = \frac{30 - 0.2}{1.2 \text{ k}\Omega} \cong 25 \text{ mA}$$

$$I_{BS} = \frac{I_{CS}}{\beta} = \frac{25 \text{ mA}}{250} = 100 \ \mu A$$

Since $I_B > I_{BS}$, the transistor is saturated and the preceding I_{CS} holds. Note that the current gain is nonlinear and a more representative gain near saturation is 100 or $I_{BS} = 250 \ \mu A$ compared to 190 μA earlier.

9.3 MANUFACTURER'S DATA

The more useful transistors are the general-purpose and Darlington although several classes of special-purpose transistors are available such as the high voltage transistor for neon bulbs. Darlington transistors are very high-gain units and will be discussed in Chap. 15.

General-purpose transistors come in the plastic or metal can variety. Figure 9.4 is a page taken from the *Motorola Small-Signal Semiconductor Catalog*. This information is similar to that given in any electronic distributor's catalog. The first thing that strikes

Bipolar Devices

Plastic-Encapsulated

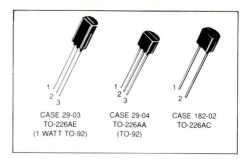

CASE 29-03	CASE 29-04	CASE 182-02
TO-226AE	TO-226AA	TO-226AC
(1 WATT TO-92)	(TO-92)	

Motorola's small-signal TO-226 plastic transistors encompass hundreds of devices with a wide variety of characteristics for general purpose, amplifier and switching applications. The popular high-volume package combines proven reliability, performance, economy and convenience to provide the perfect solution for industrial and consumer design problems. All devices are laser marked for ease of identification and shipped in antistatic containers, as part of Motorola's ongoing practice of maintaining the highest standards of quality and reliability.

In addition to the standard devices listed in the following tables, Motorola also offers special electrical selections of these devices. Please contact your Motorola Sales Representative regarding any special requirements you may have.

In each of the following tables, the major specifications of the transistors or diodes are given for easy comparison.

All transistors are available in the radial or axial tape and reel formats. Lead forming to fit TO-5 or TO-18 sockets is also available.

TABLE 1. General-Purpose Amplifier Transistors

The general-purpose transistors are designed for small-signal amplification from dc to low radio frequencies. They are also useful as oscillators and general purpose switches.

NPN	PNP	Pin Out	$V_{(BR)CEO}$ Volts Min	$f_T @ I_C$ MHZ Min	mA	I_C mA Max	$h_{FE} @ I_C$ Min	Max	mA	NF Max dB
TO-226AA										
MPS8099	MPS8599	EBC	80	150	10	200	100	300	1	—
MPSA06	MPSA56	EBC	80	100	10	50	50	—	100	—
BC546	BC556	CBE	65	150	10	100	120	450	2	10
BC546A	BC556A	CBE	65	150	10	100	120	220	2	10
BC546B	BC556B	CBE	65	150	10	100	180	450	2	10
MPS8098	MPS8598	EBC	60	150	10	200	100	300	1	—
MPSA05	MPSA55	EBC	60	100	10	500	50	—	100	—
MPS651	MPS751	EBC	60	75	50	2000	40	—	2000	—
BC182	BC212	CBE	50	200	10	100	120	460	2	10
BC237	BC307	CBE	45	150	10	100	120	460	2	10
BC239	BC309	CBE	45	150	10	100	180	800	2	10
BC547	BC557	CBE	45	150	10	100	120	450	2	10
BC547A	BC557A	CBE	45	150	10	100	120	220	2	10
BC547B	BC557B	CBE	45	150	10	100	180	450	2	10
BC547C	BC557C	CBE	45	150	10	100	380	800	2	10
BC317	BC320	CBE	45	250	10	150	110	450	2	10
2N3904	2N3906	EBC	40	300	10	200	100	300	10	5
2N4401	2N4403	EBC	40	250	20	600	100	300	150	—
2N3903	2N3905	EBC	40	250	10	200	50	150	100	6
2N4400	2N4402	EBC	40	200	20	600	50	150	150	—
MPS20	MPSA70	EBC	40	125	5	100	40	400	5	—
MPS650	MPS750	EBC	40	75	50	2000	40	—	2000	—
MPS6531	MPS6534	EBC	40	390*	50	600	10	120	100	—
MPS2222	MPS2907	EBC	30	250	20	600	100	300	150	—
2N4123	2N4125	EBC	30	250	10	200	50	150	2	—
MPS3704	MPS3702	EBC	30	100	50	600	100	300	50	—
MPS6513	MPS6517	EBC	30	330*	10	100	90	180	2	—
BC548	BC558	CBE	30	300*	10	100	120	300	2	10
BC548A	BC558A	CBE	30	300*	10	100	120	220	2	10
BC548B	BC558B	CBE	30	300*	10	100	180	450	2	10
BC548C	BC558C	CBE	30	300	10	100	380	800	2	10
2N4124	2N4126	EBC	25	300	10	200	120	360	2	—
MPS6514	MPS6518	EBC	25	480*	10	100	150	300	2	—
MPS6515	MPS6519	EBC	25	480	10	100	250	500	2	—
MPS5172		EBC	25	120*	5	100	100	500	10	—
MPS6560	MPS6562	EBC	25	60	10	500	50	200	600	—
MPS6601	MPS6551	EBC	25	100	50	1000	30	150	1000	—
BC238	BC308	CBE	25	150	10	100	120	800	2	10

FIGURE 9.4

General-purpose transistors—*Motorola Small-Signal Semiconductor Catalog. (Copyright of Motorola, Inc. Used by permission.)*

the novice designer in selecting a transistor is the multitude of choice for a given application even from a single manufacturer.

Case 29-03 TO-226AE	Style of case and pin-socket, respectively.
MPS8099 BC546 2N3904	The number 2NXXXX indicates transistor types registered by manufacturers with the Electronic Industries Association, Washington, DC. The number BXXXX is registered with the Association Internationale, Brussels, Belgium. Other designations are those of the manufacturer.
Pin-out	EBC or its permutations is the Emitter, Base, Collector as viewed from the flat on the case. There is no particular sequence from transistor to transistor. Many electronic distributors such as Newark Electronics, publish catalogs which list all transistors in numerical-alphabetical order, giving the manufacturer. Without the manufacturer's literature, the designer must test for the correct pins. This is easily accomplished with hand-held multimeters which have transistor test sockets and show the current gain when inserted in the correct order.
$V_{(BR)CEO}$	The collector-to-emitter breakdown voltage is the minimum load voltage spike which will cause the transistor to fail. It is important when switching inductive loads.
f_T	The current gain bandwidth product is simply the frequency at which the gain falls to one; thus, the transistor is ineffective as an amplifier.
I_C (max)	The maximum allowable collector or load current regardless of the load.
h_{FE}	The current gain with representative minimum and maximum values over its range.

Along with the preceding information, additional characteristics for specific transistors are listed inside the Motorola catalog. Figure 9.5 is the page for the 2N3904 transistor. Some important parameters not given previously follow:

P_D	The maximum allowable power dissipation in transistor at ambient temperature T_A or case temperature T_C. This parameter guides the active operation while I_C(max) guides the saturation operation. Transistors are rated as

Low power:	up to 500 mW
Medium power:	0.5 to 10 W
High power:	over 10 W

$V_{CE(SAT)}$	Voltage across the collector-emitter at saturation.
$V_{BE(SAT)}$	Voltage across the base-emitter at saturation.

MAXIMUM RATINGS

Rating	Symbol	Value	Unit
Collector-Emitter Voltage	V_{CEO}	40	Vdc
Collector-Base Voltge	V_{CBO}	60	Vdc
Emitter-Base Voltage	V_{EBO}	6.0	Vdc
Collector Current — Continuous	I_C	200	mAdc
Total Device Dissipation @ $T_A = 25°C$ Derate above 25°C	P_D	625 5.0	mW mW/°C
*Total Device Dissipation @ $T_C = 25°C$ Derate above 25°C	P_D	1.5 12	Watts mW/°C
Operating and Storage Junction Temperature Range	T_J, T_{stg}	−55 to +150	°C

*THERMAL CHARACTERISTICS

Characteristic	Symbol	Max	Unit
Thermal Resistance, Junction to Case	$R_{\theta JC}$	83.3	°C/W
Thermal Resistance, Junction to Ambient	$R_{\theta JA}$	200	°C/W

*Indicates Data in addition to JEDEC Requirements.

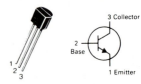

2N3903
2N3904

CASE 29-04, STYLE 1
TO-92 (TO-226AA)

3 Collector

2 Base

1 Emitter

GENERAL PURPOSE TRANSISTOR

NPN SILICON

ELECTRICAL CHARACTERISTICS ($T_A = 25°C$ unless otherwise noted.)

Characteristic		Symbol	Min	Max	Unit
OFF CHARACTERISTICS					
Collector-Emitter Breakdown Voltage(1) ($I_C = 1.0$ mAdc, $I_B = 0$)		$V_{(BR)CEO}$	40	—	Vdc
Collector-Base Breakdown Voltage ($I_C = 10$ μAdc, $I_E = 0$)		$V_{(BR)CBO}$	60	—	Vdc
Emitter-Base Breakdown Voltage ($I_E = 10$ μAdc, $I_C = 0$)		$V_{(BR)EBO}$	6.0	—	Vdc
Base Cutoff Current ($V_{CE} = 30$ Vdc, $V_{EB} = 3.0$ Vdc)		I_{BL}	—	50	nAdc
Collector Cutoff Current ($V_{CE} = 30$ Vdc, $V_{EB} = 3.0$ Vdc)		I_{CEX}	—	50	nAdc
ON CHARACTERISTICS					
DC Current Gain(1) ($I_C = 0.1$ mAdc, $V_{CE} = 1.0$ Vdc)	2N3903 2N3904	h_{FE}	20 40	— —	—
($I_C = 1.0$ mAdc, $V_{CE} = 1.0$ Vdc)	2N3903 2N3904		35 70	— —	
($I_C = 10$ mAdc, $V_{CE} = 1.0$ Vdc)	2N3903 2N3904		50 100	150 300	
($I_C = 50$ mAdc, $V_{CE} = 1.0$ Vdc)	2N3903 2N3904		30 60	— —	
($I_C = 100$ mAdc, $V_{CE} = 1.0$ Vdc)	2N3903 2N3904		15 30	— —	
Collector-Emitter Saturation Voltage(1) ($I_C = 10$ mAdc, $I_B = 1.0$ mAdc) ($I_C = 50$ mAdc, $I_B = 5.0$ mAdc)		$V_{CE(sat)}$	— —	0.2 0.3	Vdc
Base-Emitter Saturation Voltage(1) ($I_C = 10$ mAdc, $I_B = 1.0$ mAdc) ($I_C = 50$ mAdc, $I_B = 5.0$ mAdc)		$V_{BE(sat)}$	0.65 —	0.85 0.95	Vdc
SMALL-SIGNAL CHARACTERISTICS					
Current-Gain — Bandwidth Product ($I_C = 10$ mAdc, $V_{CE} = 20$ Vdc, f = 100 MHz)	2N3903 2N3904	f_T	250 300	— —	MHz

---

FIGURE 9.5

Tabular data for the 2N3904—*Motorola Small-Signal Semiconductor Catalog. (Copyright of Motorola, Inc. Used by permission.)*

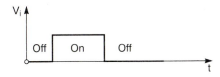

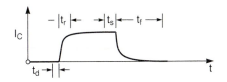

FIGURE 9.6
Transient waveform from a switched bipolar transistor.

t_d Delay time is the length of time that the transistor remains OFF after the input current is applied.

t_r Rise time is the time required for the collector current to rise to 90 percent of its final value.

t_s Storage time is the time that the collector current remains near its maximum value after the input current is cut off.

t_f Fall time is the time required for the collector current to fall to 10 percent of its maximum value.

Sooner or later every designer encounters a situation when he or she asks: How fast can the transistor switch the load current ON or OFF? Referring to Fig. 9.6,

$$\text{turn-ON time} \quad t_{ON} = t_d + t_r = 70 \text{ ns}$$
$$\text{turn-OFF time} \quad t_{OFF} = t_s + t_f = 250 \text{ ns}$$

9.4   FIELD-EFFECT TRANSISTOR (FET)

An earlier type of transistor, called the junction FET or JFET, depends upon load current control by means of a voltage rather than a base current. It consists of a channel of n-type silicon through which the current passes from drain to source [see Fig. 9.7(a)], much like the collector-emitter in bipolar transistors. The channel is surrounded by p-type silicon but the electrons flow only through the n-type, making it unipolar. When a negative voltage is applied to the gate (analogous to the base), the carrier electrons are repelled, forming a *depleted* area, and the channel resistance increases. The drain or load current is decreased. Maximum drain current occurs for zero gate voltage. Positive gate voltages destroy the operation of the device by permitting gate current. If the doping is reversed, the operation of the JFET is reversed. Current must be from source to drain. Increasing the positive gate voltage decreases the source current.

JFETs have been largely overtaken by metal oxide semiconductor FET or MOSFET, MOS for short. It includes an insulating layer of silicon dioxide between the channel and gate as shown in Fig. 9.7(b). The field is still effective in modulating the electron flow in

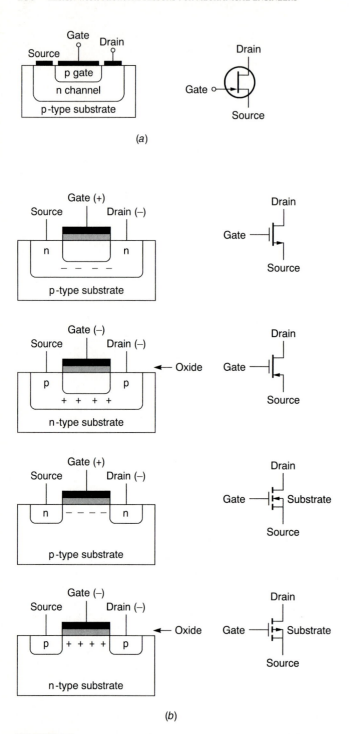

FIGURE 9.7
Types of FET transistors and their symbols. (*a*) JFET; (*b*) MOSFET.

the channel; however, the gate makes no electrical contact with the channel. Since no current can flow through the gate, positive or negative gate voltages can be applied.

There are four basic types of MOSFET structures. MOSFET can be either *n* channel (*p* substrate) or *p* channel (*n* substrate). If the channel is actually provided, it is said to be a depletion type since the applied gate voltage depletes original free carrier electrons in the channel. A zero gate voltage allows an effective drain (or load) current. Positive voltages increase the drain current while negative voltages decrease it. The wide operating range makes the depletion type useful for signal amplification. If a channel is not provided, the gate voltage induces an equal but opposite charge in the adjacent substrate. The free-moving charges behave as though a channel were present. This type is referred to as an enhancement type. Positive gate voltages are used with the *p* substrate, while negative gate voltages are used with the *n* substrate. Since an open drain-source circuit exists for the enhancement type, it is ideal for switching applications.

The performance of a MOSFET is shown in Fig. 9.8. In a confusion of terms the transistor amplifies in the *saturation region* where the drain current is nearly proportional to the gate voltage and relatively independent of the drain voltage. With the drain voltage below 5 V the device acts as a resistor, where the resistance is proportional to the gate voltage.

The major difference between the bipolar and MOSFET transistor is the infinite input impedance of the MOSFET; thus, it draws no current. This makes it ideal for amplification of small signals in high impedance circuits. MOSFET is also insensitive to temperature and cheaper to manufacture. Its higher packing densities make it ideal for integrated

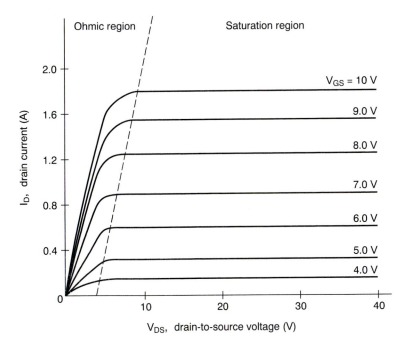

FIGURE 9.8

Drain characteristics for an *n*-channel enhancement-type MOSFET transistor.

circuits. A disadvantage is that the oxide insulation is susceptible to breakdown from static electricity. Bipolar transistors are used as an interfacing device since they can carry a much larger load current than the MOSFET. This is because the bipolar device conducts throughout its cross-sectional area while the MOSFET conducts across a narrow channel.

9.5   LOGIC FAMILIES

We are all familiar with logic gates and their different types: AND, OR, NAND, DE-CODER, etc. They are integrated circuits (ICs) constructed from a combination of transistors and resistors. A family of gates is a collection of different types made with the same technology and having the same electrical characteristics. Chips from different families can be interconnected, but this requires some care. Perhaps the most widely used family offering the greatest number of different types is the Transistor Transistor Logic (TTL) family. It is based on bipolar technology. Of growing importance in microprocessor systems is the complementary metal oxide semiconductor (CMOS) family. It is based on MOSFET technology. A designer considers two main criteria when selecting a family: power dissipation and speed.

Power Dissipation

Power dissipation is the amount of power (in milliwatts) that the IC drains from the power supply. Power dissipation is important because it drives up the scale (thus cost) of the power supply and because it increases the heat within the electronic package. Heat is a major cause of circuit failure. Power drain can be critical when the package is in remote locations (space?) or battery-powered. TTL logic will dissipate on the order of 10 mW per gate. CMOS logic draws no power unless it is in the act of switching the logic. Power dissipation increases almost linearly up to 10 mW at 1 MHz switching frequency.

Speed

The speed of a logic family is given by the propagation delay of a signal through its basic inverter. Actually, the propagation delay is the average of two close delays caused by an input change: (1) the delay t_{pHL} when the output changes from HIGH to LOW and (2) the delay t_{pLH} when the output changes from LOW to HIGH. In general, TTL logic is faster than CMOS logic although the gap is narrowing with the HCMOS (high-speed CMOS) technology built into recent microprocessors.

We will see subsequently that there is the invariable tradeoff between desired parameters. With CMOS and TTL, power dissipation goes up as the propagation delay goes down.

9.6   THE TTL FAMILY

The TTL family is available in five versions (or series): standard, low-power, high-speed, Schottky, and low-power Schottky. The standard TTL series remains popular although the low-power or LS series is often used with microprocessors.

Standard TTL was the first version in the TTL family. The basic gate was then constructed with different resistor values to produce gates with lower-power dissipation or

Table 9.1
Speed and power dissipation comparison of TTL series

Name	Propagation delay (ns)	Power dissipation (mW)	Speed-power product (pJ)
Standard TTL	10	10	100
Low-power TTL	33	1	33
High-speed TTL	6	22	132
Schottky TTL	3	19	57
Low-power Schottky TTL	9.5	2	19

higher speed. The propagation delay of a saturated transistor depends upon two factors: storage time and RC time constants. Reducing resistor values in the circuit reduces the RC time constant, thus increasing the speed (or reducing the propagation delay). However, the tradeoff is a higher-power dissipation because the lower resistance draws more current from the power supply. Schottky versions include a Schottky diode between the transistor base and the collector. This allows the transistor to be switched ON but prevents the transistor from saturating. Since the storage time is reduced, the propagation time is smaller. Of course the diode increases power dissipation. A comparison of the speed and power dissipation of the TTL series is given in Table 9.1.

Low-Power TTL

In the low-power TTL gate the resistor has been increased by a factor of 10 over the standard TTL. Power dissipation is reduced, but the propagation delay is increased. Low-power TTL is not used often in designs.

High-Speed TTL

In the high-speed TTL gate resistor values are reduced to improve the propagation delay time, but the power dissipation is increased. The speed improvements cannot match the Schottky versions, and the power dissipation is the worst of the series. It is rarely used.

Schottky TTL

Schottky TTL is a later improvement in technology which removes the transistor storage time by preventing its saturation. Since transistor storage time is the major source of the delay, the improvement is substantial. Increased power dissipation is at tolerable levels. High-speed circuits generate sharp switching wave fronts that may cause transmission line "ringing" in the longer lines connecting chips. Schottky TTL should be used in designs only when necessary.

Low-Power Schottky

Low-power Schottky TTL raises the value of the internal resistors. It sacrifices some of the speed found in Schottky but has greatly improved power dissipation (one-fifth the standard TTL). It has the best speed-power dissipation product (an important criterion for most designers), and, as a consequence, it has become the most popular version in new designs.

9.7   TTL TYPES

If the five versions of TTL are not confusing enough to the novice designer, we will add that TTL gates in all versions come in three different types of output configurations:

1. Totem-pole output
2. Open-collector output
3. Three-state (or tristate output)

Differences in the TTL versions are not in the structure of the gate or the digital function that they perform. It is in the values of resistors and the type of transistor that the gates employ. With just a subtle alteration in their structure, the way that the gate is used changes dramatically.

The totem-pole output (or construction) is the basic form of the TTL gate. It is the type used most frequently in gate design. In addition, the more complex packages such as decoders, peripheral interface adapters, and microprocessors have TTL totem-pole outputs. The open-collector output is not a necessary output in digital circuits, but it does give the user several options to be discussed later. Three-state outputs are merely totem-pole outputs with on-off switches to disconnect the signal from the circuit—an important design option when interfacing the microprocessor bus. The three types of outputs will be considered in regard to their circuit description.

Totem-pole Output

The standard TTL 2-input NAND gate is shown in Fig. 9.9. Of the resistors and four transistors, the two transistors Q3 and Q4 at the right constitute the output stage. They are referred to as a totem-pole arrangement, with one transistor sitting on top of another. The key is transistor Q2, or "phase splitter."

LOGIC LOW OUTPUT

1. When Q2 is ON, Q3 is driven into saturation (or ON).
2. The voltage drop across Q3 is $V_{CE} = 0.2$ V. This value becomes the output logic LOW.
3. The current flow to ground is *into* the gate.
4. The voltage in the collector of Q2 is the voltage drop to ground: V_{BE} of Q3 plus V_{CE} of Q2 or $0.7 + 0.2 = 0.9$ V.
5. The 0.9 V is applied to the base of Q4. This is insufficient to turn Q4 ON since V_{BE} of Q4 is one V_{BE} drop plus one diode drop D1 plus one V_{CE} (Q3) drop or $2 \times 0.7 + 0.2 = 1.6$ V.
6. The diode in the circuit is to increase the drop in the output to ensure Q4 is cut off when Q3 saturates.

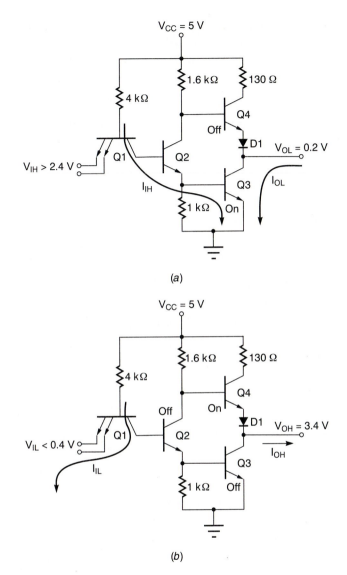

FIGURE 9.9
TTL totem-pole output NAND gate. (*a*) Logic LOW output; (*b*) logic HIGH output.

LOGIC HIGH OUTPUT

1. When Q2 is OFF, Q3 is cut off also.
2. Q4 now conducts because its base is connected to V_{CC} through the 1.6 kΩ resistor, base voltage being V_{CC}.
3. The final value of the output voltage is then 5 V minus V_{BE} drop in Q4, minus a diode drop in D1 or 3.4 V.
4. The output acts as a source with a current *out of* the gate.

Two totem-pole outputs should never be wired together; i.e., the gate output should drive another gate input. When the output of one gate is HIGH and the output of another gate is LOW, transistor Q3 is driven through the base of Q4 with the diode offering the only resistance. An excessive amount of current drawn can damage the transistors. At a minimum a valid logic LOW will not exist on the output.

The difference between various gates lies in the input section. In the NAND gate an unusual transistor Q1 having two emitters is driving Q2.

LOGIC HIGH INPUT

1. When both inputs are HIGH (greater than 2.4 V), the current from the Q1 base is forced through Q2 and Q3 turning both ON.
2. The base voltage of Q1 is 2.1 V, the transistor drop of 0.7 V through the three transistors Q1, Q2, and Q3.
3. The input is high impedance since only a very small leakage current flows *into* the gate.

LOGIC LOW INPUT

1. If either (or both) of the emitters are grounded, the voltage drop V_{BE} for Q1 will be 0.7 V, which cuts off Q2.
2. The current path is base to emitter in Q1 and *out of* the gate.

Open-Collector Output

The standard TTL 2-input open-collector (OC) NAND gate is shown in Fig. 9.10. The first thing noticed is the missing transistor Q4 and diode. Upon closer inspection we see that everything else is identical. The input characteristics are the same. A LOW gate output still passes the same input current through the saturated transistor Q3 to ground. How-

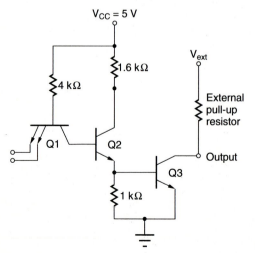

FIGURE 9.10
TTL open-collector NAND gate.

ever, an open-collector gate has no output drive capability. One is always provided in the form of an external pull-up resistor. The value of the resistor is usually several thousand ohms. When the output is HIGH, transistor Q3 is cut off. Then the output voltage is pulled up to the value of the external voltage V_{ext} tied to the resistor, providing the OC drives a high impedance load such as another gate.

Example 9.2. If an open-collector gate is driving another gate whose input voltage will drop to 3.6 V—the minimum voltage for a HIGH—while demanding a current of 0.4 mA, what is the maximum value of the pull-up resistor? Assume the pull-up voltage is 5 V.

$$R_{max} = \frac{5 - 3.6 \text{ V}}{0.4 \text{ mA}} = 3500 \ \Omega$$

Remember, the resistance must also be large enough so that Q3 will not be destroyed when a LOW appears on the output. A value of 1000 Ω is typical.

If the outputs of several open-collector TTL gates are tied together with a single external resistor, it is called wire-ANDing (sometimes inappropriately called wire-ORing). A graphic symbol for such a connection is shown in Fig. 9.11. The AND gate is drawn with lines going through the center to the common connection which distinguishes it from a regular gate. Remember, the AND function gives a HIGH only if all variables are high; otherwise the function is LOW. With the outputs of the open-collector gates connected together, a LOW on any gate output will drive the *common* output LOW.

Open-collector gates are used in three major applications:

1. Perform wire-AND logic.
2. Drive a lamp or relay.
3. Form a common data bus line.

The wire-AND connection gives additional logic with no expense or space requirement. Its drive capability takes on two aspects. First, if the pull-up resistor and voltage are removed, transistor Q3 can serve as an ON–OFF switch to ground. Any powered device connected to the output can be turned ON by a LOW for a current path to ground, or it can be turned OFF by a HIGH. Since current through the gate cannot exceed its specified value, a limiting resistor is usually included in the circuit. Second, a TTL totem-pole

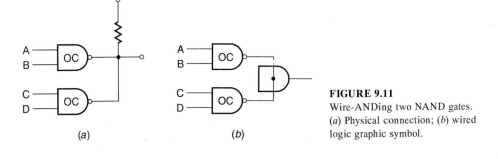

FIGURE 9.11
Wire-ANDing two NAND gates.
(*a*) Physical connection; (*b*) wired logic graphic symbol.

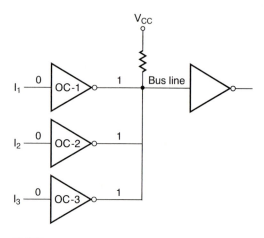

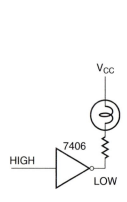

FIGURE 9.12
Using an open-collector inverter for lamp
ground.

FIGURE 9.13
Open-collector inverters forming a common bus line.

output is limited to 5 V. An open-collector is free to use any pull-up voltage to match the required interfacing device, such as a lamp, provided the pull-up resistor limits the current to the open-collector for its LOW. This use is illuminated in Fig. 9.12.

Figure 9.13 demonstrates the connection to a common bus line. Each of three inputs (any number will do) drives an open-collector inverter, and the inverter outputs are tied together to form a single bus line. The idea is to pass binary information to the bus one input at a time, say, I_3. Inputs I_1 and I_2 are held LOW so that their inverter output is HIGH. Now when I_3 is HIGH, the output of inverter 4 is HIGH. When I_3 is LOW, the output of inverter 4 is LOW. Inputs I_1 and I_2 can transmit their data in turn. Bus driving is generally accomplished by TTL three-state outputs rather than open-collector outputs.

Three-State Output

Conceptually, a three-state (or tristate) output is simple: a standard totem-pole TTL with an ON–OFF switch (enabling switch) which disconnects the line signal. Introduced by National Semiconductor in 1972, it is called a three-state because the output can be (1) logic LOW, (2) logic HIGH, or (3) very high impedance (OFF) to supply voltage V_{CC} and ground.

The symbols for four types of TTL three-state output gates are given in Fig. 9.14. There are two inputs, one being the enable. Most three-state gates are enabled by an active LOW ($\overline{E}$) as indicated by the invert circle. If several three-state devices are connected, all but one must be disabled; only the enabled output determines the logic level of the connection. If the outputs of two 3-state devices are enabled simultaneously, one may be damaged—totem-pole outputs are never wired together. At a minimum we have "bus contention" as an unknown state where the data on the bus is randomly HIGH or LOW.

Three-state TTL outputs are found just about anywhere we look. Small scale integration (SSI) packages of four gates each are available for the noninverting buffer, active

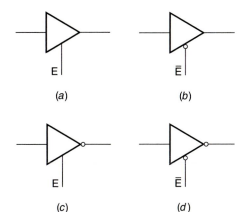

FIGURE 9.14
Types of three-state TTL outputs. (*a*) Noninverting, enable HIGH; (*b*) noninverting, enable LOW; (*c*) inverting, enable HIGH; (*d*) inverting, enable LOW.

HIGH or active LOW, as shown in Fig. 9.15. Some MSI packages, such as the 74373 octal latch in Fig. 7.3, have three-state enables. All microprocessors and their peripherals, including all memory, use three-state outputs, which allow the microprocessor to control the timing of each device's access to the buses.

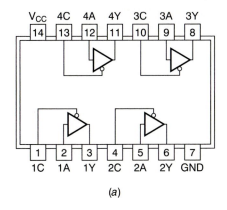

(*a*)

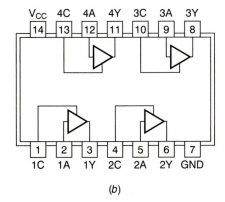

(*b*)

FIGURE 9.15
(*a*) The quad 74125 three-state buffer (active LOW); (*b*) the quad 74126 three-state buffer (active HIGH). (*Reprinted by permission of Texas Instruments.*)

9.8 TTL CHARACTERISTICS

All TTL series are within the same family, so they have almost the same electrical characteristics and can be directly interchanged without difficulty. All series operate with a $+5$ V $\pm$ 0.5 V power supply. The primary variables of interest are taken from Fig. 9.9.

1. V_{IH} is the voltage when the input is HIGH.
2. V_{IL} is the voltage when the input is LOW.
3. V_{OH} is the voltage when the output is HIGH.
4. V_{OL} is the voltage when the output is LOW.
5. I_{IH} is the current when the input is HIGH.
6. I_{IL} is the current when the input is LOW.
7. I_{OH} is the current when the output is HIGH.
8. I_{OL} is the current when the output is LOW.

Typical Characteristics

The characteristics of TTL gates are specified by the manufacturer in commonly available literature. The most widely used source is the *TTL Data Book* published by Texas Instruments Corporation. In addition to logic gates, it includes data on buffers, drivers, flip-flops, adders, decoders, encoders, multiplexers, counters, and memory, among others. A summary of typical TTL characteristics are offered in Table 9.2 for the NAND and/or inverter gate, both totem-pole output and open-collector output, although the voltage V_{OH} has no meaning for open-collectors. Other gate data will vary slightly. Table 9.3 shows the logic characteristics for gates with three-state output.

The sign convention is minus for current out of the gate regardless of input or output. See Fig. 9.9. Because of the large input impedance of a gate, attention is usually focused on the output. The gate is said to *drive* the interfacing load. When the output logic is HIGH, the gate acts as a *source* of current to the load. When the output logic is LOW,

TABLE 9.2
Logic characteristics for TTL totem-pole

	Series					
	Standard	**Low-power**	**High-speed**	**Schottky**	**Low-power Schottky**	**Units**
V_{IH} (min)	2	2	2	2	2	Volts
V_{IL} (max)	0.8	0.7	0.8	0.8	0.8	Volts
V_{OH} (min)	2.4	2.4	2.4	2.7	2.7	Volts
V_{OL} (max)	0.4	0.4	0.4	0.5	0.5	Volts
I_{IH} (max)	40	10	50	20	20	μA
I_{IL} (max)	-1.6	-0.18	-2	-2	-0.4	mA
I_{OH} (max)	-0.4	-0.2	-0.5	-1	-0.4	mA
I_{OL} (max)	16	3.6	20	20	8	mA

TABLE 9.3
Logic characteristics for TTL three-state

		Series				
	Standard	Low-power	High-speed	Schottky	Low-power Schottky	Units
V_{IH} (min)	2	—	—	2	2	Volts
V_{IL} (max)	0.8	—	—	0.8	0.8	Volts
V_{OH} (min)	2.4	—	—	2.4	2.4	Volts
V_{OL} (max)	0.4	—	—	0.5	0.5	Volts
I_{IH} (max)	40	—	—	20	20	μA
I_{IL} (max)	−1.6	—	—	−2	−0.4	mA
I_{OH} (max)	−5.2	—	—	6.5	−2.6	mA
I_{OL} (max)	16	—	—	20	24	mA
I_{OZ} (off state)	−40	—	—	−50	−20	μA

the gate acts as a *sink* for current from the load. The word "drive" encompasses both the sourcing and sinking capabilities of a gate.

An explanation of the terminology is in order:

1. V_{IH} (min) is the lowest input voltage that the gate will recognize as a HIGH, whereas V_{IL} (max) is the highest input voltage that the gate will recognize as a LOW.

2. V_{OH} (min) is the lowest guaranteed output voltage that the gate will produce as a HIGH at I_{OH} (max), whereas V_{OL} (max) is the highest guaranteed output voltage that the gate will produce as a LOW at I_{OL} (max). Expected voltages are

$$V_{OH} \text{ (typical)} = 3.4 \text{ V}$$

$$V_{OL} \text{ (typical)} = 0.2 \text{ V}$$

3. I_{IH} (max) is the largest required HIGH input current if the input voltage V_{IH} falls to V_{OH} (min) from the preceding gate.

4. I_{IL} (max) is the largest required LOW input current if the input voltage V_{IL} rises to V_{OL} (max) from the preceding gate.

5. I_{OH} (max) is the guaranteed source current at V_{OH} (typical).

6. I_{OL} (max) is the guaranteed sink current at V_{OL} (typical).

Fan-Out

The output of a gate is usually connected to the inputs of other similar gates or series in the same IC family. Each input consumes a certain amount of power from the connecting gate output so that several gates add to the load of the driver gate. *Fan-out* specifies the number of loads or gates that the output of a gate can drive without disrupting its normal operation, i.e., maintain the desired LOW or HIGH.

For a HIGH, a gate can source current I_{OH} to n gates demanding current I_{IH} each or $n = I_{OH}/I_{IH}$. Since the input impedance of a gate is large, fan-out for a HIGH is never important.

For a LOW, a gate can sink current I_{OL} with each of n gates supplying current I_{IL}.

$$n(\text{fan-out}) = \frac{I_{OL}}{I_{IL}}$$

A typical value for fan-out is 10. When heavy loads are involved, noninverting gates or buffers are usually used to provide additional drive.

EXAMPLE 9.3. How many low-power Schottky TTL loads can a standard TTL drive?

$$\text{Fan-out} = \frac{I_{OL} \ (\text{standard})}{I_{IL} \ (\text{low-power Schottky})} = \frac{16}{0.4} = 40$$

TTL Drive Capability

Gates are used to interface lamps, transistors, etc. besides driving other gates. This point raises a question: If I_{OH} (max) and I_{OL} (max) are produced at V_{OH} (typical) and V_{OL} (typical), respectively, what currents will be produced at other voltages?

Both a standard TTL inverter and a low-power Schottky TTL inverter were selected for a simple experiment. First, a logic HIGH was imposed on the output of each gate. The source current drove a variable resistor attached to ground. Source characteristics are illustrated in Fig. 9.16. Second, a logic LOW was imposed on the output of each gate. A

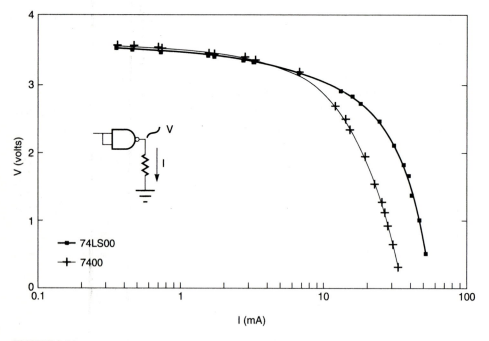

FIGURE 9.16
Source characteristics of TTL NAND or NOT.

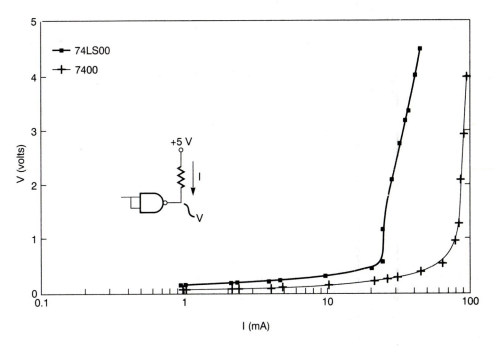

FIGURE 9.17
Sink characteristics of TTL NAND or NOT.

5-V supply generated a sink current through the variable resistor to the gate. Sink characteristics are illustrated in Fig. 9.17. Several observations can be made:

1. Higher source currents than I_{OH} (max) can be delivered, but it will be done at a lower gate voltage—an important point when driving transistors.
2. Higher sink currents than I_{OL} (max) can be accepted, but it will be done at higher gate voltages—an important point when driving a lamp.

TTL gates were designed to interface similar gates. Although we see that gates can sink much more current than I_{OL} (max), the voltage drop is much higher than 0.2 V. Excessive power dissipation and higher IC temperatures can greatly shorten the life of the system. The designer should consider buffer/drivers which can sink the higher currents at V_{OL} (typical) = 0.2 V.

EXAMPLE 9.4. Can we use the TTL 74LS04 inverter gate to drive the transistor circuit given in Fig. 9.2? Review Example 9.1.

Solution. Get the base current. Be sure the transistor is saturated. Base current and drive voltage satisfy

(*a*) $V_i = f(I_B)$, from gate drive characteristics
(*b*) $V_i = R_B I_B + V_{BE} = 10,000 I_B + 0.7$

or by trial and error

$$V_i = 3.6 \text{ V}, \qquad I_B = 0.3 \text{ mA}$$

Since the transistor base current for saturation is $I_{BS} = 0.10$ mA, the pairing will work.

Pin-Outs

The *TTL Data Book* is most frequently read for IC pin-outs. All TTL gates come with a 54XX or a 74XX number designation. They are identical except that the 54XX series is guaranteed for use from $-55°C$ to $+125°C$, whereas the 74XX series is rated for service from $0°C$ to $+70°C$. The 54XX series generally is reserved for military applications.

Pin-outs for the 7400 NAND gate are shown in Fig. 9.18 (top) just as it appears in the *TTL Data Book*. Pin-outs for other commonly used gates are given also for quick reference. The view taken for an actual IC is with the end notch (not shown) oriented to the left. A dot is sometimes placed at pin 1 although it is not necessary. The 7400 NAND gate comes in all five versions even though this is true only for a small number of ICs. The 7408 AND gate is available only in standard and both Schottky versions.

Each IC has a number stamped on the top which gives the manufacturer, gate number, version, and package type. Texas Instruments Corporation is the leading manufacturer, and the plastic dual-in-line (DIP) is the cheapest packaging.

SN74LS00N

Prefix—Manufacturer	Midfix—Version	Suffix—Package
SN—Texas Instruments	None—Standard	N—Plastic DIP
MC—Motorola	L—Low-power	J—Ceramic DIP
AM—Advanced Micro	H—High-speed	W—Flat
MM—Monolithic Memories	S—Schottky	
DM—National	LS—Low-power Schottky	
N—Signetics		

9.9   BUFFER/DRIVERS

"Buffer" and "driver" are two words that can be difficult to "pin down." Sometimes they have separate meanings; yet often they have the same meaning, i.e., buffer/driver. Buffer is another name for interfacing ICs that allow compatibility between devices. Examples of buffers serving different functions follow:

1. *7407.* Increases drive current to improve fan-out.
2. *Am2812.* Provides temporary data storage (up to 32 bytes) when data are sent to a device faster than it can process the data.
3. *74244.* Allows data to be passed from the outside world to the data bus.
4. *7407.* Eliminates the mismatch when interfacing different families.

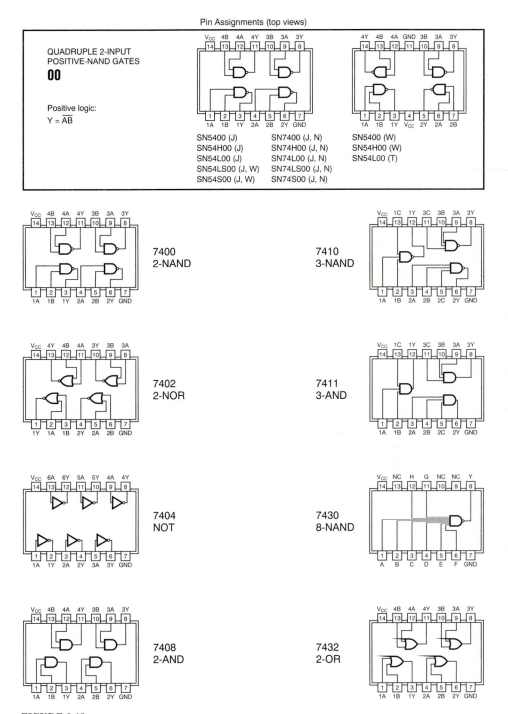

FIGURE 9.18
Pin assignments for common gates. (*Reprinted by permission of Texas Instruments.*)

Drivers supply increased power between TTL circuits and peripherals. Two samples of drivers are given:

1. *7447*. Decodes and drives a display.
2. *MC1489*. Interfaces TTL to transmission lines.

When the IC has increased current driving abilities, it is called a buffer/driver. Regular TTL can sink up to 20 mA if it is the high-speed series. Buffer/drivers can sink up to 64 mA. Beyond this level transistors must be used. Several buffer/drivers and the features which distinguish them from their cousins are offered as follows:

Totem-Pole TTL

7428 NOR	
7437 NAND (2-input)	$I_{OL} = 40$ mA
7440 NAND (4-input)	

Open-Collector TTL

7406 Inverter	$I_{OL} = 40$ mA
7407 Non-inverter	$V_{OH} = 30$ V

Three-State TTL

74241 octal buffer/driver	$I_{OL} = 64$ mA
74244 octal buffer/driver	$I_{OH} = 15$ mA
74365 non-inverter	$I_{OL} = 32$ mA
74366 inverter	$I_{OH} = 5.2$ mA

9.10   THE CMOS FAMILY

In the CMOS logic system the bipolar transistors used in TTL are replaced by enhancement-type MOSFETs. The usual NAND gate is diagrammed in Fig. 9.19. The MOSFETs occur in *complementary* pairs of p transistors and n transistors, thus the name CMOS. We should think of a gate as a simple switch. When the input is HIGH, the associated p transistor is OFF and the n transistor is ON. If both inputs are HIGH, the ouput is connected to ground by the n transistor, and the output is LOW. If either input is LOW, the associated n transistor is OFF, and the output is coupled to supply voltage V_{DD} (as opposed to accepted notation V_{CC} for TTL) by means of the p transistor. Thus, NAND logic is implemented.

The input-output characteristics are much easier to understand than the bipolar TTL technology. First, the input is blocked from the substrate by an insulating layer of silicon dioxide. So, gate input currents are practically nonexistent. When the transistor is OFF, it offers the usual very high impedance path for the current. When the transistor is switched ON, it conducts as a simple resistor following Ohm's law. An equivalent circuit for the gate output is shown in Fig. 9.20. Unlike the TTL gates, the output HIGH will be the same as the supply voltage and the output LOW will be the same as ground for an open circuit load.

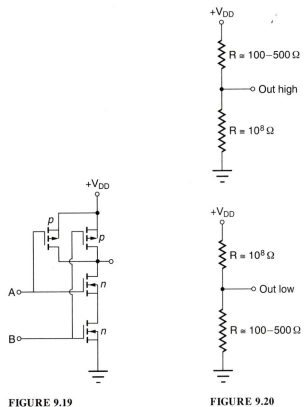

FIGURE 9.19
CMOS NAND gate.

FIGURE 9.20
Equivalent CMOS output circuit.

CMOS is widely available in two series: the 4000B series and the 74C series. The 4000B is an improvement over the original 4000A series introduced by RCA in the late 1960s. Buffers were added to sharpen the rise and fall time on both the input and output and to increase the current drive. The 74Cs were made to be pin compatible with TTL although they have the same characteristics as the B series. Interfacing rules must still be obeyed. However, we are not bound to the 5-V supply voltage of TTL. V_{DD} can range from +3 to +18 V. CMOS characteristics are given in Table 9.4. Again, these variables form the basis for logic connections.

Interfacing TTL to CMOS

Since CMOS inputs do not draw any current, the interfacing must match only the voltage. There is no problem when TTL is LOW, but a TTL HIGH output can be as low as 2.4 V, too low for a CMOS HIGH input. A pull-up resistor should be added to the TTL output that drives CMOS to raise the input voltage to 5 V. The minimum pull-up resistor ($\sim$2k) is sized so that the TTL LOW sink current is within I_{OL} (max). To interface TTL to higher voltage CMOS, the TTL 7406 or 7407 buffer/driver should be used. It has input protection against the higher voltage.

TABLE 9.4
Logic characteristics of CMOS (25°C)

| | Input voltage | | | |
	5 V	10 V	15 V	Units
V_{IH} (min)	3.5	7.0	11.0	Volts
V_{IL} (max)	1.5	3.0	4.0	Volts
V_{OH} (min at $I_{OH} = 0$)	4.95	9.95	14.95	Volts
V_{OL} (max)	0.5	0.5	0.5	Volts
I_{OH} (min)	−0.44 $V_{OH} = 4.6$ V	−1.1 $V_{OH} = 9.5$ V	−0.3 $V_{OH} = 13.5$ V	mA
I_{OH} (typical)	−0.88	−2.25	−8.8	mA
I_{OL} (min)	0.44 $V_{OL} = 0.4$ V	1.1 $V_{OL} = 0.5$ V	3.0 $V_{OL} = 1.5$ V	mA
I_{OL} (typical)	0.88	2.5	8.8	mA
Propagation delay	160	65	50	ns

Interfacing CMOS to TTL

When 5-V CMOS drives TTL, CMOS LOW is the problem. It can sink only 0.44 mA. Thus, it is capable of driving two low-power TTL loads or one low-power Schottky TTL load. Of course the added TTL can provide additional driving power. An alternative is the CMOS 4049 hex buffer, which can sink 6 mA or source 2.5 mA. An interfacing guideline can be found in Fig. 9.21.

Advantages of CMOS

1. CMOS consumes power only when switching logic levels. In low switching applications less than 10^{-6} W per gate is consumed, remarkable when compared to 1.5 mA per gate for low-power TTL and 15 mA per gate for standard TTL. These pluses are ideal for battery-driven systems. The hand-held calculator constructed from TTL would take about 4 A of power!

2. The wide supply voltage range offers greater design flexibility over TTL. TTL power supplies are very tightly regulated. In addition, analog components need a 12- to 15-V power source. A CMOS design eliminates one power supply.

3. CMOS has a higher margin for noisy signals. For a logic LOW, TTL has a range of 0 to 0.4 V compared with 0 to 1.5 V for a 5-V CMOS. At higher supply voltages both the low level and high level noise margins are improved. For example, a TTL HIGH must fall in the range 2.4 to 5 V while a 15-V CMOS will function over the range 11 to 15 V.

Disadvantages of CMOS

1. The most vulnerable feature of CMOS is the thin oxide layer insulating the input from the substrate. The infinite input impedance allows the buildup of an electrostatic charge to where a 100-V potential can damage the layer. A person walking on a carpet can easily generate several thousand volts. Internal diodes are connected between input and ground pins, which help protect it, but this is not always sufficient. It is standard practice to store all CMOS ICs on conducting foam or foil. *Be sure all loose input pins are connected to supply or ground and do not float.*

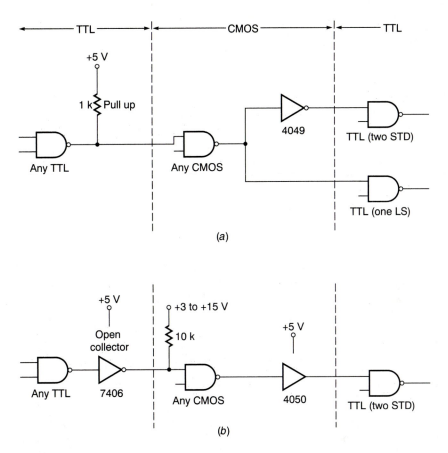

FIGURE 9.21
Interfacing CMOS and TTL. (*a*) 5-V CMOS; (*b*) 3- to 15-V CMOS.

2. CMOS is five to ten times slower than TTL. Unless we are building number crunchers, speed is often only important in timing operations. Also, CMOS transistors have large resistances, and when driving only a little capacitance, they can have large RC delay times. Just a few picofarads from other components or wiring in proximity to outputs can make the system unreliable.

3. CMOS has negligible input current demand, but it cannot supply very much source or sink current compared to TTL. CMOS has large fan-out when driving CMOS, but it can only drive one or two other TTLs.

4. TTL places a much wider variety of basic gate circuits into the hands of the designer. Most CMOS gates have a TTL equivalent although not necessarily pin compatible. Certain TTL circuits have no CMOS equivalent. An example is open-collector logic. But the converse is also true. The pure resistance nature of the switched-ON CMOS lends itself to functions not possible with TTL. An example is the 4066 analog switch. If the voltage to any of four switches is equal to V_{SS} (ground), the switch is OFF. If the voltage is equal to V_{DD}, the switch turns ON and behaves as an 80-Ω resistor.

Summary

CMOS is used in Large Scale Integrated (LSI) circuits such as memory or microprocessors, where MOS transistors are easier to fabricate and have much greater packing densities. Their outputs to the real world are usually TTL, which have superior driving characteristics. Family selection for system design involving multiple ICs is based on the user design criteria. Low-power specifications may necessitate CMOS. High-speed requirements will dictate TTL. If CMOS is contemplated, care must be exercised to avoid electrostatic charges and to isolate the CMOS from stray capacitance when assembling the system. TTL circuits are considerably more rugged than CMOS, and therefore are preferred in breadboarding and general design applications.

PROBLEMS

9.1. Make a rough sketch of the transistor collector characteristics in Fig. 9.3. Shade in the (*a*) saturation area, (*b*) area of greatest heat generation.

9.2. Verify that a CMOS gate can drive the transistor circuit in Fig. 9.2 using the base characteristics in Fig. 9.3.

9.3. For the basic circuit in Fig. 9.2, what is the maximum value for R_B which will saturate the transistor? What is the minimum value which will meet the drive specifications of a standard TTL? Assume a 5-V source for the first part of this problem.

9.4. A load resistance is to be powered by a 10-V source and switched ON by a 2N3904 general-purpose transistor. What is the maximum load that can be controlled? If the transistor is driven by a low-power Schottky TTL gate, what is the base resistance?

9.5. A DC lamp is connected to a 30-V source and should operate at 1 W of power. Select a TTL gate to drive the lamp.

9.6. Standard TTL will drive how many (*a*) standard TTL inputs? (*b*) low-power Schottky TTL inputs?

9.7. Low-power Schottky TTL will drive how many (*a*) standard TTL inputs? (*b*) Schottky TTL inputs?

9.8. If two open-collector TTL inverter outputs are tied together, what is the equivalent logic function?

9.9. Totem-pole TTL outputs should not be tied together. What sink current occurs if two standard TTL outputs are tied? What do you conclude about the result?

CHAPTER
10

INTERFACING
BINARY:
SWITCHES
AND LEDs

The microprocessor senses and responds to the outside world through peripheral devices. With personal computers, peripherals have come to mean floppy-disk drives, printers, or math coprocessors but they include the display screen (CRT) and keyboard as well. However, anything that is attached to the microprocessor, exclusive of memory and port, is considered a peripheral device. Peripherals are connected to the microprocessor through I/O ports. Ports may be simple latches (output) or three-state buffers (input), but generally the PIA is the port of choice.

The most common peripherals are mechanical switches and light emitting diode (LED) displays. Mechanical switch data entry can be manual or automatic. Manual switches include the hand-tripped type, such as a toggle switch, and the object-tripped type, such as a limit switch. Pressure, flow, and temperature switches are automatic types which trip when the specific variable reaches a prescribed value. LEDs may be a single lamp indicator or warning light variety, or they may be fashioned into dots or segments to form displays. Use of these displays in instrumentation, appliances, and other consumer products seems to be growing exponentially. LEDs often consume more power than the rest of the circuit. In applications where the supply of power is critical and CMOS ICs are the designer's choice, liquid crystal displays (LCDs) can be used.

In the selection of peripheral devices several functions often must be implemented. They can be satisfied entirely by software, eliminating some peripherals, or they can reach a compromise between software and hardware. The deciding factors are cost and reduced burden on the microprocessor. This tradeoff is nowhere more apparent than with switches and LEDs.

10.1   MECHANICAL SWITCHES

A large variety of mechanical switches are available which differ in design. The more common ones are illustrated in Fig. 10.1. All switches are referred to by the number of poles and throws. *Poles* are the number of separate circuits that can be completed by the same action. *Throws* are the number of individual contacts for each pole. Thus, in Fig. 10.2 we can identify the single-pole/single-throw (SPST), the single-pole/double-throw (SPDT), and the double-pole/double-throw (DPDT) type. All the illustrated varieties come in several types. Some companies offer combinations of poles and throws (or positions as they are often called) up to four-pole/ten-position.

A rotary switch that is used in a factory setting is the thumbwheel switch. Figure 10.3 illustrates a push variety. Any one of the 10 decimal digits can be set by an operator. The number is converted to a convenient code (such as BCD, octal, or hexadecimal) within the thumbwheel. The more commonly used BCD code shown is active HIGH if common is tied to 5 V. For example, the number 6 would register as 0110 on pins 8, 4, 2, and 1, respectively.

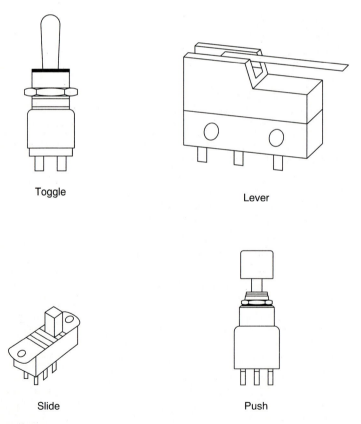

Toggle

Lever

Slide

Push

FIGURE 10.1
Mechanical switch varieties.

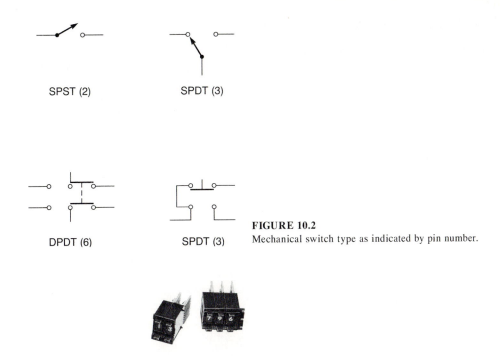

	SPST (2)	SPDT (3)
DPDT (6)	SPDT (3)	

FIGURE 10.2
Mechanical switch type as indicated by pin number.

Wiring Sequence

```
O  O  8
O  O  4
O  O  2
O  ⊖  1
O  ⊖  Common

O  O  Common
O  O  1
O  O  2
⊙  O  4
O  O  8
```

BCD & Complement of BCD

	Wheel Printing	0	1	2	3	4	5	6	7	8	9
	8	0	0	0	0	0	0	0	0		
	4	0	0	0	0					0	0
O	2	0	0			0	0			0	0
u	1	0		0		0		0		0	
t	Common	0	0	0	0	0	0	0	0	0	0
p	Common	x	x	x	x	x	x	x	x	x	x
u	1		x		x		x		x		x
t	2			x	x			x	x		
	4					x	x	x	x		
	8									x	x

FIGURE 10.3
Thumbwheel switch and BCD output. (*Courtesy of Eaton Corp., Watertown, Wis.*)

ON–OFF switches that are common in process control include flow switches, pressure switches, and temperature switches. The device becomes active when the level exceeds an upper limit or falls out of a specified range. Snap-action SPDT and DPDT are available. The trip threshold is adjusted with screws provided on the package.

Switches are generally interfaced to a peripheral interface adapter (PIA) with TTL or CMOS ports. The electrical signal is made compatible with logic circuits by using a pull-up resistor. Figure 10.4 shows the resistor for various switch types. Notice that the SPDT can be wired as a SPST. When the switch is closed, the output voltage is zero, or it

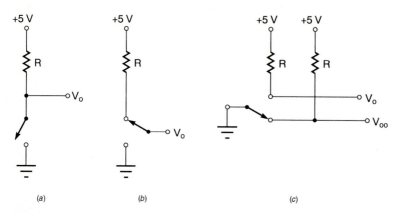

FIGURE 10.4
Switch circuits for TTL logic. (*a*) SPST; (*b*) SPDT as a SPST; (*c*) SPDT.

is active LOW. When it is open, the pull-up resistor provides a voltage $V_o = 5\text{ V} - IR$, where I is the HIGH input current to the load or port. The resistance should be large enough to limit the power dissipated when the switch is closed and should satisfy the HIGH voltage requirement. A typical pull-up resistor is 1 k.

Switch Bounce

Whenever a switch is placed in a circuit, we must be careful to avoid a condition called bounce. When a mechanical switch is engaged, the contact actually opens and closes rapidly for a short period of time before the "break" or the "make" is complete. This is not hard to understand because contacting elements are elastic and vibrate upon the switch action. The bounce for a break and make SPDT is demonstrated in Fig. 10.5. A switch contact can bounce for 5 to 50 ms, but the common design period is 20 ms. The number of bounces before the final state typically is 2 to 10.

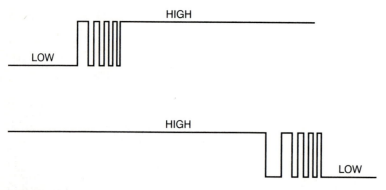

FIGURE 10.5
Switch bounce for a SPDT break and make.

When the microprocessor polls the port for the switch setting, it appears that the switch has been thrown several times. To prevent an erroneous decision by the microprocessor, the common practice is to "debounce" the switch. Debouncing can be done by using either hardware or software.

10.2   HARDWARE DEBOUNCE

SPDT Switches

There are several hardware methods to obtain a clean, sharp pulse from the switch, all based on flip-flops. Figure 10.6 shows a wiring circuit to debounce a SPDT switch using an RS flip-flop or latch. RS (reset, set) latches are available as quad (four to a package) debouncers, the 74279 IC. The RS latch can also be made from a pair of 7404 inverters in place of NAND gates. It is more cost-effective to add any unused 7404 gates on an IC than to buy new latches.

Here is how it works. The NAND gate is LOW only when the two inputs are HIGH. Thus, output V_o is HIGH when the switch is in the upper position. Upon moving the switch from the upper position nothing happens in spite of bounce because one input is locked LOW by the lower NAND gate output. When contact is made at the lower position, the first bounce LOW causes a HIGH output on the lower NAND gate. This forces the upper NAND gate output LOW. Subsequent bounces at the lower position do not change the logic configuration. A similar argument can be made for switching in the reverse direction.

A second method to debounce the switch with hardware uses a D flip-flop or dual 7474 IC package. Figure 10.7 shows the wiring diagram. The circuit ignores the input and clock but activates the preset (top) and clear (bottom). A LOW on the preset sets a HIGH

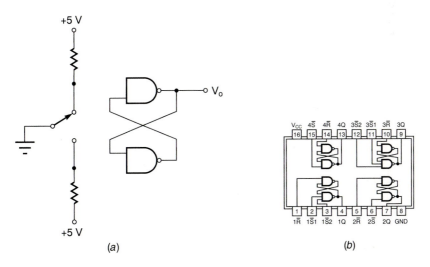

(a)   (b)

FIGURE 10.6
SPDT switch debounce using an RS latch. (*a*) Basic circuit; (*b*) 74279 latch. [(*b*) *Reprinted by permission of Texas Instruments.*]

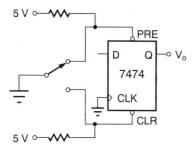

FIGURE 10.7
SPDT switch debounce using a D flip-flop.

output. The bounce that occurs after leaving the upper position does not affect the output. When the switch engages the lower position, the first bounce clears the flip-flop or sets a LOW on the output. Subsequent bounces are ignored; the flip-flop must be reset for a HIGH.

SPST Switches

SPST switches have only a single contact and cannot be debounced by the previous circuits. One method uses a D flip-flop as shown in Fig. 10.8. Since the output does not change until a position-edged clock signal is imposed, the switch is debounced by choosing a clock period that is greater than the period of the bounce or less than 50 Hz.

An alternative method uses a Schmitt trigger, Fig. 10.9. A Schmitt trigger converts a noisy or slowly varying analog signal into the clean digital signals needed in digital cir-

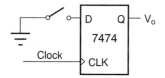

FIGURE 10.8
SPST switch debounce using a D flip-flop.

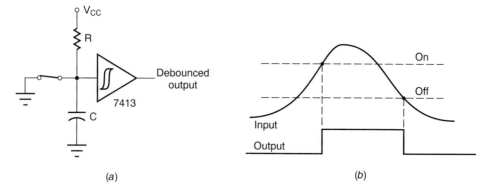

(a)   (b)

FIGURE 10.9
SPST switch debounce using a Schmitt trigger. (a) Basic circuit; (b) Schmitt trigger action.

cuits. As the input to the trigger rises from a low value, the trigger ignores the input, maintaining a LOW output. When the varying input signal reaches the ON threshold, the output "snaps" to a HIGH. It remains HIGH until the input signal falls to the OFF threshold. At that point the Schmitt trigger snaps LOW. Schmitt triggers find many applications in digital circuits where signals must be sharpened for TTL/CMOS.

The debounce scheme depends upon charging and discharging an *RC* network. When the switch is opened, the input to the Schmitt trigger remains LOW for a period of time as the supply voltage charges the capacitor. The bounces only serve to lengthen the charge time. The output of the trigger goes HIGH when the threshold voltage is reached. When the switch is closed, the first bounce discharges the capacitor, and the output goes LOW because the capacitor does not have time to charge up to the threshold. The *RC* time constant should be longer than the switch bounce time.

10.3   SOFTWARE DEBOUNCE

Both SPST and SPDT switches can be debounced with software instead of hardware. Usually, software is cheaper than hardware, but the microprocessor must poll the inputs continually. There are several steps to the debounce routine:

1. Detect if the switch is closed.

2. Wait 20 ms.

3. Check the switch to be certain the same reading occurs.

4. Repeat step 1 if the bounce is still present.

Example 10.1. The following program debounces a single switch. It detects if the active LOW switch at address 6000 is closed. If the switch is closed, the program jumps to a 20-ms delay subroutine. It verifies if the switch is still closed and repeats if the switch is open.

Main Program

```
LOOP    LDA A   6000
        CMP A   #FF         Detect closure.
        BEQ     LOOP

        JSR     DELAY       Jump to delay subroutine.

        LDA A   6000
        CMP A   #FF         Check closure again.
        BEQ     LOOP
```

Subroutine

```
DELAY   LDX     #09C3
BACK    DEX                 Delay 20 ms.
        BNE     BACK
        RTS
```

Multiple Switches

When several switch or key closures are to be detected, the data can be entered in a parallel bank of single keys. In Fig. 10.10 each bit of port B is connected to an input key. When the key is closed, the corresponding line is pulled LOW. To determine which key is pressed, the program must identify the key LOW in the byte at port B. It does this by continuously shifting the byte into the carry bit until a zero is registered. If the number of shifts are counted, the count represents the key number pressed. The shift routine occurs after the key is debounced. Therefore, the following step is added to the single key program.

5. Identify the switch.

> **Example 10.2.** Write a program to debounce and identify multiple switches.
>
> The following program uses a reset PIA at addresses 6000–6003 to interface eight active LOW switches. The PIA is initialized. A LOW is detected at port B. A delay subroutine idles the program for 20 ms. If the switch is still LOW, the port byte is rolled into the carry until the zero is found. The roll count is stored in memory location called BIT.

Main Program

BIT	DB		Address for key number.
INIT	LDA A	#04	Initialize PIA port B.
	STA A	6003	
LOOP	LDA A	6002	
	CMP A	#FF	Detect closure.
	BEQ	LOOP	
	JSR	DELAY	Jump to delay subroutine.
	LDA A	6002	
	CMP A	#FF	Check for closure again.
	BEQ	LOOP	
	LDA B	#07	
IDENT	ROL A		
	BCC	END	Identify the switch.
	DEC B		
	BRA	IDENT	
END	STA B	BIT	

Subroutine

DELAY	LDX	09C3	
BACK	DEX		Delay 20 ms.
	BNE	BACK	
	RTS		

This program can function only when one switch is active at a time. If two or more bits are active, the program finds the one nearest the PB7 input bit and ignores the lower bits. Thus, it is imperative that the subsequent control action silences the demand switch. There are some applications where several switches can be active at once, and all the in-

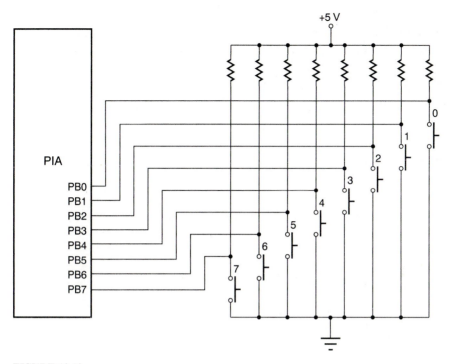

FIGURE 10.10
Interfacing independent switches to the PIA.

puts enter into the control decision process. A program to store all active bit numbers in memory would replace the "identify" section in Example 10.2 with:

```
            LDX     #BIT7
            LDA B   #07
IDENT       ROL A                    Identify all active switches.
            BCC     STOR
            DEC B
            BEQ     CONTNU
            DEX
            BRA     IDENT
STOR        STA B   0,X
            DEC B
            DEX
            BRA     IDENT
CONTNU      —
            —
```

A block of 8 bytes BIT0 to BIT7 is set aside in memory for the switches. A roll accumulator to the left starts the search with the switch at PB7. Each time the carry bit is clear, the roll count or switch number is stored at an address BITN pointed to by the index register. Any inactive switch will leave the original zero in its proper location.

10.4 ENCODING SWITCHES

When multiple switches are to be processed in addition to other inputs, the required number of PIA ports can be excessive. For example, a given design might require a 12-bit analog-to-digital converter plus 16 switches for three PIAs. A cheaper and more efficient design can be made from an encoder. An *encoder* is an IC that takes many inputs (switches here) and binary codes them into a reduced number of outputs.

The 74148 8-to-3 priority encoder and the 74147 10-to-4 priority encoder are good candidates. Priority encoders have an advantage over regular encoders (simply made from OR gates and not sold as an individual IC) by ranking the incoming requests. For the 74148, input 7 has the highest priority. If it is LOW, all other requests are ignored and the output is a binary 0. Input 6 has the next priority. If it is LOW, all requests but 7 are ignored and the output is binary 1. The ranking continues down to input 0, which will be recognized only if all other inputs are inactive (HIGH). The priority scheme is also ideal for ranking interrupt requests for which the priority encoder is intended.

A wiring diagram for a 74148 encoder is shown in Fig. 10.11. A 3-bit binary code for the input is loaded directly into the accumulator. There is no need for the switch identification section of the code. Another advantage is the GS strobe. This line goes LOW when any switch is active. Thus, it serves as an interrupt request if the microprocessor must be dedicated to other tasks rather than polling.

10.5 KEYPADS

The linear arrangement of switches (keys) without encoders is a disadvantage for a large number of keys because there is an input for each key. A better arrangement is the matrix organization for keypads. A *keypad* is a keyboard with 16 or fewer keys. Figure 10.12 shows a microwave oven "touch" keypad and a 4 × 4 key matrix. It now takes a single PIA port instead of two ports for 16 separate keys. Larger keyboards would have encoders for the key matrix.

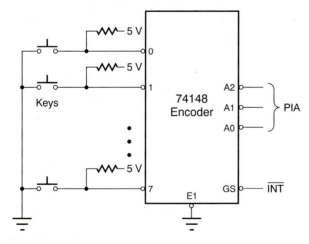

FIGURE 10.11
Encoding switch inputs with a 74148 priority encoder.

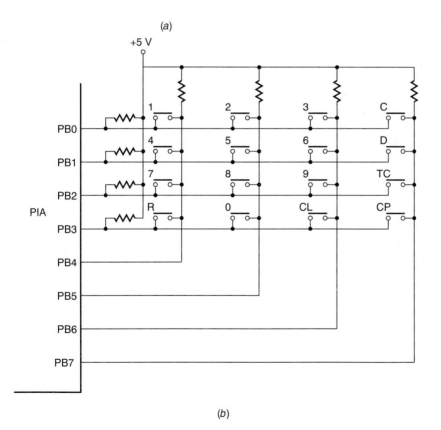

1	2	3	COOK
4	5	6	DEFROST
7	8	9	TEMP CONTROL
RESET	0	CLOCK	COOK PROGRAM

(a)

(b)

FIGURE 10.12
Keypad matrix input to PIA. (a) Microwave oven keypad; (b) key matrix circuit.

A key is characterized by its row and column number. The idea is to make the rows inputs and the columns outputs. When a particular column is grounded by a LOW (remaining columns HIGH), any key closed in that column will cause a LOW on its row. All inputs have a pull-up resistor so that all other rows are HIGH. Since the closed key grounds the row, two resistors in parallel are tied to row bit and column bit. Row input current I_{IL} is small. Therefore, the resistor is sized so that each carries less than one-half the TTL current I_{OL} to the column bit.

To find the pressed key, a LOW is placed on bit PB4, column 1, and a HIGH is placed on the remaining three columns. If no key in the column is closed, the four input bits will read HIGH. Proceed to a LOW on the second column. If the key on row 2 is closed, a LOW will appear on row 2; otherwise all inputs are HIGH. Successive LOWs are placed on each column until a LOW is found on one of the four rows.

Once a row LOW is found, the port is read. Remember, PIA bits programmed as an output can be read by the MPU. But it will read the last bits latched on a STA instruction, i.e., the four column bits producing the row LOW. The input nibble and the output nibble combine to form the port byte. In Fig. 10.12(b) key 5 closure (second row, second column) would read 1101 1101. All key closure readings are given in Table 10.1 along with their hexadecimal code.

Port readings are now stored in an MPU accumulator. We know to which key it is pointing by looking at the table. But the program doesn't know! The program identifies the key by its position in the table. To find this position we implement a look-up table. The complete program would contain the following steps:

1. Place the look-up table bytes in memory starting at location TABLE.

2. Initialize the PIA. Set PB0–BP3 as input bits and PB4–PB7 as output bits.

3. Send LOWs on all outputs (columns). Read the inputs (rows). Loop until a LOW input is found indicating a pressed key.

TABLE 10.1
Key closure reading on port B

Key	Port B PB7–PB0	Hex
1	1110 1110	EE
2	1101 1110	DE
3	1011 1110	BE
C	0111 1110	7E
4	1110 1101	ED
5	1101 1101	DD
6	1011 1101	BD
D	0111 1101	7D
7	1110 1011	EB
8	1101 1011	DB
9	1011 1011	BB
TC	0111 1011	7B
R	1110 0111	E7
O	1101 0111	D7
CL	1011 0111	B7
CP	0111 0111	77

4. Jump to the debounce delay. Repeat step 3 to be certain a key has been pressed.

5. Start by sending a LOW to column 1. Repeat for successive columns. After each LOW check the input for any LOW bit. When a LOW is found, we have the byte for the pressed key.

6. Search the table for the numerical position of the port byte. This number represents the key in subsequent program operations. Store the position in memory location KEY.

Example 10.3. Write a program to read a matrix keypad by using the previous steps.

```
KEY      —
TABLE    EE, DE, BE, 7E
         ED, DD, BD, 7D
         EB, DB, BB, 7B
         E7, D7, B7, 77

INITPIA  LDA B    #F0
         STA B    6002
         LDA B    #04
         STA B    6003

LOOP     LDA A    #F0
         STA A    6002
         STA A    KEY        Find any pressed key.
         LDA A    6002
         CMP A    KEY
         BEQ      LOOP
         JSR      DELAY
         LDA A    #F0
         STA A    6002
         STA A    KEY        Be certain the key is
         LDA A    6002       pressed.
         CMP A    KEY
         BEQ      LOOP

BACK     LDA A    #FE
NEXT     STA A    KEY        Send out LOWS on
         LDA A    6002       successive columns
         CMP A    KEY        until the key is found.
         BNE      FOUND
         ROL A
         BCS      NEXT
         BRA      BACK

FOUND    LDX      #TABLE
         CLR B
LOOK     CMP A    0,X        Find the key position in the
         BEQ      DONE       look-up table and
         INC B               store at KEY.
         INX
         BRA      LOOK
DONE     STA A    KEY
```

Rollover

In some situations typing, e.g., is very desirable to detect only a single key pressed, followed by its release. *Rollover* is the problem caused when more than one key is pressed at the same time, generating wrong codes.

The priority encoder will solve the problem since the 3-bit output code represents the highest priority key. However, it is not functional if the first key pressed takes precedence.

Two methods are used to prevent these problems inherent in rollover: two-key rollover and *N*-key rollover. In the two-key-rollover scheme the keypad (or keyboard) is ignored until the pressed key is processed. If a second key is pressed before the first key is released, it will be recognized only after the first key is released. *N*-key rollover handles the problem of more than two keys pressed at the same time.

Although lengthy software can be written to circumvent rollover, hardware is generally the accepted means. The MM74C923 keyboard encoder is available for two-key-rollover protection. This IC handles up to 20 SPST keys. Additional circuitry eliminates key bounce. In addition, a "key pressed" signal can be used to interrupt the microprocessor and to free it from the key detect polling task.

10.6 DISPLAYS

An integral part of any microprocessor-based system is light indicators to give the system status. Indicators can be single ON–OFF lights (lamps) or alphanumeric displays. A wide range of technologies is used to implement and control these devices. Table 10.2 lists four types of indicators and their traits.

Neon depends upon a high voltage-low current discharge through a gas, with gas ionization providing the light. The indicators are not common and are expensive, but they can be powered directly from line voltage. The light comes only in red wavelengths.

Incandescent is powered over a wide range of voltages but demands the highest currents (nominally 80 to 200 mA). The light is white so that lenses can form any desired light color. Some incandescent displays are manufactured which equal LEDs in voltage and current but are five times the cost. Their chief advantage is brightness.

Liquid crystal displays (LCDs) work on the principle of light polarization so that they reflect light and produce none of their own. LCDs require an AC voltage but no current, which makes them a natural for CMOS logic. Cost is approximately 10 times LEDs.

Light emitting diodes (LEDs) are by far the most common indicator with microprocessor-based systems. They are cheap and power compatible with TTL logic ICs. Low-current demand makes them a candidate, along with LCDs, for battery-powered systems.

TABLE 10.2
Comparison of light indicators

	Neon	Incandescent	LCD	LED
Voltage (V)	≈ 120 AC/DC	1–30 DC	12–30 AC	1.5–3
Current (mA)	0.5–3	> 20	None	5–25
Brightness (nf)	10^2	10^4	Reflected	10^2
Color	Red	White	Reflected	Red, yellow, green
Life (h)	10^3	10^4	10^5	10^5

10.7   LIGHT-EMITTING DIODES (LEDs)

For diodes made with certain materials, a significant fraction of the electrical power input is converted into light energy centered around the visible spectrum. LEDs can be designed to emit light from ultraviolet to infrared. The most common and most efficient LEDs emit a red light using a gallium arsenide phosphide (GaAsP) chip. Yellow (GaAsP on GaP) and green (GaP) light is also available.

The construction of LEDs is illustrated in Fig. 10.13. A small chip is epoxied to the cathode pin, also making electrical contact with the anode pin. It is encapsulated at the bottom of a plastic cone-shaped segment, which diffuses and amplifies the light. A tinted glass epoxy fills the cone to give a uniform appearance to the diffused light. The symbol for an LED is also shown.

Single-lamp LEDs are found in a wide variety of sizes and shapes. Numeric and alphabetic displays come in two basic types as shown in Fig. 10.14: segmented and dot matrix. There are several physical arrangements of the display elements, referred to as fonts. See Fig. 10.15. The 7-segment display is by far the most common display, and it is easy to utilize from an electrical standpoint. However, it is limited to numeric and a small number of alphabetic characters (includes hexadecimal characters). The 16-segment display can duplicate all alphabet letters but has not gained any acceptance in the field. The 5 × 7 dot matrix display is the choice when a full alphanumeric capability is required; however, it involves a more complex circuitry and/or software. A modified 4 × 7 dot matrix breaks up the segments of the 7-segment display and gives a more pleasing shape to the hexadecimal letters.

10.8   LED EQUIVALENT CIRCUIT

The DC forward current through each segment or dot LED is limited to a maximum DC rating, typically 20 mA per segment or dot. However, the maximum rating should be

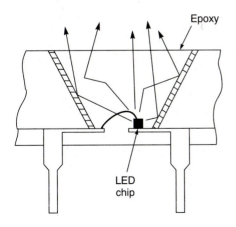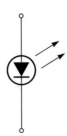

FIGURE 10.13
LED construction and symbol.

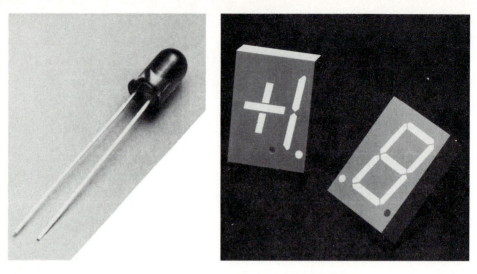

FIGURE 10.14
LED types. (*a*) Lamp; (*b*) displays. (*Reproduced with permission of Hewlett Packard Co.*)

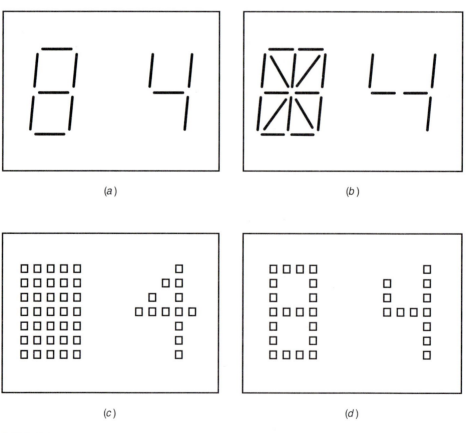

(a)

(b)

(c)

(d)

FIGURE 10.15
Physical arrangement of LED display elements. (*a*) 7-segment; (*b*) 16-segment; (*c*) 5 × 7 dot matrix; (*d*) modified 4 × 7 dot matrix.

verified from the manufacturer's literature on optoelectronic devices. Several ratings are reported here:

Hewlett Packard 5082-76XX	7-segment displays: 20 mA
Hewlett Packard 3530-3730	7-segment displays: 30 mA
Texas Instruments TIL 302-304	7-segment displays: 30 mA
Texas Instruments TIL 305	dot matrix: 10 mA

The displays are visible at current levels much less than the maximum rating. For a 20-mA maximum rating with the observer standing at 5 m, the values in Table 10.3 are suggested as a starting point in the design. While the current sinking ability of TTL falls within the range, it is standard practice to use drivers which can source or sink much higher currents than the DC rating (see Sec. 10.13). An advantage of drivers is that the single IC has 8-bit outputs. Regardless, a current limiting resistor is required for each segment or dot to achieve the design current. It is placed between the driver and display. The magnitude of the resistor depends upon the LED equivalent circuit.

The current-to-voltage behavior of an LED is given in Fig. 10.16. As with all diodes, a "turn-ON" voltage must be reached before there is a current. Above this value

TABLE 10.3
Design currents for seven-segment displays

Lighting	High-efficiency red	Yellow	Green or standard red
Dim (home)	4 mA	6 mA	8 mA
Moderate (office)	8	10	12
Bright (outdoors)	10	16	20

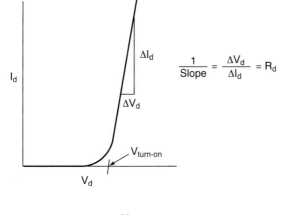

$$\frac{1}{\text{Slope}} = \frac{\Delta V_d}{\Delta I_d} = R_d$$

FIGURE 10.16
LED equivalent circuit.

TABLE 10.4
Equivalent circuit parameters

Color	R_d(min)	R_d(typical)	R_d(max) Ω	V(turn-ON) (V)
Standard red	—	3	7	1.55
High-efficiency red	17	21	33	1.55
Yellow	15	25	37	1.60
Green	12	19	29	1.65

the diode shows a small resistance. An equivalent circuit consists of a turn-ON voltage battery and resistor. Therefore, the diode voltage drop is

$$V_d = V_{\text{turn-ON}} + R_d I_d \tag{10.1}$$

Some variation can occur in the parameters. Table 10.4 gives the turn-ON voltage and resistance range found from lot to lot. The typical resistance is used in Eq. (10.1).

Example 10.4. Determine the limiting resistor when interfacing a 5082–7650 high-efficiency red display to the 74373 latch in Fig. 10.17.
The 74373 latch can sink 24 mA and maintain a logic LOW. However, the display must be limited to 20 mA:

$$5 \text{ V} - (V_{\text{OL}})_{\text{max}} = V_{\text{turn-ON}} + (R_d + R)I_d$$
$$5 \text{ V} - 0.4 = 1.55 \text{ V} + (21 + R)(20 \text{ mA})$$
$$R \approx 130 \text{ } \Omega$$

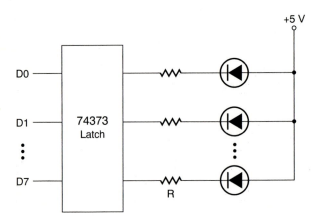

FIGURE 10.17
Using current-limiting resistors with LEDs.

10.9 LED 7-SEGMENT DISPLAYS

Seven-segment LED displays (cost: 50¢ to $2.00) can be ordered in two versions: common anode and common cathode. In Fig. 10.18 all common anode segments have a common 5-V supply connected to the anode or input of the diode. A particular LED is turned on by a logic LOW on the driver bit. Common cathode circuits tie each diode cathode to a common ground. The driver now supplies a source current through a current limiting resistor to each diode anode. The LED is activated by a logic HIGH on the driver bit. Both versions are offered to give the designer some selection for his or her circuits. However, common anode is usually the choice because current sinks in the appropriate current range are readily available with TTL.

Pin-Out

Hewlett Packard offers a series of high-efficiency red, yellow, and green 7-segment LEDs. From Fig. 10.19 we see that each series comes in the anode or cathode version. In addition, the displays are further broken down into three fonts: left-hand decimal (package A), right-hand decimal (packages B and C), and overflow ∓ 1 (package D). There is a choice of three numeral sizes: 0.3, 0.43, and 0.8 in.

As we can see, each segment has a letter assigned to it; start at the top segment, a, and proceed alphabetically in clockwise manner. This lettering is universal among all manufacturers for most 7-segment displays (some differences occur for the overflow display). Of the 14 pins on the DIP display, several are not used (or even missing). The common anode for a 5-V supply is connected to 2 pins, similarly for the common cathode ground. Figure 10.20 shows the pin-outs for each of the four packages. Pin-outs vary among manufacturers.

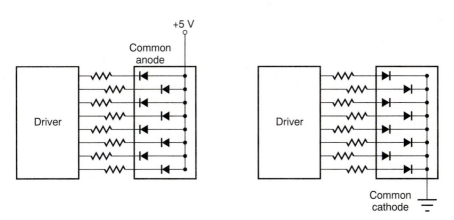

FIGURE 10.18
Common anode and common cathode versions of 7-segment LEDs.

Devices

Red 5082-	AlGaAs[1] Red HDSP-	HER[1] 5082-	Yellow 5082-	Green HDSP-	Description	Package Drawing
7730		7610	7620	3600	7.6 mm common anode left-hand decimal	A
7731		7611	7621	3601	7.6 mm common anode right-hand decimal	B
7740		7613	7623	3603	7.6 mm common cathode right-hand decimal	C
7736		7616	7626	3606	7.6 mm universal ±1. Overflow right-hand decimal[2]	D
7750	E150	7650	7660	4600	10.9 mm common anode left-hand decimal	E
7751	E151	7651	7661	4601	10.9 mm common anode right-hand decimal	F
7760	E153	7653	7663	4603	10.9 mm common cathode right-hand decimal	G
7756	E156	7656	7666	4606	10.9 mm universal ±1. Overflow right-hand decimal[2]	H

NOTES:
1. These displays are recommended for high ambient light operation. Please refer to the HDSP-E10X AlGaAs and HDSP-335X HER data sheet for low current operation.
2. Universal pin-out brings the anode and cathode of each segment's LED out to seperate pins. See internal diagram H.

(a)

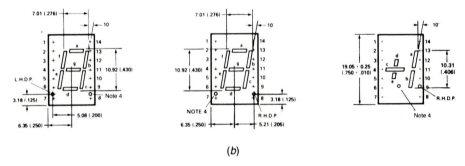

(b)

FIGURE 10.19
Hewlett Packard catalog listing with 7-segment LED display variations. (*a*) Devices; (*b*) packages. (*Reproduced with permission of Hewlett Packard Co.*)

Interfacing a 7-Segment Display

Seven-segment LED displays require a certain set of 7 bits to be applied to the pins for each character formed. The hexadecimal number representing those bits is called the code. A typical microprocessor system will produce output information in the form of BCD data. Thus, three functions need to be performed:

1. *Latch* the data from the bus.
2. *Decode* the BCD data for the display character.
3. *Drive* the display.

A display user can purchase one, some, or all the functions in a single IC. In addition, decoding can be accomplished with software (look-up table). Price, availability, system architecture, and power limitations must be taken into account when deciding how to

Internal Circuit Diagram

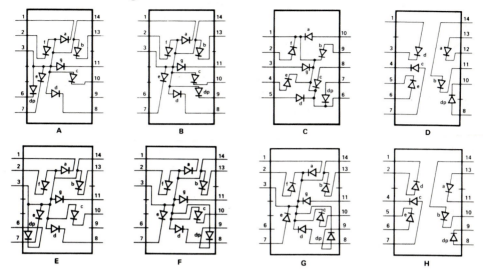

	Function			
Pin	A	B	C	D
1	Cathode-a	Cathode-a	No pin	Anode-d
2	Cathode-f	Cathode-f	Cathode[6]	No pin
3	Anode[3]	Anode[3]	Anode-f	Cathode-d
4	No pin	No pin	Anode-g	Cathode-c
5	No pin	No pin	Anode-e	Cathode-e
6	Cathode-dp	No conn.[5]	Anode-d	Anode-e
7	Cathode-d	Cathode-e	No pin	Anode-c
8	Cathode-d	Cathode-d	No pin	Anode-dp
9	No conn.[5]	Cathode-dp	Cathode[6]	No pin
10	Cathode-c	Cathode-c	Anode-dp	Cathode-dp
11	Cathode-g	Cathode-g	Anode-c	Cathode-b
12	No pin	No pin	Anode-b	Cathode-a
13	Cathode-b	Cathode-b	Anode-a	Anode-a
14	Anode[3]	Anode[3]	No pin	Anode-b

	Function			
Pin	E	F	G	H
1	Cathode-a	Cathode-a	Anode-a	Cathode-d
2	Cathode-f	Cathode-f	Anode-f	Anode-d
3	Anode[3]	Anode[3]	Cathode[6]	No pin
4	No pin	No pin	No pin	Cathode-c
5	No pin	No pin	No pin	Cathode-e
6	Cathode-dp	No conn.[5]	No conn.[5]	Anode-e
7	Cathode-e	Cathode-e	Anode-e	Anode-c
8	Cathode-d	Cathode-d	Anode-d	Anode-dp
9	No conn.[5]	Cathode-dp	Anode-dp	Cathode-dp
10	Cathode-c	Cathode-c	Anode-c	Cathode-b
11	Cathode-g	Cathode-g	Anode-g	Cathode-a
12	No pin	No pin	No pin	No pin
13	Cathode-b	Cathode-b	Anode-b	Anode-a
14	Anode[3]	Anode[3]	Cathode[6]	Anode-b

FIGURE 10.20

Pin assignments for Hewlett Packard 7-segment LED displays. (*Reproduced with permission of Hewlett Packard Co.*)

design the interfacing subsystem. Three basic subsystem approaches are explained in the next three sections.

10.10   INTERFACING A 7-SEGMENT DISPLAY WITH SOFTWARE DECODING

A subsystem for interfacing a 7-segment display without a decoder is presented in Fig. 10.21. The correct code for each character must appear on the data bus. Since we are using a common anode display, a LOW on the pin will light the corresponding segment. For example, numeral 7 must have a LOW on pins PB0, PB1, and PB2. The code is 1111 1000 or F8.

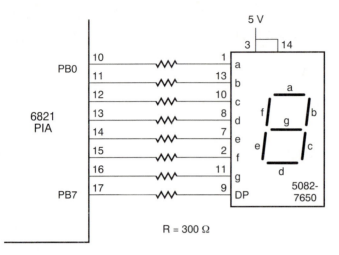

FIGURE 10.21
Driving an anode display using software decoding.

Data can be latched with a 74LS373/74LS374 octal latch or a 6821 PIA. The 74LS373/74LS374 latch (cost: $.80) can serve as its own driver with a sink current to 24 mA. However, the PIA (cost: $1.30) is used here. It can sink ~9 mA with $V_{pin} = 0.5$ V, which calls for 300 Ω resistors to limit the current to 9 mA. Pin voltage will rise to 1.25 V if the current is allowed to be 20 mA.

Since software will be used to generate data, each character must have its corresponding code stored in memory. Table 10.5 gives the code for all possible characters that a 7-segment display can form. Now the common anode requires LOWs for each lit segment while the common cathode requires HIGHs for each lit segment. Thus, the two codes for a given character are complementary.

Example 10.5. Write a program to display two BCD numbers which have been packed into one byte and stored at memory address 0000. Two PIA ports will be used, and they are located at 6000 and 6002 in the memory map. Each is connected to a separate driver and common anode display. The program starts at address 0200. A look-up table for the display code starts at address 0300:

```
0200      LDA A    #FF
          STA A    6000        Initialize the PIA
          STA A    6002        with two output
          LDA A    #04         ports.
          STA A    6001
          STA A    6003

          LDA B    0000
          AND B    #0F         Output least
          STA B    BLANK1      significant nibble
          LDX      #0300       to second display.
          LDA A    0,X
```

TABLE 10.5
Hexadecimal code for the 7-segment display

Display	Common anode	Common cathode
0	C0	3F
1	F9	06
2	A4	5B
3	B0	4F
4	99	66
5	92	6D
6	82	7D
7	F8	07
8	80	7F
9	98	67
A	88	77
C	C6	39
E	86	79
F	8E	71
G	82	70
H	89	76
I	F9	06
J	E1	1E
L	C7	38
O	C0	3F
P	8C	73
U	C1	3E
Y	99	66
b	83	7C
c	A7	58
d	A1	5E
h	8B	74
n	AB	54
o	A3	5C
r	AF	50
u	E3	1C
–	BF	40
?	AC	53
.	7F	80
Blank	FF	00

```
BLANK1   —
         STA A    6002

         LDA B    0000
         LSR B            Output most
         LSR B            significant nibble
         LSR B            to first display.
         LSR B
         STA B    BLANK2
         LDX      #0300
         LDA A    0,X
```

```
BLANK2    —
          STA A    6000
          WA1
```

The look-up table shows

```
0300   C0
0301   F9
0302   A4
   ·
   ·
   ·
```

Example 10.6. Write a program to flash the message: HELLO. Use a single common anode display, and flash each character in sequence for one second followed by a one-second blank display after each character. Store the characters starting at address 0000.

```
0000      89, FF, 86, FF
0004      C7, FF, C7, FF
0008      CD, FF

START     LDA A   #FF          Initialize the
          STA A   6002         single PIA
          LDA A   #04          output port.
          STA A   6003

          CLR B
LOOP      STA B   NUMBER
          LDX     #0000
          LDA A   0,X
NUMBER    —
          STA A   6002
          INC B
          JSR     DELAY
          BRA     LOOP
```

Delay Subroutine

```
DELAY     LDX     #F423
AGAIN     DEX                  Delay one second.
          BNE     AGAIN
          RTS
```

Example 10.7. When the seven-segment display is a common cathode, all segment outputs share a common ground and a voltage HIGH to a pin lights a particular segment. Open-collector buffers (7407) with pull-up resistors is one method to drive the display. Figure 10.22 shows another method using a 74LS244 buffer/driver which sources 15 mA at 5 V.

Find the magnitude of the seven current limiting resistors.

$$5 \text{ V} - V_{LED} = I_{LED}R$$

$$5 \text{ V} - [1.55 + I_{LED}(21)] = I_{LED}R \qquad \text{where } I_{LED} = 15 \text{ mA}$$

$$R = 209 \cong 200$$

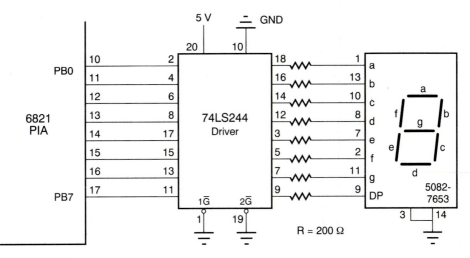

FIGURE 10.22
Driving a cathode display using software decoding.

10.11 INTERFACING A 7-SEGMENT DISPLAY WITH HARDWARE DECODING

Software decoding allows some versatility in programming, but it increases both the program length and execution time. Perhaps the simplest method of driving a 7-segment display is to have one decoder/driver for each digit. Decoder/drivers are single ICs which take a BCD logic number, decode it for the display, and provide ample current drive for the segments. They introduce some rigidity into the design since decoder/drivers can only decode for the 10 decimal numbers. There are many BCD to 7-segment decoder/drivers on the market. They are divided into common cathode drivers which source current and common anode drivers which sink current. Table 10.6 lists a few.

TABLE 10.6
BCD to 7-segment decoder/drivers

Device	Type	Current (mA)	Comment	Source
4039	CA	20		MOT
4511	CC	25	CMOS with latch	MOT
7447	CA	40		MOT, TI
7448	CC	4		TI
74247	CA	40		TI
74C48	CC	50	CMOS	NS
8857	CC	60		NS
9368	CC	22	Constant current with latch	FAIR
9370	CA	25	With latch	FAIR

MOT = Motorola; NS = National Semiconductor;
TI = Texas Instruments; FAIR = Fairchild.

The 7447 Decoder/Driver

The most popular decoder/driver for 7-segment displays is the 7447 (cost: $.90). The pinout and the function table are shown in Fig. 10.23. It receives a BCD number at inputs ABCD. The seven outputs, a to g, are to be connected directly to the common anode segments, a to g. Current limiting resistors must be placed between each display segment and the 7447 output. The function table uses the word "ON" to denote that the open-collector output is on. However, the output bit is active LOW. For example, to illuminate the number 0, BCD input DCBA is 0000, which generates the 7-bit output 0000001. This binary number lights all segments but g. If a number greater than 9 is applied to the BCD input, an unintelligible symbol appears on the LED display. The number 15 (FF) will blank all segments.

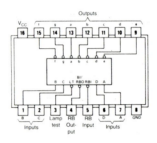

Decimal or function	Inputs						BI/RBO	Outputs							NOTE
	LT	RBI	D	C	B	A		a	b	c	d	e	f	g	
0	H	H	L	L	L	L	H	ON	ON	ON	ON	ON	ON	OFF	
1	H	X	L	L	L	H	H	OFF	ON	ON	OFF	OFF	OFF	OFF	
2	H	X	L	L	H	L	H	ON	ON	OFF	ON	ON	OFF	ON	
3	H	X	L	L	H	H	H	ON	ON	ON	ON	OFF	OFF	ON	
4	H	X	L	H	L	L	H	OFF	ON	ON	OFF	OFF	ON	ON	
5	H	X	L	H	L	H	H	ON	OFF	ON	ON	OFF	ON	ON	
6	H	X	L	H	H	L	H	OFF	OFF	ON	ON	ON	ON	ON	
7	H	X	L	H	H	H	H	ON	ON	ON	OFF	OFF	OFF	OFF	
8	H	X	H	L	L	L	H	ON	ON	ON	ON	ON	ON	ON	1
9	H	X	H	L	L	H	H	ON	ON	ON	OFF	OFF	ON	ON	
10	H	X	H	L	H	L	H	OFF	OFF	OFF	ON	ON	OFF	ON	
11	H	X	H	L	H	H	H	OFF	OFF	ON	ON	OFF	OFF	ON	
12	H	X	H	H	L	L	H	OFF	ON	OFF	OFF	OFF	ON	ON	
13	H	X	H	H	L	H	H	ON	OFF	OFF	ON	OFF	ON	ON	
14	H	X	H	H	H	L	H	OFF	OFF	OFF	ON	ON	ON	ON	
15	H	X	H	H	H	H	H	OFF	OFF	OFF	OFF	OFF	OFF	OFF	
BI	X	X	X	X	X	X	L	OFF	OFF	OFF	OFF	OFF	OFF	OFF	2
RBI	H	L	L	L	L	L	L	OFF	OFF	OFF	OFF	OFF	OFF	OFF	3
LT	L	X	X	X	X	X	H	ON	ON	ON	ON	ON	ON	ON	4

H = high level, L = low level, X = irrelevant

FIGURE 10.23
Pin assignments and function table for the 7447 BCD to a 7-segment display driver. (*Reprinted by permission of Texas Instruments.*)

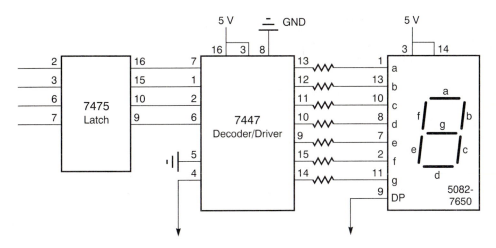

FIGURE 10.24
Using a 7447 decoder/driver to interface a 7-segment display.

A BCD nibble on the data bus can be latched with a 7475 quad latch (cost: $.50). Alternatives are the 74LS373/74LS374 octal latch and the 6821 PIA. These ICs can latch two and four BCD numbers, respectively. A circuit with the 7475 quad latch is shown in Fig. 10.24. The programmer merely places a packed BCD byte on the bus for each pair of displays. Two latches are enabled for the same address. Decimal points, DPs, must be latched separately if they are needed.

Ripple Blanking

Ripple blanking is a function of decoder/drivers which allows them to blank out the leading 0s. For example, if the number 20 is the output for a four-digit display, it is rather distracting to see 0020. We wish to blank out the first two zeros and to leave the last zero in place. Notice that the 7447 has a ripple blanking input (RBI) and a ripple blanking output (RBO). If the RBI is LOW and the inputs ABCD are all LOW, the display is blanked. When the BCD input to the 7447 is not zero, the RBO is HIGH regardless of RBI. Figure 10.25 shows how the scheme works. The first display has the RBI *grounded* and its RBO is connected to the RBI of the next display. A zero will blank the first display and the RBO will be LOW. This RBO will drive the RBI of the next stage. The first nonzero digit will propagate an RBO HIGH through the remaining stages and they will not be blanked regardless of the digit.

10.12   LED DISPLAYS WITH LOGIC

An alternative approach to the discrete component 7-segment display subsystem is an on-board IC display (cost: $15.00). This IC contains all the subsystem functions on the display: latch, decoder, driver. They are referred to as displays with logic. The IC cost is high

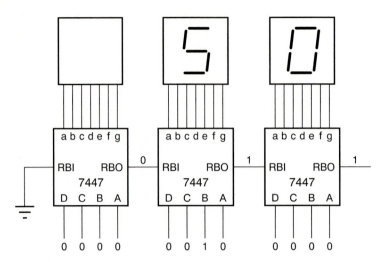

FIGURE 10.25
Ripple blanking with the 7447 decoder/driver.

compared to the total cost of the individual function ICs. However, there are several *advantages* to using displays with logic:

1. Reduced assembly time and cost
2. Reduced space requirements
3. Reduced power consumption
4. Increased reliability

The Texas Instruments TIL308 and TIL309 series of 7-segment displays are shown in Figure 10.26. The 16-pin device is available with right- or left-hand decimal point. BCD data and decimal points are latched into the IC by the latch strobe input pin. When the strobe is LOW, the data in the latches follow the data on the inputs. When HIGH, the data in the latches is held constant and is unaffected by new data on the inputs. BCD data stored in the latches is available at the latch output pins. A LOW on the blanking input will blank (turn OFF) the entire display. Thus, it must be HIGH for normal operation. When the LED test input is LOW, it will turn ON the entire display. The test input must be HIGH for normal operation. In addition to the BCD code for numerical displays, a binary 1011 will generate a minus sign.

The TIL306 and TIL307 series of 7-segment numeric displays with logic do not have a BCD 4-bit input but are driven by a clock pulse. A clear input resets the count to zero. Each positive-going transition of the clock input will increment the count. A maximum count output will go LOW when the digit is at 9. It will return HIGH when the digit changes to zero. Therefore, the maximum count output serves as the clock input for the next display.

The TIL311 is designed to display decimal and hexadecimal data. It accepts 4-bit data corresponding to the character and displays it on a 4 × 7 dot matrix; the characters are shown in Fig. 10.27. Pin-out is identical to the BCD 7-segment display with logic.

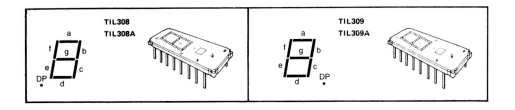

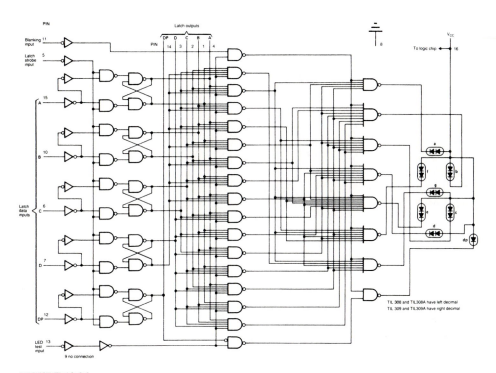

FIGURE 10.26

The TIL308/309 series of 7-segment displays with logic (BCD input). (*Reprinted by permission of Texas Instruments.*)

10.13 MULTIPLEXING 7-SEGMENT DISPLAYS

Each display, driven by a decoder/driver with latched data, is economical up to approximately four displays. However, the number of ports and drivers increases in direct proportion to the number of displays. It has become common practice to multiplex a large number of displays where a single decoder/driver is used for all displays. In multiplexed displays each display is only driven for a portion of the time, but the eye perceives the character as though it were on constantly. Multiplexing is not only efficient in terms of drive circuitry, it is the most efficient way of lighting an LED. For a given average power dissipation, the light output is much greater when pulsed with a higher current. Average current levels equal to 70 percent of the DC current level will produce the same lighting

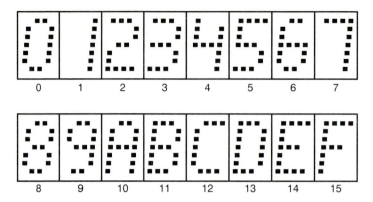

FIGURE 10.27
Characters for the TIL 311 hexadecimal display with logic.

effect. A disadvantage of multiplexing is that the data can no longer be latched and forgotten; the microprocessor must be more dedicated.

Six-digit multiplexed displays are illustrated in Fig. 10.28. The displays and driver are the common cathode type. The BCD data is latched at port A, and the character code is presented to all displays. The decoder/driver is a NS8857 which will source up to 60 mA. An open-collector buffer (7407) drives the decimal point. Each display has its common cathode tied to ground through a transistor, called a digit driver. The display cannot light unless its corresponding transistor is turned ON. The transistors are con-

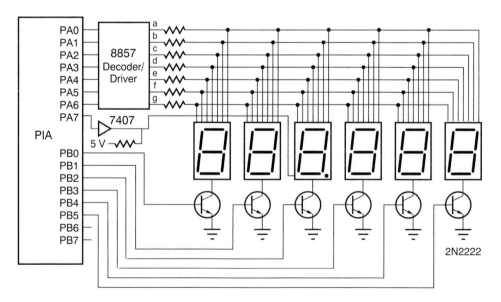

FIGURE 10.28
Circuit for multiplexing six 7-segment displays.

trolled by the contents of port B. Only one transistor at a time is turned ON, the one whose character code is on the driver output.

Each display is turned ON, or refreshed, at a frequency called the refresh rate. The refresh rate should produce a flicker-free character. A minimum practical refresh rate is about 100 Hz. To optimize the overall performance of a 7-segment display, the refresh rate should be at least 1 kHz. Given N displays refreshed at f Hz, the maximum ON time t_0 for each character is

$$t_o = \frac{1}{Nf} \tag{10.3}$$

Since the ON time is cycled continually from one display to the next, the ratio of time that the display is ON, called the duty factor, is simply $1/N$.

Suppose we select an LED color and lighting conditions from Table 10.3 such that the DC current should be 10 mA. A more efficient multiplexed display will need an average current of 7 mA. Because the display is ON for one-sixth the time, the peak current (ON time current) is 42 mA. All catalogs for LED displays will list a maximum peak current allowable. For the HP 5082-76XX series, the rating is 60 mA. Each general-purpose transistor must be selected to sink the peak current for a maximum of eight LEDs (7-segments plus a decimal point) or 336 mA. If 1 kHz is the refresh rate chosen, the ON time for each display is 0.167 ms. Any concern for the LED response time is unfounded. It is ~0.1 μs. Current limiting resistors R can be found from

$$V_{OH} - V_{LED} - V_{TR} = 0 \tag{10.4}$$

or

$$V_{OH} - [1.55\ V - I_{LED}(21 + R)] - 0.2\ V = 0$$

$$5\ V - [1.55\ V - 42\ mA(21 + R)] - 0.2\ V = 0$$

$$R = 56\ \Omega$$

Peak currents for multiplexed displays exceed the average allowable currents. If a display is left ON for an extended time, it will be damaged. Care should be exercised when debugging the design. In the event of some system failure the displays can be protected by detecting an extended transistor ON time and interrupting the microprocessor.

Example 10.8. Write a program to multiplex six 7-segment displays. The digits are stored in memory starting at address DIGIT. They are loaded in order, starting with the least significant digit and proceeding to the most significant digit. The segment-driven address at port A is 6000. The digit driver address at port B is 6002:

```
DIGIT     DB, DB, DB,
          DB, DB, DB

INITPIA   CLR A
          COMA                PIA initialized.
          STA A    6000       Port A = output
          STA A    6002       Port B = output
          LDA A    #04
          STA A    6001
          STA A    6003
```

```
ANEW    LDX     #DIGIT      Point to digit.
        LDA B   #20         Turn on driver.
        CLC
LOOP    STA B   6002

        LDA A   0,X         Send digit code.
        STA A   6000
        JSR     DELAY
        INX
        ROR B               Get new digit driver.
        CMP B   #01
        BNE     LOOP
        BRA     ANEW

DELAY   LDA A   #1B
COUNT   DEC A               Delay 0.167 ms.
        BNE     COUNT
        RTS
```

After the mandatory PIA initialization the subroutine starts by loading the index register with the address of the six character codes. It then loads accumulator B with binary 00100000. Notice the solo one is located N places to the left, where N is the number (6) of displays. After clearing the carry the binary number is output to port B, which turns ON the transistor for the least significant display. Its character code is the first byte stored at address DIGIT. Using the index mode, the byte is output to port A and held for 0.167 ms. The next digit is pointed to. Accumulator B is rolled (with carry) to the right, turning on the next transistor to the left. Its character code is output. The process continues until the solo one occupies the Nth place to the right in accumulator B.

Of course the program cycles continuously, but it can be written as a subroutine in which the six digits are refreshed once each pass through the main program. This method will force a change in the refresh rate, delay time, and peak current. A third method can rely on an external interrupt driven by a timer (see Chap. 14) for the refreshment cycle. In all methods the average LED current is constant.

10.14 DOT MATRIX DISPLAYS

Seven-segment displays are capable of a limited character set. By increasing the number of elements (dots), it is possible to achieve any two-dimensional character. In practice, the 5×7 dot matrix array gives satisfactory resolution for all alphanumeric characters. Figure 10.29 shows the LED array and pin-out for the Texas Instruments TIL305 red display. The maximum peak current rating is 100 mA vs. 10 mA for the maximum average current. The array consists of five columns, each connecting seven LED anodes in the column. Each row connects to the cathode of five LEDs. Thus, to turn ON a given LED, power is applied to its column and its row is grounded.

The generation of a character in a 5×7 array is more complex than forming a 7-segment character. With 7-segment displays the character is formed by one byte to the driver. In a 5×7 array each column of seven LEDs is similar to a 7-segment display and the dot matrix display takes on the appearance of five multiplexed 7-segment displays. Let us see how it works. Suppose we want to display the letter H as illustrated in Fig. 10.30.

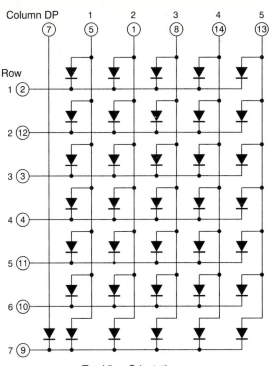

Column DP   1   2   3   4   5

Row

1

2

3

4

5

6

7

Top View Orientation

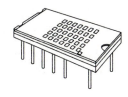

1FIGURE 10.29
Pin assignments for the TIL 305 5 × 7 dot matrix display. (*Reprinted by permission of Texas Instruments.*)

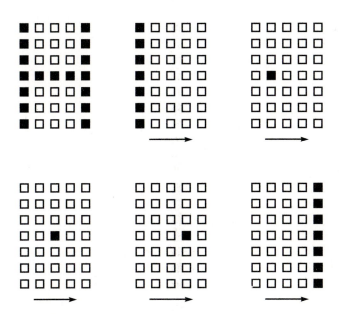

FIGURE 10.30
Strobing the columns of a 5 × 7 dot matrix display to form the letter H.

Each column must be turned ON or strobed in succession while the remaining columns are OFF. The task is accomplished with software just like multiplexed 7-segment displays, or we can use hardware such as shift registers. When column 1 is ON, all seven rows must be ON. For column 2, only row 4 is ON, etc., until column 5 is reached. The pattern is repeated continuously at the refresh rate, which is the same as for multiplexing 7-segment displays.

Each character is stored in memory as a block of 5 bytes. For example,

Character	Column No.	Row code
H	1	7F
	2	08
	3	08
	4	08
	5	7F

If the data is in the accumulator as an ASCII hexadecimal code, the code is used in a look-up table search for the starting address of the character's 5 bytes to be multiplexed in the array. The ASCII code would require 640 bytes of memory.

A circuit diagram for driving two TIL305 arrays is shown in Fig. 10.31. The column drivers are MPSA29 Darlington drivers (see Chap. 15) and row drivers are 2N3904 general-purpose transistors. Each array is addressed separately by ANDing the row byte with an enable signal from PB7. The data would be stored in memory with the PB7 bit zero, automatically addressing the second array. A software change would enable the first array. If the DC current per diode is 5 mA and there are 10 columns each time sharing for the two arrays, each diode must get 50-mA peak current during the multiplex ON time. Each column transistor must be able to switch current for up to seven LEDs or a 350-mA peak current, while each row transistor only needs to switch a 50-mA peak current. The current limiting resistor is found from the following equation:

$$V_{CC} - V_{LED} - IR - 2V_{CE} = 0 \tag{10.5}$$

Using the TIL305 and MSPA29 specifications yields

$$12 \text{ V} - 1.65 \text{ V} - (50 \text{ mA})R - 2(0.4) = 0$$

or
$$R = 192 \ \Omega$$

A program to display two characters in the circuit of Fig. 10.31 follows. Five bytes of data for the right-hand character are placed in memory at address 1CHAR. The left-hand character bytes start at 2CHAR. The routine is almost identical to the program for multiplexing 7-segment displays. The only difference between the two character loading operations is the ORA (OR accumulator) instruction for character 2. This operation places a one in bit 7 of port B to enable the left-hand array for each of its column strobes.

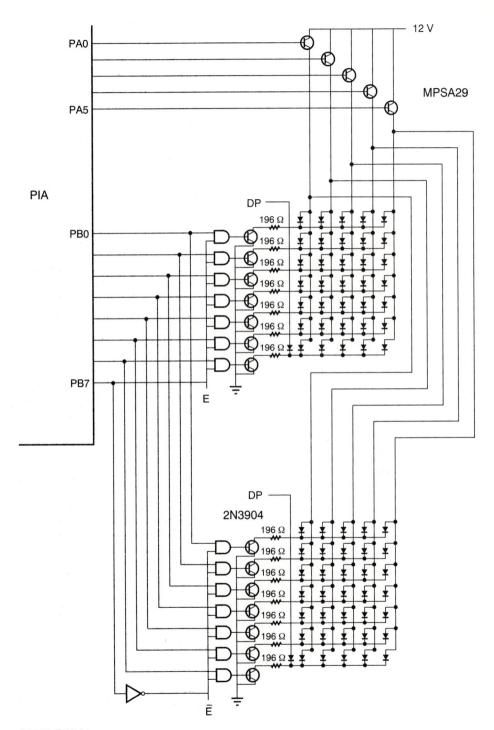

FIGURE 10.31
Circuit for multiplexing two 5 × 7 dot matrix displays.

Dot Matrix Program

```
1CHAR    DB, DB, DB, DB, DB
2CHAR    DB, DB, DB, DB, DB

INITPIA  CLR A
         COMA
         STA A   6000
         STA A   6002
         LDA A   #04
         STA A   6001
         STA A   6003

ANEW     LDX     #1CHAR
         LDA A   #10
         CLC
LOOP1    STA A   6000
         LDA B   0,X
         STA B   6002
         JSR     DELAY
         INX
         ROR A
         CMP A   #01
         BNE     LOOP1

         LDX     #2CHAR
         LDA A   #10
         CLC
LOOP2    STA A   6000
         LDA B   0,X
         ORA B   #80
         STA B   6002
         JSR     DELAY
         INX
         ROR A
         CMP A   #01
         BNE     LOOP2
         BRA     ANEW

DELAY    PSH A
         LDA A   #10       Delay 0.100 ms
COUNT    DEC A             for refresh rate = 1 kHz.
         BNE     COUNT
         PULA
         RTS
```

In those applications where program memory space is a premium the multiplexing of columns for multiple arrays can be handled with hardware. A 7496, 5-bit shift register is used to cycle the ON bit continually among the columns. An EPROM stores the 640-byte

FIGURE 10.32
The Hewlett Packard HDSP-2000 5 × 7 dot matrix LED alphanumeric display. (*Reproduced with permission of Hewlett Packard Co.*)

code for the ASCII characters. Circuitry details are in the Texas Instruments *Optoelectronics and Image-Sensor Data Book*.

The Hewlett Packard HDSP-2000 in Fig. 10.32 has four dot matrix arrays in one package. The column code data is entered serially with each HIGH-LOW transition of an external clock. On-board is a serial-in, parallel-out 7-bit shift register for each digit which controls the row drivers. Packages may be tied together for up to 80 digits.

10.15   LIQUID CRYSTAL DISPLAYS (LCDs)

Liquid crystal displays (LCDs) are ideally suited to designs which have a limited supply of power although they appear more frequently in instrument front panels. They are teamed with CMOS logic in battery-operated devices such as calculators, watches, and computer games. Figure 10.33 shows a Beckmann hand-held thermocouple calibrator with LCD readout. Battery life is 500 h.

LCDs do not create any light; instead, they reflect surrounding light for the front-lit version or transmit light from a back-lit version. The LCD is a sandwich of an organic compound between a clear glass plate with transparent indium-oxide electrodes in the shape of a 7-segment display and a back glass plate electrode. When the electrodes are activated by a voltage field, the long cylindrical organic molecules align themselves with the segment electrodes. The sandwich has a front vertical light polarizer and a back horizontal light polarizer followed by a reflector. Reflected light cannot pass through the aligned molecules and the activated segment appears dark.

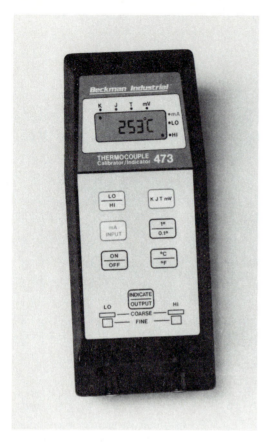

FIGURE 10.33
Beckman industrial digital thermocouple calibrator. (*Courtesy of Beckman Industrial.*)

LCDs are powered by low level AC voltages. A *segment driver* is shown in Fig. 10.34(*a*). It consists of a CMOS exclusive-OR gate whose output goes to the segment electrode. The oscillator or clock input cycles the exclusive-OR gate and the backplane electrode. Since the device functions as a capacitor, it is important not to cycle the signal at high frequency. The lower reactance will cause a current drain. If the frequency is too low, the display will flicker. Typical clock frequencies range from 40 to 60 Hz. The segment control input is similar to the a to g input of LED 7-segment displays. Each segment driver is commonly built into an LCD driver chip that comes with the display.

The LCD voltage level falls within the range of 3 to 12 V RMS (root mean square). To understand what this means, waveforms for the operation are illustrated in Fig. 10.34(*b*). The clock oscillates between supply V_{DD} and ground V_{SS}. When a logic 1 is applied to the control input, the voltage across the electrode sandwich varies from $+(V_{DD} - V_{SS})$ to $-(V_{DD} - V_{SS})$. Ground is typically zero, so the voltage is $\pm V_{DD}$. For a square wave, the RMS voltage is supply voltage. It is important that the DC offset voltage is less than 50 mV. LCDs last almost indefinitely under normal conditions but DC offset will rapidly deteriorate the material.

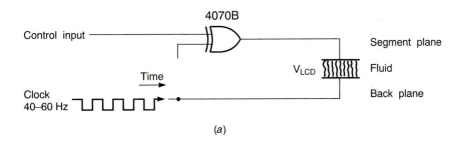

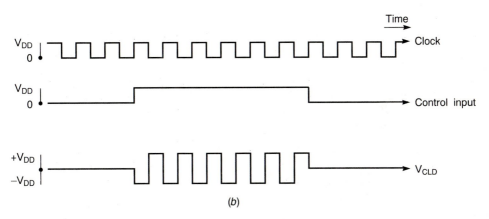

FIGURE 10.34
Driver circuit for an LCD segment. (*a*) Circuit; (*b*) timing ($V_{SS} = 0$).

Each LCD display can be driven by a CMOS BCD-latch/decoder driver such as the 4543 in Fig. 10.35. The BCD input is obtained directly from the bus and is latched by the enable E. A timer generates the clock signal. Displays are usually found four to eight per package. A 4543 driver can be used for each digit, but it is more convenient to drive the package with a four-digit LCD driver such as National Semiconductor's MM74C945 counter/latch/decoder driver.

Philips makes an assortment of 7-segment and 5×7 dot matrix LCD displays. The latter have up to 80 characters in one to four lines. The module receives ASCII characters from the MPU and latches the characters into its 80-character RAM. A ROM-character generator produces the dot matrix pattern. Also, on-board is the LCD clock. The only interface needed is shown in Fig. 10.36. Module initiation is completed with a series of four codes (38, 0F, 01, 06) each followed by a 1-μs or more pulse. The RS line should be LOW when the code is sent. A blinking cursor identifies the home position. ASCII characters are sent in succession with the RS line HIGH. The cursor shifts to the right after each transaction.

LCDs warrant consideration when power must be conserved and adequate background lighting is present. However, they do have several drawbacks. The response time is slow, on the order of 200 ms, so that segment images appear to fade out rather than snap out. Also, operating temperatures are limited at the low end to about 20°F. This precludes such applications as automotive use.

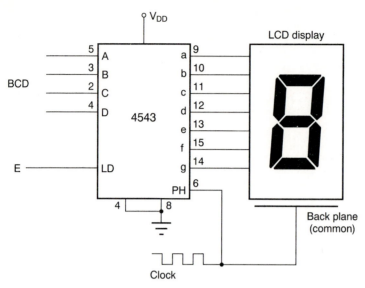

FIGURE 10.35
Driving a 7-segment LCD display with a CMOS BCD latch/decoder/driver.

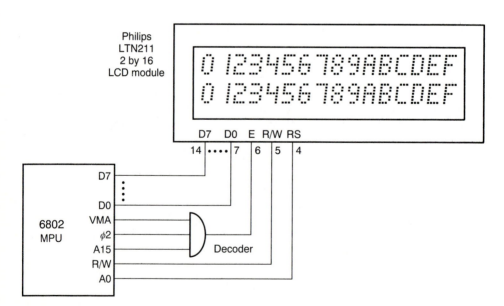

FIGURE 10.36
Driving a two-line multiple character 5 × 7 dot matrix LCD module.

PROBLEMS

10.1. Change the debounce routine in Example 10.1 to detect a closed switch HIGH.

10.2. A switch will bounce up to 10 times upon a break or make contact. Write a program to count the number of bounces in your switch.

10.3. You have decided to reduce your PIA inputs by using a priority encoder. Rewrite the program in Example 10.2 to identify a switch closure when using an encoder.

10.4. Write a program to identify a debounced switch when the PIA in Fig. 10.10 is configured for 16 inputs using two ports.

10.5. The program in Example 10.2 stores the active switch number in memory. Assume you have four switches with each requiring a service routine when active. Modify the program so that active switch N will jump to its service routine at 10N0.

10.6. An array of eight switches is to be examined to see if any one of them is set. Write a program which will check the array for this condition and will cause a lamp to flash if affirmative.

10.7. A 7-segment display is to show the characters (*a*) 8, (*b*) 6, (*c*) F. Verify the hexadecimal code for common cathode displays.

10.8. A circuit is to use four 7-segment displays to flash the message: HELP. Sketch the circuit and write a program to accomplish the task.

10.9. Redesign the hardware in Fig. 10.22 for a 20-mA current through the LEDs. Use a 74LS244 driver. What is the current limiting resistance?

10.10. The 7448 common cathode decoder/driver has the same pin-out as the 7447. The 7448 sources 4 mA, but it sinks 6 mA. If it is wired with a pull-up resistor (to 5 V) on each output line, the 7448 can be used to turn OFF a segment by sinking the current from a pull-up resistor while allowing the 5-V source to drive the segment when it is HIGH.
(*a*) Sketch a circuit for the 7448 in this arrangement.
(*b*) Find the maximum source current possible and the value of the pull-up resistor.

10.11. Find the value of the current limiting resistor in Fig. 10.25.

10.12. Each of four 7-segment displays is driven by a separate 7447 decoder/driver. The data is output from a single 6821 PIA.
(*a*) Show the circuit diagram.
(*b*) Write a program to output the number 9876.

10.13. Port A of the PIA interfaces an 8857 common cathode decoder/driver for a 7-segment display. Determine the value of the current limiting resistor if (*a*) a single display is used, (*b*) four displays are multiplexed. A desirable DC current is 10 mA.

10.14. A 6821 PIA is to be used to multiplex four common anode 7-segment displays. The average current is 4 mA.
(*a*) Select a decoder/driver. Show a circuit diagram.
(*b*) Write a program to implement.

10.15. In Example 10.8 the program to multiplex six 7-segment displays does nothing else. Suppose the program was only part of a larger program in which the microprocessor had other tasks to perform. Assume assembly language with 200 machine cycles precedes the multiplexing program and 150 cycles follow it before looping to the beginning of the main program. How would you change the multiplexing refresh rate, delay time, peak current, and limiting resistor?

CHAPTER
11

INTERFACING
STEP
MOTORS

In many instances a microprocessor must control the motion of an object. While a few cases are handled with solenoids, the universal actuator is a motor. Over 100 electric motors can be found in an average American home, including hair dryers, refrigerators, fans, dishwashers, clothes dryers, VCRs, cassette tape players, garbage disposals, etc. Many of these motors which are AC are ideal for applications that require constant speed with little variation in load. AC motors have some advantages over DC motors. They use line current and give more power per size. Control over AC motors is exercised by simply turning them on or off. However, designs often require control of position or speed under changing loads. Control motors for these tasks are invariably DC motors.

The most widely used control motor today is the step (stepper or stepping) motor. Step motors are hybrid DC motors with unique construction and functional features that distinguish them from their DC cousins. We will follow the nomenclature which refers to step motors as a separate category. Thus, any further reference to DC motors excludes step motors.

When comparing step motors and DC motors, there are certain advantages and disadvantages associated with their use in designs:

1. *Step motors are open loop.* They rotate a fixed number of degrees for each input pulse. Most DC control motors are operated closed loop. They require encoders to sense shaft position and often tachometers to sense velocity. This increases components and price.
2. *Step motors are easily controlled with microprocessors.* However, logic and drive electronics are more complex.

242

3. *Step motors are brushless.* Brushes contribute several problems. They wear and must be replaced. Brushes spark, which can be dangerous in flammable or explosive environs. Brushes cause electrical transients that can lead to glitches in the electronics.

4. *DC motors have a continuous displacement and can be accurately positioned.* Step motor motion is incremental and its resolution is limited to the step size.

5. *Step motors can slip if overloaded and the error can go undetected.* For this reason a few step motors use closed-loop control.

6. *Feedback control with DC motors gives a much faster response time compared to step motors.*

Because of improvements in permanent magnets and the development of large-scale integration (LSI) electronics, step motors can compete effectively with DC motors over most design areas except the high torque-speed range. More will be discussed on this topic in Sec. 12.9.

The first application of step motors was in 1935. Early models had poor performance and were inefficient. Great strides have been made in their performance over the last 10 years, paralleling development of the microprocessor. Today step motors can be found in computer peripherals, robotics, chart recorders, *x-y* plotters, pumps, clocks, drafting tables, valves, machine tools, medical equipment, automotive devices, small business machines, and scanners to name a few applications. Figure 11.1 shows the magnetic head

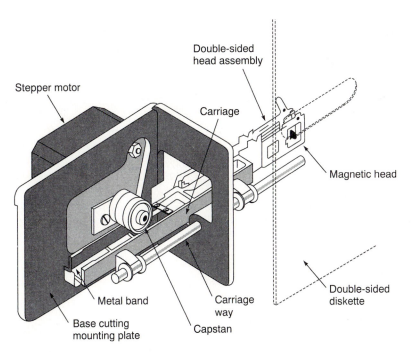

FIGURE 11.1
Head actuator for 8-in floppy disk drive. (*Design Engineering, May 1981.*)

actuator for a double-sided 8-in floppy disk drive. A flexible metal band translates stepper motion into a linear read/write head action. Track-to-track access time is 3 ms.

11.1 STEP MOTOR TYPE

Step motors are classified into three types according to their rotor structure: (1) variable reluctance, (2) permanent magnet, and (3) hybrid. In addition, these types can fall into one or both of two construction methods. In method I, Fig. 11.2(*a*), the rotor has regular projections or teeth. The stator or stationary part has similar projections which hold the windings. In method II, Fig. 11.2(*b*), rotor and stator end projections are more teethlike and more numerous. The advantage of more teeth is smaller step angles.

Variable Reluctance

Variable reluctance step motors have a *soft iron* rotor which rotates when teeth on the rotor are attracted to the electromagnetic stator teeth. This action is similar to a solenoid. Iron rotors have lower inertia than other types. This permits a faster response. However, since the rotor contains no magnet, there is no residual torque when the motor is deenergized and the rotor is said to be free wheeling. Generally, step angles of variable reluctance steppers are 7.5° or 15°.

Permanent Magnet

Permanent magnet step motors contain a *permanent magnet* rotor which has a holding torque when the motor is not energized. Each permanent magnet tooth is axially oriented with alternating north and south poles. Some steppers have magnets inserted into the stator

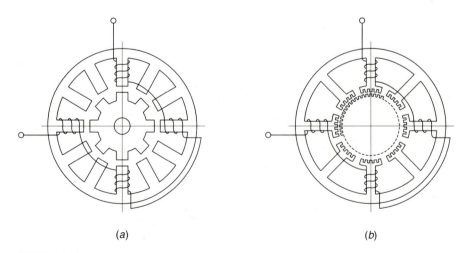

(*a*)   (*b*)

FIGURE 11.2
Construction methods. (*a*) Single teeth; (*b*) multiple teeth. (*Figures courtesy of Oriental Motor.*)

to enhance the electromagnetic field and to provide higher torque. Magnets are made from alnico (an aluminum, nickel, cobalt alloy) or newer rare earth materials (samarium-cobalt) which have half again as much magnetic energy. Permanent magnet step motors require less power to operate than other types. They also have better response damping characteristics. Step angles can be found over the whole range of standard angles, including 1.8°, 7.5°, 15°, 30°, 45°, and 90°.

Hybrid

Hybrid step motors combine rotor features of the variable reluctance stepper and permanent magnet motor. A smaller permanent magnet surrounds the shaft. It differs from the permanent magnet stepper by having one rotor end a circumferential north pole and the opposite end a south pole. Rotor teeth are cut into two iron core cups which fit over each end. The hybrid stepper uses construction method II only. See Fig. 11.3. Hybrid step motors can entertain more rotor teeth and have higher torque. Typical step angles are 0.9° and 1.8°.

Type Summary

Ten years ago the most popular type was the variable reluctance stepper. Most recent improvements in performance have pertained to permanent magnet steppers and hybrid steppers so that today the 1.8° step angle of either type is the most widely used. Accuracy for most steppers is about 3 percent of the step angle regardless of the number of steps; thus, accuracy is improved by going to smaller angles.

Any performance advantages are small, and comparisons among the three types of motors do not provide general rules. They compete effectively by making slight changes in motor size or control drive characteristics. Manufacturers concentrate on one type of motor, usually over a specific size range.

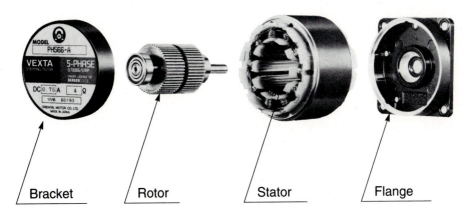

Bracket   Rotor   Stator   Flange

FIGURE 11.3
Exploded view of a five-phase hybrid stepping motor. (*Courtesy of Oriental Motor, Vexta Brand stepping motor.*)

11.2   HOW STEPPERS WORK

To see how a step motor works, consider a simple permanent magnet rotor with a single north–south (N–S) pole and a four-tooth stator driven by a pair of windings, $A_1 - A_2$ and $B_1 - B_2$. When $A_1 - A_2$ (and $B_1 - B_2$) are connected to a DC supply voltage and ground, respectively, the top tooth (and right-side tooth) becomes a north pole and the opposite tooth becomes a south pole as shown in Fig. 11.4. This pushes the rotor into a stable position at $+45°$, where rotor poles line up between opposite poles of two windings. When polarity of winding $A_1 - A_2$ is reversed by switching supply voltage and ground, the rotor advances $90°$ to a new stable position at $-45°$. This means it has a $90°$-step angle. By carrying out the four-step sequence in Table 11.1, the hypothetical motor will rotate one full revolution. When the step sequence is repeated, the motor advances another four steps. If the step sequence is reversed, motor direction is reversed. The number of sequence steps and switching pattern will vary with motor construction and the manufacturer.

The preceding sequence is called the *full-step mode* because each step advances the stepper by the amount specified as its step angle. A step motor can be made to rotate one half-step angle by operating it in the *half-step mode*. Consider Fig. 11.4 again. If only one

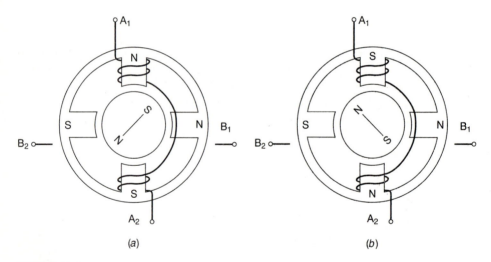

(a)   *(b)*

FIGURE 11.4
Energization sequence for step motor. B_1 to B_2 winding not shown. (*a*) Step 1; (*b*) step 2. (*Figures courtesy of Oriental Motor.*)

TABLE 11.1
Full-step mode

Step	A_1	A_2	B_1	B_2
1	H	L	H	L
2	L	H	H	L
3	L	H	L	H
4	H	L	L	H
5	Repeat 1 to 4.			

TABLE 11.2
Half-step mode

Step	A_1	A_2	B_1	B_2
1	H	L	H	L
2	L	L	L	H
3	L	H	H	L
4	H	L	L	L
5	L	H	L	H
6	L	L	H	L
7	H	L	L	H
8	L	H	L	L

winding is energized, the motor will come to equilibrium directly under its opposite pole, halfway between the two full steps shown. Our 90°-step angle motor is now a 45°-step angle motor. In the half-step mode only one coil is energized every other step to give an eight-step sequence listed in Table 11.2. The effect of a strong step followed by a weak step is lower torque at low speed but a slightly higher torque at a high speed.

Phase

Step motor phase refers to the number of *independent* windings on the stator. For example, Fig. 11.2(a) shows a three-phase motor. One winding is drawn; two identical windings can be made on the remaining stator teeth. Similarly, Fig. 11.2(b) shows a two-phase motor. In general, the number of teeth on the rotor and stator are related by

$$N_S = N_R \pm P \tag{11.1}$$

where N_S = number of stator teeth
$\quad N_R$ = number of rotor teeth
$\quad\quad P$ = number of stator teeth per phase

Step angle θ_0 in degrees per step is given by

$$\theta_0 = \frac{360}{N} \tag{11.2}$$

where N is the number of steps per revolution:

$$N = \left| \frac{N_R N_S}{N_S - N_R} \right| \tag{11.3}$$

For the preceding three-phase motor, the step angle is calculated to be 15°.

A step motor is often classified by the number of connector wires protruding from the case, which is not related to the number of phases. Figure 11.5 shows various wiring schemes for two-, three-, and four-phase step motors.

Two-phase motors with four wires are called *bipolar* motors because the steppers require the winding polarity to reverse in order to generate a four-step sequence. This can be accomplished with four switches and a bipolar ($\pm$) source, although it is typically done with eight switches and a unipolar (+) source (see the next section). If the source is

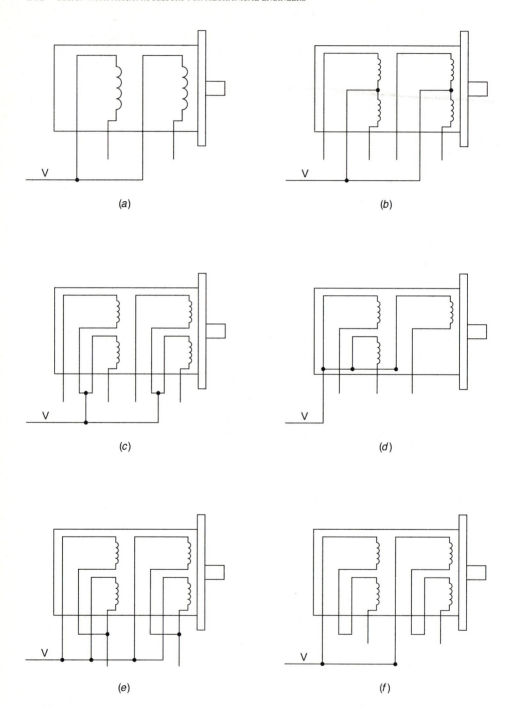

FIGURE 11.5
Motor phase and connector wires relationship. (*a*) Two-phase: four wire (bipolar); (*b*) two-phase: six wire (unipolar); (*c*) two-phase: eight wire (unipolar); (*d*) three-phase: four wire; (*e*) four-phase: eight wire (parallel); (*f*) four-phase: eight wire (series).

applied to the center-tap of each winding, there is no need to switch polarity since current flow is opposite in each winding half. Such an arrangement is called a *unipolar or bifilar* motor. Connector wires may total five, six, or eight, depending on how the source wires are handled. Two-phase bifilar steppers are erroneously called four-phase motors. Both have four (apparent or otherwise) coils, but the physical layout on the stator would be different. For a unipolar motor to have the same number of turns per winding as a bipolar motor, the wire diameter must be decreased and therefore resistance must be increased. Thus, unipolar steppers have 20 to 40 percent less torque at low step rates. However, at high step rates performance of the unipolar stepper is superior. Two-phase steppers are commonly used in light-duty applications where switching elements are not expensive.

Three-phase motors have one 4-wire arrangement. It requires the fewest switches: three. Three-phase windings are used exclusively in variable reluctance steppers.

Four-phase motors have eight wires, a pair for each coil. Leads may be wired externally, in parallel, or in series. The parallel wiring gives a better performance at high step rates, whereas the series arrangement has higher torque at low step rates. High-power steppers are primarily of the four-phase type.

11.3 STEP MOTOR CONTROL

Now that we have examined how step motors operate, we will discuss how these motors are controlled. Torque, speed, and acceleration characteristics are very much a function of control circuitry. The electronics have become more complex, but there is improvement in stepper performance, particularly at higher step rates. At the same time circuitry is a major cost item in the step motor system, often exceeding the cost of the motor. We will discuss the fundamental components of stepper control. LSI chips with all components on-board are available for interfacing steppers. These packages will be covered in Sec. 11.5.

Basic Switch Circuit

As mentioned before, each phase of the stepper must be switched ON and OFF in sequence to advance the rotor. The particular ON-OFF pattern varies with the motor and manufacturer. Airpax Corporation makes a series of small low-cost step motors with the switching sequence shown in Fig. 11.6. The bipolar motor has four color-coded leads. Although color coding is consistent for a given manufacturer, it differs among manufacturers. Either bipolar or MOSFET switching transistors can be used to reverse the phase current. They are not integrated into the motor package but are furnished by the customer. Care must be taken in microprocessor software so that transistors in series do not short the power supply by coming on at the same time. The eight-wire unipolar motor has one-half the number of switches and no short problem, which makes it superior for laboratory work.

Back EMF Suppression

Coils of wire in any motor (or solenoid) are inductors as well as resistors. Inductors store energy in its surrounding magnetic field. When the current is switched OFF, the magnetic field begins to collapse and return the energy that was put into it. In that brief instant after cutting the power from an inductor circuit voltage will build up until there is sufficient

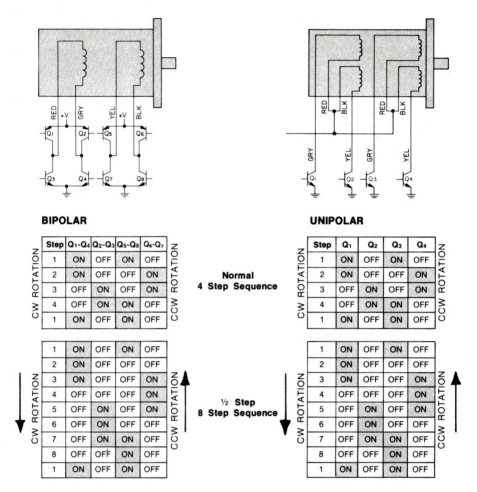

BIPOLAR

Normal
4 Step Sequence

Step	Q₁-Q₄	Q₂-Q₃	Q₅-Q₈	Q₆-Q₇
1	ON	OFF	ON	OFF
2	ON	OFF	OFF	ON
3	OFF	ON	OFF	ON
4	OFF	ON	ON	OFF
1	ON	OFF	ON	OFF

(CW ROTATION / CCW ROTATION)

½ Step
8 Step Sequence

Step	Q₁-Q₄	Q₂-Q₃	Q₅-Q₈	Q₆-Q₇
1	ON	OFF	ON	OFF
2	ON	OFF	OFF	OFF
3	ON	OFF	OFF	ON
4	OFF	OFF	OFF	ON
5	OFF	ON	OFF	ON
6	OFF	ON	OFF	OFF
7	OFF	ON	ON	OFF
8	OFF	OFF	ON	OFF
1	ON	OFF	ON	OFF

UNIPOLAR

Step	Q₁	Q₂	Q₃	Q₄
1	ON	OFF	ON	OFF
2	ON	OFF	OFF	ON
3	OFF	ON	OFF	ON
4	OFF	ON	ON	OFF
1	ON	OFF	ON	OFF

Step	Q₁	Q₂	Q₃	Q₄
1	ON	OFF	ON	OFF
2	ON	OFF	OFF	OFF
3	ON	OFF	OFF	ON
4	OFF	OFF	OFF	ON
5	OFF	ON	OFF	ON
6	OFF	ON	OFF	OFF
7	OFF	ON	ON	OFF
8	OFF	OFF	ON	OFF
1	ON	OFF	ON	OFF

FIGURE 11.6
External transistor circuitry for switching bipolar and unipolar steppers. (*Courtesy of Airpax Corp.*)

potential to exceed the maximum breakdown voltage $V_{CE(max)}$ of the transistor. This voltage can be suppressed by several methods. Each depends upon creating a current loop to dissipate the energy while the transistor is OFF.

A model of the phase winding is simply an inductance L and resistance R in series, Fig. 11.7(*a*). The simplest method of suppression is to place a (fast) *diode* in parallel with the winding, Fig. 11.7(*b*). When the winding is energized, current cannot flow through the diode because it is reversed bias. After switching OFF the transistor, an emf voltage greater than supply voltage V will short circuit the current through the diode and winding resistance. The diode is called a "flyback" diode or "free-wheeling" diode, describing its function.

A better method connects a *resistor and diode* in series. See Fig. 11.7(*c*). The loop time constant is reduced by the additional resistance and energy dissipates faster. How-

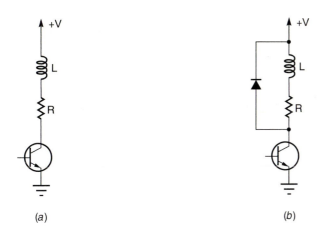

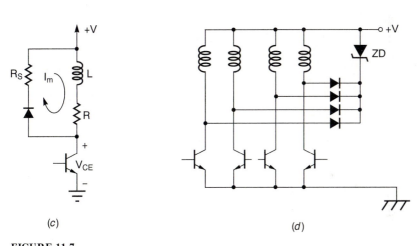

FIGURE 11.7
Back emf suppression methods for steppers. (*a*) Equivalent winding circuit; (*b*) diode suppression; (*c*) diode + resistance suppression; (*d*) diode + zener suppression.

ever, back emf will be higher and must not exceed the maximum breakdown voltage. Maximum resistance R_S is

$$R_{S(max)} = \left[R\frac{V_{CE(max)}}{V} \right] - 1 \tag{11.4}$$

Using a *zener diode and diode* will make the short-circuit current damp faster than the diode or diode + resistor method. The zener diode breakdown voltage is selected near the transistor breakdown voltage rating. At this voltage zener diodes will conduct all current necessary to maintain its breakdown voltage. Most of the energy is dissipated in the zener diode. In practice, there is no need to supply a zener diode for each winding. Figure 11.7(*d*) shows the correct method for using one.

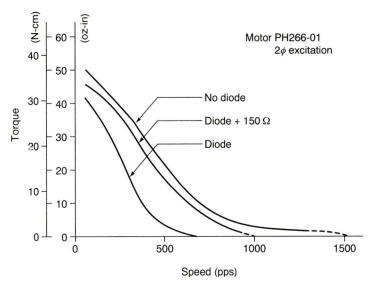

FIGURE 11.8
Torque-speed curves of an Oriental stepping motor showing effect of suppression methods. (*Courtesy of Oriental Motor, Vexta Brand stepping motor.*)

Because back emf voltage works as a brake on the motor, operating torque will drop. Although the magnitude of the back emf is important, it is the length of time (dissipation time) over which an average value acts which is critical. Figure 11.8 displays torque-speed curves for each suppression method. The zener diode + diode curve will fall inside the diode curve.

Series Resistance Drive

A problem with the previous drive circuits is slow stepper speed, caused by inductance. Inductance resists buildup of energy as current is changing. Once steady current is reached, inductance has no effect. Suppose it takes 0.01 s after power is switched ON to reach steady current. Remember, torque is proportional to current. At 1000 steps per second the motor coil will be energized for 0.001 s. This is only one-tenth the time needed to reach maximum torque. As a result, the actual torque from the stepper is severely limited. We must reduce the motor winding time constant.

An energized coil is governed by the following equation:

$$L\frac{dI}{dt} + RI = V_S \tag{11.5}$$

$$I(0^+) = 0$$

The solution is simply

$$I(t) = \frac{V_S}{R}[1 - e^{-t/\tau}] \tag{11.6}$$

where the time constant $\tau = L/R$. One method to speed up the current buildup is to *increase* the resistance by connecting a resistor in series with the winding resistance R. The series resistor is always some multiple of R. This method is called *series resistance drive* or *L/nR* drive. The higher winding resistance of the unipolar stepper explains its better high-speed torque when compared to bipolar steppers. Figure 11.9 shows the circuit for an *L/4R* drive. An additional resistance $(3R)$ must be placed in series with each winding that is on during a step sequence. A medium-size step motor (Airpax Corporation) is shown in Fig. 11.10(*a*). Its running torque curves for *L/R* drive (no external resistance) and *L/4R* drive can be compared in Fig. 11.10(*b*).

Series resistance drive is simple to wire and cheap (two resistors). But added voltage drop across the series resistance increases the needed power supply voltage to a factor of 4 over the operating voltage. The method is also inefficient because series resistance must absorb three times the winding power loss.

Example 11.1. The Airpax unipolar step motor in Fig. 11.10 uses the wiring diagram in Fig. 11.9. Determine the series resistance, transistor, and zener diode.

Solution. Series resistance is simply three times the winding resistance. A winding is interpreted as the coil section between supply and ground. Thus,

$$3R = 312\Omega$$

The steady-state current in each winding is

$$I_{SS} = \frac{V_S - V_{SAT}}{R} - \frac{24 - 0.2}{104} = 0.23 \text{ A}$$

Select a transistor with a current rating of 0.23 A or better. Any β greater than 200 will give a switching current of 1 mA or less. Then we can use a PIA or gate to switch the transistor, avoiding a buffer. A 2N3903 would do nicely.

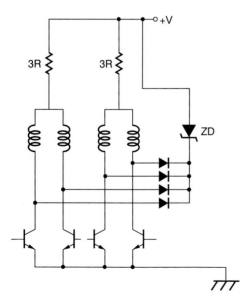

FIGURE 11.9
External resistance for *L/4R* drive.

(a)

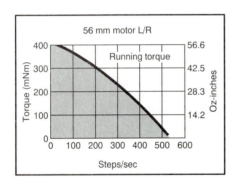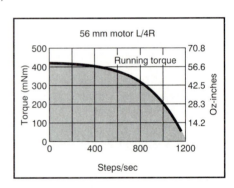

(b)

DC operating voltage = 24 V   Rotor moment of inertia = 1×10^{-2} g-m^2

Resistance per winding = 104 Ω Detent torque = 7 oz-in

Inductance per winding = 89 mH   Step angle = 1.8° ± 5%

Holding torque = 50 oz-in   Weight = 16 oz

(c)

FIGURE 11.10
Medium-size stepper. (a) Airpax model 45H-24A46S (uinpolar); (b) torque-speed curves; (c) specifications. (*Courtesy of Airpax Corp.*)

The transistor has a breakdown voltage

$$V_{CE(max)} = 40$$

The zener diode breakdown voltage must be *less* than the transistor breakdown voltage minus the power supply voltage or

$$V_z = 40 - 24 \text{ V} = 16 \text{ V}$$

Also, only one transistor is switched from ON to OFF in any two-phase step sequence. At the moment of switching this current is I_{SS} and will decay exponentially through the zener diode. Since inductance energy must equal dissipated energy,

$$\frac{1}{2}LI_{SS}^2 = V_z \int_0^\infty I \, dt$$

where

$$I = I_{SS}e^{-t/\tau} \cong \frac{V_S}{R}e^{-t/\tau}$$

Solving this equation, we have the time constant of current decay

$$\tau = \frac{0.5LV_S}{RV_z}$$

If the time of current decay is approximately 4τ, an average dissipated power in the zener diode is

$$\overline{P}_z = \frac{1}{2}\frac{LI_{SS}^2}{4\tau} = \frac{V_S V_z}{4R}$$

The zener diode is

$$V_z = 16 \text{ V}$$

$$\overline{P}_z = 2 \text{ W} \qquad \text{(safety factor of 2)}$$

Chopped-Voltage Drive

We can see in Eq. (11.6) that an alternate route, different from the time constant, is available to speed up winding current rise time, namely, supply voltage. Chopped-voltage drives apply much higher voltages than rated voltage to the step motor. High voltages cause current to rise quickly with torque following the current. As the current exceeds rated current (V_S/R), a current-sensing network switches OFF the voltage. When winding current decays below the rated current, the voltage is switched ON. The effect is a chopped-winding current with an average current equal to the rated current. Figure 11.11 illustrates the

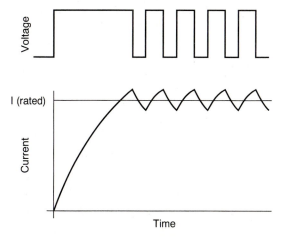

FIGURE 11.11
Voltage and current waveform for chopper drive.

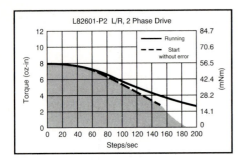 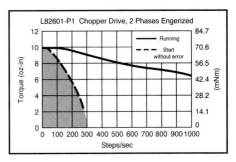

FIGURE 11.12
Comparison of *L*/*R* drive and chopper drive for Airpax motor. (*Courtesy of Airpax Corp.*)

voltage and current relationship. Motor inertia is sufficiently large to average the chopped torque and to give a smooth motion.

Figure 11.12 compares an *L*/*R* drive (no series resistance) with a chopper drive. The overall improvement in torque-speed curve is significant and much greater than can be expected with a series resistance drive. Furthermore, chopper drives are highly efficient. They are also more complex and costly than alternatives and are now built into stepper logic-driver LSI chips, discussed in Sec. 11.5. Today chopper drives are the most popular method to improve stepper performance.

11.4 STEP MOTOR PERFORMANCE

Static (Holding) Torque

After being energized the rotor settles into an equilibrium position between a pair of magnetic poles. Any attempt to displace the rotor in either direction will result in an opposing torque, called the holding torque. See Fig. 11.13. The holding torque is given by

$$T = -T_{\mathrm{H}} \sin\left(\frac{360°}{\theta_0}\theta\right) \tag{11.7}$$

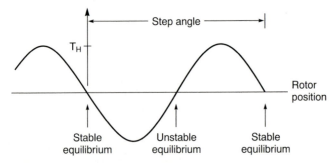

FIGURE 11.13
Holding torque for single phase.

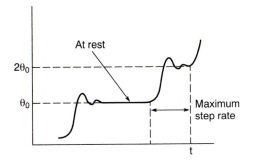

FIGURE 11.14
Motor displacement vs. time for single steps.

where θ_0 is the step angle. Magnetic stiffness $(dT/d\theta)$, coupled with the mechanical inertia J, produces a second-orderlike response governed by

$$J\ddot{\theta} = T \tag{11.8}$$

where small internal damping has been neglected. If the step motor is pulsed at a slow enough rate, it comes to rest at the end of each step and oscillates as shown in Fig. 11.14.

The nonlinear behavior of Eq. (11.8) results in a *band of resonant frequencies* between

$$\omega_{N1} = \left[\frac{360T_H}{\theta_0 J}\right]^{1/2} \quad \text{and} \quad \omega_{N2} = \left[\frac{720T_H}{\pi\theta_0 J}\right]^{1/2} \tag{11.9}$$

Operating a stepper at frequencies within this band leads to erratic behavior, including large oscillations and missed steps.

Several techniques may be successful in reducing resonance: (1) Choose a less effective emf suppressor since back emf damps the oscillation. (2) Place a mechanical damper on the shaft. (3) Step with a smaller step angle. (4) Increase load inertia to lower resonant frequency. (5) Use the half-step mode.

Example 11.2. Calculate the resonant band for Airpax motor 4SH-24A46S.

From specifications listed in Fig. 11.10, $T_H = 50$ oz-in and $J_m = 0.546$ oz-in². Assume that the motor is not driving a load; otherwise load inertia must be added to the motor inertia:

$$\omega_{N1} = \left[\frac{360 \times 50 \text{ oz-in} \times 386 \text{ in/s}^2}{1.8 \times 0.546 \text{ oz-in}^2}\right]^{1/2} = 2659 \text{ rad/s}$$

$$= 84{,}600 \text{ steps/s}$$

Both step rates are beyond the range for steppers. Small, large step angle motors would not fair so well.

Dynamic Torque

As stepping rate increases, the motion changes from discrete steps to a continuous motion referred to as *slewing* (slew: to twist or turn about). We can see in Fig. 11.15 that stepper velocity in steady-state slewing goes through small oscillations about the average because the rotor does not fully come to rest.

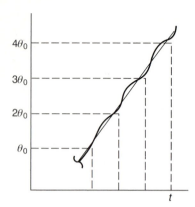

FIGURE 11.15
Motor displacement vs. time for steady-state slewing.

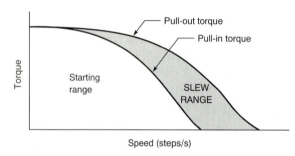

FIGURE 11.16
Starting and running torque range.

The most important performance parameter for steppers is steady torque output. Figure 11.16 shows a typical torque-speed plot. The curves do not define specific operating points but outline regions where the motor will operate satisfactorily. Steppers develop their highest torque at standstill. As the step rate is increased, winding inductance prevents the current from reaching its steady-state value and torque decreases with the step rate.

Step motors function by virtue of an electronically switched power source which rotates the magnetic field. Each pulse sequence steps the stator field by a fixed angle. At low pulse rates the field decays and builds up in its new orientation, pulling the magnetically coupled rotor. The rotor must be able to accelerate from zero to full speed in the time it takes a stator pole to move by. *Pull-in torque* or start without error torque is the maximum torque at which the stepper will start from rest (or stop without loss of a step) when operated at the given step rate. Since pull-in torque data includes the rotor inertia torque, pull-in torque is the steady frictional (or gravitational) torque that the motor will drive from rest.

Pull-in torque is not the maximum torque delivered by steppers. Part of the drive torque is used to accelerate motor inertia. Once running speed is reached (one or two steps), inertia torque is available for friction torque. *Pull-out torque* or running torque is the maximum frictional torque that can be applied to the motor while running at a steady rate. The difference between the pull-in torque curve and the pull-out curve at a fixed step rate is the torque to overcome motor inertia. The area between the two torque curves is called the slew range.

There are two aspects to the design problem which need to be discussed before using torque-speed curves. First, load frictional torque is known (fixed). Therefore, its intersection with the pull-in torque curve gives the maximum step rate to move the load from

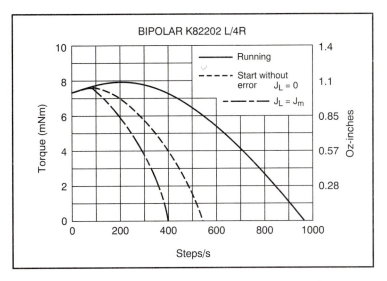

FIGURE 11.17
Starting torque curves with and without an equivalent load inertia. (*Courtesy of Airpax Corp.*)

rest. Any lesser step rate is also acceptable. The design torque intersection with the pull-out torque curve gives the maximum step rate (slew rate) possible after the motor is running at pull-in step rate. However, the motor must be carefully accelerated to this speed and decelerated to a stop again if any steps are not to be missed. Stepping rates outside the pull-out torque curve will cause the motor to stop, oscillating about its fixed position. Second, torque-speed curves do not account for load inertia. Suppose the stepper is coupled to a disk spindle which has the same moment of inertia as the motor. We would not expect the pull-in torque curve to be very helpful although the pull-out torque curve is still valid. There is a simple solution to this problem. Since the vertical distance between the two curves is a motor inertia torque, a reflection of this torque on the downside of the pull-in torque gives the new pull-in torque curve for the combined moment of inertia. The construction is demonstrated in Fig. 11.17. This method is valid for any multiple of the motor moment of inertia.

> **Example 11.3.** The small Airpax step motor in Fig. 11.17 is used in a medical device to meter medicine. Frictional load is measured with a torque wrench and found to be 2.5 mNm. Load inertia is negligible. What is the maximum *constant* step rate to be programmed into the microprocessor?
>
> At a 2.5-mNm torque the maximum start without error speed is 450 steps per second. The motor will operate at 825 steps per second, but the program must start at a rate of 450 steps per second and slowly increase the rate until running speed is achieved.

> **Example 11.4.** The motor in Example 11.3 is considered for a design where the friction torque is 1.0 mNm and the load moment of inertia is twice the motor moment of inertia. Determine the start speed and the running speed.
>
> A curve must be constructed by subtracting double the vertical distance between the original start without the error line and the running line. At a 1.0-mNm torque intercept the start speed is 320 steps per second. Maximum running speed remains at 910 steps per second.

Ramped Step Rate

When step motors are used in the start-stop region, they can be started or stopped immediately. For a given frictional torque load T_f, step rate S_S on the pull-in curve is the maximum step rate for this region. Torque T_m, used to accelerate motor inertia J_m, is the vertical distance between pull-in and pull-out torque curves at S_S. To reach the running step rate, the motor must be accelerated from S_S to the running speed S_R in time t_0. When slowing down, the motor must be decelerated at a similar rate, or it will run past the intended next step. Between the two torque-speed curves the torque available for acceleration is not known, but it will decrease from T_m to zero at the running speed. We will give an approximate method for step rate ramp-up/ramp-down.

Motor acceleration (deceleration) obeys the equation

$$T - T_f = J\frac{d\omega}{dt} = JK\frac{dS}{dt} \tag{11.10}$$

where J = total moment of inertia

$$K = \frac{2\pi \text{ rad/rev}}{360°/\theta_0 \text{ steps/rev}}$$

Proposing a linear change in step rate

$$S = \frac{(S_R - S_S)t}{t_0} + S_S \tag{11.11}$$

with an average acceleration torque of $T_m/2$, the ramp time is

$$t_0 = \frac{2JK(S_R - S_S)}{T_m} \times \text{(safety factor of 2)} \tag{11.12}$$

The microprocessor delivers each sequence pattern with a delay S^{-1} between each step. Thus, the delay will decrease from S_S^{-1} for the first step to S_R^{-1} at time t_0, continuing the latter delay until the motor is ramped down. The average delay is

$$t_a = S_S^{-1} - 0.5(S_S^{-1} - S_R^{-1}) \tag{11.13}$$

Therefore, the number of steps to reach the running speed S_R is $\delta = t_0/t_a$.

Example 11.5. Outline a program to ramp-up a stepper:

1. Set loop count equal to δ.
2. Output the first sequence pattern.
3. Delay for S_S^{-1} seconds.
4. Decrease delay time by t_a.
5. Output next pattern.
6. Decrease count.
7. Is count zero?
8. If not, return to step 4.
9. Continue.

11.5 MICROPROCESSOR INTERFACING OF STEP MOTORS

After a description of step motors and their control interfacing steppers is relatively simple. The proper logic pattern is written to a suitable latch, such as a PIA, which is interfaced with motor control transistors. Since switched inductive windings generate a very large

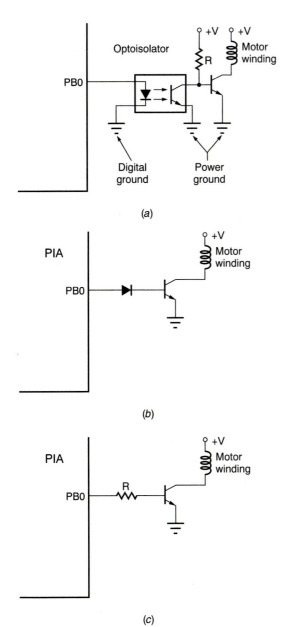

FIGURE 11.18
Three ways to isolate PIA from stepper.
(*a*) Optoisolator; (*b*) diode; (*c*) resistance.

back emf, the PIA (and microprocessor system) must be protected in case a control transistor should short out. Figure 11.18 shows three ways to *isolate* the PIA from the motor:

1. *Optoisolator (optocoupler).* This device contains an LED and a phototransistor in a single chip. The PIA drive current activates an LED, causing the phototransistor to conduct. There is no direct coupling (only light) to allow reverse currents. Most commercial driver boards have optoisolators. A discussion of optoisolators will follow in Chap. 15.
2. *Diodes.* A diode offers a cheap, simple interface. It has negligible resistance in the forward mode and will not load the PIA. It prevents current in the reverse mode, becoming a high resistance when its breakdown voltage is exceeded.
3. *Resistance.* Using resistors is the least desirable way to isolate the PIA. It does not completely isolate the PIA and must be sized to assure sufficient drive power for the control transistor. However, it does limit the transistor base power consumption.

Some thought should be given to the selection of control transistors. PIA port B is able to provide 1 mA of current. This level can drive standard bipolar switching transistors. If the current gain β is on the order of several hundred, a transistor collector current of 0.2 to 0.4 A is sufficient to drive small steppers. But many step motors require winding currents of 1 A or more. Then bipolar Darlingtons or power MOSFETs must be used for control transistors. These items will also be reviewed in Chap. 15.

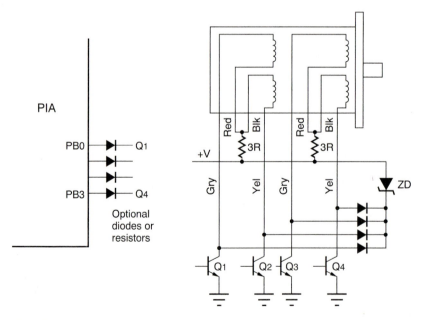

FIGURE 11.19
Circuit diagram for stepper and PIA interface.

A complete step motor interfacing system is shown now in Fig. 11.19. An Airpax series 82400 stepper is chosen. It has an $L/4R$ drive circuit for the unipolar control. Specifications are:

DC voltage:	5 V
Resistance per winding:	15.5 Ω
Inductance per winding:	13.5 mH
Step angle:	7.5°

Example 11.6. Software can be written to displace the rotor a fixed number of steps or run the motor at a given step rate. Write a program to run the Airpax series 82400 at a steady 100 rpm.

$$100 \; \frac{rev}{min} \times 360 \; \frac{deg}{rev} \times \frac{1}{60} \; \frac{min}{s} \times \frac{1}{7.5} \; \frac{step}{deg} = 80 \; \frac{steps}{s}$$

The delay between steps is $\frac{1}{80}$ s. The sequence pattern for unipolar control is 05, 09, 0A, 06, repeat.

Program

```
              LDA A   #FF      Set up PIA port B
              STA A   6002     at address 6002
              LDA A   #04      for output.
              STA A   6003

      START   LDA A   #05      Advance first step.
              STA A   6002
              JSR     0100     Jump to delay subroutine.

              LDA A   #09
              STA A   6002     Advance second step.
              JSR     0100

              LDA A   #0A
              STA A   6002     Advance third step.
              JSR     0100

              LDA A   #06
              STA A   6002     Advance fourth step.
              JSR     0100

              JMP     START
```

Step Motor IC Controllers

Integrated circuits (ICs) for step motor control are available which greatly simplify interfacing hardware and software. They are divided into logic circuits and power circuits, either on separate chips or combined in a single IC. Sequencing logic is on-board, and the motor steps or changes direction with a pulse input. Many ICs have high performance chopper drives. Power circuits are essentially switching transistors. Emf suppression diodes may be an integral part. Bipolar or unipolar (bifilar) control is built into the power

TABLE 11.3
Step motor ICs

Vendor	Number	Maximum volts	I (pk) mA/ phase	I (avg) mA/ phase	Drive mode	Power	Logic	Type
Cybernetics	CY512	7	80	—	L/R	No	Yes	Unipolar
Motorola	1042	20	500	300	L/R	Yes	Yes	Bipolar
Oriental	OMD240	28	400	250	L/R	Yes	Yes	Unipolar
RCA	CA3169	17	2500	600	L/R	Yes	No	
Rifa	PBL3717	45	1000	800	C	Yes	Yes	Bipolar
SGS	L293	40	1500	1000	L/R	Yes	Yes	Bipolar
SGS	L297	5	—	—	—	No	Yes	Bipolar
SGS	L298	45	3000	2500	C	Yes	No	Bipolar
Signetic	SAA1027E	18	500	350	L/R	Yes	Yes	Unipolar
Sil, General	SG36535	35	5000	2000	C	Yes	No	Bipolar
Sil, General	SG293D	40	1500	1000	L/R	Yes	Yes	Bipolar
Sil, General	SG3637	40	2500	2000	L/R	Yes	No	Bipolar
Sprague	UCN4202A	15	800	500	L/R	Yes	Yes	Unipolar
Sprague	UDN2849Z	28	1000	800	C	Yes	No	Bipolar
Sprague	UDN2976W	60	5000	4000	—	Yes	Yes	Bipolar
Texas Instruments	376CNE	18	800	500	C	Yes	Yes	Unipolar
Unitrode	PIC930	45	1000	800	C	Yes	Yes	Bipolar
Unitrode	PIC900B	60	5000	5000	C	Yes	No	Bipolar

circuit with the logic stepping sequence identical for both. Table 11.3 gives a summary for a number of control ICs. Two of these ICs will be discussed.

Signetics SAA1027

The SAA was an early driver IC. It is a single chip system for driving small steppers with winding currents up to 350 mA. See Fig. 11.20. Three inputs are controlled by applying

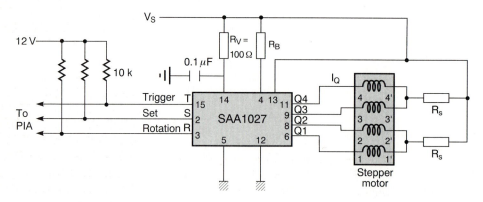

FIGURE 11.20
Interfacing a stepper to the SAA1027 controller IC. (*Adapted from Airpax stepper motor handbook.*)

HIGH or LOW voltage levels to the terminals. *Trigger input T* is normally held HIGH when not being stepped. The motor will step on a positive going edge, in other words, on the LOW-HIGH of a HIGH-LOW-HIGH pulse. A HIGH on *direction input R* will cause the motor to step CCW while a LOW gives a CW rotation. *Set input S* returns the logic state Q1 to Q4 to the initial step pattern LOW (winding on), HIGH (winding off), LOW, HIGH by momentarily applying a LOW to pin 2 when *T* is held HIGH. Pin 2 is normally held HIGH. The supply voltage is 9 to 18 V, nominally 12 V.

Control logic for inputs is 12 V (and 25 mA). To drive the inputs with the PIA, 10 k pull-up resistors should be tied from the pin(s) to a 12-V supply.

If 12-V motors are driven, series resistors R_s are eliminated in Fig. 11.20 to give an *L/R* drive. For 5-V motors, R_s resistors (25 Ω, 5 W) step-down the voltage to give an *L/nR* drive.

Bias resistor R_B for the SAA1027 limits the motor average winding current to specifications and protects the motor from being overdriven. The average winding current I_Q, which is on half the time, is found from the motor voltage V_m and winding resistance:

$$I_Q = \frac{V_m}{2R}$$

For the 5-V Airpax 82400 unipolar drive, $I_Q = 5V/2(15.5) = 0.161$ A. The relationship between I_Q and the bias current I_B at pin 4 is

$$I_Q = 5I_B - 0.05 \text{ A}$$

or $I_B = 0.042$ A for the previous motor. Another relationship between the bias current and the bias voltage at pin 4 is

$$V_B = \left(\frac{1000}{15}\right)I_B + \left(\frac{2}{3}\right)$$

or $V_B = 3.5$ V. Then the bias resistance is obtained from the equation

$$R_B = \frac{V_s - V_B}{I_B}$$

or

$$R_B = 200 \ \Omega$$

SGS L297/L298

One of the most popular two-chip logic-driver sets today is the SGS L297/L298. The L297 chip generates four-phase TTL logic signals for two-phase bipolar and (four-) phase unipolar step motors. The L298 is a high-current, full bridge driver designed to accept standard TTL logic levels and drive inductive loads such as relays, solenoids, DC motors, as well as step motors. Figure 11.21 shows the internal circuitry of each chip and a wiring diagram for the pair driving a stepper.

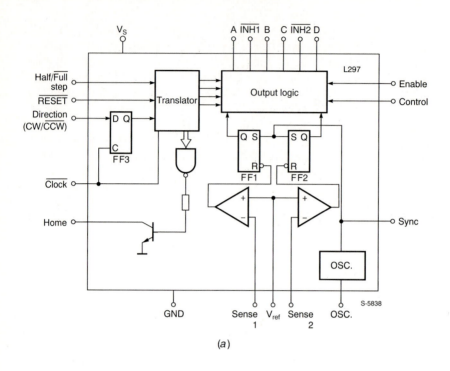

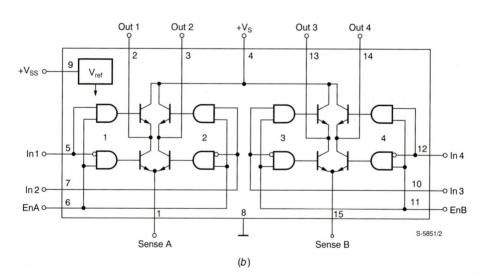

FIGURE 11.21
L297/L298 stepper motor controller. (*a*) L297 block diagram; (*b*) L298 block diagram; (*c*) combined circuit. (*Diagram by permission. SGS-Thomson Microelectronics.*)

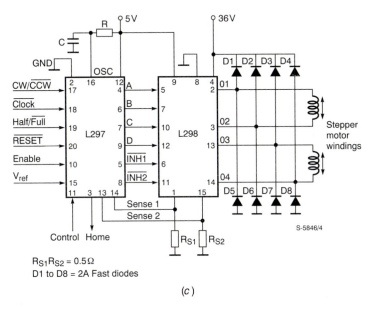

RS1RS2 = 0.5 Ω
D1 to D8 = 2A Fast diodes

(c)

FIGURE 11.21 (continued)

L297s have four inputs under microprocessor control which determine how the windings are energized:

Half/$\overline{\text{Full}}$	When HIGH, selects half-step mode. When LOW, selects full-step mode.
$\overline{\text{Reset}}$	An active LOW restores logic pattern to home state (ABCD = 0101).
Direction (CW/$\overline{\text{CCW}}$)	An active LOW rotates motor counterclockwise; HIGH rotates it clockwise.
$\overline{\text{Clock}}$	An active LOW on this pin advances the motor one step. Step occurs on the signal LOW to HIGH transition.

Three additional inputs have varied functions:

Enable	When LOW, ABCD = 0000. Must be HIGH for motor to operate.
Control	Must be LOW for chopper to activate $\overline{\text{INH1}}$ and $\overline{\text{INH2}}$ (see discussion).
V_{ref}	Reference voltage for chopper circuit determines peak winding current. Set to match given stepper current rating (see discussion).

The translator generates motor phase sequences and a dual pulse-width-modulated (PWM) chopper circuit regulates current in the motor windings. An on-chip oscillator drives the duel chopper. It supplies pulses at the chopper rate which sets two flip-flops, FF1 and FF2. When winding current reaches the prescribed peak current, the voltage across a sense resistor (RS1 and RS2) equals set voltage V_{ref}, and the corresponding parator resets the flip flop, interrupting the drive current. If the sense resistors are set at 0.50 Ωs, a motor winding current of 2 A will cause the sense resistor voltage drop to be 1.0 V ($E = IR$). For a reference voltage of 1.0 V the voltage comparator will switch and reset the flip-flops. At the same time the output logic will generate appropriate inhibit outputs ($\overline{INH1}$ and $\overline{INH2}$) to turn off the transistors in the L298, thus current flows through the winding. The oscillator (OSC) will turn on the phase pattern ABCD for the next cycle when current flow is again high enough to trigger the comparator.

Oscillator frequency is set by an external RC network wired to pin 16. Oscillation frequency is found from

$$f \cong \frac{1}{0.69RC}$$

This frequency should be at least two or three times the motor step rate.

HOME output is an open-collector output that indicates when the L297 is in its initial state (ABCD = 0101). The transistor conducts when this signal is active.

The L298 circuit contains eight power switching transistors in the configuration required for bipolar control. Consider the left half only with $\overline{INH1}$ (pin 6) HIGH, allowing the transistors to switch normally. The preceding initial state corresponds to step 3 in Fig. 11.6. Thus, the pattern 0101 will place a logic LOW on pin 5 and a logic HIGH on pin 7. We see that transistors $Q_1 - Q_4$ are indeed OFF and transistors $Q_2 - Q_3$ are ON. The reader may confirm the right half of L298 by completing step 3.

PROBLEMS

11.1. The actuator head in Fig. 11.1 must span 77 tracks spaced 0.0208 in apart. Access time is 3 ms track-to-track. If the stepper angle is 1.8°, determine the capstan radius.

11.2. The step motor in Fig. 11.4 is a (a) one-, (b) two-, (c) three-phase motor.

11.3. Construct a step motor based on a pair of permanent magnet rotor poles. Select the minimum number of stator teeth and phase windings for a viable motor. Give the step angle.

11.4. The motor in Fig. 11.3(b) has 50 rotor teeth and 40 stator teeth. Find the step angle.

11.5. A step motor has the specifications shown. Determine the (a) external winding resistance, (b) zener diode, and (c) transistor.

Operating voltage = 12 V
Resistance per winding = 20.5 Ω
Inductance per winding = 45 mH

11.6. A step motor has the specifications shown. Determine if a resonance problem may occur if driving a 10^{-4} g-m² load.

Holding torque = 0.57 oz-in
Operating voltage = 5 V
Step angle = 15°
Rotor inertia = 5×10^{-5} g-m²

11.7. The step motor in Fig. 11.17 accelerates to a running speed of 200 steps per second in t_0 s. The step angle is 7.5°. The load inertia is equivalent to J_m. Find acceleration time t_0.

11.8. Repeat Example 11.4 if the load inertia is twice the motor inertia.

11.9. Write the assembly language for Example 11.5.

11.10. Determine the supply voltage in Fig. 11.19 if the motor voltage is 5 V.

11.11. Write a program to advance the motor in Example 11.5 by 90°, pause one second, and then return. Maximum step rate is 200 steps per second.

11.12. The proposed ramp-up rate is linear. A slightly superior method is an exponential ramp-up rate $S = S_R(1 - e^{-t/\tau})$.

(*a*) Determine time constant τ.

(*b*) Outline a program to achieve this step rate.

11.13. Assume your stepper is driven by an SAA1027 IC. Write a program to advance the motor in steps of 36° with a pause of one second after each step. At the 180° point the motor reverses to the start point.

CHAPTER
12

INTERFACING PERMANENT MAGNET DC MOTORS

The basic operation of a simple conventional DC motor is illustrated in Fig. 12.1. It features a rotating *armature* with one (shown) or more separate coil windings. Each winding terminates at a split ring (commutator) with power being transferred to the commutators through brushes. Between the commutators there is an insulator so that the ring acts as a double-pole, double-throw switch. As the armature rotates, the commutator constantly switches the current so that the armature magnetic field remains fixed. The armature is propelled by opposing electromagnetic forces from a separately wound stationary magnet, called the *field*. Fields of permanent magnet (PM) motors are generated by permanent magnets.

In the past a motor required an electromagnetic field if it were to deliver any significant power. Permanent magnet fields using ferrite magnets cannot supply a sufficiently intense magnetic flux. Alnico magnets greatly improved the power of permanent magnets but they offer little resistance to demagnetization forces as do the ferrites. Magnet material is magnetized during manufacture by placing it in a strong electromagnetic field. But high armature currents, approximately eight times the stated operating current, can demagnetize the stator magnet. This situation occurs while reversing high-speed motors. Alnico is used in relatively low-torque applications.

The development of rare earth (samarium cobalt) magnets has revolutionized the DC motor. They are three times stronger than Alnico; yet the material virtually defies demagnetization. Samarium-cobalt magnets are more expensive than Alnico, but they are used in most high-performance PM motors.

270

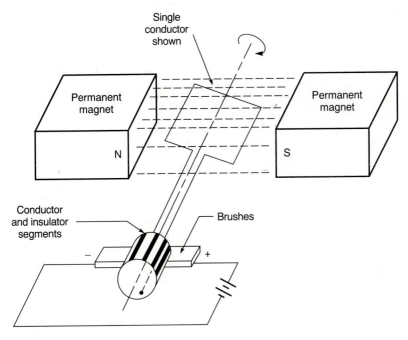

FIGURE 12.1
DC motor components.

Permanent magnet motors have several advantages over conventional wound field DC motors. First, PM motors have a linear torque-speed curve. This makes them ideal for control motors. Second, elimination of the wound field reduces heat generation within the motor. Temperature places an upper bound on the allowable current rating, thus the torque. *Today virtually all control motors are DC permanent magnet types.*

12.1 DC MOTOR TYPES

There are three types of PM motors: (1) iron core, (2) disk, and (3) cup. The name describes the rotor construction. Brushless DC motors have permanent magnets, but they are never called PM motors because of their distinct drive characteristics. Brushless DC motors are discussed in a separate section. Torque motors are PM or brushless DC motors which have been optimized for torque rather than power. This translates into shorter length and larger diameter motors.

Iron Core Motor

Iron core PM motors are simply conventional DC motors with permanent magnet fields. The armature is a laminated iron structure with slots containing separate windings. Iron provides a low reluctance path for the field flux. Each of approximately one dozen separate windings is terminated by a pair of commutators. This arrangement increases the torque and reduces cogging—a tendency for torque output to ripple as the brushes transfer power to

FIGURE 12.2
Cutaway of an iron core motor. (*Photo supplied by Leeson Electric Corp., Grofton, WI.*)

successive commutators. Iron core motors deliver the highest power and have a high starting torque. But they have large rotor inertia. Therefore, rotors are built with small diameters and long lengths to decrease the mechanical time constant. Figure 12.2 shows a cutaway.

Disk Motor

In many applications it is necessary to accelerate or decelerate the load quickly. The disk or printed circuit motor was designed to fill this need. Originally, the "pancake" shaped armature was formed by etching copper away from a printed circuit board to form the wires. Today armature wires are stamped from flat sheets of copper and laminated together with insulation between the sheets. The laminations are connected to form a continuous wire. Current flows radially to the circumference and between circumferentially placed field magnets. Figure 12.3 shows an exploded view of the components.

This ironless type of design has several advantages over an iron core motor. The lower rotor inertia allows faster acceleration. Peak currents are higher because of the non-magnetic armature and arrangement of the field magnets. Low "winding" inductance gives small electrical time constants and longer brush life since there is little arcing. Cogging is no longer a problem. However, disk motors cannot deliver the speed or power of iron core motors. Winding current is limited due to the lack of a heat conducting path, unlike iron armatures. Also, the design leads to a more expensive motor.

FIGURE 12.3
Disk motor construction. (*Courtesy of PMI Motors.*)

Cup Motor

The cup (bell, shell, or basket wound) armature motor was developed as a solution to the problem of making a high-torque, low-inertia motor. Figure 12.4 illustrates the hollow cup-like structure of the armature. The shell is formed by wire held together with epoxy or fiberglass. An extremely light weight, small diameter rotor gives the smallest inertia among the three types, and thus the fastest acceleration. These motors are generally used in low-power applications.

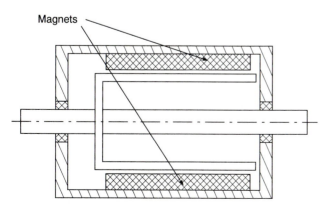

Magnets

FIGURE 12.4
Illustration of a cup motor.

12.2 PERMANENT MAGNET (PM) MOTOR PERFORMANCE

The most fundamental performance characteristic for electric motors is the speed-torque curve. Most manufacturers supply this curve along with several other performance specifications. Figure 12.5 gives the speed-torque curve and several specifications for a pair of small disk motors by PMI Motors. The linear curves are characteristic for all PM motors. Motor speed is maintained by changing the input DC voltage to meet any torque demand. Continuous stall torque is the maximum torque delivered by the motor when operated continually without cooling. Heat, generated by winding currents, is the chief limiting factor for motor power. If the user is willing to air cool the motor, its steady power can be increased. If the motor operates for short duty cycles, which occur when accelerating inertial loads, the current and peak torque can be increased substantially over the continuous values.

(a)

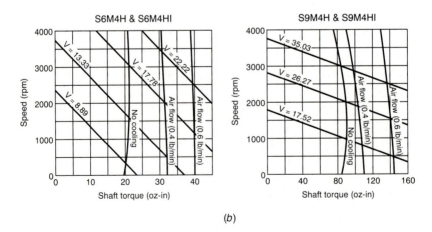

(b)

FIGURE 12.5
Typical PM motor characteristics. (a) Disk motors by PMI; (b) speed-torque curves; (c) specifications. (*Courtesy of PMI Motors.*)

25°C Ambient unless specified

Maximum Performance		S6M4H S6M4HI	S9M4H S9M4HI	
Peak torque	TP	217 153	935 660	oz-in N-cm
Continuous stall torque	TS	20 14	85 60	oz-in N-cm
Peak current	IP	51	79	A
Continuous stall current	IS	4.8	7.5	A
Peak acceleration (no load)	AP	256	167	krad/s²

Intrinsic Motor Constants

Torque constant	KT	4.26 3.01	11.91 8.41	oz-in/A N-cm/A
Back EMF constant	KE	3.15	8.80	V/krpm
Terminal resistance	RT	1.207	0.85	Ω
Armature resistance	RA	0.940	0.66	Ω
Average friction torque	TF	0.9 0.6	4.0 2.8	oz-in N-cm
Viscous damping constant	KD	0.16 0.11	1.32 0.93	oz-in/krpm N-cm/krpm
Moment of inertia	JM	0.00085 0.060	0.0056 0.396	oz-in s² kg-cm²
Armature inductance	L	<100	<100	μH

(c)

FIGURE 12.5 (continued)

Motor constants can be related to the speed-torque curve as well as to a mathematical model for the motor/load. This model is necessary to implement control of speed or position.

Motor Constants

The dynamic equation for DC motor armatures is given by

$$V = L\frac{dI}{dt} + RI + E \tag{12.1}$$

where V is the applied voltage to the armature, R and L are the winding resistance and inductance, and E is the back emf. E is given by

$$E = K_E\omega \tag{12.2}$$

where K_E is called the back emf constant or *voltage constant* and ω is the angular velocity (speed) of the rotor. Back emf power is the same as the mechanical power produced by the motor. Thus,

$$P = EI = T\omega \tag{12.3}$$

Substituting Eq. (12.2) into Eq. (12.3), we obtain the relationship between the torque and the current:

$$T = K_T I \tag{12.4}$$

where K_T is called the *torque constant* and is equal to K_E. In English units they will have different numerical values because they are expressed in different, but equivalent, units:

$$K_T \text{ (oz-in/A)} = 1.3524 K_E \text{ (V/krpm)}$$

$$K_T \text{ (Nm/A)} = 9.5493 \times 10^{-3} K_E \text{ (V/krpm)}$$

$$K_T \text{ (Nm/A)} = K_E \text{ (V/rad s}^{-1})$$

Another important performance measure is the amount of torque developed in relation to the amount of power dissipated as heat in the armature windings. *Motor constant* K_m is defined by

$$K_m = \frac{T}{\sqrt{RI^2}} = \frac{K_T}{\sqrt{R}} \tag{12.5}$$

A larger motor constant indicates a higher torque is generated and less power is lost as heat. The motor constant increases with motor size, but it also depends upon the quality of motor materials and how well they have been utilized.

Steady-State Performance

Let us consider Eq. (12.1) under *steady state conditions*. Substituting Eq. (12.2) for E and Eq. (12.4) for I yields

$$T = \frac{K_T}{R} V - K_m^2 \omega \tag{12.6}$$

This linear torque equation is the basis for the speed-torque curves shown earlier. It provides a simple method for predicting the motor performance from a single given design. Stall torque is

$$T_S = \frac{K_T}{R} V \tag{12.7}$$

and the theoretical no-load (torque) speed is

$$\omega_0 = \frac{K_T V}{R K_m^2} = \frac{V}{K_E} \tag{12.8}$$

Thus,

$$T_S = K_m^2 \omega_0 \tag{12.9}$$

When the motor is rotating at constant speed, it produces mechanical power $P = T\omega$, or

$$P = \left[\frac{K_T V}{R} - K_m^2 \omega \right] \omega \tag{12.10}$$

The power curve is a parabola with maximum power delivered at one-half the maximum speed ω_0. Maximum power is given by

$$P_m = \frac{K_m^2 \omega_0^2}{4} \tag{12.11}$$

Dynamic Model

Now the mechanical load consists of the inertia J and the constant torque T_f due to friction or gravity. Therefore,

$$T = J\frac{d\omega}{dt} + T_f \tag{12.12}$$

where the drive inertia J includes the motor inertia J_m and the load inertia J_L reflected through the gear ratio $N = \theta_m/\theta_L$:

$$J = J_m + \frac{1}{N^2}J_L \tag{12.13}$$

Using Laplace notation and substituting current for torque into Eq. (12.12), we obtain

$$K_T I(s) = Js\omega(s) + T_f(s) \tag{12.14}$$

Rewriting Eq. (12.1), we have

$$V(s) = LsI(s) + RI(s) + K_E\omega(s) \tag{12.15}$$

The preceding current is replaced by Eq. (12.14). After several steps of algebra

$$\frac{V(s)}{K_E} = (\tau_e\tau_m s^2 + \tau_m s + 1)\omega(s) + \frac{1}{K_m^2}(\tau_e s + 1)T_f(s) \tag{12.16}$$

where the electrical time constant τ_e and the mechanical time constant τ_m are

$$\tau_e = \frac{L}{R} \tag{12.17}$$

$$\tau_m = \frac{J}{K_m^2} \tag{12.18}$$

For most DC motors, the electrical time constant is negligible and

$$(\tau_m s + 1)\omega(s) = \frac{V(s)}{K_E} - \frac{T_f(s)}{K_m^2} \tag{12.19}$$

Equation (12.19) is a *dynamic* model for the motor/load combination. The solution for speed in the time domain depends upon the applied voltage and constant load torque.

12.3 OPEN-LOOP CONTROL OF PM MOTORS

The dynamic model for motor/load can be arranged into an open-loop block diagram with voltage and frictional (or gravitational) load torque as inputs and speed as the output. See Fig. 12.6. The "loop" includes a voltage amplifier with gain K_A. Part of the control voltage V_R goes to accelerating the inertia load. The steady-state speed is proportional to the control voltage and if T_f is small, $\omega = (K_A/K_E)V_R$. Thus, open-loop control of PM motors is speed control, although position control can be inferred from the time integral of the control

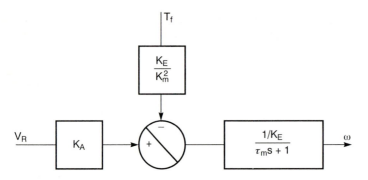

FIGURE 12.6
Open-loop speed control for DC motor.

voltage; i.e., applying a known voltage $V_R(t)$ for a fixed time will give a corresponding rotation θ.

There are two methods to control the motor by means of the amplifier: (1) linear transistors and (2) switching transistors. Linear transistors are power transistors that operate in their linear range, motor voltage being proportional to V_R. Figure 12.7 shows a T-type configuration for driving motors with complementary power transistors. When the control voltage is positive, the motor runs clockwise; negative voltages reverse the motor. A single transistor can be used for unidirectional control.

An alternative drive method using switching transistors is more suitable to digital control. These Darlington and MOSFET transistors are discussed in Chap. 15. There are two approaches: (1) pulse-width-modulated (PWM) and (2) pulse-frequency-modulated (PFM). PWM control voltages work by switching the full voltage of the DC power supply to the motor on and off at a fixed frequency. By varying the duty cycle (on portion of the

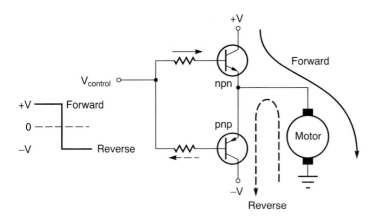

FIGURE 12.7
T circuit with linear amplifiers for open-loop speed control.

cycle), the average voltage seen by the motor varies in proportion to the duty cycle. Switching frequency is above the audible range, on the order of 16,000 cycles per second, much higher than the motor time constant so that the motion is smooth and continuous. The PFM control voltage switches the DC supply on for a fixed period of time (fixed width) but varies the spacing between the pulses or repetition rate. For low average voltages as seen by the motor, the frequency is low. Voltages are increased by increasing the switching frequency. PFM control is less popular but common for hobby servos used in radio controlled models. An H-bridge configuration, Fig. 12.8, consists of four switching transistors. Motor direction is controlled by which input receives the PWM voltage. In the forward mode, e.g., transistors Q1 and Q4 are ON, and current flow is left-to-right through the motor. A system that is more compatible with microprocessors is shown in Fig. 12.9. A PWM voltage is applied to a single input, and the motor direction is controlled with a HIGH/LOW voltage on a second input. Either T or H configurations can employ switching or linear amplifiers. The H bridge is preferable because it takes a single unipolar power supply, but it is more difficult to control with the linear method.

Linear transistors are confined to low-to-medium performance (less than 5 A) motors because of heat and power loss. Power dissipation in transistors is the product of current and voltage drop. Linear transistors have large voltages across them, whereas switched transistors have a voltage drop on the order of 0.2 V. If the excessive heat generated is not removed, motor current (torque) must be reduced. Linear transistors are susceptible to breakdown at peak current levels, which occur when motors are overloaded. For this reason current limiting circuits are employed.

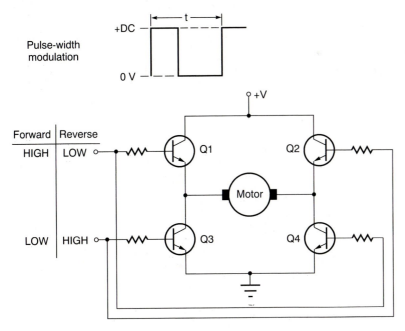

FIGURE 12.8
H circuit with switching amplifiers (PWM) for open-loop speed control.

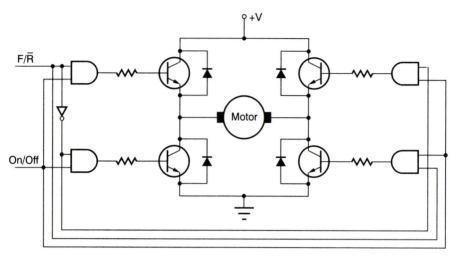

FIGURE 12.9
Microprocessor interface for H circuit.

12.4 CLOSED-LOOP CONTROL

Open-loop speed control inherently assumes that all parameters will remain constant as the system operates. The moment of inertia J and the steady-state load torque T_f are relatively constant for some systems such as peripheral drive spindles, but they can change drastically in other systems such as robotics. The gravitational load torque T_f of a manipulator changes with position, and joint inertias can vary by a factor of 10 over the range of movement. Independent of the dynamic load is the effect of heat on the motor constants. Temperature can be related to the constants from the following:

$$R = R_0[1 + 0.004(\theta - \theta_0)]$$
$$K_T = K_{T_0}[1 - 0.002(\theta - \theta_0)]$$

where nominal values are for a magnet temperature $\theta_0 = 25°C$. Thus, a 25 percent or more change is likely at operating temperatures. For these reasons closed-loop control for speed is the rule, and closed-loop control for position is a necessity with PM motors.

12.5 CLOSED-LOOP SPEED CONTROL
WITH DC MOTORS

With closed-loop control the motor speed is sensed with a tachometer. A tachometer is similar to a PM motor running as a generator. Since the device does not produce power, it is constructed to give an output voltage that is accurately proportional to the rotor speed (slight cogging can be eliminated with a low-pass filter). "Tachs" are housed in a separate case, which is piggybacked onto the motor. Manufacturers often sell the pair as a package. Tach feedback voltage goes to a comparator which subtracts the voltage from reference voltage V_R. The difference drives the gain amplifier. Figure 12.10 is a block diagram of the system.

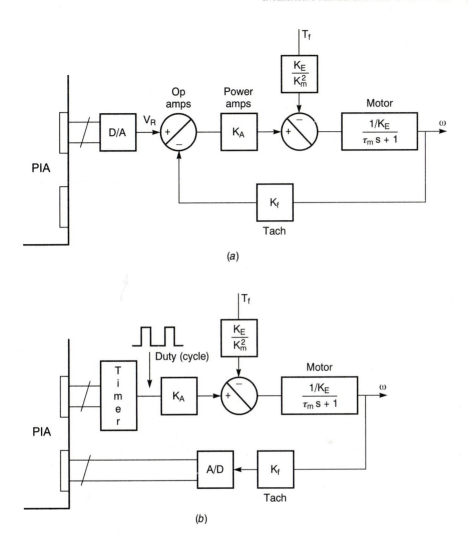

FIGURE 12.10
Two methods for closed-loop speed control. (*a*) Linear (analog) system; (*b*) digital (PWM) system.

We can see that there are two drive approaches to using microprocessors:

1. *Analog.* The microprocessor merely supplies a reference byte to PIA port B, which is converted to a voltage with a digital-to-analog converter (DAC). An op-amp comparator circuit drives linear transistors having gain K_A. The loop is analog.
2. *Digital.* Switching transistors require a microprocessor to close the loop. Some manufacturers supply an amplifier box unit that inputs analog voltages and outputs a PWM signal for switching transistors. For breadboard designs, tachometer voltages are converted to digital data for processing with an analog-to-digital converter (ADC). The microprocessor compares actual speed with desired speed located in memory. In many cases, including robotics, the microprocessor controls a single motor (distributed

control). A central computer supplies an updated desired speed to the PIA. The difference byte goes to a 6840 programmable timer which controls the signal duty cycle. Gain K_A is not a physical device but represents the duty cycle fraction of the switched voltage, the gain being 0 to 100 percent of full scale.

If we apply block diagram algebra to the control loop, output speed is expressed as

$$\omega(s) = \left[\frac{\alpha}{\alpha\tau_m s + 1}\right]\left[\frac{K_A}{K_E}V_R(s) - \frac{1}{K_m^2}T_f(s)\right] \qquad (12.20)$$

where $\alpha = \dfrac{1}{1 + K_f K_A/K_E}$.

Comparison of Eq. (12.19) with Eq. (12.20) shows similar forms of the system equations, but the effect of tachometer feedback is to reduce the time constant. However, assuming no change in the desired speed, a constant torque disturbance will produce a steady-state velocity error in this system:

$$E_{SS} = \frac{\alpha T_f}{K_m^2}$$

Increasing system gains K_A and K_f will reduce the effect of the torque disturbance. For applications where the load inertia is variable, the only system change will be in the time constant.

Steady-state velocity errors can be eliminated by introducing a proportional plus integral (PI) controller. See Fig. 12.11. Proportional + integral action is easily constructed with op amps for the analog closed-loop shown. With the digital system proportional + integral action becomes part of the software. Referring to the more general three-term or PID (Proportional + Integral + Derivative) controller, output c is expressed by

$$c = K_p e + K_i \int e\, dt + K_d \frac{de}{dt} \qquad (12.21)$$

or

$$c = K_p\left[e + \frac{1}{T_I}\int e\, dt + T_d \frac{de}{dt}\right] \qquad (12.22)$$

since conventional analog controllers specify the integral and derivative action times (T_I, T_d), which are more meaningful to the control engineer. Representing Eq. (12.22) in discrete format for software implementation, it can be shown that

$$c_N = c_{N-1} + A_0 e_N - A_1 e_{N-1} + A_2 e_{N-2} \qquad (12.23)$$

where A_0, A_1, and A_2 are constants defined by

$$A_0 = K_p\left[1 + \frac{\Delta t}{T_I} + \frac{T_d}{\Delta t}\right]$$

$$A_1 = K_p\left[1 + 2\frac{T_d}{\Delta t}\right]$$

$$A_2 = K_p\frac{T_d}{\Delta t}$$

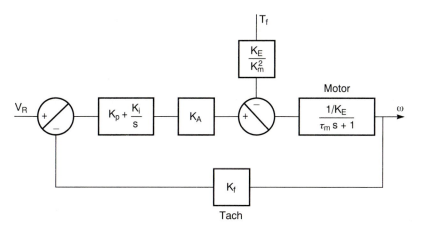

FIGURE 12.11
Closed-loop speed control with a PI controller.

Again, employing block diagram algebra, the PI system reduces to

$$\omega(s) = \frac{\left[\dfrac{K_p}{K_i}s + 1\right]\omega_R - \dfrac{sT_f}{K_m^2 K_f K_i K_A}}{\dfrac{s^2}{\omega_N^2} + \dfrac{2\zeta}{\omega_N}s + 1} \tag{12.24}$$

where $\omega_R = K_f V_R$

$$\omega_N = \left(\frac{K_f K_i K_A}{\tau_m K_E}\right)^{1/2}$$

$$\zeta = \frac{K_E + K_p K_f K_A}{[2K_f K_i K_A K_E \tau_m]^{1/2}}$$

We see that the response is now second order. Integral gain and tachometer gain increase the natural frequency, and thus decrease the response time. Damping is controlled by the proportional gain. A disadvantage of this system is that it can no longer tolerate dynamic changes in the load inertia J which affects the damping and system oscillation.

12.6 POSITION SENSING

DC motors are used to control position more than velocity. Feedback transducers for position are either resolvers or optical encoders. A *resolver* resembles a motor, having an armature and field windings. The single armature winding is energized with an AC voltage. Two field windings are offset 90° from each other. See Fig. 12.12. The two outputs will generate sine and cosine waves whose voltage levels vary depending upon the position of the resolver rotor. An electronic circuit (resolver converter) compares the two waves and converts the field voltages to an analog signal proportional to the rotor angle. Accuracy of commercial resolvers is typically 2 to 20 min of arc.

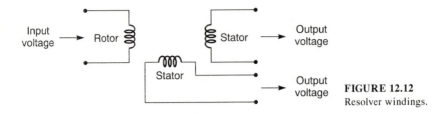

FIGURE 12.12
Resolver windings.

The most widely used position transducer is an *optical encoder*. It is price-competitive with resolvers, and its binary output makes the optical encoder ideal for microprocessor interfacing. Optical encoders have four major components: light source, optical disk, light sensor, and conditioning electronics. An optical disk (glass or plastic) is made with alternating opaque and transparent bands. The disk is placed between an LED and phototransistor. When the disk rotates, light passes through the transparent portions and is blocked by the opaque portions of the disk. Conditioning electronics produce clearly defined pulses when light is transmitted through the disk. There are two types of optical encoders: incremental and absolute.

Incremental encoders have one, two, or three tracks—an LED and phototransistor for each track. Figure 12.13(*a*) shows a two-track incremental encoder. A single track simply outputs a series of pulses as the disk rotates but cannot determine the direction of rotation. Its output is the same for both directions of rotation. Two-track incremental encoders can determine direction. One track is constructed with one-quarter cycle shift in relation to the other. A cycle is defined as one light plus one dark band. The result is a 90° shift in pulses from each track as illustrated in Fig. 12.13(*b*). Track B pulses will lead track A pulses for a clockwise (CW) rotation. Track A pulses will lead track B pulses for a counterclockwise (CCW) rotation. We can also increase the pulse count per revolution by a factor of 4 with two tracks. Special circuits can generate pulses in response to rising and falling edges of the sensor outputs. Encoders are available with single track increments of 5000 per revolution.

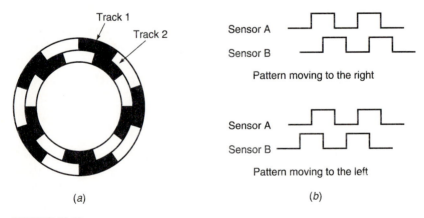

(*a*)

(*b*)

FIGURE 12.13
Two-track incremental encoder. (*a*) Bands; (*b*) output signals.

In addition to two primary tracks, an optical encoder may have a third track that has a single transparent band. Its output is one pulse per revolution, called the zero reference or index pulse.

Incremental encoders can be used to determine velocity by counting pulses for a specified time period. In feedback loops the time period must be small to give a short-term average of the "instantaneous" velocity. Since the revolution is ±1 count, large errors occur with a corresponding small number of counts. This is why tachometers are preferred for velocity measurement.

Example 12.1. A 256-count per revolution incremental encoder supplies the feedback position in a control loop. Thus each count represents 1.4°. There are two tracks, A and B, wired to bits A0 and A1 of a PIA port. Write a program to store the number of complete revolutions at address 0000 and the position binary number at address 0001. Allowances should be made for both CW and CCW motions.

As the rotor moves CW, the port bit pattern for each count will be 00, 01, 11, 10, repeat. For a CCW rotation, the pattern will be 00, 10, 11, 01, repeat. See Fig. 12.13(*b*). The program must detect 00, then use the next bit pair to establish direction. If the direction is CW, increase the count. If the direction is CCW, decrease the count. Set up a 16-bit count; the HIGH-byte is rotations.

```
            0000    Rotations
            0001    Count
ENCODE  CLR     0000
        CLR     0001
GET00   LDA A   PORT      Isolate bits 0,1
        AND A   #03       and find pattern 00.
        CMP A   #00
        BNE     GET00

DIR     LDA A   PORT
        AND A   #03       Determine the
        CMP A   #01       direction.
        BEQ     CW
        CMP A   #10
        BEQ     CCW
        BRA     DIR

CW      CLR B
        ADD A   #01       Increment count.
        CLC
        ADD A   0001
        ADC B   0000
        BRA     GET00

CCW     CLR B
        ADD A   #01       Decrement count.
        CLC
        SUB A   0001
        SBC B   0000
        BRA     GET00
```

As an alternative to the preceding software, incremental encoder pulses can be converted directly to a binary number by using the dual 4-bit 74LS393 binary counter. A short program must be written to keep track of the total number of complete revolutions.

One disadvantage of incremental encoders is false counts being generated by electrical transients or outside disturbances. Large errors can also result from a power interruption because the count is lost. These errors can be eliminated with absolute encoders.

Absolute encoders differ from incremental encoders by including many more tracks (up to 10) on the disk. Each track with a LED-phototransistor pair form a bit in a binary parallel output to define the shaft position. Figure 12.14 shows a *four*-track absolute encoder with a resolution of one in $(2^4 - 1)$. Each bit is a position change of 24°.

Some absolute encoders have a Gray code output rather than a binary output. See Table 12.1. The Gray code changes from one number to the next by changing only one bit. When counting is done in standard binary, it is not uncommon for several bits to change

TABLE 12.1
Comparison of binary to Gray code

Decimal	Binary	Gray
0	0000	0000
1	0001	0001
2	0010	0011
3	0011	0010
4	0100	0110
5	0101	0111
6	0110	0101
7	0111	0100
8	1000	1100
9	1001	1101
10	1010	1111
11	1011	1110
12	1100	1010
13	1101	1011
14	1110	1001
15	1111	1000

Photocells

FIGURE 12.14
Four-track absolute encoder.

at one time. For example, changing binary 7 (0111) to binary 8 (1000) changes all 4 bits. For the Gray code, changing code 7 (0100) to code 8 (1100) changes only the most significant bit (MSB).

Using the Gray code significantly reduces any error that may occur. If the binary encoder is read at the instant of transition from 7 (0111) to 8 (1000), the first 3 bits of 8 may be read as 000 and the MSB also read as 0 because the disk may not have moved far enough to change the MSB. The number read would be 0000, not 1000. This represents an error that is the weight of the MSB. Transition from 7 to 8 with the Gray code can only change the MSB, the difference between a single digit. When Gray codes are used, the largest error that can be made is the weight of the least significant bit (LSB).

An instruction sequence to convert 8-bit Gray code output to binary is

```
GR-BIN   LDA A   PORT
         STA A   MEM
         CLR A
LOOP     EOR A   MEM
         LSR     MEM
         BNE     LOOP
```

Hewlett Packard makes high-performance optical encoders which are easy to use. The *HEDS-5500* is a quick assembly version that comes with its own hex wrench to mount onto the motor shaft. See Fig. 12.15. The encoder contains an LED light source and a code wheel which rotates between the LED emitter and detector. Integrated circuitry provides a TTL compatible output for the two-channel (A and B) quadrature signals. Specifications are:

Power:	5 V at 40 mA (max)
Size:	1.62×1.18 in
Moment of inertia:	8×10^{-6} oz-in-s^2
Resolutions available:	96, 100, 192, 200, 256, 360, 400, 500, 512 (cycles/rev)
Shaft diameters:	2 mm, 3 mm, $\frac{1}{8}$ in, $\frac{5}{32}$ in, 4 mm, $\frac{3}{16}$ in, 5 mm, $\frac{1}{4}$ in

Hewlett Packard supplies quadrature decoder/counter interface ICs to utilize the optical encoder easily. Channels A and B are tied directly to the chip, which outputs a binary count of the cycle resolution for the microprocessor. HCTL-2000, shown in Fig. 12.16, has a 12-bit output while the HCTL-2020 extends to a 16-bit output:

Pin-Out (HCTL-2000)

V_{DD}	5 V
V_{SS}	GND
CH A, CH B	To the optical encoder.
D0–D7	Data lines to the MPU.
CLK	To the $\phi2$ clock of MPU for internal timing.
RST	Resets the internal counters to zero. Can be tied to MPU RST.

(a)

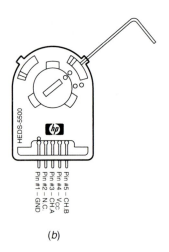

(b)

FIGURE 12.15
Hewlett Packard incremental optical encoder Model No. HEDS-5500. (a) Full view; (b) pin-out. (*Reproduced with permission of Hewlett Packard Co.*)

$\overline{\text{OE}}$	Output enable (or chip select) tied to address decoder for direct interface to MPU.
SEL	When SEL is HIGH, the LOW-byte of the count is on data pins. When SEL is LOW, HIGH-byte (with 0000 at HIGH nibble) is on data pins. Connect to A0.

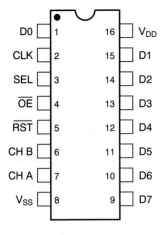

FIGURE 12.16
Hewlett Packard HCTL-2000 quadrature decoder interface for an optical encoder. (*Reproduced with permission of Hewlett Packard Co.*)

12.7 CLOSED-LOOP POSITION CONTROL WITH DC MOTORS

Shaft position is obtained by integrating the velocity output of the speed control loop. By adding a position feedback loop and position comparator, we arrive at the classical closed-loop block diagram for position control with DC motors. See Fig. 12.17. Carrying out the necessary steps for loop reduction using block diagram algebra, position θ is given by

$$\theta = \frac{V_R/K_B - K_E T_f/K_m^2 K_A K_B K_C}{(s^2/\omega_N^2) + (2\zeta s/\omega_N) + 1} \tag{12.25}$$

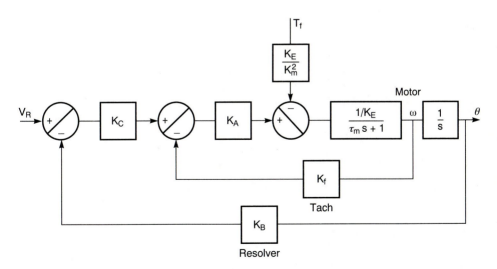

FIGURE 12.17
Position control with tachometer feedback.

where $\omega_N = \sqrt{\dfrac{K_A K_B K_C}{K_e \tau_m}}$

$$\zeta = \frac{1}{\alpha \tau_m \omega_N}$$

$$\alpha = \frac{1}{1 + (K_f K_A / K_E)}$$

Gains are chosen so that the damping factor ζ is 0.7 and the natural frequency is large for broad bandwidth (i.e., fast speed of response). Damping is controlled by velocity feedback gain K_f.

Since the command position is $\theta_R = V_R / K_B$, there is no steady-state error when the torque disturbance is zero. Again, a torque disturbance on the system produces a steady-state position error, but the error can be minimized by selecting a high gain $K_A K_C$. Position error from a torque disturbance can be eliminated by a proportional + integral controller in place of K_C or included within the inner loop. This design must be done carefully since the characteristic equation is of third order rather than second order, and inappropriate selection of controller gains will cause an unstable response. The problem is compounded when the system moment of inertia changes over the range of operation. A more complex controller scheme with microprocessor software sometimes is used to deal with large inertia variations.

An alternative position control system is shown in Fig. 12.18. It consists of a single feedback loop for position but has a more complex control network, namely, a lead-lag compensator. Either system can be controlled with a microprocessor. Given the possible combinations of drive systems, control loops, analog, digital, software, and hardware, many approaches to motor control can be found.

The classical system, Fig. 12.17, generally is an analog system with the microprocessor supplying the command voltage V_R. The motor is driven with linear transistors (or power op amps), and the comparators are op-amp circuits which also set the gains. A resolver mea-

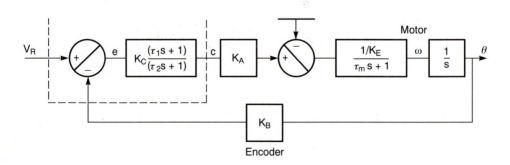

FIGURE 12.18
Position control with lead-lag compensator.

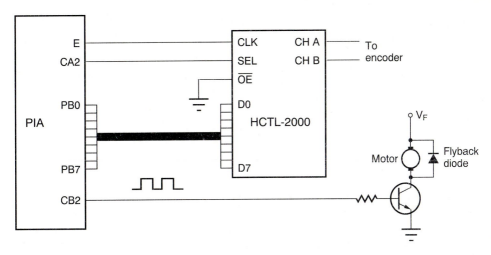

FIGURE 12.19
Connection of a PIA to the Hewlett Packard HCTL-2000 decoder/counter interface IC.

sures position. If an encoder is used, the first comparator becomes part of the microprocessor software. If control is PWM, microprocessor software encompasses both comparators.

The system in Fig. 12.18 is more suitable for microprocessor control. Compensation is easily implemented with software by converting the block to finite difference form:

$$c_N = B_0 e_N - B_1 e_{N-1} + B_2 c_{N-1} \tag{12.26}$$

where constants $B_0 = K_C[(\tau_1/\Delta t) + 1]\gamma$
$B_1 = K_C \tau_1/(\Delta t \gamma)$
$B_2 = \tau_2/(\Delta t \gamma)$
$\gamma = (\tau_2/\Delta t) + 1$

The error signal e is computed by subtracting the encoder count from a set-point value. The latter can be stored in memory or supplied to the MPU system port by a higher level computer.

Figure 12.19 shows the hardware to drive a simple unidirectional DC motor. An HCTL-2000 provides a position count to port B of the PIA. Line CA2 is configured for the output mode to select the proper count byte. Once the motor control voltage is determined, line CB2 can be used to output a PWM signal to the motor power transistor. Chapter 15 gives a thorough discussion of power transistors.

Example 12.2. Write a program to generate a PWM signal on PIA line CB2. Full motor voltage V_F is applied to the power transistor. Let the 8-bit binary number $(N)_2$ stored at address (0000) represent the control voltage V_C, where

$$V_C = \left[\frac{(N)_{10}}{256}\right] V_F$$

If CB2 is ON during a timing loop initialized by $(N)_2$, and if CB2 is OFF during a timing loop initialized by $FF-(N)_2$, the correct PWM signal will be generated. Place the PIA at address space 4000–4003 and initialize CRB for programmed output:

```
        ⋮
        INIT PIA
        ⋮
PWM     LDA A   #3C       Send a HIGH.
        STA A   4003
        LDA A   0000
        INC A             Delay count (N)₂
ON      DEC A
        BNE     ON
        LDA A   #34       Send a LOW.
        STA A   4003
        LDA A   #FF
        SUB A   0000
        INC A             Delay count FF-(N)₂.
OFF     DEC A
        BNE     OFF
        BRA     PWM       Another pulse.
```

Pulse period is the sum of two timing loops, or approximately 1.5 ms.

All these approaches require a dedicated microprocessor. Just recently several general-purpose motion control ICs have come on the market which free the microprocessor. Configuration bytes are written to several on-board registers, including the set-point register. All that is needed for a complete servo system is a host processor to specify commands, an amplifier, and a motor with an incremental encoder. The Hewlett Packard HCTL-1100 motor controller provides position and speed control for DC, DC brushless, and stepper motors. Its block diagram is given in Fig. 12.20. A host processor views the motor controller as a block of memory through an 8-bit multiplexed address/data bus interface. The input command (set point) and feedback register are 24-bit counters keeping track of position. The HCTL-1100 can execute any one of four DC control algorithms selected by the user:

1. Position control
2. Proportional speed control
3. Integral speed control
4. Trapezoidal profile control for point-to-point moves

Desired position (or speed) is compared to actual position (or speed) and the signal multiplied by a lead compensator (digital filter). Minimum programmable sample period is 128 μs with a 1 MHz clock. The output is available at the motor command port as a byte and at the PWM port as a PWM signal. The HCTL-1100 can also provide electronic commutation for DC brushless and stepper motors. The commutator is programmable to enable the correct phase sequence for most motor encoder combinations.

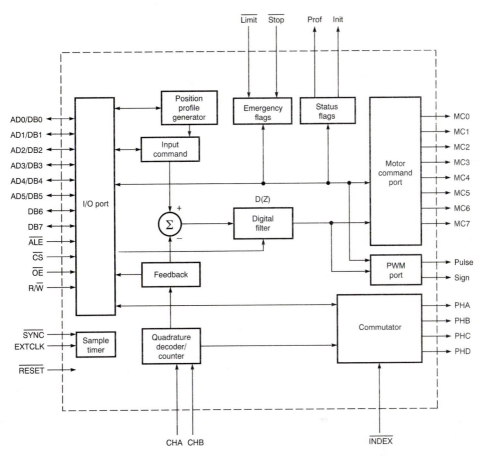

FIGURE 12.20
Hewlett Packard HCTL-1100 motion controller. (*Reproduced with permission of Hewlett Packard Co.*)

The National Semiconductor LM628/LM629 motor controller is similar to the HCTL-1100 for DC motor control. On board is a PID compensator. The LM628 has an 8-bit output to drive an 8-bit or 12-bit DAC and power amplifier. The LM629 provides the PWM output magnitude and sign for directly driving H switches. It can be interfaced to an LM621 brushless motor commutator to drive power switches for a brushless DC motor.

Presently, these motion controller ICs are very expensive, but time and competition quickly change the equation in electronics.

Example 12.3. A DC motor is to be used for position control. It comes with a tachometer, and an encoder (256 counts per revolution) is to be added. Determine the appropriate gains in Fig. 12.17. Specifications are listed as follows:

$K_T = 6.2$ oz-in/A $\tau_e = 2.1$ ms
$K_E = 4.6$ V/krpm $\tau_m = 35$ ms
$R_T = 2.2\ \Omega$ $K_f(\text{tach}) = 4$ V/krpm
$J_m = 4 \times 10^{-3}$ oz-in-s^2 $(1/N^2)J_L = 2 \times 10^{-3}$ oz-in-s^2
Rated voltage: 30 V

The motor time constant is always specified with no load. Adjusting for J_L yields

$$\tau_m = \frac{(4+2)(35)}{4} \text{ ms}$$

$$K_E = \frac{4.6(60)}{1000} \frac{\text{V} - \text{s}}{\text{rev}}$$

$$K_B = 1 \text{ byte/rev}$$

$$K_f = \frac{4(60)}{1000} \left[\frac{\text{byte}}{5 \text{ V}} \right]$$

where the bracketed term is the ADC for the tachometer. Set point is stored in memory as one byte with each bit representing $360°/256°$. The output byte to a programmable timer generates the PWM control signal. The duty cycle is 100 percent for a full byte (FF) to the timer. Then

$$K_A = 30 \text{ V/byte}$$

Solve Eq. (12.25) for K_C with $\zeta = 0.7$. Tachometer feedback gain can be increased with an op amp if larger K_C is desired.

12.8 BRUSHLESS DC MOTORS (BLDCs)

Permanent magnet DC motors described up to this point all have brushes to transmit power to the armature windings. Brush arcing causes electronic noise and maintenance problems from excessive wear. A brushless DC motor (BLDC) has been developed to overcome these problems. It substitutes electronic commutation for the conventional mechanical brush commutation. Because the electronic commutation in BLDCs exactly duplicates the brush commutation in conventional DC motors, it exhibits the same linear torque-speed curve. Furthermore, the BLDC has the same motor constants and obeys the same performance equations.

Actually, the BLCD has been around since the 1960s, but cheap electronics and the advent of microprocessors for which it is ideally suited has made these motors a very competitive design alternative today. BLDCs have several advantages over their counterparts:

1. *High reliability.* The life of BLDCs is almost indefinite. Bearing failure is the most likely weak point.
2. *Quiet.* A lack of mechanical noise from brushes makes it ideal for a people environment. An added advantage is that there is no mechanical friction.
3. *High speed.* Brush bounce limits DC motors to 10,000 rpm. BLDCs have been designed for speeds up to 100,000 rpm, limited by the mechanical strength of the PM rotors.
4. *High peak torque.* BLDCs have windings on the stator housing. This gives efficient cooling and allows for high currents (torque) during low duty cycle, stop-start operations. Peak torques are more than 20 times their steady ratings compared to 10 times or less for conventional DC motors. Maximum power per unit volume can be 5 times conventional DC motors.

The biggest hurdle to overcome using BLDCs is the relatively high cost. Choice is restricted because there are few manufacturers. However, price is usually worth it when considering complex machinery where normal downtime and maintenance are not only costly in itself but often unacceptable.

Figure 12.21 shows the standard BLDC with the windings on the stator. This design is sometimes referred to as an inside-out DC construction because it is opposite the typical DC PM motor with the windings on the armature. BLDCs are also made with the stator on the inside, the rotor shell surrounding the windings. Inside rotors have less inertia and are better suited for the start-stop operation. Outside rotors are better for constant load, high-speed applications. Here is how it works.

The operation of a BLDC is similar to the operation of a step motor. A PM rotor with pairs of N-S poles aligns itself with the N-S field of a wound stator. Several separate windings (phases) surround the stator. Electronic switching of power to the phases advances the field around the stator (electric commutation) with the rotor following. BLDC design stresses continuous rotation as opposed to "step" motion.

FIGURE 12.21
Brushless DC motor with windings on the stator. (*Courtesy of Motion Systems, Kimco Magnetics Division.*)

To determine the instant when to switch the field, "Hall effect" sensors are strategically located around the stator and near the rotor. These sensors are small semiconductor slabs of indium/antimony. A voltage is generated across its output terminals proportional to the magnetic flux from the rotor poles. A HIGH is generated as long as the N-pole is across from a sensor.

Most BLDCs have three phases and two or four poles. Figure 12.22(a) is an illustration of a *two-pole* motor. We can see that two power transistors are required for each of three terminals. Wye-connected stator windings are alternately shared when the field commutates, and the current through the various phases changes direction as in conventional brush commutated armatures. Three Hall sensors are located near each phase 120 mechanical degrees apart. In position 1 we find that the rotor N-pole has just passed sensor *a* but has not yet reached sensor *b*. See Table 12.2. Thus, the sensor output is $\bar{a}bc$, and phases *A*, *B* are activated by transistors A^+ and B^-. The *A* side of the winding is an N-pole and the *B* side of the winding is an S-pole, thus propelling the rotor in a CW direction. After the rotor moves 60° to position 2, sensor *b* turns on as well as phases *A*, *C*. Continuing for 360°, sensor logic and phase combinations change six times. Each sensor is ON for a 180° rotation and OFF for a 180° rotation.

ON transistors are governed by a unique sensor logic combination. For example,

$$A^+ = \bar{a}\bar{b}c + \bar{a}bc = \bar{a}c$$

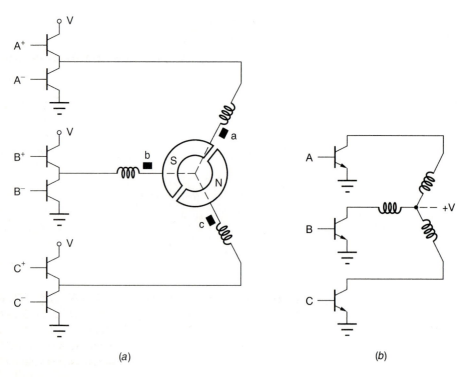

(a)   (b)

FIGURE 12.22
Operation of a two-pole, three-phase brushless DC motor. (a) Six transistors; (b) three transistors.

TABLE 12.2
Commutation logic and phase switching for two-pole BLDC

	Sensor logic (120 mechanical degrees)			
Position	*a*	*b*	*c*	ON transistors
1 (0°)	0	0	1	A^+ B^-
2 (60°)	0	1	1	A^+ C^-
3 (120°)	0	1	0	B^+ C^-
4 (180°)	1	1	0	B^+ A^-
5 (240°)	1	0	0	C^+ A^-
6 (300°)	1	0	1	C^+ B^-

Logic for all transistors is

$$A^+ = \bar{a}c \qquad B^+ = b\bar{c} \qquad C^+ = a\bar{b}$$
$$A^- = a\bar{c} \qquad B^- = \bar{b}c \qquad C^- = \bar{a}b$$

Rotor direction can be reversed by simply reversing the logic (negate signals *a*, *b*, *c*).

It is typical to operate wye-connected motors with the voltage applied to the common phase center as in Fig. 12.22(*b*). Current flow is now in one direction only, and three transistors are required. Four-pole (or more) magnets can be used with either connection. With four-pole rotors, Hall sensors can be located 120 mechanical degrees apart or grouped 60 mechanical degrees apart. In many references this spacing is expressed in electrical degrees, where

$$\text{Electrical degrees} = \text{mechanical degrees} \left(\frac{\text{No. rotor poles}}{2} \right)$$

In either case sensor logic changes every 30 mechanical degrees, and each sensor is ON for a 90° rotation and OFF for a 90° rotation. See Table 12.3. When the Hall sensors are placed 60° apart, phase logic is

$$A = b\bar{c} \qquad B = \bar{a}c \qquad C = a\bar{b}$$

TABLE 12.3
Commutation logic and phase switching for four-pole, common phase center BLDC

	Sensor logic (mechanical degrees)						Phase		
	60°			120°					
Position	*a*	*b*	*c*	*a*	*b*	*c*	*A*	*B*	*C*
1 (0°)	1	0	1	1	1	0	OFF	OFF	ON
2 (30°)	1	0	0	1	0	0	OFF	OFF	ON
3 (60°)	1	1	0	1	0	1	ON	OFF	OFF
4 (90°)	0	1	0	0	0	1	ON	OFF	OFF
5 (120°)	0	1	1	0	1	1	OFF	ON	OFF
6 (150°)	0	0	1	0	1	0	OFF	ON	OFF
Repeat									

By using three transistors instead of six, the circuit is simplified and less costly. But the motor constant K_m is reduced by approximately 20 percent at lower speeds to 40 percent at higher speeds.

BLDC Controllers

The electronics to drive BLDCs is considerably simplified when the engineer uses a controller/driver IC. These devices are primarily intended to operate open loop but are versatile enough to adapt for closed-loop control. Both the Motorola MC 33034 controller and the Unitrode UC3620 controller are similar in operation. The Motorola unit has outputs which can drive either MOSFET or bipolar power transistors supplied by the user. The Unitrode unit has power transistors (and suppression zeners) on-board, so we will discuss it.

A block diagram for the UC3620 is shown in Fig. 12.23. It contains commutation and drive, error amplifier, and chop-mode modulator. Commutation and drive consists of a decoder and the six power transistors. Their function is to receive positional information from the Hall sensors and to commutate the three motor phases in a proper sequence. The emitters of the three bottom transistors are connected to ground (pin 1) through resistor

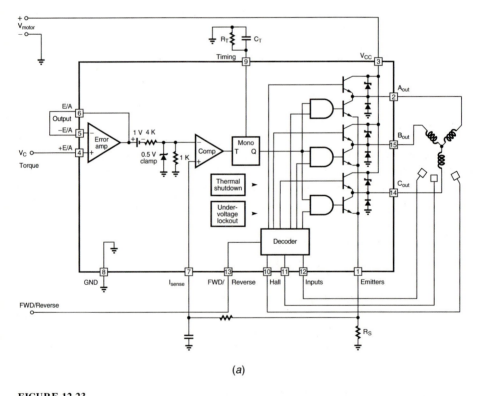

(a)

FIGURE 12.23
Unitrode's UC3620 3-phase brushless DC motor controller IC. (*a*) Open-loop speed control; (*b*) closed-loop speed control. (*Courtesy of Unitrode Integrated Circuits.*)

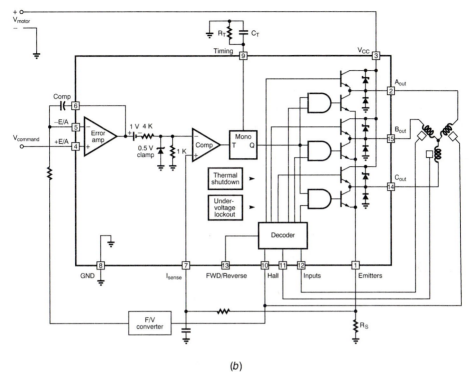

(b)

FIGURE 12.23 (continued)

R_S. Since all the motor current flows through this resistor, a voltage proportional to the motor current will be generated. This current sense voltage serves as a feedback signal for switching OFF the drive in the chopped mode. Feedback to pin 7 is through an RC filter to remove voltage spikes. An internal voltage comparator compares the current sense voltage to the positive voltage output of the error amplifier. When the current sense voltage rises above the error voltage, a monostable vibrator disables the output. The OFF time is set by

$$T_{\text{OFF}} = 0.916R_T C_T$$

If pin 7 voltage is less than the error voltage, the motor is turned ON, the ON time being proportional to the difference between the comparator voltages. A 0.5-V clamp zener on the error voltage input assures that the motor voltage never exceeds the current limit since an emitter voltage greater than 0.5 V would disable the drive. Peak motor current for the UC3620 is limited to 3 A.

In the open-loop mode a command voltage for speed is applied to pin 4. Pins 5 and 6 are connected externally to form a voltage follower. In the closed-loop mode the external pins allow several gain options—a proportional or an integral gain. As Fig. 12.23(b) shows, a feedback capacitor forms integral action within the inner velocity loop. A feedback voltage proportional to speed can be obtained from one of the Hall sensors (pin 10), thus eliminating the need for a tachometer. Sensor output, consisting of two pulses per revolution for a four-pole rotor magnet, is converted to a voltage for pin 5. There are several circuits to accomplish the task. One option is the MC 33039 closed-loop BLDC adapter specifically made for this purpose. Closed-loop position control is the same as DC brush motors.

The UC3620 can handle any number of rotor magnet poles or Hall sensor locations. The rule is, multiply the number of poles by the mechanical angle between Hall devices. If the product is 240 or greater, the UC3620 will handle it. If less, simply invert one Hall line.

12.9   MOTOR SELECTION

We have reviewed several qualitative features about each motor type as well as motor specifications, which present a start in the motor selection process. It should be emphasized that there is no infallible guide to selecting the best motor. There are always several workable configurations. Constraints can often eliminate several designs. For instance, lack of space can limit the motor diameter or the positioning resolution can rule out step motors.

The engineer must make the motor drive system work both electrically and mechanically. He or she should look at the motor-to-load interface before looking at the electrical drive-to-motor interface. In a typical motion control application the requirement will be to overcome some load frictional force and move a mass through a certain distance in a specified time. Therefore, the designer should weigh the following: (1) moment of inertia, (2) torque, (3) power, and (4) cost.

Load Inertia

For optimum system performance, the load moment of inertia should be similar to the motor inertia. When gear reducers intervene between motor and load, the reflected load inertia is J_L/N^2, where N is the gear ratio. If the motor inertia J_m is equal to the reflected load inertia, the fastest load acceleration will be achieved, or, conversely, the torque to obtain a given acceleration will be minimized. Therefore, matched inertias are best for fast positioning. On the other hand, peak power requirements are minimized by selecting the motor inertia so that the reflected load inertia is 2.5 times as large as the motor inertia. The torque will be increased but the maximum speed will be further reduced. A load inertia greater than 2.5 times the motor inertia is less than ideal, but it should not present any problems if the ratio is less than 5. A larger motor inertia implies that the same performance can be achieved at a lower cost by selecting a smaller motor.

Only 10 years ago PM motors were limited in the choice of type. A cup or basket motor (often called an ironless motor although, strictly speaking, a disk motor is ironless also) dominated the low inertia region, whereas the iron core motor dominated the high-inertia region. There was little overlap. Figure 12.24 shows the range of motor inertias on the market today. Overlap is extensive. An engineer can virtually consider any type of motor including brushless and stepper, at the first stage of the design-inertia match.

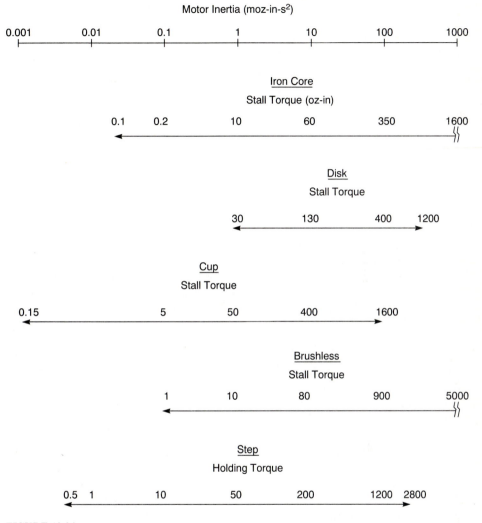

FIGURE 12.24
Range of motor inertia and typical stall torques for commercially available motors. Read vertically to match inertia and torque.

Torque

The motor must supply sufficient torque T_m to overcome the load friction and to accelerate the load over a distance (radians) s in time τ. Torque and acceleration at the motor are given by

$$T_m = \frac{T_L}{N} + J_m\alpha_m$$

$$\alpha_m = N\alpha_L$$

where the load torque is

$$T_L = T_f + J_L\alpha_L$$

Substituting the load torque in the preceding equation, we obtain

$$T_m = \frac{1}{N}[T_f + \alpha_L(J_L + N^2 J_m)]$$

For linear acceleration over distance s in time τ,

$$\alpha_L = \frac{2s}{\tau^2}$$

For damped ($\zeta = 0.7$) second-order response over distance s,

$$(\alpha_L)_{max} = \omega_N^2 s$$

Allowances should always be made for variations in load and bearing behavior as well as motor production variations. A designer should aim for a 50-percent torque allowance for most industrial applications.

An initial design should be planned without a gear reducer. In many cases direct drive is not possible because load torque requirements far exceed the torque delivered by a motor of reasonable size. Critical needs on space or weight can lead to gear reducers for otherwise perfectly matched motor/load systems. The problem with gear reducers is gear backlash. If gears mesh too tightly, there is severe sliding friction between teeth which can cause lockup. Thus, the teeth spacing, backlash, is a tradeoff between reducing the power loss within the gears (loose fit) or improving the position accuracy (tight fit) of the load. Resolution of step motors can be enhanced with gear reducers, but their accuracy remains within the bounds of backlash. Direct drive robots give a repeatability or maximum positioning error, which is an order-of-magnitude better than traditional robots with gear reducers.

Figure 12.24 presents the stall torque for each type of motor over its moment of inertia range. These values represent a composite of several commercially available motors at each designated inertia. Each manufacturer limits its line to a few types and sizes. There are several factors which have considerable effect on the stall torque for a given size motor: (1) the strength of the field PMs which is reflected in weight, not size, and (2) the air cooling of the motor which allows larger stall currents through the windings. For example, the data in Table 12.4 was obtained from two sources.

TABLE 12.4
The effect of magnet strength and core cooling on the stall torque of cup motors ($J_m = 1$ moiss)

Manufacturer #1		Manufacturer #2	
Stall torque	Weight	Stall torque	Fan static pressure drop
30 oz-in	1.7 lbf	55 oz-in	0 in H_2O
55	7.5	78	10
170	20	142	20

TABLE 12.5
Comparison of design parameters among motor types (J_m = 1 moiss)

Type	Cost	Size (D × L)	Peak torque/ stall torque (oz-in)	Time constant (ms)	Efficiency (%)
Iron core	1	2" × 3"	5–10	20	50–75
Cup	2	4 × 5	5–10	2	50–75
Disk	3	3 × 2	10–15	8	50–75
Brushless	1	2 × 1	5–25	20	75–90
Step	1	2 × 2	1–2	—	25–40
Encoder	1	—	—	—	—

Motor types are competitive over their inertia range with regard to stall torque. As expected, the ironless core motors do have better torque magnitudes. However, Table 12.5 shows that these motors are more expensive and tend to be larger in size. Small mechanical time constants combined with higher torques make them ideal for fast response systems.

Stall current, thus stall torque, is limited by the maximum permissible core temperature. If the design calls for short bursts of acceleration, peak currents—thus peak design torques—can far exceed their stall counterparts as long as the average power remains the same. All PM motors have excellent peak torque to stall torque ratios, with brushless motors having the highest ratio. Although step motors have exceptional stall torque characteristics, their peak torque capability is poor, and they are not good candidates for quick load accelerations.

Power

Besides maximum torque requirements, torque must be delivered over the load speed range. The product of torque and speed is power. Total power P is the sum of the power to overcome friction P_f and the power to accelerate the load P_a, the latter usually the dominant component:

$$P = T_f\omega + J\alpha\omega$$

Peak power required during acceleration depends upon the velocity profile. If the load is linearly accelerated over distance s in time τ, the maximum power is

$$(P_a)_{max} = \frac{4Js^2}{\tau^3}$$

If the load undergoes a damped ($\zeta = 0.7$) second-order response over distance s, the maximum power is

$$(P_a)_{max} = (0.146)J\omega_N^3 s^2$$

It is interesting that to accelerate a load in one half the time will require eight times the power.

Now the torque-speed curve for DC PM motors is a linear line from stall torque to no-load speed. Therefore, the maximum power produced by the motor is the curve midpoint or one-fourth the stall torque and maximum speed product. The maximum speed for various motors is:

Step motor:	200–400 steps/s
Step motor (L/nR drive):	400–800 steps/s
Step motor (chopper drive):	10,000 steps/s
Permanent magnet motor:	10,000 rpm
Brushless motor:	> 20,000 rpm

As a starting point, choose a motor with double the calculated power requirement.

Cost

Among several designs the single most important criteria is cost. Although it may be more prudent to choose the first workable design when only several units are involved, high-volume applications demand careful study of the economic tradeoffs. For example, a motor with a given inertia size can deliver a wide range of torques, depending upon the magnet strength. Price can vary by a factor of three or more over this torque range. Yet, by going to a larger motor with a lower strength magnet the same torque can be achieved at little or no increase in cost.

Table 12.5 shows the relative cost for motor types. Except for ironless motors, direct motor cost is similar. However, permanent magnet DC motors operate closed loop. The cost of encoders can equal if not exceed the cost of the motor itself. In addition, stepper and brushless motors have electronic expenses greater than brush motor electronic expenses. Are motor controllers to be used or is a greater burden of control placed on the software? Are you going to buy a motor control board? A step motor can be controlled for less than $10 worth of electronics. Add PIAs, optoisolators, chopper drive, and overload features; place them on a board for personal computers (PCs), and $100 worth of components can cost $1000.

PROBLEMS

12.1. Outline the programming steps necessary for implementing PI controller action in Eq. (12.23).

12.2. Outline the programming steps to control the motor in Fig. 12.11. The comparator and controller are in software. Input and output are binary through analog converters.

12.3. Write an assembly language code for reading an incremental encoder if dual 4-bit 74LS393 binary counters are placed between the encoder and PIA.

12.4. Explore the software problem of using an incremental encoder with 1024 pulses per revolution.

12.5. Write the assembly language subroutine for a lead-lag controller.

12.6. Outline the programming steps to control the motor in Fig. 12.18. How does the program change if the encoder pulses trigger an interrupt instead of being read at a port?

12.7. An HCTL-2000 decoder interface IC is to be connected directly to the 6802 MPU. The decoder occupies address space 4000–7FFF. Show a wiring diagram. Write the assembly language line(s) which will load the 12-bit count into addresses 1000–1001.

12.8. Write the complete assembly language code for the motor position system shown in Fig. 12.18. Replace the lead-lag compensator with a simple proportional gain K_c. Drive the motor with the PWM system in Fig. 12.19.

12.9. A spindle is to be controlled to position a mechanism at $5°$ intervals, $\mp 1°$. The driving DC motor, having characteristics given in Fig. 12.5, is powered by linear amplifiers. The mechanical time constant is 10 ms, and its tachometer gain is 1 V/krpm. Assume the load is a constant 10 oz-in with negligible inertia.

(*a*) Select an encoder pulse count per second.

(*b*) Calculate appropriate loop gains.

(*c*) Give the assembly language for its operation.

CHAPTER
13

INTERFACING
ANALOG

The microprocessor manipulates data in *digital* (binary) form: HIGH (logic 1) or LOW (logic 0), corresponding to discrete ranges of voltage. For TTL logic the range recognized is 2.0 to 5.0 V for a HIGH and 0 to 0.8 V for a LOW. When the microprocessor interfaces the physical world, we have observed that a number of peripherals are compatible with binary logic. Input switches are simply ON/OFF and optical encoders supply a pulsed TTL output. Seven-segment LEDs and stepping motors respond to bytes of binary output. Personal computer peripherals, such as keyboards, CRTs, and printers are binary interfaced.

When the microprocessor is used as a mechanism (or process) controller or as a data gathering tool, the interfacing signals are commonly voltages (or currents) which are *continuous in time.* These signals are called *analog,* meaning an entity similar to another entity. In mathematics it refers to the identical governing equations for dependent variables from different physical phenomena. In the real world it is the continuous electrical signal generated by a transducer sensing some physical parameter such as force, displacement, velocity, flow, pressure, strain, temperature, etc. A transducer simply converts one form of energy to another, an electrical signal generally. The physical principles on which the process is based include resistive, capacitive, inductive, photoconductive, piezoelectric, and electromagnetic elements.

13.1 ANALOG COMPONENTS

Analog signals from the transducer must be converted to a digital code to interface the microprocessor. The conversion process is carried out by an *analog-to-digital converter (ADC).* Analog signals are more complex than binary signals. Typical transducer outputs are in the millivolt (or less) range. This level must be amplified by operational amplifier circuits to TTL or CMOS levels before interfacing. Ground and impedance matching

306

problems must be circumvented. It may be necessary to follow the amplifier with a filter. If the analog signal changes rapidly, a *sample-and-hold* amplifier must be employed to give the ADC time to complete the conversion process. When multiple transducers must be read, the analog signals pass through a *multiplexer,* which is nothing more than a multiposition switch that connects each analog signal in turn to the sample-and-hold amplifier prior to entering the ADC. The microprocessor controls which signal is multiplexed through to the converter and what specific point in time the sample is taken. Each analog input path is a channel. Multichannel systems are called data acquisition systems. See Fig. 13.1.

When analog outputs are required, *digital-to-analog converters (DACs)* generate piecewise continuous signals from digital code. These converters can be used for real-time signal generation and signal processing applications. However, their main function is to generate control voltages for two actuator types: (1) solenoids and (2) DC motors for speed or position control. If several analog devices are controlled, a separate DAC is usually used for each line, or a single DAC can be used with a *demultiplexer.* The analog demultiplexer simply switches the single converter output between several output channels as shown in Fig. 13.2. Each demultiplexer output must interface a *sample-and-hold* amplifier to latch the particular channel analog signal. These low-level current signals are then boosted in power to drive the actuator.

All the aforementioned analog devices are found in the manufacturers' literature under *linear* (analog) ICs. We will discuss each, concentrating on the converters and the operational amplifiers. The latter are an integral part of the converters as well as the signal conditioning circuits.

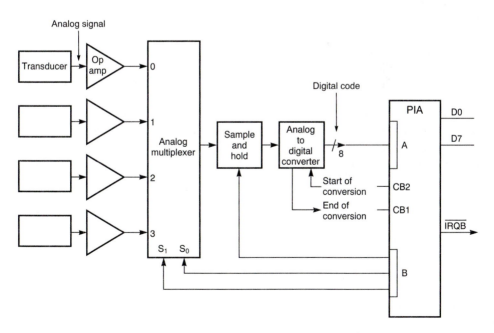

FIGURE 13.1
Analog input system to the microprocessor system.

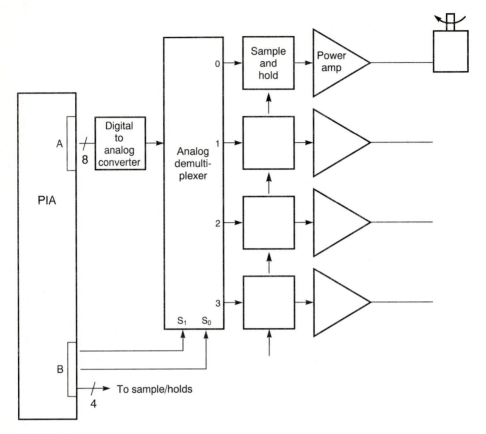

FIGURE 13.2
Analog output system from the microprocessor system.

13.2 IDEAL OPERATIONAL AMPLIFIERS

Operational amplifiers (op amps) are the basic ingredient in virtually all aspects of analog signal handling. Op amps are available as separately packaged linear circuits or are found as components in larger IC packages such as ADCs or DACs. The op amp itself is symbolized by a triangle. See Fig. 13.3. There are two inputs, one marked with a plus (+)—a

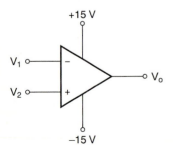

FIGURE 13.3
Symbol for an open-loop op amp: Comparator.

noninverting input—and one marked with a minus (−)—an inverting input—and a single output. The device is normally powered by dual supplies (±), although some single supply versions are available. Rated power is generally ±15 V with some op amps responding to a range: ±5 to ±20 V. An ideal op amp has the following characteristics:

1. Output voltage is proportional to the voltage *difference* between the noninverting input (+) and the inverting input (−). Thus,

$$V_o = A(V_+ - V_-) \qquad (13.1)$$

where A ideally is infinite. In practice, A is on the order of 10^6.

2. Input impedance at both inputs is infinite. An impedance of 1 MΩ is a more realistic value.

3. Output impedance is zero. It is typically 1 to 10 Ω.

4. Bandwidth is infinite; i.e., the gain will not fall off for rapidly changing inputs.

The output of an op amp is limited to a voltage less than the power supply and saturates at about ±12 V. It is apparent that an op amp in the preceding configuration will quickly saturate for anything but the smallest input signals as shown in Fig. 13.4. This arrangement is called an op-amp *comparator* since it compares the relative inputs and responds with a positive or negative saturation accordingly. Op amps can be made compatible with TTL logic systems by adding a pair of diodes to a 5-V source and to ground. See Fig. 13.5. The output now saturates to 0 and +5 V. Specifically designed op amps, also called comparators, such as the AD790, have TTL compatible outputs. Comparators are the basis for ON/OFF control systems. The two inputs are set-point voltage and sensor

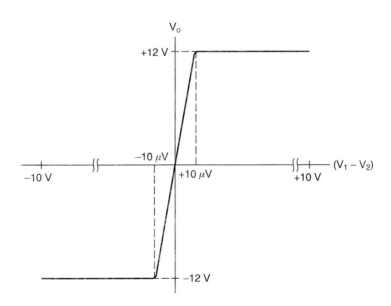

FIGURE 13.4
Output of a comparator.

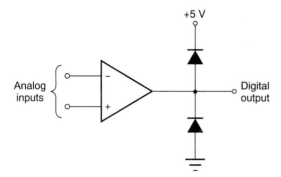

FIGURE 13.5
Comparator with 0 and 5 V output.

feedback voltage. The output drives a power transistor which switches the controlling power device ON or OFF, maintaining a constantly changing output about the set point.

13.3   COMMON OP-AMP CIRCUITS

The op amp would not be very useful if it were connected only as a comparator. Its strength lies in the concept of feedback. Consider Fig. 13.6(*a*), which shows an op amp with an added input impedance Z_i and a feedback impedance Z_f. The common node at the noninverting input is called the summing junction. Remember that the open-loop gain A attempts to drive the input voltage difference to zero; see Eq. (13.1). Since V_+ is ground, summing junction voltage $V_{SJ} = V_-$ is practically zero and is called a virtual ground. Current into the op amp is zero so that $I_i = I_f$. Then

$$\frac{V_i - V_{SJ}}{Z_i} = \frac{V_{SJ} - V_o}{Z_f}$$

We know that $V_{SJ} \approx 0$, so that

$$V_o = -\frac{Z_f}{Z_i} V_i \tag{13.2}$$

Practical output voltages are limited to supply voltage, and output currents cannot exceed the 10 to 20 mV range. Therefore, loads should be larger than approximately 1000 Ω. In addition, loads should be smaller than 10 MΩ so that currents do not fall into the noise range. Voltage gains of 1 to 1000 are typically found, whereas loads for the lower gains are around 10 kΩ.

Figure 13.6 illustrates several standard op-amp feedback configurations that are useful in signal conditioning and control. The reader is no doubt familiar with most of these configurations. For many transducers the magnitude of the input resistance is the same order as the transducer. Thus, the op amp will load the measured circuit and alter the very voltage it is trying to amplify. Table 13.1 lists common transducer output impedances and their signal levels to be amplified. To isolate a transducer from an op-amp amplifier, a buffer in the form of a *unity gain follower,* Fig. 13.6(*d*), can be used, although the noninverting circuit, Fig. 13.6(*e*), will do as well. In each case the op-amp resistance as seen by the transducer is infinite. Wheatstone bridge circuits accomplish this isolation by nulling any current through the sensor.

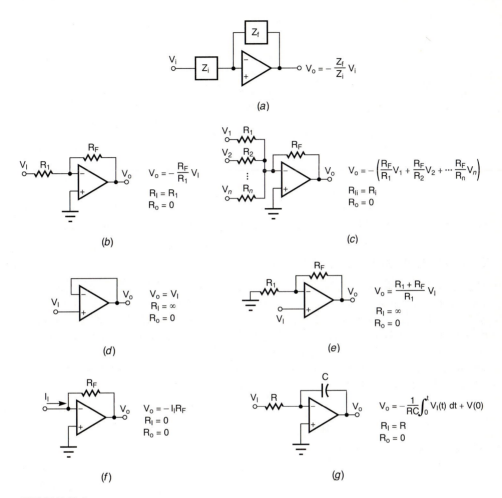

FIGURE 13.6
Common op-amp circuits. (*a*) Generalized circuit; (*b*) inverter; (*c*) inverter summer; (*d*) follower;
(*e*) noninverter; (*f*) current-to-voltage converter; (*g*) integrator.

TABLE 13.1
Characteristics of common transducers

Transducer	Typical signal range	Output impedance (Ω)
Thermocouple	0–50 mV	—
Thermistor	1–100 mV	10^3–10^7
Resistance (platinum) thermometer	—	10^2–10^3
Strain gauge	0.1–10 mV	10^4
Piezoelectric crystal	1–100 mV	10^4
Photodiode	1 nA–1 mA	10^2
Photomultiplier	1–1000 nA	10^8
pH electrode	−1 to +1 V	10^9

The high input impedance of op amps makes it convenient for them to be regarded as voltage amplifiers, and their low output impedance maintains a required output voltage over a wide range of output currents (0 to 20 mA). Input resistors simply serve to establish the currents to be drawn from their respective source voltages. If the desired input is already a current, an input resistor is unnecessary and a direct connection to the summing junction can be made. This *current-to-voltage converter,* Fig. 13.6(*f*), is a useful configuration for instruments in which small currents are to be measured.

In all the preceding cases the amplifier has one input, usually the noninverting input, connected to ground. It is said to be operating in a ground reference or single-input mode. In contrast, a differential mode has both inputs active. An example is the *subtractor circuit* or the differential circuit shown in Fig. 13.7. The output of this circuit is

$$V_o = \frac{R_2}{R_1}(V_2 - V_1) \tag{13.3}$$

The differential amplifier is ideal for low-level signals in which the signal on each wire is measured with respect to ground. In fact, connecting one lead to a noisy ground in single mode fashion can affect the differential signal. Most low signal level transducers generate a differential signal. In addition, the circuit is useful when it is desirable to amplify small differences between two large voltages because it reduces common mode error (see Sec. 13.4).

The subtractor circuit offers the same input resistance as regular amplifier circuits and will load low-level transducers. In addition, common mode error will not be eliminated unless the resistors are perfectly matched. *Instrumentation amplifiers* solve these problems; its circuit is shown in Fig. 13.8. Two op-amp followers prevent the source from being loaded. The second stage amplifier is a subtractor. Output of the circuit is given by

$$V_o = \left(1 + \frac{2R_2}{R_1}\right)(V_2 - V_1) \tag{13.4}$$

Several manufacturers produce instrumentation amplifiers in a single package. They have better specifications in terms of gain, noise, stability, and common mode error. On-chip resistors are carefully matched. A single external resistor is used to set the gain. As with most carefully constructed hybrids, they can be quite expensive. But the expense is justi-

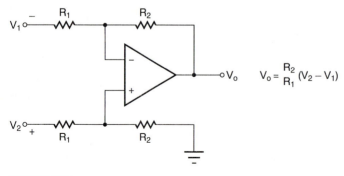

FIGURE 13.7
Subtractor op-amp circuit.

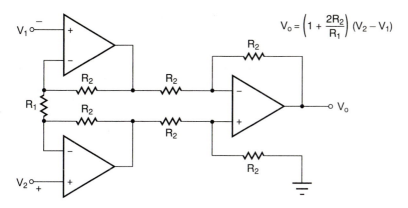

$$V_o = \left(1 + \frac{2R_2}{R_1}\right)(V_2 - V_1)$$

FIGURE 13.8
Instrumentation amplifier from three op amps.

fied if the application calls for very high gain, high precision, and low noise. It is ideally suited for strain-gauge-type transducers used in the measurement of stress/strain, pressure, force, and torque.

The AD524 precision instrumentation amplifier is shown in Fig. 13.9(*a*). It has pin programmable gains of 1, 10, 100, and 1000. Simply connect a lead from pin 3 to the desired gain pin. Furthermore, the AD524 can be configured for gains other than the present ones. Just connect an external resistor R_G between pin 3 and pin 16. Its value is found from

$$R_G = \frac{40 \text{ k}\Omega}{G - 1} \tag{13.5}$$

A typical bridge application is illustrated in Fig. 13.9(*b*).

An instrumentation amplifier for thermocouples is the AD594/595. See Fig. 13.10. The chip contains an ice point reference and is precalibrated to produce a high level (10 mV/°C) output. The °C version has an accuracy of ±1°C. The AD594 is precalibrated for type J (iron-constantan) thermocouples, whereas the AD595 is precalibrated for type K (chromel-alumel) inputs. Other types can be interfaced by adding two or three external resistors.

Example 13.1. A thermistor measures temperature over the range 0°C to 100°C. Its resistance over the range decreases nonlinearly from 12432 to 1384 Ω. The circuit in Fig. 13.11 will generate a corresponding voltage of 0 to 10 V, which will interface an ADC. Determine appropriate values for the resistors.

Solution. Drive the thermistor with the op-amp power supply if possible—±15 V. Small thermistor resistance values can lead to self-heating and altered resistance. Voltage V_1 is given by

$$V_1 = V_S\left(\frac{R}{R + R_T}\right)$$

Select R so that V_1 extends over much of the supply voltage range as the temperature changes full scale:

$$-11.75 \leq V_1 \leq -4.30 \qquad \text{for } R = 5 \text{ k}\Omega$$

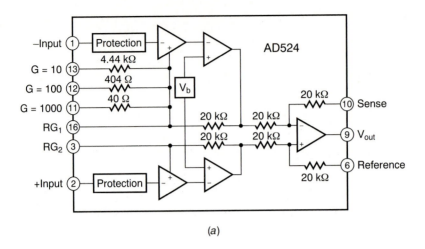

(a)

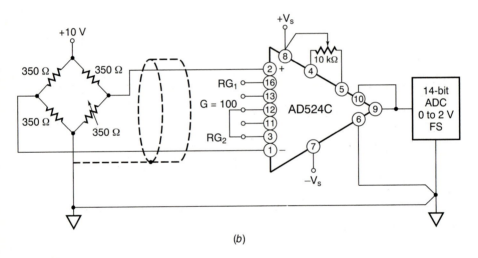

(b)

FIGURE 13.9
The Analog Devices AD524 precision instrumentation amplifier. (a) Pin assignments; (b) bridge application. (*Courtesy of Analog Devices.*)

A unity-gain voltage follower A_1 isolates the temperature transducer from amplifier A_2, or $V_2 = V_1$. For amplifier A_2, output voltage is given by

$$V_3 = \frac{R_F}{R_1} V_2 - \frac{R_F}{R_2} (+15 \text{ V})$$

Satisfying the 0 and 10 V output for full-scale temperature, we obtain

$$R_1 = 10 \text{ k}\Omega \qquad R_2 = 12.77 \text{ k}\Omega \qquad R_F = 13.42 \text{ k}\Omega$$

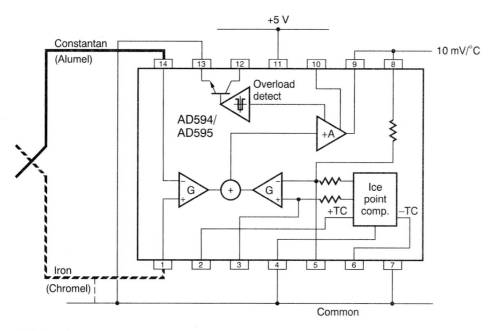

FIGURE 13.10
The Analog Devices AD594 instrumentation amplifier for thermocouples. (*Courtesy of Analog Devices.*)

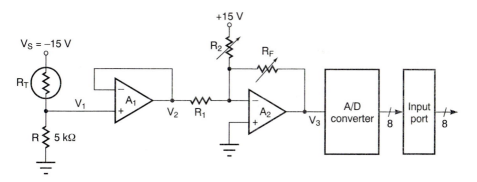

FIGURE 13.11
Circuit for interfacing a thermistor to a microprocessor system.

13.4   OPERATIONAL AMPLIFIER CHARACTERISTICS

So far we have discussed ideal op-amp behavior and, in passing, mentioned their less than ideal open-loop gain and output impedance. There are a number of real op-amp characteristics that must be understood if we are to select the proper op amp for a given application. Fortunately, these characteristics have a negligible effect on the most widely used parameter: the closed-loop gains given in Fig. 13.6. Real op-amp characteristics do affect noise, speed, and accuracy. The following parameters are found in most manufacturers' specifi-cations of op amps.

Open-Loop Voltage Gain

The open-loop voltage gain A is the ratio of output voltage to the voltage difference be-tween the two inputs. A practical open-loop gain is never infinity but varies from 10^4 to 10^7. Of course, op amps are not used in the open-loop mode, and voltage gains for the feedback or closed-loop mode are much lower, normally 1 to 1000.

Unity-Gain Bandwidth

Input signal frequency can attenuate the closed-loop gain G if the open-loop gain is not large enough. Figure 13.12 shows a Bode plot for a typical open-loop gain. At a frequency (10 Hz) called the full gain bandwidth the open-loop gain rolls off at a rate of 20 db per decade. The unity-gain bandwidth is the range of frequencies from DC to the point where the open-loop gain becomes unity (0 db). We see that the closed-loop gain is unaffected until its magnitude is similar to the open-loop gain. Closed-loop gain bandwidth can easily be determined from the unity-gain bandwidth. For example, a closed-loop gain $G = 10$ (20 db) has a bandwidth (10^5 Hz), one decade removed from the unity-gain bandwidth.

Rated Output

Rated output is the peak values of output voltage and current that can be supplied by the amplifier. The maximum output voltage V_{max} is limited by the power supply. Maximum

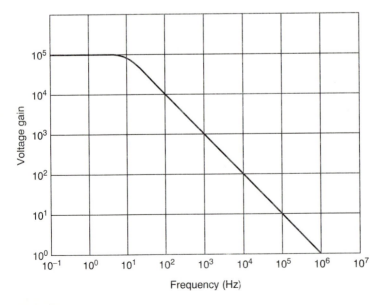

FIGURE 13.12
Open-loop frequency response of a typical op amp.

output current I_{max} is limited by the output transistors and is 20–25 mA for most op amps. At 10 V the load on the output must be greater than 400 Ω.

Output Resistance

Output resistance R_o is the effective resistance of the amplifier in the open-loop mode when viewed as a source. When not given, R_o can be estimated from V_{max}/I_{max}. Of greater interest is the closed-loop output resistance

$$R_{oc} = \frac{R_o}{1 + A\beta} \tag{13.6}$$

where $\beta = R_i/(R_i + R_f)$. This resistance is generally less than one.

Slew Rate

Slew rate is the maximum rate at which the output voltage can change. See Fig. 13.13. Assume an op amp is amplifying sine waves with small amplitudes that fall within its bandwidth for the closed-loop gain. As we increase the input amplitude, a point is reached at which we can no longer increase the output amplitude. This occurs when the rate of change of the output voltage exceeds the slew rate. The limitation imposed by the slew rate in its most severe form is expressed by the *full-power bandwidth*, the maximum frequency an output can oscillate when the output voltage is the rated voltage. Full-power bandwidth may be two orders of magnitude smaller than its small signal bandwidth.

Settling Time

Settling time is defined as the elapsed time from the application of a step input to the time when the amplifier output has settled within a specified error band. See Fig. 13.13. This band is usually 0.1 percent of the final value.

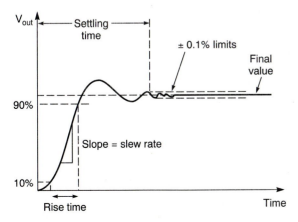

FIGURE 13.13
Transient response of an op amp.

Input Offset Voltage

When both inputs are connected to ground (0 V), a small voltage appears at the op-amp output. Input offset voltage V_{os} is defined as the voltage required at the input to drive the output to zero. Offset voltage is subject to the same voltage gain as the difference signal applied to the two inputs. Thus, it is one of the major sources of error in op-amp circuits. Some op amps are provided with two pins which can be used to balance or "null" the offset with a single preset potentiometer. It is usually preferrable to purchase a better amplifier than attempt to correct an inferior one. Another problem is that the offset voltage varies with temperature (input voltage drift, specified in mV/°C). If the op amp is driving a low impedance load, the output current may be large enough to heat the package. To minimize drift problems high-gain amplifiers should have output loads of 10 kΩ or more.

Input Bias Current

Suppose we ground both inputs and null the offset voltage. A small current, called the input bias current I_B, flows into the negative input (and out of the positive input). If an input resistor is placed between ground and the op amp as in Fig. 13.14, the bias current causes a voltage drop across the resistor. Input voltage difference $(V_+ - V_-)$ is now finite and the difference is amplified, appearing on the op-amp output. This bias current error is superimposed onto the offset voltage error. The total (drift) error on the output can be found from

$$V_o = -\frac{R_f}{R_i}\left[V_i + \underbrace{\frac{V_{os}(R_f + R_i)}{R_f} + I_B R_i}_{\text{(drift) error}}\right] \tag{13.7}$$

where input signal V_i is no longer zero.

Error due to bias current can be minimized if a resistor with the value

$$R_B = \frac{R_f R_i}{R_f + R_i} \tag{13.8}$$

is added between the positive input and ground. Drift error is now found from

$$V_o = -\frac{R_f}{R_i}\left[V_i + \frac{V_{os}(R_f + R_i)}{R_f} + I_{os} R_i\right] \tag{13.9}$$

where I_{os} is the *input offset current,* the difference between the two bias currents.

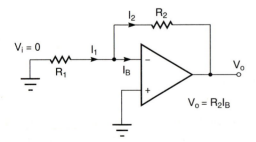

FIGURE 13.14
Model for output voltage due to input bias current.

Common Mode Rejection Ratio (CMRR)

Op amps are also subject to what are called common mode errors. These errors are most serious when comparing two large signals (subtractor circuit) which have almost the same voltage. An ideal op amp gives zero output when V_+ and V_- have identical values. In practice, the (+) and (−) inputs are not perfectly symmetrical, and a nonzero output occurs when a common mode voltage V_{CM} is applied:

$$V_{CM} = \frac{V_+ + V_-}{2} \tag{13.10}$$

The common mode error is represented by $V_{CM}/CMRR$, where CMRR is the common mode rejection ratio specified for every op amp. CMRR is usually expressed in decibels and is typically in the range 60 to 100 db, which translates into 10^3–10^5. Common mode errors for single-ended input circuits (inverters, noninverters, and followers) are on the order of one part in CMRR or less. However, the subtractor circuit (and instrumentation amplifier) is a different story. If input V_2 differs by a small amount v from input V_1, the output voltage is given by

$$V_o = \frac{R_2}{R_1}\left(v + \frac{V_{CM}}{CMRR}\right) \tag{13.11}$$

where common mode voltage $V_{CM} \cong V_1$. Now the error can be significant if an amplifier with poor CMRR is selected.

13.5   OPERATIONAL AMPLIFIER TYPES

One of the first monolithic [total integrated circuit (IC) on a silicon chip] IC operational amplifiers was the 741 introduced by Fairchild Semiconductor in 1968. The 741 is the prototype general-purpose op amp and is still widely used today. It is said to be a (internally) compensated device as opposed to an uncompensated one (301). Compensation refers to the presence of a capacitor built into the IC which makes the op amp stable against oscillations when feedback is applied. The 301 uses an external capacitor which enables one to extend the bandwidth for AC signals. Most op amps are compensated.

Early monolithic op amps were constructed with bipolar transistor technology. In the mid-1970s FET transistors, as the first stage of the amplifier, were successfully marketed. These BiFETs, as they are called, have both advantages and disadvantages relative to totally bipolar op amps.

Bipolar advantages	BiFET advantages
1. Low offset voltage	1. High input resistance
2. Low noise	2. Low bias current
3. Low drift with temperature	3. Low offset current

Specification improvements over the years have tended to blur some distinctions, although the cost of BiFETs remains higher. Both bipolar and BiFET op amps are popular today.

The choice of an op amp depends upon which class of specifications is most critical to the designer. His or her requirements generally fall into one of several catagories: (1) general-purpose amplifiers, (2) precision amplifiers, (3) low-input current amplifiers, and (4) high-speed amplifiers. Table 13.2 gives the key specifications for several op amps in each category. Literally hundreds of op amps are available with a slightly different mix on the specifications.

General-Purpose Op Amps

General-purpose op amps are the workhorse among the different types. Their characteristics are adequate for most applications, and they are cheaper. In fact, cost is the single most important item in identifying an op amp as general purpose. Later general-purpose op amps offered such a substantial improvement in specifications that they were called *high performance* op amps. These include the AD741L and the LF356.

Pin-outs for the AD741 are given in Fig. 13.15, but *all general-purpose op amps (bipolar or BiFET) have the same pin-out.* Offset null pins are optional, using an external precision potentiometer to zero the offset voltage. Nulling can affect the temperature drift problem and will add cost plus complication. It may be better in the long run to choose a higher performance amplifier if the offset voltage magnitude is undesirable. Pin 8 is not connected (NC).

Precision Op Amps

Precision op amps are also called premium op amps and instrument-grade op amps (not to be confused with instrument amplifiers). As the name implies, they have very high performance levels at a price. There are two groups which emphasize specifications pertinent for two basic uses:

1. *Low bias current and high impedance.* These traits are inherent in BiFET op amps. Applications include integrators, current-to-voltage converters, and very high-impedance transducers such as radiation detectors.

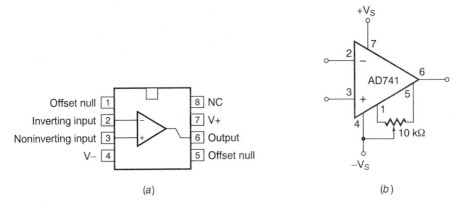

(a) (b)

FIGURE 13.15
Pin assignments for general-purpose op amps. (a) Top view; (b) connection for offset null.

TABLE 13.2
Specifications for several commercial op amps

		General-purpose			Precision		Low-current		High-speed	
		AD741C	AD741L	LF356A	ADOP-07	AD547L	AD515A	AD548	AD509	AD381
Input offset voltage	mV	6	0.5	0.5	0.025	0.25	3	0.5	8	1
Technology		Bipolar	Bipolar	BiFET	Bipolar	BiFET	BiFET	BiFET	Bipolar	FET
Input voltage drift	μV/°C	15	5	5	0.6	2	50	5		15
Input bias current	NA	500	50	0.05	2	0.025	3×10^{-4}	0.01	200	0.02
Input offset current	NA	200	5	0.01	2	0.015	—	0.005	25	0.005
Input impedance	MΩ	2	2	10^{6}	80	10^{12}	10^{13}	10^{12}	100	10^{12}
CMRR	db	90	90	90	110	80	80	82	90	76
Unity-Gain bandwidth	MHz	1	1	4	0.6	0.25	1	1	20	5
Slew rate	V/μs	0.5	0.5	10	0.17	3	0.3	1.8	120	30
Cost	$	0.25	2.00	5.00	7.00	15.00	18.00	—	—	—
Output current	mA	25	25	25	—	—	—	—	—	—
Power dissipation	mW	500	500	570	500	—	—	—	—	—
Settling time	μs	30	30	1.5	80 (est)	6 (est)	50 (est)	8	0.2	0.7

2. *Low offset voltage.* This trait is inherent in bipolar op amps. Additional characteristics include low-voltage drift, low noise, and high CMRR. We are after the high-voltage accuracy necessary for low level DC transducer circuits, instrumentation, and control systems. Most precision op amps are bipolar.

Pin-outs for the AD 0P–07 are given in Fig. 13.16. All precision op amps use the same pin-outs and circuit. Again, offset nulling (pins 1 and 8) is optional. The 0.01 μF capacitors are recommended to minimize the effects of power supply noise.

Low-Input Current Amplifiers

Low-input current amplifiers are called electrometer op amps. They rely on BiFET technology where the bias currents follow a characteristic doubling for every 10°C temperature rise above the rated value of 25°C. Every effort should be made to minimize operating temperature, e.g., restricting the output loads to at least 10 kΩ. Electrometer op amps are best suited for low-current output transducers, such as photodiodes and oxygen sensors, and for most biomedical instrumentation. They are also ideal for long-term precise integration.

High-Speed Amplifiers

High-speed op amps are characterized by fast settling times, high slew rates, and wide bandwidths. They are important as components in applications with rapidly changing or switched analog data such as analog buffers, multiplexers, DACs, and ADCs. It is also the choice for amplification of AC signals with broad bandwidth. The ultrahigh frequency AD 5539 for video circuits has a settling time of 12 ns.

> **Example 13.2.** An op-amp integrator is constructed with a 1-MΩ input resistor and a 0.1-μF feedback capacitor so that $RC = T = 10^{-1}$ s. If it is desirable to integrate for one second, how reliable is a general-purpose op amp for this application?

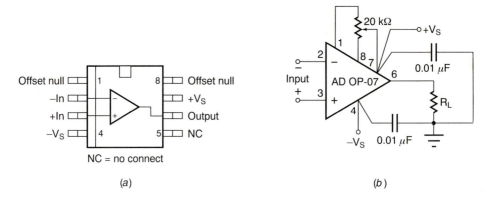

FIGURE 13.16
Pin assignments for precision op amps. (*a*) Top view; (*b*) connections for offset null.

Solution. Drift error caused by bias current and offset voltage can be found from current conservation at the summing junction:

$$\frac{dV_o}{dt} = \frac{I_B}{C} \qquad \text{(bias current error)}$$

$$\frac{dV_o}{dt} = \frac{V_{os}}{RC} \qquad \text{(offset error)}$$

using a high-performance AD741L, we see that

$$\frac{dV_o}{dt} = \frac{5 \times 10^{-8} \text{ A}}{10^{-7} \text{ F}} = 0.5 \text{ V/s}$$

$$\frac{dV_o}{dt} = \frac{0.5 \times 10^{-3} \text{ V}}{10^{-1} \text{ s}} = 0.5 \times 10^{-2} \text{ V/s}$$

Bias current error in a full-scale 10-V output is 5 percent—too high. Select a precision BiFET instead.

13.6 DIGITAL-TO-ANALOG CONVERTERS (DACs)

A digital-to-analog converter (DAC or D/A converter for short) is a device that receives a binary word from the microprocessor and converts it to a scaled analog voltage: e.g., 0 to 10 V. In Fig. 13.17 a 3-bit word incrementing in time appears as an increasing voltage where each level corresponds to a given digital word. There exists only a finite number of voltage levels, and the output is a staircase approximation of the desired continuous voltage. A smooth output can be obtained with filtering.

DACs are nothing more than op amps whose gains can be programmed digitally. An inverting op amp has a fixed resistance ratio which scales a varying input voltage.

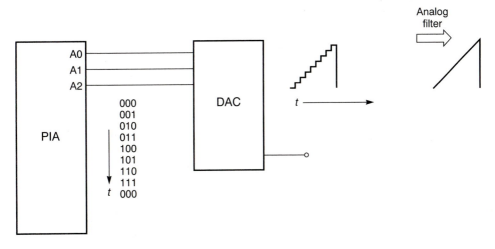

FIGURE 13.17
DAC input/output.

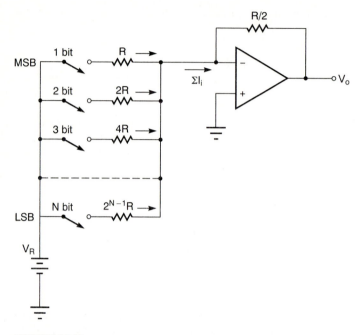

FIGURE 13.18
Binary weighted ladder DAC.

DACs fix the input voltage but switch a series of input resistors to vary the gain and voltage output level. Consider Fig. 13.18. All input voltages are the reference voltage V_R. Each input resistor is twice as big as the one preceding it. Thus, the construction is called a *binary weighted ladder*. The output voltage from each input is one-half the value produced by the preceding input:

$$\text{1-bit only} \qquad V_o = -\frac{V_R}{2}$$

$$\text{2-bit only} \qquad V_o = -\frac{V_R}{4}$$

$$\text{3-bit only} \qquad V_o = -\frac{V_R}{8}$$

$$\vdots \qquad\qquad \vdots$$

$$N\text{-bit only} \qquad V_o = -\frac{V_R}{2^N}$$

The minimum step size, called the *resolution,* is produced by the LSB or bit N:

$$\text{Minimum step size} = \frac{V_R}{2^N} \qquad\qquad (13.12)$$

Full-scale output occurs when all bits are closed:

$$V_o = V_R\left(\frac{1}{2} + \frac{1}{4} + \frac{1}{8} + \cdots + \frac{1}{2^N}\right)$$

$$= -V_R\left(1 - \frac{1}{2^N}\right) \tag{13.13}$$

As the number of bits increases, the full-scale output approaches the reference voltage but is never equal to it. Generalizing for an arbitrary binary input, we have

$$V_o = -V_R \sum_{i=1}^{N} \frac{b_i}{2^i} \tag{13.14}$$

where b_1 = MSB and b_N = LSB. For example, byte = 10011001 gives

$$V_o = -V_R\left(\frac{1}{2} + \frac{0}{4} + \frac{0}{8} + \frac{1}{16} + \frac{1}{32} + \frac{0}{64} + \frac{0}{128} + \frac{1}{256}\right)$$

$$= -\frac{153}{256} V_R$$

Binary weighted ladders are not used in practice for several reasons: (1) Resistance values require precision trimming. They must be accurate to less than one part in 2^N for the R_N input to be meaningful. This is difficult to do on ICs. (2) To get better resolution by going to more bits, $R_N = R \times 2^N$ will be very large. The resulting small currents in the branch can approach current noise levels. (3) Most wiring designs have stray capacitance in the pF range. Although the level is small, combining it with very large resistances can cause undesirable RC time constants or slow conversion times.

Example 13.3. Determine the voltage output levels for the staircase in Fig. 13.17. Assume a 10-V reference.

Binary	Analog (V)
000	0
001	1.25
010	2.50
011	3.75
100	5.00
101	6.25
110	7.50
111	8.75

R/2R Ladder Network

The *R/2R ladder* network, Fig. 13.19, resolves the problems of the binary weighted ladder. Today it is the most popular single package DAC. There are only two resistor values, R and $2R$. Observe that each binary switch either connects (LOW input) to true ground or

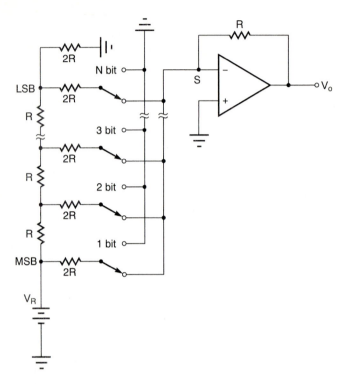

FIGURE 13.19
R/2R ladder DAC.

connects (HIGH input) to the op-amp virtual ground. Therefore, the current in each $2R$ branch remains constant. Current through the switch is either shunted to true ground or shunted to the op amp. Since there are no voltage transients, the network responds faster. When current from the reference approaches the first branch node, it sees a $2R$ resistor toward the binary switch and a $2R$ equivalent resistance toward the remaining network. This splits the current with half going to the MSB switch. The current at *any* branch node sees a $2R$ resistor toward the binary switch and an equivalent $2R$ resistance toward the remaining network. The explanation is apparent if we start at the LSB $2R$ resistor and its end companion. This pair of parallel resistors can be combined to form a single resistance R. The formed R is in series with the network R for an equivalent $2R$ resistance as viewed from the next node. The pattern is repeated for successive nodes.

Output voltage V_o from the $R/2R$ ladder obeys the same equations—(13.12) to (13.14)—as does the binary weighted ladder.

Multiplying DACs

We can see that the DAC output voltage is proportional to the reference voltage V_{ref}. A conventional DAC has an internal V_{ref}, which is derived from the fixed power supply voltage to the chip. A *multiplying DAC* has an external reference voltage. Figure 13.20(*a*) shows

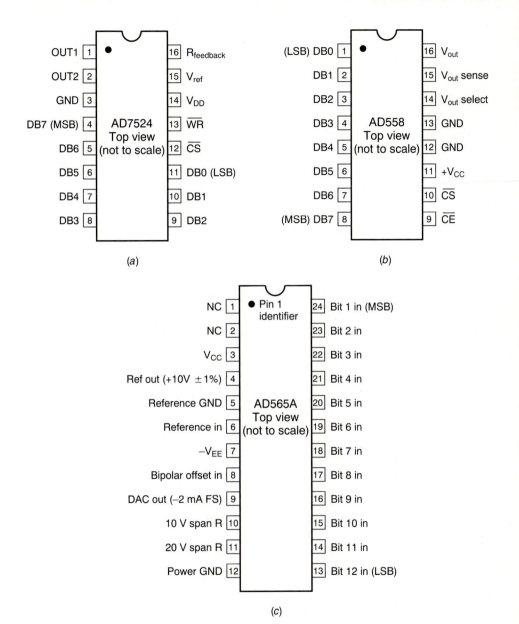

FIGURE 13.20
External and internal references for Analog Devices DACs. (*a*) External; (*b*) internal; (*c*) internal with external trim. (*Courtesy of Analog Devices.*)

the pin assignment for the AD7524 multiplying DAC. V_{ref} is clearly marked by a separate pin (15). The conventional low-cost AD558 in Fig. 13.20(*b*) has only the supply voltage pin (11), whereas the conventional high-speed AD565A in Fig. 13.20(*c*) has an additional two pins (4 and 6), called ref-in and ref-out. They are not to be confused with V_{ref}. A 100-Ω

trim potentiometer between pins 4 and 6 is used to adjust the internal reference voltage so that the DAC output is exactly full scale.

Multiplying DACs offer several advantages. First, the external reference can be a time-varying analog voltage which *multiplies* a binary function; see Eq. (13.14). This is a subtle but important difference. In conventional DACs the internal V_{ref} is fixed, and time-varying bytes to the DAC produce the programmed discrete staircase output. With the multiplying DAC, V_{ref} can be a variable analog signal and a fixed programmable byte scales the *continuous* output. See Fig. 13.21. Multiplying DACs are applicable to programmable gain control of AC signals.

Second, the external reference can also be fixed. A major source of error in the output voltage of a DAC is reference voltage drift with temperature. A precision reference voltage IC is almost always more accurate than the internal reference. Figure 13.22 shows the pin-out for an AD586 high precision 5-V reference. It consists of a "buried" zener diode reference and op amp to buffer the output. The recommended 12- to 15-V supply voltage (pin 2) can be as high as 36 V. An optional trim potentiometer is used to set the output level to exactly 5 V.

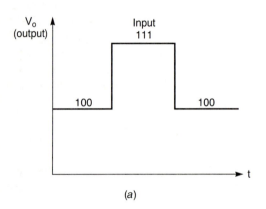

(a)

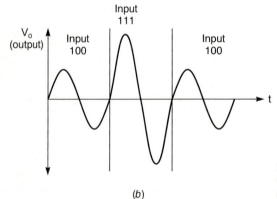

(b)

FIGURE 13.21
Conventional DAC vs. a multiplying DAC.
(a) V_{ref} = constant; (b) $V_{ref} = A \sin \omega t$.

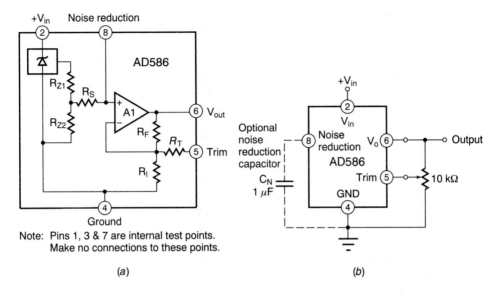

FIGURE 13.22

The Analog Devices AD586 voltage reference. (*a*) Block diagram; (*b*) connection. (*Courtesy of Analog Devices.*)

13.7 DAC ERRORS

So far we have presented a somewhat idealized behavior for the DAC. However, there are numerous errors in the output voltage to which the user must be alerted. For an N-bit converter there are 2^N different analog output voltages equally spaced over the full-scale output. The spacing, 2^{-N} of the full-scale span, is called the *resolution* of the DAC. The resolution can be expressed in several ways. See Table 13.3. The number of millivolts corresponds to the output voltage step size or output of the LSB. Resolution and accuracy are often given in percent full scale (% FS) or parts per million (ppm). Bit fraction can be converted to log-magnitude (dB) by the formula

$$dB = 20 \log_{10}(2^{-N})$$

TABLE 13.3
Analog converter resolution

N	Fraction (2^{-N})	mV at $V_R = 10$ V	% FS	ppm	dB
6	0.01562	156.2	1.56	15,620	−36.1
8		0.003906		39.1	0.39	3,906
						−48.2
10	0.0009766	9.8	0.098	977	−60.2
12	0.0002441	2.4	0.024	244	−73.3
14	0.0000610	0.61	0.0061	61	−84.3
16	0.0000153	0.15	0.0015	15	−96.3
18	0.0000038	0.038	0.00038	4	−108.4

Output errors include (1) gain error, (2) offset error, and (3) nonlinearity error (often called linearity). Gain error is a slope deviation from the ideal gain and is manifested by a less than full-scale output for a given reference voltage. Offset error is the output when a binary zero is applied to the input. It is often expressed as a fraction of the LSB. Nonlinearity is the failure of the output to fall on a straight line. In general, accuracy is the difference between the ideal output and the actual output for a given binary input, and it is expressed as a % FS.

A DAC calibration is carried out to eliminate offset and gain error. Offset error is eliminated by applying a zero digital input and adjusting an external trim potentiometer on either the voltage reference or the op amp. The gain is set to give full-scale output by adjusting a gain trim potentiometer on the DAC. Nonlinearity is a failure to meet tolerances in the internal components and cannot be eliminated.

All the stated error measurements are made at room temperature (25°C). In practice, the device functions over a wide temperature range, depending upon environment and circuit board power dissipation rates. Three temperature ranges are used for analog converters:

Military range:	$-55°C$ to $+125°C$
Industrial range:	$-25°C$ to $+85°C$
Commercial range:	$0°C$ to $+70°C$

Unfortunately, parameters of interest (including gain, linearity, and offset) will change with the IC temperature after it is calibrated. This "drift" is expressed in terms of parts per million (ppm) per degree rise—ppm/°C—and should be weighed against the resolution. The predominant temperature drift is gain drift and is often the only drift cited in the specifications. The main source of gain drift is the internal reference voltage. References incorporated within a converter rarely provide accuracy over the temperature range compatible with the resolution itself. Drift problems can be reduced by using a multiplying DAC with an external high precision voltage reference.

Example 13.4. You are to choose between a 12-bit DAC with an external reference and a 16-bit DAC with an internal reference. The maximum operating temperature is 125°C. Cost is not a factor:

12-bit DAC:	Linearity = 0.012% at 25°C
	Drift = 5 ppm/°C
Reference:	Drift = 1 ppm/°C
16-bit DAC:	Linearity = 0.003% at 25°C
	Drift = 10 ppm/°C.

Maximum error (after calibration at 25°C) over the temperature range is defined by:

$$\text{Error} = (\text{linearity error}) + (\text{drift})(\Delta T)(10^{-4})$$

For the 16-bit DAC,

$$E = 0.003 + (10 \text{ ppm/°C})(100°C)(10^{-4})$$

$$= 0.103\% \text{ FS}$$

$$\equiv 1030 \text{ ppm or slightly worse than 10-bit accuracy}$$

For the 12-bit DAC,

$$E = 0.012 + (5 \text{ ppm/°C} + 1 \text{ ppm/°C})(100)(10^{-4})$$

$$= 0.072\% \text{ FS}$$

$$\equiv 720 \text{ ppm or slightly better than 10-bit accuracy}$$

Therefore, choose the 12-bit DAC.

13.8   DAC SPECIFICATIONS

There are literally hundreds of DAC families on the market today, each with several models. The reason for this variety is choice for the designer. The basic structure of all conventional DACs is to provide a network of precision resistors and a set of binary switches. Then some decisions must be made. Do you want an internal or external reference? Is the output to be current (without op amp) or voltage (with op amp)? Is an I/O port provided, or is it interfacing the DAC directly to the bus? What overall accuracy and speed do you want? To minimize cost, the engineer should select the lowest level of accuracy and speed to get the job done. Table 13.4 gives the specifications for several commercially available DACs. Most DACs today are both TTL and CMOS compatible.

Resolution

DACs are made in 8-, 10-, 12-, 14-, 16-, and 18-bit versions. Of course, the resolution gets smaller as the number of bits increases, but it becomes more difficult to meet the required tolerances. Eight-bit versions provide enough resolution for most microprocessor control circuits, and they are cheap. Twelve-bit DACs are very popular in data acquisition because they have a good balance between accuracy and cost. Sixteen-bit converters are available but relatively expensive. Eighteen-bit DACs are rare and command a corresponding price.

Linearity

Linearity, often called accuracy in a restricted meaning of the word, is the maximum deviation from the expected value over the full range of the output at room temperature (25°C). It is expressed as a percent of full scale. From Table 13.3 this percentage has an equivalent LSB and ppm. For example, the Motorola MC1408 has a linearity of 0.19%, which is equivalent to $\frac{1}{2}$LSB or 1953 ppm. Although linearity is the worst case over the full range, a potential error of 50% exists for a binary input of $(01)_{16}$.

As the DAC reaches steady-state temperature, accuracy deteriorates. If the specifications give an operating range of 0°C to 70°C, full-scale drift at 70°C is measured and expressed in ppm/°C. Proportionality over the temperature range is assumed. Thus, the AD767 has a linearity of $\frac{1}{2}$LSB plus an additional error at 70°C of $(30)(70° - 25°)/$ $(224 \text{ ppm}) = 6 \text{ LSB}$.

Settling Time

Settling time is the time it takes the DAC to reach within (usually) $\frac{1}{2}$ LSB of its new voltage after a binary change. This time limits conversion rates. The major contribution to

TABLE 13.4
Specifications of several commercially available DACs

DAC	Manufacturer	Bits	Power (V)	Linearity (% FS)	Drift (ppm/°C)	Settling time	Multiplying or internal ref.	Internal op amp	Latch	Cost ($) (Sm. quant.)
MC1408(L8)	Motorola	8	+5, −15	0.19	20	300 ns	M	No	No	1.50
MC3412	Motorola	12	+5, −15	0.012	15	200 ns	I	No	No	5.00
DAC811	Burr-B.	12	+5, +15 −15	0.006	10	4 μs	I	Yes	Yes	25.00
DAC7541	Burr-B.	12	+5 to +15	0.024	5	1 μs	M	No	No	13.00
DAC702	Burr-B.	16	+5, +15 −15	0.0015	10	350 ns	I	No	No	65.00
DAC729	Burr-B.	18	+5, +15 −15	0.00075	5	8 μs	I	Yes	No	179.00
AD7524	Analog D.	8	+5 to +15	0.19	10	250 ns	M	No	Yes	—
AD767	Analog D.	12	±15	0.012	30	4 μs	I	Yes	Yes	—

Bit	8	10	12	14	16
Resolution (ppm)	3906	977	244	61	15

settling time is the op-amp circuit. Current output DACs have times from 10 ns to 1 μs with several hundred nanoseconds being common. When op amps are added for a voltage output, settling time slows to the 1 μs to 10 μs range.

Settling time for DACs is not so important in microprocessor-based designs. The 16/8-bit microprocessors can write continuous data to a DAC without having to wait for the conversion to take place. For example, the STA instruction alone takes 5 μs to execute. This is not true for ADCs. They are relatively slow, and rapid data transfers must be coordinated.

Reference

Most designs with DACs will require a fixed reference. If accuracy is a problem, then multiplying DACs with a voltage reference should be used. Typical voltage reference costs and temperature drifts are given in the following table. Again, accuracy is rarely critical for 16/8-bit microprocessor-based control circuits. The internal reference is recommended until proven otherwise.

Voltage reference ICs

Reference	Output (V)	Drift (ppm/°C)	Cost ($)
AD586J	5	25	2.95
AD586L	5	5	6.50
AD588CD	10	1.5	38.80

Output Type

The analog output of basic DAC resistive networks is current. The simplest way to obtain a voltage output is to drive a pure resistive load R_L. However, this has its limitations. First, binary switches of the $R/2R$ ladder send the reference current fractionals directly to R_L, not to ground. Thus, the output voltage is a function of the load unless R_L is much greater than the ladder resistance R. Second, output power will be small, and the voltage range will be restricted.

An ideal current-to-voltage transducer is an op amp. It has HIGH-input impedance and LOW-output impedance, so loading effects are negligible. External resistors can be connected to the op amp for gain adjustment. Furthermore, offset voltages can be added to the summing junction for additional signal conditioning. DACs can be purchased with op-amps on-board (voltage output) or with the resistive network only (current output). The purchaser must provide an op amp for the latter. This would be desirable if a faster DAC settling time was necessary.

Latched

If the DAC has an internal latch (sometimes called a buffer), it may be interfaced directly to the data bus. A memory-mapped address is given to the DAC as though it were a PIA. Its chip select (or enable) pin is activated by a decoded signal.

13.9   INTERFACING DACs

So far we have assumed straight binary input to the DAC, and output voltages are between zero and the reference voltage. This output is called *unipolar*. Many outputs must have both a positive and negative sign or *bipolar* output. Analog control of DC motors is one example. Arithmetic operations by the microprocessor would also be using 2s complement binary. Bipolar outputs should conform to the signed binary input. To compound the situation further most commercial DACs call for an offset binary input. A comparison of these binary inputs and corresponding voltage outputs for a hypothetical 4-bit DAC is given in Table 13.5. Outputs are proportional to the reference voltage and the straight binary weight of the input. Notice that the offset binary and 2s complement binary are very similar. In fact, *offset binary is identical to 2s complement binary except that the MSB is inverted.*

In most cases our DAC will require a unipolar output. However, all commercial DACs can be wired to produce a bipolar output. We will examine several commercial DACs and their interfacing to the microprocessor.

Motorola MC1408

A very popular 8-bit DAC over the years has been the Motorola MC1408 multiplying DAC because of its flexibility and cost. Figure 13.23 shows a block diagram. It is available in several varieties. The MC1408 series operates in the commercial temperature range, whereas the MC1508 series operates in the military temperature range. The MC1408 also

TABLE 13.5
DAC input/output coding

	Unsigned arithmetic			Signed arithmetic		
				DAC input		
Decimal	DAC input	DAC output	Decimal	Offset binary	Two's complement	DAC output
0	0000	$+V_R (0/16)$	+7	1111	0111	$+V_R (7/8)$
+1	0001	$+V_R (1/16)$	+6	1110	0110	$+V_R (6/8)$
+2	0010	$+V_R (2/16)$	+5	1101	0101	$+V_R (5/8)$
+3	0011	$+V_R (3/16)$	+4	1100	0100	$+V_R (4/8)$
+4	0100	$+V_R (4/16)$	+3	1011	0011	$+V_R (3/8)$
+5	0101	$+V_R (5/16)$	+2	1010	0010	$+V_R (2/8)$
+6	0110	$+V_R (6/16)$	+1	1001	0001	$+V_R (1/8)$
+7	0111	$+V_R (7/16)$	0	1000	0000	$+V_R (0/8)$
+8	1000	$+V_R (8/16)$	−1	0111	1111	$-V_R (1/8)$
+9	1001	$+V_R (9/16)$	−2	0110	1110	$-V_R (2/8)$
+10	1010	$+V_R (10/16)$	−3	0101	1101	$-V_R (3/8)$
+11	1011	$+V_R (11/16)$	−4	0100	1100	$-V_R (4/8)$
+12	1100	$+V_R (12/16)$	−5	0011	1011	$-V_R (5/8)$
+13	1101	$+V_R (13/16)$	−6	0010	1010	$-V_R (6/8)$
+14	1110	$+V_R (14/16)$	−7	0001	1001	$-V_R (7/8)$
+15	1111	$+V_R (15/16)$	−8	0000	1000	$-V_R (8/8)$

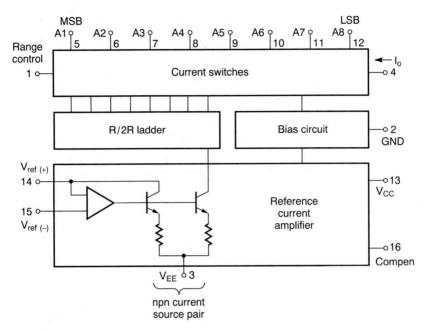

FIGURE 13.23
The Motorola MC1408 multiplying DAC. (*Reprinted with permission of Motorola, Inc.*)

comes with 8-bit accuracy (MC1408L8), 7-bit accuracy, or 6-bit accuracy (MC1408L6); but all are 8-bit devices.

When using the MC1408 DAC, there are three requirements to be met:

1. Unlike most multiplying DACs the V_{ref} (pin 14) is not a true voltage input. Within the IC is a reference current amplifier driving the $R/2R$ ladder network. Thus, the input is a nominal 2-mA reference current (5-mA maximum). A high precision resistance R_{ref} should be inserted between the voltage reference V_{ref} and pin 14. Pin 15 is normally tied to ground although it is used as the LOW input when a differential pair source voltage is used in the multiplying mode.
2. There is no internal op amp. Thus, a current-to-voltage converter circuit must be used on the DAC output (pin 4) if an output voltage rather than a current is desired. Maximum output current is 2.1 mA.
3. An 8-bit output port or PIA must be used with the MC1408. It has no internal latch to save the input from the bus.

Unipolar Connection

An MC1408, external op amp, and PIA wiring diagram for unsigned outputs is shown in Fig. 13.24. The theoretical equation for output voltage is

$$V_o = \frac{V_{ref}}{R_{ref}}(R_o)\left(\frac{B7}{2} + \frac{B6}{4} + \frac{B5}{8} + \frac{B4}{16} + \frac{B3}{32} + \frac{B2}{64} + \frac{B1}{128} + \frac{B0}{256}\right) \quad (13.15)$$

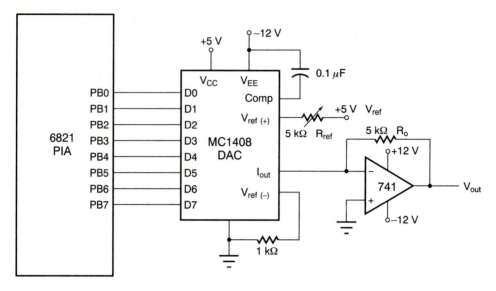

FIGURE 13.24
Connection for unsigned outputs (0 to 10 V) from the MC1408 DAC.

Thus, for the wired circuit and maximum binary input 11111111,

$$V_o = \frac{5\ V}{2.5\ k}(5\ k)\left(\frac{1}{2} + \frac{1}{4} + \frac{1}{8} + \frac{1}{16} + \frac{1}{32} + \frac{1}{64} + \frac{1}{128} + \frac{1}{256}\right)$$

$$= 10\ V\left(\frac{255}{256}\right) = 9.961\ V$$

For most applications the maximum output is adjusted to the theoretical value with the reference potentiometer.

The chip calls for two separate power supply voltages: $V_{CC} = +5.0$ V at pin 13 and $V_{EE} = -5.0$ to -15 V at pin 3. A 37-pF compensation capacitor (pin 16) is required to prevent oscillations only if the DAC is coupled to very high-speed amplifiers.

Bipolar Connection

Suppose your control application requires that the DAC supply a signed output with a 10-V range or -5 to $+5$ V. How can the previous circuit be altered for the bipolar range? An obvious approach is to add a fixed voltage (-5 V) into the op-amp summing junction. Equation (13.15) is changed to

$$V_o = 10\ V\left(\frac{B7}{2} + \frac{B6}{4} + \frac{B5}{8} + \frac{B4}{16} + \frac{B3}{32} + \frac{B2}{64} + \frac{B1}{128} + \frac{B0}{256}\right) - 5\ V$$

Substituting several binary inputs yields

$$11111111: \quad V_o = 4.961\ V$$
$$10000000: \quad V_o = 0\ V$$
$$00000000: \quad V_o = -5\ V$$

These voltages are exactly the values for an 8-bit offset binary input. Since the microprocessor is operating with 2s complement binary, we must invert the MSB input line to the DAC or B7. When the *input is 2s complement binary,* the theoretical equation for output voltage is

$$V_o = \frac{V_{ref}}{R_{ref}}(R_o)\left(\frac{\overline{B7}}{2} + \frac{B6}{4} + \frac{B5}{8} + \frac{B4}{16} + \frac{B3}{32} + \frac{B2}{64} + \frac{B1}{128} + \frac{B0}{256}\right) - V_{ref}\frac{R_o}{R_B}$$

$$(13.16)$$

Here we are using the reference voltage and input resistor R_B source for the offset current. Since the output is zero for a binary zero input,

$$\frac{R_B}{R_{ref}} = 2$$

A wiring diagram for signed outputs is shown in Fig. 13.25.

Analog Devices AD7524

The AD7524 is a TTL/CMOS compatible 8-bit device that is designed for direct interface to the microprocessor bus. As a multiplying DAC it has the more conventional reference voltage input to the $R/2R$ ladder as shown in Fig. 13.26(a). Options include $\frac{1}{2}$ LSB, $\frac{1}{4}$ LSB, and $\frac{1}{8}$ LSB linearity for each of the three temperature ranges.

Figure 13.26(b) shows the circuit connection for the unipolar output. Multiplying DACs with unipolar connections are often referred to as two-quadrant connections when the reference voltage is a variable; i.e., the references can take on positive or negative

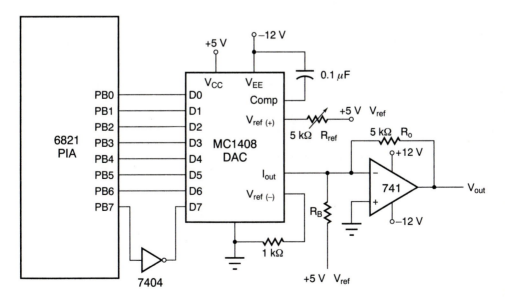

FIGURE 13.25
Connection for signed outputs (-5 to $+5$ V) from the MC1408 DAC.

signs, whereas the product binary term in Eq. (13.15) is positive only. The power supply voltage (pin 14) is +5 to +15 V depending upon the CMOS system logic level (or +5 V for TTL inputs). The reference resistance (R1) serves a different function from R_{ref} for the MC1408. It is a small (compared to the $R/2R$ ladder resistors) potentiometer to trim the true reference voltage for full-scale output. *Both R1 and R2 are used only if gain adjustment is necessary.* The output voltage from an external op-amp A1 obeys Eq. (13.15) except that the output is inverted or negative.

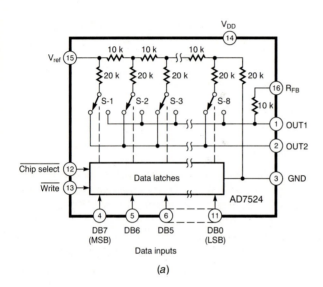

(a)

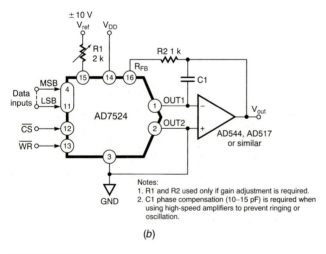

Notes:
1. R1 and R2 used only if gain adjustment is required.
2. C1 phase compensation (10–15 pF) is required when using high-speed amplifiers to prevent ringing or oscillation.

(b)

FIGURE 13.26
The Analog Devices AD7524 microprocessor compatible DAC. (*a*) Block diagram; (*b*) unipolar connection; (*c*) bipolar connection. (*Courtesy of Analog Devices.*)

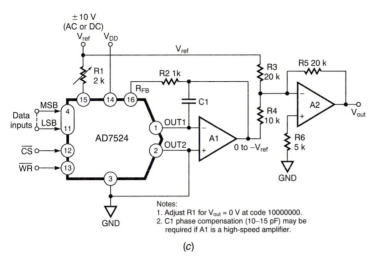

Notes:
1. Adjust R1 for V_{out} = 0 V at code 10000000.
2. C1 phase compensation (10–15 pF) may be required if A1 is a high-speed amplifier.

(*c*)

FIGURE 13.26 (continued)

Figure 13.26(*c*) shows the circuit connection for the bipolar output or four-quadrant output. In this case op-amp A2 inverts the output from op-amp A1 as well as adds the negative offset voltage. The compensation capacitor is 10 pF–15 pF if it is needed. Bipolar outputs follow the offset input convention. Again, the MSB data input to the AD7524 should be inverted for the 2s complement convention.

Data transfer is controlled by $\overline{WR}$ and $\overline{CS}$. When $\overline{CS}$ and $\overline{WR}$ are both LOW, the AD7524 analog output follows the data input in a nonlatched fashion. When either $\overline{CS}$ or $\overline{WR}$ is HIGH, the output holds the last value prior to $\overline{WR}$ or $\overline{CS}$ assuming the HIGH state. The manufacturer recommends using the address decoding circuit output for $\overline{CS}$ and VMA · $\phi2$ for $\overline{WR}$.

Analog Devices AD767

Another set of specifications is found on the AD767, Fig. 13.27(*a*). It is a 12-bit DAC designed for use with fast microprocessors. Power is supplied to pin 8 (+12 V) and pin 10 (−12 V). An internal "buried" zener reference (10 V) is typically used although an external reference can be applied instead to pin 7. The AD767 is a complete voltage output DAC. An input latch allows direct interfacing to a 12-bit bus. Figure 13.27(*b*) and (*c*) illustrate wiring connections for the 0- to 10-V unipolar output and the ±5-V bipolar output, respectively. Trim resistor R1 is the zero offset voltage adjustment. Usually, it is not needed for the unipolar output, and pin 4 can be connected to ground. Trim resistor R2 adjusts the full-scale gain. For the bipolar output, trim resistor R1 serves a valuable purpose by adding the −5-V offset (output adjusted to −5 V with all bits OFF).

The most likely data bus for the 12-bit DAC will have 8 bits. This presents a minor problem. When the data bus is used to write the first 8 bits, the DAC will be enabled to latch the data, and at the same time a voltage will appear based on the new 8 bits and the old 4 bits. After a short time elapses the DAC will be enabled for the last 4 bits of the new word, and the voltage will change accordingly. The undesirable sudden shift in the voltage is called a "glitch." A solution to the problem is double-buffering (latching). That is, the system must latch the entire 12-bit data before it is latched by the DAC. Then only a single voltage appears on the DAC output. In Fig. 13.28 a PIA is the first latch. The least significant 8 bits are sent to port A. The most significant 4 bits are sent to port B. Remember that port B control register can be configured so that a low pulse is generated on the CB2 output control line when a write to port B takes place. CB2 enables the 12-bit DAC for the new 12-bit word on the PIA.

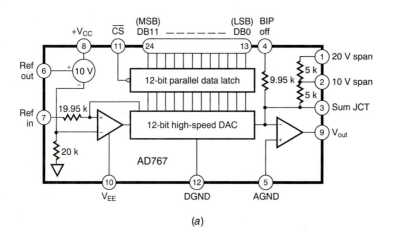

(a)

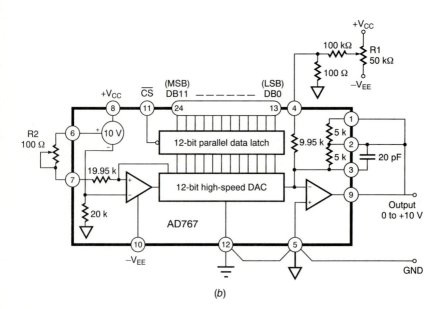

(b)

FIGURE 13.27
The Analog Devices AD767 voltage output DAC. (a) Block diagram; (b) unipolar connection; (c) bipolar connection. (*Courtesy of Analog Devices.*)

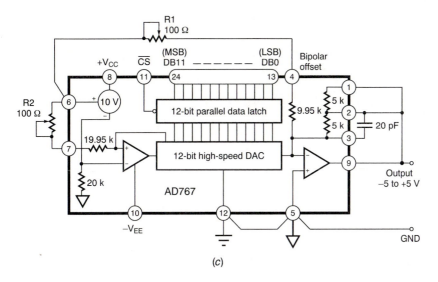

(c)

FIGURE 13.27 (continued)

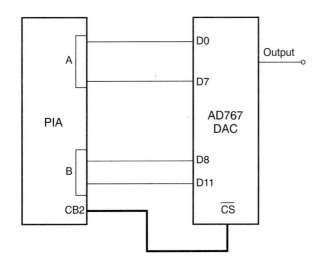

FIGURE 13.28
Interfacing a 12-bit DAC to an 8-bit data bus.

13.10   ANALOG-TO-DIGITAL CONVERSION

The counterpart to the DAC is the analog-to-digital converter, also called the ADC and A/D converter. Input/output (I/O) relationships are simply reversed; i.e., a continuous analog input is represented by a digital output (N lines) that scales the input in 2^N discrete steps. The output for a 3-bit ADC is shown in Fig. 13.29. Since the output of the ADC can be only one of eight possible states to represent the analog input, there will be a range of inputs for which the output code will not change. For example, with an input voltage of $\frac{5}{8}V_{FS}$ the output is $(101)_2$. Yet the output is the same for a range of inputs called the quantization interval Q, equivalent to the DAC step size. For an ideal ADC, the *quantization*

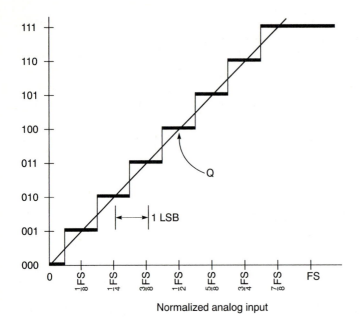

FIGURE 13.29
Ideal ADC conversion.

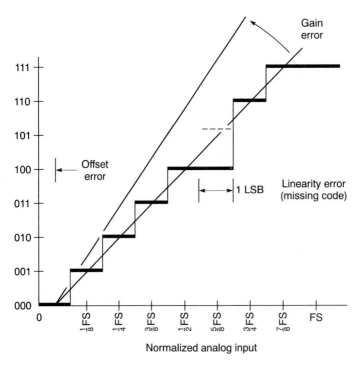

FIGURE 13.30
Real ADC conversion.

error is zero when the input is centered over the interval. On either side of the center the maximum error is equal to one-half the interval or $Q/2 = \pm(V_{FS}/2^N)/2$ or $\pm\frac{1}{2}$ LSB.

So far we have been discussing an ideal ADC. Other errors parallel the digital-to-analog converter (DAC). If the first transition does not occur at exactly $+\frac{1}{2}$ LSB or $\frac{1}{16}$ FS for the 8-bit converter, an offset error results. See Fig. 13.30. Gain error is the difference between the slope of the ideal I/O curve of the ADC and the actual curve. *Differential linearity error* is the amount of deviation of any step from its ideal size of $V_{FS}/2^N$. If the differential linearity error is greater than one LSB, the possibility of missing a binary code exists.

Time of Conversion

The time required for the ADC to measure the input voltage and generate the proper output code is called the conversion time T_{oc}. This time is much greater than the time to complete a DAC operation. We will see later that some form of handshaking with the microprocessor may be required since data can be written to the ADC faster than it can be converted. Additionally, analog signals with various frequency components will place two bounds on the conversion time.

First, consider a pure sine wave of frequency f. It takes several data samples (conversions) within the period $(1/f)$ to reproduce the wave. If the ADC samples at a much slower rate, we get a false image of the real event. See Fig. 13.31. This phenomenon is called aliasing. Nyquist's Theorem states that data must be sampled at a rate greater than twice the frequency of the highest frequency component to avoid aliasing. Then

$$T_{oc} < \frac{1}{(2f)} \qquad (13.17)$$

High-speed data acquisition systems should include a low-pass filter to eliminate any frequencies greater than half the sampling frequency.

Second, the analog signal will change by an amount ΔV during the conversion time. If we take too long to determine the reading, the signal will change by an amount

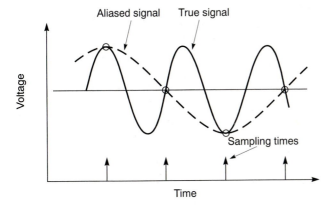

FIGURE 13.31
Effect of slow ADC read rate of fast changing data.

exceeding the resolution. The uncertainty when dependent upon the conversion time is undesirable. In this connotation the conversion time is called the aperature time. We want to establish the maximum allowable aperature time that will still give the ADC resolution. The worst time to sample a sinusoid is where the rate of change is a maximum:

$$\Delta V = \left(\frac{dV}{dt}\right)_{max} T_{oc} \tag{13.18}$$

Therefore, the maximum conversion time is found from

$$T_{oc} < \frac{\Delta V/V}{2\pi f} \tag{13.19}$$

This restriction is far greater than the one for aliasing. If the ADC is not fast enough to make the conversion in this time, a sample-and-hold amplifier must be used to hold the analog signal until conversion is completed.

> **Example 13.5.** Suppose we want to read a 100-Hz periodic signal to 12-bit accuracy. An ADC is available which has a conversion time of 10 μs. Is the ADC up to the job?
> From Table 13.2, $\Delta V/V = 1/2^N$ or 0.000244. Therefore, $T_{oc} \leqq 0.4$ μs. Use a sample and hold with the ADC or a faster ADC.

13.11   A/D CONVERTER TYPES

The two most important quantities when selecting an ADC are speed and resolution. Unfortunately, these factors are not independent. There are three widely used ADC technologies today: (1) flash or parallel converters, (2) dual slope converters, and (3) successive approximation converters. Flash converters are extremely fast, but they are not very accurate. They are applied in real-time signal processing such as oscilloscopes, electronic warfare, and communications. Dual slope converters are very slow, but they are highly accurate. Their averaging technique makes them immune to noise. They can be found in instruments (multimeters), test equipment, and measurement systems (data acquisition). By far the most popular ADCs are the successive approximation type. They offer the best trade-off among speed, resolution, and cost. Table 13.6 gives a comparison between ADC types.

Flash Converters

A typical flash converter is shown in Fig. 13.32. Suppose we wish to convert an analog input to a 3-bit output. A string of eight equal value resistors divides a 10-V reference into equal decreasing steps: 10, 8.75, 7.50, 6.25, 5.00, 3.75, 2.50, 1.25. Eight comparators subtract the split reference voltage from the analog input, producing an output of 0 or 1 depending upon whether the result is negative or positive. Thus, an input of 3 V would produce 0s from comparators 7-2 and 1s from comparators 1-0. This pattern is then encoded into a 3-bit number by a 74148 priority encoder. The encoder gives a binary output (010) established by the first input (comparator 2) to be at the zero level. In general, an N-bit converter must be constructed with 2^N comparators and resistors. Therefore, the resolution is limited by the number of comparators we can afford. Flash converters over 8 bits are very expensive.

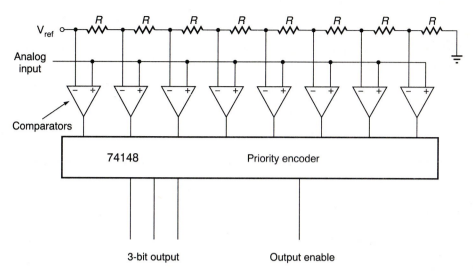

FIGURE 13.32
A flash ADC based on parallel conversion.

TABLE 13.6
Comparison of ADC converters

	Flash	Dual slope	Successive approximation
Speed	4–1000 ns	50–1000 ms	1–100 μs
Resolution	4–10 bits	10–22 bits	8–16 bits
Cost (approx.)	$150 (6 bit)	$15 (10 bit)	$200 (16-bit)
	$3000 (10 bit)	$40 (16 bit)	$40 (12-bit)
			$5 (8-bit)

Dual Slope Converters

There are several versions of the integration method: single slope, dual slope, and multiple slope. We confine our discussion to dual slope converters since they are by far the most common. A circuit for the dual slope converter is diagrammed in Fig. 13.33(a). An unknown voltage is applied to the input where an analog switch connects it to an integrator. The integrator drives a comparator; its output goes HIGH as soon as the integrator output is more than several millivolts. When the comparator output is HIGH, an AND gate passes clock pulses to a binary counter. The binary counter counts pulses until the counter overflows. This time period T_1 is fixed by the clock frequency f_c and counter size M, and is independent of the unknown voltage. However, the integrator output voltage is proportional to the unknown voltage:

$$V_1 = \left(\frac{T_1}{RC}\right)V_{\text{in}} = \frac{MV_{\text{in}}}{f_c RC} \tag{13.20}$$

When the counter overflows, it resets to zero and sends a signal to the analog switch, which disconnects the unknown voltage and connects a reference voltage. The polarity of

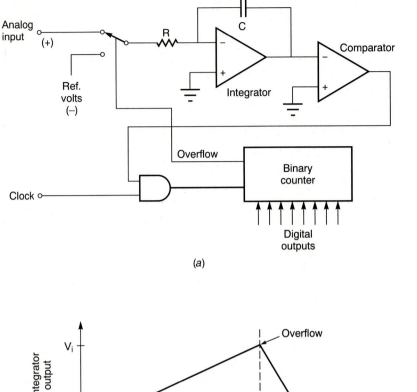

(a)

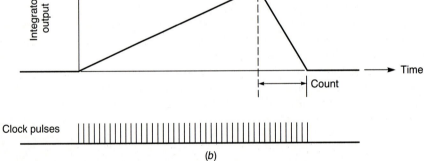

(b)

FIGURE 13.33
ADC dual slope converter. (a) Circuit diagram; (b) timing diagram.

the reference voltage is opposite that of the unknown voltage. The integrator voltage decreases at a rate proportional to the reference voltage. In the meantime the counter has started counting again from zero at the instant the reference voltage is applied to the integrator. When the integrator output reaches zero, the comparator goes LOW, bringing the AND gate LOW. Clock pulses no longer pass through, and the counter stops counting at m.

$$V_i = \left(\frac{T_2}{RC}\right) V_{\text{ref}} = \frac{m V_{\text{ref}}}{f_c RC}$$

The triangle base V_i in Fig. 13.33(b) is common to both; therefore, the count is

$$m = \left(\frac{M}{V_{\text{ref}}}\right) V_{\text{in}}$$

Accuracy of the dual slope method depends only upon V_{ref}. Noise rejection is excellent because the integral action of the converter averages out random negative and positive contributions over the sampling period T_1. Accuracy of ADCs is often quoted in (base 10) digits. Thus, a $3\frac{1}{2}$-digit converter has an accuracy 1 in $10^{3.5}$ or 1 in 3162. This is equivalent to 12-bit (actually $11\frac{1}{2}$-bit) accuracy.

Successive Approximation Converters

The successive approximation method resembles weighing an object on a chemical balance with a set of binary weights. If the set consists of three weights ($\frac{1}{2}$, $\frac{1}{4}$, $\frac{1}{8}$ lbf), there are eight combinations to weigh an object. The largest weight (MSB) is placed on the balance. If the unknown is heavier, the scales do not tip and the $\frac{1}{2}$ lbf weight is left ON. The middle weight is added. If the combination tips the scales, the middle weight is left OFF. Finally, the smallest (LSB) weight is added and is left ON if the scales do not tip. The binary reading (101) or $\frac{5}{8}$ lbf represents the fraction of full scale or 1 lbf with a resolution of 1 LSB. The highest reading is full scale minus 1 LSB or $\frac{7}{8}$ lbf.

A successive approximation circuit is shown in Fig. 13.34(a). The heart of the converter is a DAC. The control logic, sometimes called the successive approximation register (SAR), brings the MSB HIGH on the first clock pulse; all remaining bits are LOW. The binary number is stored and sent to the DAC. The DAC converts the number to a voltage, and the output is compared to the unknown voltage. If the output of the DAC is greater than the unknown voltage, the comparator goes LOW and the bit is reset to zero. If the output is less than the unknown, the MSB remains at 1.

The next MSB is set to 1 and a comparison is made again. It is retained if the output of the DAC is less than the unknown; otherwise it is reset. The sequence of setting and testing bits continues until all the bits have been tested. At this point the SAR outputs an end-of-conversion (EOC) signal to the microprocessor (or PIA). The number stored in the SAR is transferred to the output register of the SAR and the process is complete. The bit status for a 5-bit ADC is represented in Fig. 13.34(b).

13.12 INTERFACING THE ADC

Analog-to-digital converters are selected primarily by application needs: conversion, speed, and resolution. Beyond these traits all characteristics and specifications parallel their mirror function, digital-to-analog converting. Offset and linearity errors are defined similarly, specified on the digital side. They may be purchased with on-chip latches for direct interfacing to the data bus. Most later models come with an internal voltage reference. ADCs will accept unipolar or bipolar voltage inputs and produce straight binary or offset binary outputs. However, unlike DACs, this option is not a matter of external circuitry but must be selected with the converter.

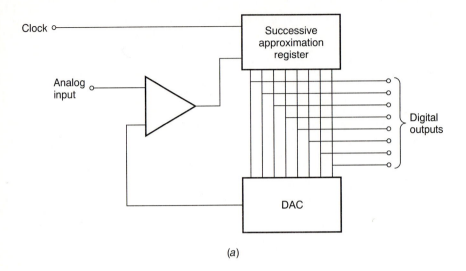

(a)

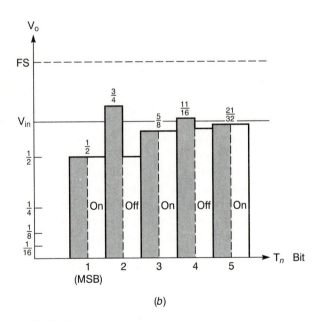

(b)

FIGURE 13.34
Successive approximation ADC. (a) Circuit; (b) bit status—shaded columns indicate trial.

National Semiconductor ADC0809

The National Semiconductor ADC0809 successive approximation ADC is an older, very successful IC that can be bought for only a few dollars. In addition, it is versatile and relatively easy to use. It is an 8-bit device which is sufficiently accurate for control purposes. A block diagram is shown in Fig. 13.35.

Start   Clock

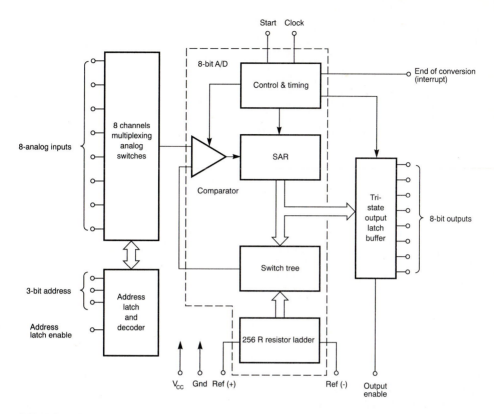

FIGURE 13.35
The ADC0809 block diagram. (*Reprinted with permission of Motorola, Inc.*)

The ±REF pins are external references for the digital-to-analog conversion phase of the process. A $256R$ ladder network approach was chosen over the conventional $R/2R$ ladder to guarantee no missing codes. Furthermore, output symmetry is achieved so that external zero and full-scale trim adjustments are eliminated. The ADC0809 needs less than a milliamp of supply current. Supply voltage should be a nominal +5 V with a maximum rating of 6.5 V.

After the conversion is completed, the 8-bit result is placed into an internal latch until another conversion is completed. The latch is tristated so that the conversion does not appear on the output until the latch tristates are enabled by a HIGH on the output enable pin OE. This allows the user to connect the output lines directly to the data bus. It is said to be microprocessor compatible. When the ADC0809 is interfaced to a PIA, the data is again latched by the PIA. Consequently, the output enable should be permanently tied to +5 V.

Notice that there are eight analog input lines. This is an added feature because most ADCs come with a single analog input (for example, the ADC0804). An internal multiplexer allows you to switch any one of the eight inputs for conversion. Figure 13.36 shows the pin-out for the ADC0809. Input channels are labeled IN0 through IN7. A particular channel is selected by using the 3-bit address decoder labeled ADD A, ADD B, and ADD

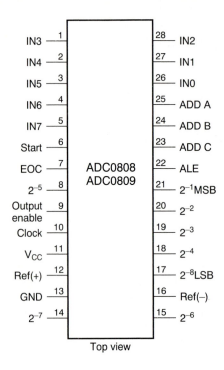

Top view

FIGURE 13.36
Pin assignments for the Motorola ADC0809.
(*Reprinted with permission of Motorola, Inc.*)

TABLE 13.7
Truth table for analog input select

Input channel	Address line		
	C	**B**	**A**
IN0	0	0	0
IN1	0	0	1
IN2	0	1	0
IN3	0	1	1
IN4	1	0	0
IN5	1	0	1
IN6	1	1	0
IN7	1	1	1

C. Table 13.7 lists the input states that are necessary to select a desired analog channel. The channel is latched into the decoder with a LOW-to-HIGH transition signal on the address latch enable (ALE) pin. Analog inputs are 5 V full scale. Of course, the binary outputs 00000000 to 11111111 represent the analog inputs 0 to $(255/256)V_{\text{ref}}$.

Recall that an ADC is much slower than the microprocessor can read. Thus, most ADCs have two control lines for communication (handshaking). With the ADC0809 one line is labeled START to initialize the conversion process. Other manufacturers may label this line CONVERT or RUN. When the conversion is complete, the ADC signals the mi-

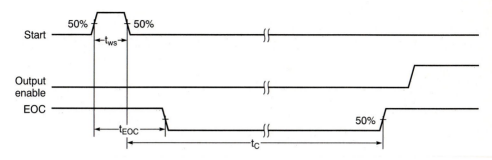

FIGURE 13.37
The Motorola ADC0809 timing diagram.

croprocessor with the end-of-conversion or EOC line, sometimes called the STATUS or data valid (DV) line. A timing diagram is shown in Fig. 13.37. A pulse is applied to the START pin; minimum pulse width should be $t_{ws} = 200$ ns. At time $t_{EOC} = $ (eight clock periods $+ 2 \mu s$) after the rising edge of the START pulse the EOC pin goes LOW and the conversion begins. The conversion process itself lasts for $t_C = 90$ to $116 \mu s$ ($100 \mu s$ typical). When the conversion is complete, the EOC pin goes HIGH. At this point the OUTPUT ENABLE pin can be switched to HIGH by a READ command, placing the latched data on the data bus.

Example 13.6. You are to interface the ADC0809 directly to the data bus. Memory space A800–ABFF has been set aside for the ADC. The eight analog inputs are to be addressed at A800–A807. Show the circuit diagram. Use the interrupt capability of the microprocessor to detect the EOC. Give sufficient software to show how a binary representation of IN3 can be placed in accumulator B.

The hardware is shown in Fig. 13.38. An 8-input NAND is used for basic decoding of the memory space. Control line $R/\overline{W}$ allows the same address to be used for starting conversion (STA instruction) and for reading the result (LDA instruction). Multiplexing with the three LSB address lines selects a particular input.

Address	Content	Mnemonic		Comment
FFF8	01			Load start of
FFF9	00			interrupt routine.
0000		PROGRAM	CLI	
		·		
		·		
		·		
		STA A	A803	Start conversion
		·		on IN3.
		·		
		·		
0100		INTRPT	LDA B A803	
			RTI	

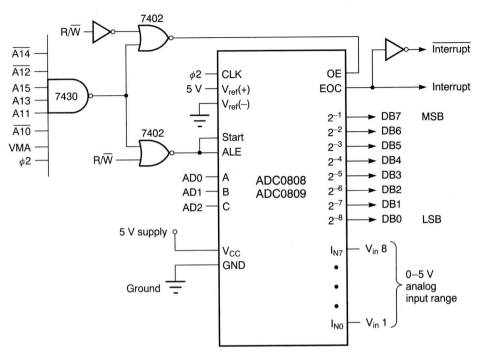

FIGURE 13.38
Hardware for Example 13.6.

Example 13.7. The ADC0809 can be easily interfaced to the PIA. Show a circuit diagram for the interfacing. Assume that the address space for the PIA is 7000 to 73FF and that the analog data is on IN3. Use interrupts. Write the assembly code to store the conversion in accumulator B.

The complete circuit is shown in Fig. 13.39. PIA port A is the input port for the binary result from the ADC. Port B is used to send the multiplex code to the ADC, selecting the analog input channel. CB2 signals the start of conversion with a pulse. The pulse is generated by sending the proper code to control register B (a review of Chap. 8 may be in order). CB1 is set up as an interrupt channel for a LOW-to-HIGH transition from EOC. The software is given in the following table.

Address	Content	Mnemonic			Comment
FFF8	01				Interrupt
FFF9	00				vector.
0000		INIT PIA	CLR A		
			STA A	7001	
			STA A	7003	Set up port A
			STA A	7000	(7000) as input
			COM A		and port B
			STA A	7002	(7002) as output.
			LDA A	#04	
			STA A	7001	
			STA A	7003	

```
PICK IN3  LDA A  #03    Select channel
          STA A  7002   IN3.
START     LDA A  #37
          STA A  7003   Send LOW-
          LDA A  #3F    HIGH-LOW
          STA A  7003   out CB2
          LDA A  #37
          STA A  7003
CONTNU           —
0100      INTRPT LDA B  7000
          RTI
```

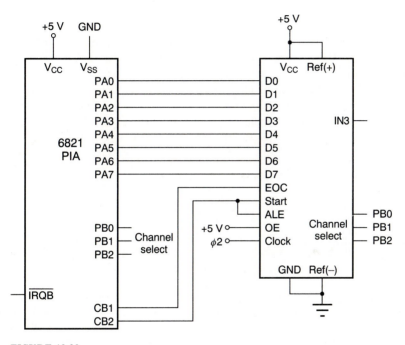

FIGURE 13.39
Hardware for Example 13.7.

Example 13.8. The interrupt method is advantageous because the microprocessor can perform other tasks during the conversion process. Also, a port bit is not needed to detect the EOC, which may save a whole port in some cases. If the microprocessor is dedicated to reading the ADC, an alternate method is polling. Suggest hardware and software changes in Example 13.6 to poll port B for the EOC and to store the conversion in accumulator B.

A simple hardware change is to move the EOC line to PB7, setting up bit 7 as an input. After the START pulse is delivered, we must delay the polling for 10 μs until the EOC goes LOW and starts the conversion. EOC is detected when this line goes HIGH:

```
INIT PIA   PB0–PB2 output,
           PB7–input
PICK IN3
START
```

```
DELAY    NOP
         NOP
         NOP
         NOP
         NOP
POLL     LDA A 7002
         AND   #80
         BEQ   F9
READ     LDA B 7000
         WA1
```

Analog Devices AD574A

Analog Devices AD574A is one of the more popular ADCs today, primarily for data acquisition systems. It is a 12-bit successive approximation converter with a three-state buffered output for direct interface to the data bus. A high precision voltage reference and clock are included on-chip. The block diagram and pin configuration are shown in Fig. 13.40.

Power is supplied to pin 1 (+5 V), pin 7 (+15 V), and pin 11 (−15 V). For *unipolar* inputs two ranges are possible: 0 to +10 V (pin 13) or 0 to +20 V (pin 14). The

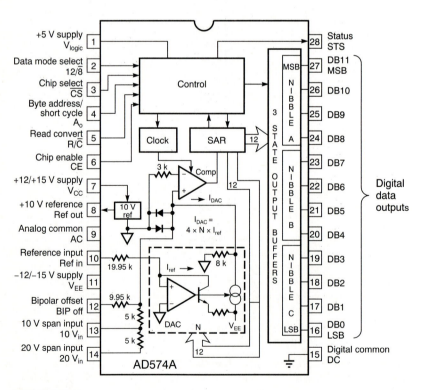

FIGURE 13.40
The Analog Devices AD574A ADC block diagram and pin assignments. (*Courtesy of Analog Devices.*)

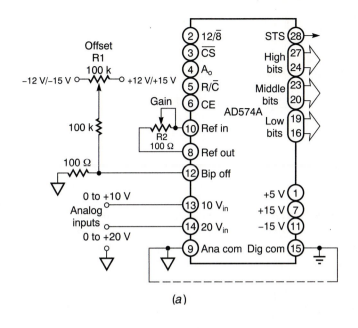

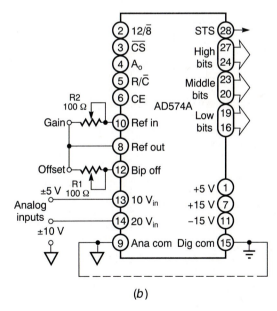

FIGURE 13.41
The Analog Devices AD574A connections. (*a*) Unipolar; (*b*) bipolar. (*Courtesy of Analog Devices.*)

unipolar connection is given in Fig. 13.41(*a*). External trim potentiometers are used to calibrate the offset error and full-scale error. A typical offset error is ± 1 LSB and a full-scale error is ± 2 LSB. In most applications no calibration trimming is necessary. Therefore, pin 12 for offset trim can be connected directly to ground (pin 9). If full-scale trim is not required, replace the trim potentiometer with a 50-Ω resistor. For bipolar inputs two ranges are possible: ±5 V (pin 13) or ±10 V (pin 14). The bipolar connection is given in Fig. 13.41(*b*). Again, the trims may not be necessary, and the two potentiometers can be replaced with 50-Ω resistors. Output is offset binary; invert the MSB if 2s complement numbers are desired.

Control logic is simplified if dedicated input ports, such as a PIA, are available. Both enables (CE and $\overline{\text{CS}}$) must be active before the ADC will operate. The READ/CONVERT or R/$\overline{\text{C}}$ input serves to start the conversion process and to read the result. A conversion is started when R/$\overline{\text{C}}$ goes HIGH to LOW. This allows operation with either a high pulse or a low pulse. The three-state buffers for the output are enabled when R/$\overline{\text{C}}$ is HIGH. However, once a conversion cycle has begun, it cannot be stopped or restarted and data cannot be read until conversion is completed. Conversion time is 35 μs. When conversion starts, the STATUS (STS) pin goes LOW to HIGH, remains HIGH for the conversion period, and signals the EOC with a HIGH to LOW transition. This signal can be polled by the processor through one of the PIA port bits. It can also be used to generate an interrupt. But the conversion time is so short that it makes more sense just to delay the 35 μs by inserting a sufficient number of "do-nothing" instructions or by using a delay subroutine.

The AD574A includes internal logic, determined by the 12/$\overline{8}$ pin input, to permit a direct interface to an 8-bit data bus. In the previous application, 12/$\overline{8}$ pin HIGH, all 12 data lines are connected to the PIA. When this pin is hard wired LOW, the 12-bit output is established by two read operations. First, connect pin A0 to the LSB of the address bus, and connect the data bus to DB0–DB7. The even address (A0 LOW) contains the 8 MSBs (DB11 through DB4). The odd address (A0 HIGH) contains the 4 LSBs (DB3 through DB0) in the upper half-byte, followed by four zeros in the lower half-byte.

13.13   SAMPLE-AND-HOLD AMPLIFIERS

Analog-to-digital converters are relatively slow devices that cannot follow fast changing analog signals; i.e., the analog signal will change by more than 1 LSB over the conversion cycle. Sample-and-hold amplifiers (S/H or SHA) are fast devices (0.1 to 1 μs) that are able to hold an instantaneous voltage upon command until the ADC can convert it. The basic circuit consists of an electronic analog switch, a capacitor, and an op-amp follower as shown in Fig. 13.42(*a*). When the switch is closed, the voltage at the input appears across the capacitor and the output voltage will equal the input voltage. As the input changes, the capacitor charges and discharges, and the output follows. When the switch is opened, the capacitor retains its charge, and the output voltage remains equal to the input voltage at the instant the switch was opened.

Figure 13.42(*b*) shows how the ideal SHA performs. When the SHA follows the signal, it is said to be tracking. In fact, many SHAs are called track-and-hold amplifiers. An actual SHA cannot perform as cleanly as illustrated. The dynamics of switching from sample to hold or from hold to sample introduce a number of specifications that are peculiar to SHAs. The actual performance is shown in Fig. 13.42(*c*).

Acquisition time. The time t_{ac} required for the capacitor to charge within a given error band of the tracking voltage after it is switched from hold to sample. Acquisition time is given by the product of the capacitance and ON-state switch resistance.

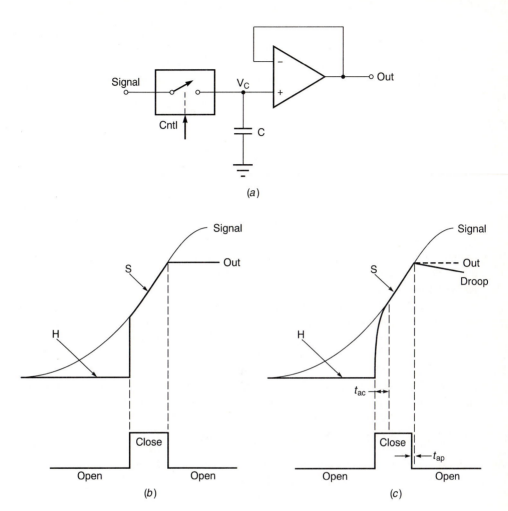

FIGURE 13.42
Sample-and-hold amplifier. (*a*) Basic conversion; (*b*) ideal performance; (*c*) actual performance.

Aperature time. The time t_{ap} required to switch from sample to hold. The delay is usually measured in tens of nanoseconds.

Droop rate. The voltage decay at the output when the SHA is in the hold mode. It is due to switch leakage current.

There are improvements that can be made in the basic SHA. First, adding a voltage follower in front of the switch gives the circuit a high-input impedance. This is important in some applications since a high-impedance transducer may be the source, and it should not be loaded. Second, offset voltage on the output can be eliminated by moving the hold capacitor to a feedback position on the output voltage follower. The accepted circuit today is shown in Fig. 13.43.

A commercial SHA selection is based on acquisition time, accuracy, and internal/external hold capacitor. Accuracy is the (non)linearity expressed as a percent of full scale

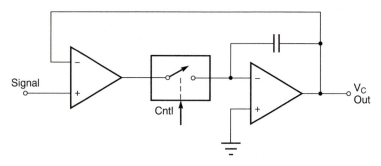

FIGURE 13.43
Commercial version of the SHA.

($\pm 0.01\%$ typical). Hold capacitors are generally on-chip, but several SHAs from every manufacturer are available with pins for an external capacitor supplied by the user. The advantage of an external capacitor is the choice of a smaller capacitor to improve acquisition time. However, accuracy is compromised with smaller capacitors. The hold capacitor should be a high-quality polystyrene or Teflon type. Capacitors as small as 100 pF may be used if less than 12-bit accuracy is acceptable.

Analog Devices AD585 is a high-speed precision sample-and-hold amplifier with an internal hold capacitor. It is recommended for 10- and 12-bit data acquisition systems. The acquisition time is 3 μs with a maximum offset of 2 mV for a 10-V input. Droop rates less than 1 mV/ms, using the on-chip hold capacitor, can be improved by adding a larger external hold capacitor. A block diagram is shown in Fig. 13.44. On-chip precision resistors can be externally connected to provide signal gains of +1 (shown), −1, or +2. The control signal for the HOLD command can be either active HIGH or active LOW. Outputs are protected against damage from accidental short circuits.

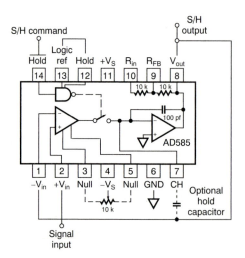

FIGURE 13.44
The Analog Devices AD585 sample and hold.
(*Courtesy of Analog Devices.*)

13.14   ANALOG MULTIPLEXERS

Analog data from transducer/sensors must be converted to digital data for the microprocessor. A single ADC coud be used for each analog signal, but this solution is not cost-effective. Only one ADC is required if an analog multiplexer functions as a selector switch, connecting one voltage at a time to the ADC. We saw that the ADC0809 had a built-in eight-channel multiplexer. However, this type of ADC is not common because the user would like some flexibility in choosing his or her own multiplexer-ADC match to solve his or her design problem.

Analog multiplexers are available as 2-, 4-, 8-, or 16-channel devices. The applied digital logic (channel address) selects the desired input channel through an internal decoder circuit. Some multiplexers come with on-chip latches to facilitate interfacing to the microprocessor bus. The enable line must be active before any channel is passed to the output. Switching times are less than 1 μs.

Most analog multiplexers are connected internally for single-ended inputs, one input line for each channel with all inputs referenced to a common ground. Often transducer signals are differential where the signal is the difference between two voltages, floating and isolated from ground. One solution is a differential op-amp interface to the ADC. However, for these applications multiplexers are also available with differential inputs. These multiplexers may or may not have the differential op-amp on-chip for the single-ended output.

The Analog Devices analog multiplexers are shown in Fig. 13.45. The ADG506A is a 16-channel single-ended input multiplexer. Four binary addresses switch the signal to a common output. The ADG507A switches one of eight differential inputs to a common *differential* output.

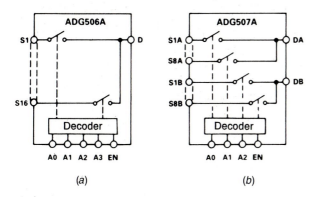

(a)   (b)

FIGURE 13.45
The ADG506A/507A analog multiplexers. (a) Single-ended; (b) differential. (*Courtesy of Analog Devices.*)

PROBLEMS

13.1. Figure 13.1 represents a basic data acquisition system. Write a program that will collect data from eight transducers and store the binary number at addresses 0000–0007. Assume the analog-to-digital converter (ADC) converts automatically and always has the number available at port A.

13.2. What is the difference between a sample-and-hold amplifier and a latch?

13.3. Sketch a possible circuit for an ON/OFF control system. Assume that the system is a water tank with a heater where the temperature must be maintained at or near set point. Use a comparator.

13.4. An op amp drives a load which has a variable resistance between 10 and 1000 Ω. When the load is infinite (open circuit), the output voltage is 120 V. Assuming the input to the amplifier does not change, determine the effect of load on the output voltage. State any assumptions.

13.5. A differential amplifier uses 10-k and 50-k resistors for a gain equal to 5. It is amplifying the difference between two signals at 9.8 and 10.0 V. The CMRR is 80 db. What error do you expect?

13.6. An inverter op amp is used to boost the input voltage (tens of millivolts). Output voltage should be in the tenths of volts range. Would you recommend an AD741C, an LF356A, or an AD0P–07A?

13.7. List the DAC output levels for a 3-bit input device with a 5-V reference.

13.8. The MC1408(L6) digital-to-analog converter (DAC) has 6-bit accuracy although it is an 8-bit device. Explain.

13.9. Which has better accuracy at 70°C, the DAC811 or the DAC7541? Express their accuracy in LSBs.

13.10. Figure 13.26(*b*) gives the pin connections for the microprocessor compatible AD7524 DAC. Show an interface circuit for address space 3XXX. How does the interface change if a PIA is used?

13.11. A 12-bit DAC is interfaced to a PIA. See Fig. 13.28. You wish to generate a 0 to full-scale voltage ramp in 0.1 s. Write the assembly code if the PIA address space is 4XXX.

13.12. How should you configure the PIA to DAC interface in Fig. 13.28 if the DAC has no on-chip latch?

13.13. Construct an input/output (I/O) graph similar to Fig. 13.29 for an ADC bipolar connection.

13.14. Write software for Example 13.6 to place 256 bytes of data into memory, one byte every 0.1 ms.

13.15. What is the fastest rate at which we can collect data for Prob. 13.14? Suggest software changes to accelerate the data-taking process.

13.16. Show hardware and software for interfacing the AD574A ADC to the PIA.

13.17. The AD574 ADC needs a sample and hold if the analog signal changes more than $\frac{1}{2}$ LSB over the conversion time interval. Find the maximum signal frequency for which a SHA will not be needed.

CHAPTER 14

INTERFACING
TIMERS

Timers are integrated circuits (ICs) which deliver continuous or single pulse outputs to serve as clocks for other ICs or to time external events. Some examples that have appeared in this text include (1) SPST switch debounce, (2) dot-matrix display drivers, (3) liquid crystal displays (LCDs), (4) dedicated DC motor controllers, and (5) pulse-width-modulated (PWM) control of DC motors. Many control applications do not require feedback but only the "timely" ON/OFF switching of control elements. These include (1) appliances, (2) process control, (3) robots, (4) automobiles, and (5) machine tool control.

In some cases, such as LCD inputs, the clock frequency is fixed permanently. In other cases, such as PWM control, the pulse character must be continually changed through programming from a microprocessor. Figure 14.1 shows the system for maintaining the air/fuel ratio at or near stoichiometry in a fuel-injected engine. The exhaust gas oxygen (EGO) sensor measures the amount of oxygen in the exhaust gas. Its voltage output is read by a microprocessor which determines the correct PWM signal for the fuel injectors. In Fig. 14.2 automobile speed is controlled with a vacuum-operated throttle actuator. Throttle position is varied by changing the average pressure in the actuator chamber. This is done by rapidly switching the pressure control valve between outside air and the manifold pressure, which is near vacuum. The control voltage is a PWM signal which controls the vacuum duty cycle.

Time delays and pulses can be generated directly by the microprocessor/PIA. But while the MPU is polling, it cannot handle other assignments. An even more difficult situation occurs when two timing operations must be running simultaneously. Consider the microprocessor control of a four-cylinder fuel-injection engine. The crankshaft position (and velocity) must be determined while fuel injector on-time is being controlled. The MPU cannot do both jobs alone, and yet monitor temperature and manifold pressure. The solution is to give these timing tasks to a timer IC.

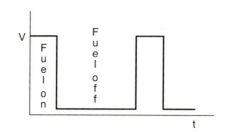

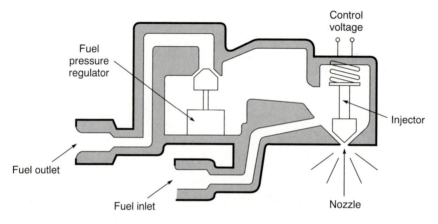

FIGURE 14.1
Pulse-width-modulated (PWM) fuel injection system.

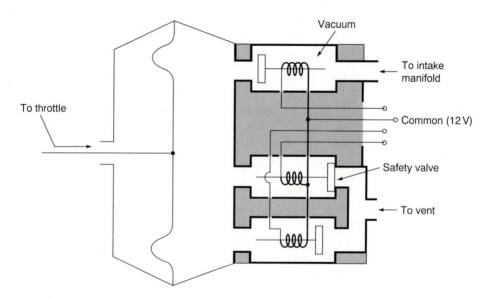

FIGURE 14.2
Throttle actuator for an automobile speed control system.

This chapter essentially discusses the Motorola 6840 programmable timer module (PTM) and its applications. The PTM can generate an assortment of pulse signals. It also has an input capability which can be used to measure the duration of events. Although the MC 6840 can handle permanent pulse modes easily, they can be generated more cheaply with fixed timers such as the classic 555 timer which requires only an external resistor/capacitor to set the frequency. The 555 timer will be reviewed as well.

14.1 THE MC 6840 PROGRAMMABLE TIMER MODULE (PTM)

The MC 6840 programmable timer is designed to operate with the Motorola 6800 family, but it can be used with other MPUs. It is interfaced directly to the data bus and has two chip selects for memory-map decoding much like the MC 6821 PIA. The PTM is fully TTL compatible and requires only a single 5-V supply. Power dissipation is about 500 mW.

The PTM has three timers which operate independently from the microprocessor once they are programmed. Each timer can be programmed for one of four basic tasks. See Fig. 14.3:

1. *Continuous output.* The wave form is a continuous pulse with programmed frequency. In addition, it can be programmed to give *pulse-width-modulated* (PWM) signals. When the timer counts down to zero, it generates an interrupt signal that can be used for "scheduling" external control duties. The term "time-out" or TO is given to this event.

2. *Single pulse output.* The single pulse or "one-shot" is similar to the continuous output except that only a single pulse is produced. A single time-out is generated at the pulse end. The pulse width and its delay are programmable.

3. *Period (frequency) input measurement.* The timer counter can be made to stop and start by applying an external signal to the gate input pin. Signal transition from HIGH-to-LOW stops the counter. By reading the counter content one can determine the exact time or period between the two triggers. Frequency is simply the reciprocal of the period.

4. *Pulse width input measurement.* Pulse width measurement is similar to period measurement. In this case the counter is stopped by a LOW-to-HIGH transition of the input signal. The time of any external event can be measured by generating the proper transition signal at the beginning and end of the event.

The MC 6840 PTM has 10 programming registers, addressed similar to the PIA. Each timer has a counter register and each counter has a latch register. Individual control registers determine how each timer is to be used. An overall status register keeps track of timer interrupts. Figure 14.4 shows the timer registers and I/O lines. Each timer has an output line, a gate line for the input mode, and an optional external clock input.

Latches

Data (the count to be decremented) is loaded into the latch from the MPU via the bidirectional data bus. The latch is 16 bits wide while the data bus is only 8 bits wide; therefore,

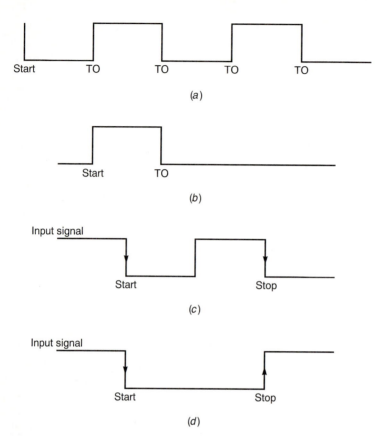

FIGURE 14.3
The MC 6840 timer programming modes. (*a*) Continuous pulse output; (*b*) single-pulse output; (*c*) frequency measurement; (*d*) pulse measurement.

the data is entered in two parts. The most significant byte (MSByte) is entered first and stored temporarily in a buffer. When the LSByte is written, both are transferred simultaneously to the latch. The data must be loaded from the 16-bit MPU index register (or stack pointer). Transfer is accomplished with the STX instruction. Thus, the latch is a *write-only register*. This write operation is called the (counter) *latch initialization*.

Counters

Data must then be transferred from the latch to the counter register. This process is called *counter initialization*. Conditions or commands for counter initialization will be reviewed shortly. Once the counter receives the data it will not operate until the *counter enable* condition has been met, which is also reviewed later. The counter is decremented by one for each clock cycle. If the internal clock is selected, the clock is the MPU $\phi2$. An external clock can be selected if a different decrement rate is wanted. In fact, the output of one timer can be used to supply an external slow clock to an adjacent timer. Timer counters are *read-only registers*. They can be read by the MPU "in progress" or read in a hold

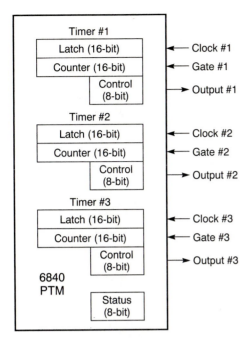

FIGURE 14.4
The MC 6840 timer registers and I/O lines.

state due to a stop signal on a respective gate line. Since counters are 16-bit registers, we must read them into a 16-bit MPU register such as the index register. The LDX instruction serves this purpose.

Control Registers

Control registers are 8-bit *write-only registers.* Their function is to control operating modes and related actions of the timer. Register bit 0 performs a different function on each of the three timers. However, bit 1 through bit 7 of each control register perform the same function. Control registers will be central to the PTM initialization process later. A summary of bit functions is given in Fig. 14.5.

Bit 0. In control register #1, bit 0 determines whether the counters are held or allowed to operate. When this bit is logic zero, *all* timers are allowed to operate. Writing a one into this bit constitutes a *software reset* (see Sec. 14.3). Counters will not start the count or stop if the count is in progress.

Bit 0 of control register #2 determines whether control register #3 or control register #1 is selected when addressed by the PTM. Each share the same address. Control register #1 will be written into if bit 0 is set; control register #3 will be written into if bit 0 is cleared. Bit 0 functions in the same manner as bit 2 of the PIA control register when selecting the data direction register or the output port.

In control register #3, bit 0 gives the option of choosing a slower clock. When the bit is cleared, the clock driving timer #3 only is divided by 1. When the bit is set, the clock is divided by 8. The ÷8 prescaler is placed between the clock input and the input to counter #3. Thus, the prescaler can be applied to either the internal clock or an external clock.

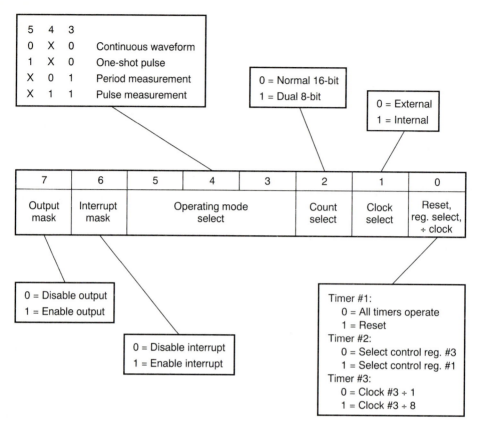

FIGURE 14.5
Control register bit function.

Bit 1. This bit determines whether the internal or external clock will act as the clock for that counter. If the bit is set, the internal clock will be selected. This clock derives from the E pin and is usually the $\phi2$ signal from the microprocessor. A cleared bit enables the external clock tied to the clock input line of the counter.

Bit 2. This bit controls the counting mode. If the bit is cleared, the associated timer will operate with a 16-bit binary number in the counter. A set bit will operate the timer with two 8-bit numbers in the counter. Since two separate numbers are decremented to zero instead of one, slight variations in the basic mode can be achieved. The counter is said to be operating in the 16-bit mode or the dual 8-bit mode. For example, the dual 8-bit mode is used to generate PWM output signals other than a 50% duty cycle.

Bits 3, 4, and 5. These bits determine the counter operating mode and are the interrupt control bits for the input modes. Together with bit 2, they make up the 12 variations of the four basic modes for each timer. Each mode will be discussed in detail later. Table 14.1 lists the programming for bits 3, 4, and 5.

TABLE 14.1
6840 PTM mode selection. (*Reprinted with permission of Motorola, Inc.*)

CRX3	CRX4	CRX5	
0	0	0	Continuous operating mode: Gate ↓ or write to latches or reset causes counter initialization
0	0	1	Frequency comparison mode: Interrupt if gate ↑‾‾‾↑ is < counter time out
0	1	0	Continuous operating mode: Gate ↓ or reset causes counter initialization
0	1	1	Pulse width comparison mode: Interrupt if gate ↑‾‾‾↑ is < counter time out
1	0	0	Single shot mode: Gate ↓ or write to latches or reset causes counter initialization
1	0	1	Frequency comparison mode: Interrupt if gate ↑‾‾‾↑ is > counter time out
1	1	0	Single shot mode: Gate ↓ or reset causes counter initialization
1	1	1	Pulse width comparison mode: Interrupt if gate ↑‾‾‾↑ is > counter time out

Bit 6. Bit 6 is the timer interrupt enable or disable bit. Recall that the timer will generate an interrupt signal when the timer count reaches zero or the timer is stopped. If the bit is cleared, the interrupt is masked; it has no effect on the interrupt flag in the status register. The bit must be set to enable the interrupt.

Bit 7. This bit controls the timer output. A bit value of zero masks the output from appearing at its output pin. If the bit is set, the output is enabled.

Example 14.1. The contents of timer #1 control register is A2. How is the timer configured?

LSB	0	All timers operate.
	1	Internal clock.
	0	16-bit count.
	0 ⎫	
	0 ⎬	Period input mode.
	1 ⎭	
	0	Interrupt disabled.
MSB	1	Output enabled.

Status Register

The MC 6840 has a *read-only* status register which contains the status of the interrupt flags. See Fig. 14.6. Bits 0, 1, and 2 are assigned to timers #1, #2, and #3, respectively. Bit 7 is a composite flag which will be set if *any* of the individual interrupt flags are set. The remaining 4 bits are not used, and they default to zero when read. No interrupt flag will be set unless bit 6 of the corresponding control register is set.

When bit 7 of the status register goes HIGH, the 6840 PTM interrupt pin goes LOW, which interrupts the microprocessor if it's connected to IRQ. The MPU service subroutine reads the status register to determine which timer caused the interrupt. The interrupt flags are cleared by one or more of the following methods:

1. Hardware (external) reset.

2. Software reset.

3. A read of the status register followed by a read of the timer *counter* which caused the interrupt.

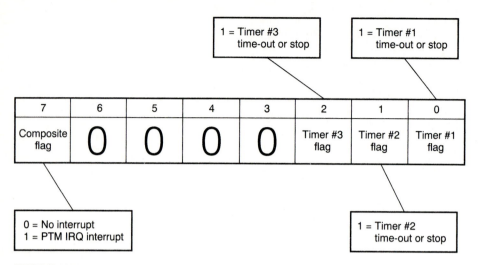

FIGURE 14.6
Status register bit function.

4. A write to the latches when control register bits 3 and 4 are zero.

5. The external gate line goes LOW (and control register bit 3 is zero).

14.2 MC 6840 PIN-OUT AND INTERFACING

The MC 6840 programmable timer is a 28-pin IC. A pin assignment is shown in Fig. 14.7, and a complete interface to the 6802 is shown in Fig. 14.8.

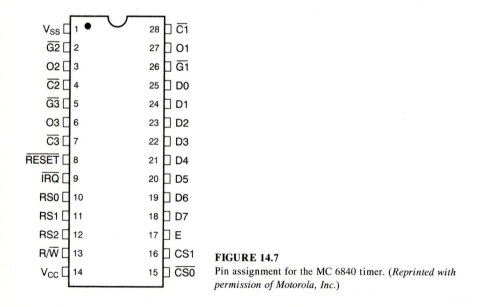

FIGURE 14.7
Pin assignment for the MC 6840 timer. (*Reprinted with permission of Motorola, Inc.*)

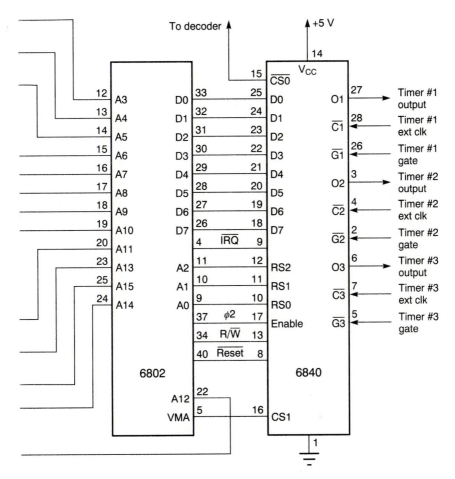

FIGURE 14.8
The MC 6840 timer interface to the 6802 microprocessor.

Data (D0–D7). The bidirectional data lines (D0–D7) allow transfer of data between the MPU and PTM. Outputs are three-state devices which remain in the high impedance OFF-state except when the MPU performs a read.

Read/Write (R/W̄). An input signal from the MPU to control the direction of data transfer.

Enable (E clock). The E clock signal synchronizes data transfer between the MPU and PTM as well as clock resets and gate inputs. Normally, this signal is the 1.0-MHz $\phi2$ clock from the 6802 MPU. For other 6802 processors, the faster 1.5-MHz MC 68A40 PTM or the 2.0-MHz MC 68B40 PTM should be used.

Interrupt request (ĪRQ). The interrupt request line goes LOW when the composite interrupt flag (bit 7) in the status register is set. This pin is tied directly to ĪRQ on the MPU.

$\overline{\text{Reset}}$. A LOW on this pin causes a hardware reset. The ramifications will be discussed in the initialization section.

Chip selects ($\overline{\text{CS0}}$, CS1). Two chip selects are used to activate the data bus interface and allow data transfer from the MPU. CS1 should be connected to the VMA line from the MPU. $\overline{\text{CS0}}$ should come from the decoder for memory-mapped I/O.

Output (01, 02, 03). Output lines for timers #1, #2, and #3, respectively.

Gates ($\overline{\text{G1}}$, $\overline{\text{G2}}$, $\overline{\text{G3}}$). Input lines for timers #1, #2, and #3, respectively. A HIGH-to-LOW TTL compatible signal acts as an optional trigger for the output mode, initializing the counter. For the input mode a gate transition will stop the counter. Gate transition is recognized by the PTM on the fourth cycle of the clock. Gate levels must be stable for at least one clock cycle.

Clock inputs ($\overline{\text{C1}}$, $\overline{\text{C2}}$, $\overline{\text{C3}}$). Optional external clock inputs which will accept TTL voltage levels to decrement the counters. A counter is not decremented until the fourth internal clock pulse. This does not affect the input frequency, but it will create a delay between the clock input transition and internal recognition by the PTM.

Register select lines (RS0, RS1, RS2). The register select lines allow access to the PTM registers. These lines are connected to the three least significant MPU address lines A0, A1, and A2, respectively. There are eight possible addresses. Since each register is read or write only, the additional R/$\overline{\text{W}}$ line gives a combination of 16 possible registers that can be addressed. However, the three latches and three counters are 16-bit registers and occupy two addresses each. This leaves 10 locations for the 10 registers in address space XXX0–XXX7. Table 14.2 gives the address for each register assuming the PTM is decoded for memory space 5000–5007. Of course, the decoding circuit can be made to map the PTM into any address space. The A0, A1, and A2 connections will assure that the first eight addresses of the decoded space will select the registers in the order shown in Table 14.2.

TABLE 14.2
PTM register addresses for decoded space 5XXX

Address	Read operation	Write operation
5000	NONE	CONTROL REG. #1 OR #3
5001	STATUS REG.	CONTROL REG. #2
5002	TIMER #1 COUNTER	TIMER #1 LATCH
5004	TIMER #2 COUNTER	TIMER #2 LATCH
5006	TIMER #3 COUNTER	TIMER #3 LATCH

Example 14.2. A 16-bit number A06B is to be loaded into the counter latch for timer #3. How can this be done? Assume the 6840 PTM is at address 4000–40FF.

From Table 14.2 the latch is at address 4006. A write can be accomplished two ways:

```
Program 1:   LDA A   #A0
             STA A   4006
             LDA A   #6B
             STA A   4007
Program 2:   LDX     #A06B
             STX     4006
```

Program 2 is more efficient. The reader should be aware that he or she cannot use index register instructions like increment or decrement. Among other reasons this type of instruction uses both the read and write cycle.

14.3   MC 6840 PTM INITIALIZATION

Before any programs are written, a certain sequence of events must be carried out. First, the PTM undergoes a hardware reset similar to the PIA and the microprocessor itself. Second, the latches are loaded with data (latch initialization). Third, the data is transferred to the counters (counter initialization). Fourth, the counters are enabled (decremented). These steps may be software driven or initiated by gate signals. Their details will now be discussed to learn how the PTM functions and to use it effectively.

Resets (Hardware)

The $\overline{\text{RESET}}$ pin on the PTM is normally tied to the microprocessor $\overline{\text{RST}}$ pin. The reset signal is asserted during power up and whenever the system "Reset" button is pushed. The pin should be brought LOW for at least 10.25 μs. When the pin is returned HIGH, several events will have occurred within the IC during the LOW:

1. All counter clocks in progress are disabled, and all outputs are reset to zero.
2. All status register bits (interrupt flags) are cleared.
3. All latches are preset to their maximum value FFFF (65,535).
4. All counters are preset to the latch contents.
5. All control register bits are cleared with the exception of bit 0 of control register #1, which is set. This bit is the software internal reset bit.

Resets (Software)

The internal software reset bit is set to one by the hardware reset. It is also set by writing a HIGH to bit 0 of control register #1. Upon being set the following conditions take place:

1. All counter clocks in progress are disabled, and all outputs are reset. Counters maintain their last value at the time the count is stopped. Latches can be written to, and counters can be read. But they will not count.

2. All interrupt flags in the status register are cleared.

3. All counter latches retain the last data written to them.

4. All remaining control register bits are unchanged.

Latch Initialization

The latches are initialized by simply writing a 16-bit data word directly to the latch. It is also initialized by a hardware reset and will assume the default value (FFFF) unless a new word is written. The write operation was given previously in Example 14.2:

LDX #XXXX
STX (XXX2) or (XXX4) or (XXX6)

Counter Initialization

Data is transferred from the (counter) latch to the counter register in one of four ways:

1. Bit 0 of control register #1 is changed from one to zero (reset). This method is normally used in the output mode.

2. The gate associated with the counter receives a HIGH-to-LOW transition signal. Bit 0 of control register #1 should be zero and the corresponding control register should be programmed for a gate input. This method is valid for either the output mode or the input mode. No individual interrupt flag should be asserted for the input mode.

3. A data word is written to the latches, provided bits 3 and 4 of the control register are both zero. This is true regardless of bit 0 in control register #1.

4. The counters automatically reinitialize themselves when they count down to zero in the continuous mode.

Counter Enable

The counter is released and allowed to count under the following conditions. For the output mode, bit 0 of control register #1 is cleared and the signal gate is LOW. Thus, the gate should be tied LOW if it is not used. Gate level makes no difference for the single shot mode. For both the input and output modes, the gate pin goes HIGH-to-LOW, bit 0 of control register #1 is already cleared, and there is no individual interrupt flag asserted. If at any time a period or pulse width measurement is in progress, a write to the (counter) latch will disable the counter.

Program Initialization

Up to now we have discussed the conditions which lead to the operation of the three MC 6840 timers (counters). How do we translate this information into an efficient program to initialize the PTM? There are two major steps:

1. Write the beginning count to each timer latch. Do this by using index register instructions

> LDX #
> STX

for each latch. The order of the latches is not important. If a value is not written to a given timer, the beginning count will default to FFFF.

2. A control word is written to each timer control register. These registers are 8 bits wide, so use the sequence

> LDA A #
> STA A

The registers should be loaded in the following order:

> Write to control register #3.
> Write to control register #2.
> Write to control register #1.

The reason for this order is simple efficiency. Remember that bit 0 of control register #2 is used to select between control register #3 and control register #1. Since the hardware reset clears all bits except bit 0 of control register #1, control register #3 is already programmed to be addressed. Then control register #2 is addressed, which includes changing bit 0 to address control register #1. Control register #1 is a natural for last since its bit 0 is cleared to start all timers.

14.4 OUTPUT MODE: CONTINUOUS WAVEFORM

The PTM can be programmed to generate a continuous rectangular TTL signal on the output pin. Waveform pulses can have a variable duty cycle. The counter value establishes the frequency. For a single timer the frequency is relatively high, but timers can be cascaded together to obtain much lower frequencies. Periods from 2 μs to 282 years can be generated with the three timers in a single PTM using the $\phi 2$ clock.

There are two types of continuous output programming modes:

> Bit 2 = 0   16-bit counting mode
> Bit 2 = 1   dual 8-bit counting mode

The 16-bit mode generates a symmetrical waveform (50 percent duty cycle). When the counter reaches zero (time-out), a signal transition occurs and the counter is reset. The counter decrements twice for each period. With the dual 8-bit mode the count is divided into a MSB and a LSB. The MSB decrements once for every full countdown of the LSB and an asymmetrical PWM signal can be generated.

There are two further options with regard to bit 4:

> Bit 4 = 0   Write to latches OR gate transition initializes counter.
> Bit 4 = 1   Gate transition initializes counter.

When a gate signal starts the counter, set bit 4 to prevent a false start from erroneous software.

The following notation is used throughout:

N = *decimal* number in the 16-bit latch

M = decimal number of the MSB in N

L = decimal number of the LSB in N

T = clock period. Unless stated otherwise, the internal clock of 1 MHz is assumed. $T = 1\ \mu s$

P = waveform period

Case I. 16-Bit Counting Mode

Control register format

b₇	b₆	b₅	b₄	b₃	b₂	b₁	b₀
X	X	0	0	0	0	X	X

Pulse width W = space width $S = (N + 1)T$

Period $P = W + S = 2(N + 1)T$

Frequency $= 1/P$

Example 14.3. Set up timer #3 for a square wave output with a frequency of 30 Hz. PTM is at address 5000–5007.

$$P = \tfrac{1}{30}\ s$$
$$W = \tfrac{1}{60}\ s$$
$$(N + 1)(10^{-6}\ s) = \tfrac{1}{60} \qquad N = (16{,}666)_{10}$$
$$= 411A$$

Control register #3

7	6	5	4	3	2	1	0
1	0	0	0	0	0	1	0

Bit 0: Divide clock by 1.

Bit 1: Use internal clock.

Bit 2:   16-bit counting mode (symmetrical output).
Bits 3, 4, 5:   Continuous output.
Bit 6:   Interrupt disabled.
Bit 7:   Output enabled.

Control register #2

7	6	5	4	3	2	1	0
0	0	0	0	0	0	0	1

Bit 0:   Select control register #1 for a write.
Bits 1–7:   Don't care.

Control register #1

7	6	5	4	3	2	1	0
0	0	0	0	0	0	0	0

Bit 0:   Start all counters.
Bits 1–7:   Don't care.

The following program to implement Example 14.3 is complete. Timer #3 will start counting immediately after the write to control register #3.

Program

LDX	#411A	Load timer #3 latch with count.
STX	5006	
LDA A	#82	Configure register #3.
STA A	5000	
LDA A	#01	Prepare to write register #1.
STA A	5001	
LDA A	#00	Start the counter.
STA A	5000	

Example 14.4. The larger the latch word (FFFF maximum) is, the lower the timer output frequency will be. The lowest frequency for a single timer using the internal clock (1 MHz) is 7.6 Hz. The ÷8 clock option will lower this frequency to approximately 1 Hz. If lower frequencies are desired, timers should be cascaded as in Fig. 14.9. The output of timer #3 goes to the clock input of timer #2. The output of timer #2 goes to the clock input of timer #1 (if needed). Configure that PTM for a continuous 0.1-Hz waveform from timer #2.

The waveform frequency from timer #3 is

$$F_3 = \frac{1}{2(N_3 + 1)T} = \frac{1}{T_3}$$

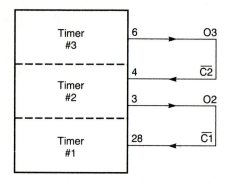

FIGURE 14.9
Cascading timers for lower output frequency.

The desired frequency is

$$F_2 = \frac{1}{2(N_2 + 1)T_3}$$

Therefore,

$$4(N_3 + 1)(N_2 + 1)T = \frac{1}{F_2} = 10 \text{ s}$$

$$T = 10^{-6} \text{ s}$$
$$N_3 = (65,535)_{10} = \text{FFFF}$$
$$N_2 = (37)_{10} = 25$$

Control register #3

7	6	5	4	3	2	1	0
1	0	0	0	0	0	1	0

Bit 0:   Divide clock by 1.
Bit 1:   Use internal clock.
Bit 2:   16-bit counting mode.
Bits 3, 4, 5:   Continuous output.
Bit 6:   Interrupt disabled.
Bit 7:   Output enabled.

Control register #2

7	6	5	4	3	2	1	0
1	0	0	0	0	0	0	1

Bit 0:   Select control register #1 for a write.
Bit 1:   Use external clock.
Bit 2:   16-bit counting mode.
Bits 3, 4, 5:   Continuous output.

Bit 6:   Interrupt disabled.
Bit 7:   Output enabled.

Control register #1

7	6	5	4	3	2	1	0
0	0	0	0	0	0	0	0

Bit 0:   Start all counters.
Bits 1–7:   Don't care.

Program

```
LDX     #FFFF     Load timer #3 latch.
STX     5006
LDX     #0025     Load timer #2 latch.
STX     5004
LDA A   #82       Configure register #3.
STA A   5000
LDA A   #81       Configure register #2.
STA A   5001
LDA A   #00       Start counters.
STA A   5000
```

Note: The load of timer #3 latch can be omitted since FFFF is loaded automatically if no write is received.

Case II. Dual 8-Bit Counting Mode

Control register format

b_7	b_6	b_5	b_4	b_3	b_2	b_1	b_0
X	X	0	0	0	1	X	X

Pulse width $W = LT$

Space width $S = [M(L + 1) + 1]T$

Period $P = W + S = (M + 1)(L + 1)T$

Duty cycle $= \dfrac{W}{P} \times 100\% = \dfrac{L}{(M + 1)(L + 1)} \times 100\%$

A special condition exists if $L = 0$. In this case the counter will revert to the 16-bit mode with each pulse and space equal to $(M + 1)T$. If $M = L = 0$, the counters do not change but the output toggles at a rate of one-half the clock frequency.

Example 14.5. Set up timer #1 so that the output has a period of 10^{-3} s (1000 Hz) and a 10 percent duty cycle.

$$\text{Duty cycle} = \frac{LT}{P} \times 100\% = 10\%$$

$$L = (100)_{10} = 64$$

$$\text{Period } P = (M + 1)(L + 1)T = 10^{-3}\text{ s}$$
$$M = 9$$

Control register #3

7	6	5	4	3	2	1	0
N	O	T		U	S	E	D

Control register #2

7	6	5	4	3	2	1	0
0	0	0	0	0	0	0	1

Bit 0:   Select control register #1 for a write.
Bits 1–7:   Don't care.

Control register #1

7	6	5	4	3	2	1	0
1	0	0	0	0	1	1	0

Bit 0:   Start all timers.
Bit 1:   Use internal clock.
Bit 2:   Dual 8-bit mode.
Bits 3, 4, 5:   Continuous output.
Bit 6:   Interrupt disabled.
Bit 7:   Output enabled.

Program

```
LDX     #0964     Load latch #1.
STX     5002
LDA A   #01       Prepare to write register #1.
STA A   5001
LDA A   #86       Start counter.
STA A   5000
```

In most problems the period is specified for the PWM signal. Some casual thought will quickly reveal just how much restriction this places on the duty cycle. For Example 14.5 the maximum duty cycle is 25 percent. The next section will explain a method that gives us a precise duty cycle over its whole range.

14.5   OUTPUT MODE: SINGLE PULSE WAVEFORM

The single-pulse (one-shot) mode is similar to the continuous waveform in several respects. The internal counting mechanism remains cyclical, and a time-out occurs each half period. Bit 2 configures the timer for a 16-bit count or a dual 8-bit count. Counter initialization is identical, occurring with a "write to the latch" if bit 4 is clear. There are several differences:

1. The output is enabled for only one pulse and remains LOW until another counter enable, and thus the name one-shot.
2. The gate does not need to be held LOW to enable the timer. Use a software reset only. However, a HIGH-to-LOW transition on the gate pin will cause a one-shot if bit 0 of timer #1 is clear, i.e., after a software reset. In this case a write to the appropriate control register during program initialization occurs *after* a write to control register #1. Any gate trigger must be HIGH at least four clock cycles to be valid.
3. If the latch count equals zero ($N = 0$), then the output line will go LOW and stay LOW until a number is put into the latches.

Case I.   16-Bit Counting Mode

Control register format

b_7	b_6	b_5	b_4	b_3	b_2	b_1	b_0
X	X	1	0	0	0	X	X

$$\text{Pulse width } W = NT$$
$$\text{Pulse space } S = T$$
$$\text{Period } P = W + S = (N + 1)T$$

Example 14.6. Set up timer #2 to deliver a single pulse with a 1 ms width. Assume the PTM occupies address space 5000–5007.

$$N(10^{-6}) = 10^{-3} \text{ s}$$
$$N = (1000)_{10} = 03E8$$

Control register #3

7	6	5	4	3	2	1	0
N	O	T		U	S	E	D

Control register #2

7	6	5	4	3	2	1	0
1	0	1	0	0	0	1	1

Bit 0:	Select control register #1 for write.
Bit 1:	Use internal clock.
Bit 2:	16-bit counting mode.
Bits 3, 4, 5:	Pulse output.
Bit 6:	Interrupt disabled.
Bit 7:	Ouput enabled.

Control register #1

7	6	5	4	3	2	1	0
0	0	0	0	0	0	0	0

Bit 0:	Start all counters.
Bits 1–7:	Don't care.

Program

```
LDX     #03E8     Load timer #2 latch.
STX     5004
LDA A   #A3       Configure register #2.
STA A   5001
LDA A   #00       Start counters.
STA A   5000
```

Case II. Dual 8-Bit Counting Mode

Control register format

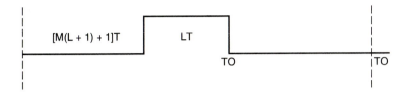

b_7	b_6	b_5	b_4	b_3	b_2	b_1	b_0
X	X	1	0	0	1	X	X

$$\text{Pulse width } W = LT$$

$$\text{Space width } S = [M(L + 1) + 1]T$$

$$\text{Period } P = (L + 1)(M + 1)T$$

Example 14.7. The continuous dual 8-bit counting mode does not give a full-range PWM output. A better method uses one timer to generate a single pulse with the correct width and a second timer to generate a continuous waveform for gating the single pulse repeatedly. Configure the PTM for a PWM signal with a duty cycle equal to 78 percent and a period of 1 ms.

A sketch of the system is given in Fig. 14.10. Timer #1 is programmed for a continuous output 01 with period $P_1 = 2(N_1 + 1)T$. Timer #2 is programmed for a pulse output 02

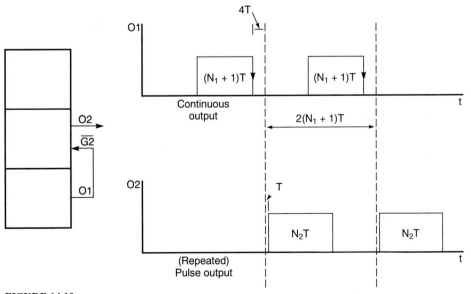

FIGURE 14.10
Full range duty cycle by cascading the continuous and pulse modes.

with width $W_2 = N_2T$. However, the output 01 is connected to timer #2 gate $\overline{G2}$. Each HIGH-to-LOW transition of the continuous waveform triggers a new pulse. Although the trigger is not recognized by timer #2 for four clock cycles, and the pulse output is delayed for one clock cycle, the period and pulse width are not affected. Thus,

$$\text{Duty cycle} = \frac{W_2}{P_1} \times 100\% = \frac{N_2}{2(N_1 + 1)} \times 100\%$$

$$\text{Period} = 10^{-3} = 2(N_1 + 1)(10^{-6})$$

$$N_1 = (499)_{10} = 01F3$$

$$N_2 = (780)_{10} = 030C$$

Control register #2

7	6	5	4	3	2	1	0
1	0	1	0	0	0	1	1

Bit 0:	Select control register #1 for write.
Bit 1:	Internal clock.
Bit 2:	16-bit mode.
Bits 3, 4, 5:	Pulse output.
Bit 6:	Disable interrupt.
Bit 7:	Enable output.

Control register #1

7	6	5	4	3	2	1	0
1	0	0	0	0	0	1	0

Bit 0:	Start counters.
Bit 1:	Internal clock.
Bit 2:	16-bit mode.
Bits 3, 4, 5:	Continuous output.
Bit 6:	Disable interrupt.
Bit 7:	Enable output.

Program

```
LDX     #01F3
STX     5004
LDX     #030C
STX     5002
LDA A   #82
STA A   5000
LDA A   #A3
STA A   5001
```

14.6 INPUT MODE: PERIOD MEASUREMENT

b_7	b_6	b_5	b_4	b_3	b_2	b_1	b_0
X	X	X	0	1	X	X	X

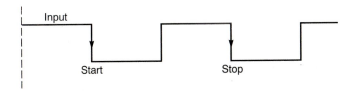

For the input mode a gate transition signal can cause several events within the PTM. The counter is started upon a HIGH-to-LOW signal transition, provided the counter has all other conditions for an enable. The period measurement mode uses a second HIGH-to-LOW signal transition to stop or reinitialize the counter. Furthermore, interrupts may be generated on the second or subsequent transitions or upon counter time-out. Inter-relationships depend upon bit 5. Three basic measurements are possible: (1) the period between the first and second transition, (2) a comparison of the gate period (frequency) with the counter time-out period (frequency), (3) event counting. The key to understanding the input mode is to focus on the interrupt.

Case I. Interrupt upon Gate Transition

7	6	5	4	3	2	1	0
X	X	0	0	1	X	X	X

When zero is written to bit 5 in the control register, an interrupt is generated upon gate transition but only if the transition occurs *before* counter time-out (TO). This fact has two consequences.

First, in Fig. 14.11(*a*) the gate signal period is less than the counter time-out period. Upon the first gate transition after the counter starts, an interrupt signal is generated and the counter is disabled (stopped). Therefore, the count difference multiplied by the clock period gives the gate signal period. This is by far the most important feature of the input mode. Alternatively, the counter final value can be ignored, and the interrupt service subroutine merely increases a memory location by one, counting external events. Subsequent clearing of the interrupt flag permits the PTM to register the next count (see the status register discussion).

Second, Fig. 14.11(*b*) shows a gate signal period that is greater than the counter time-out. Since the count reaches zero (TO) before gate transition, no interrupt occurs, not even for TO. Upon gate transition the counter is reinitialized, and the process continues

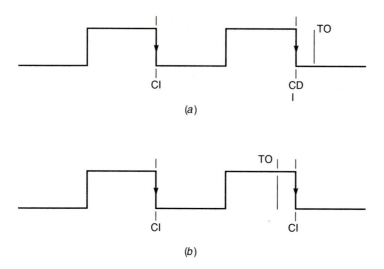

FIGURE 14.11
Frequency input mode: Bit 5 cleared. CI = counter initialized; CD = counter disabled; I = interrupt. (*a*) Transition before time-out (TO); (*b*) transition after time-out (TO).

indefinitely or until the gate signal frequency changes. From this perspective we have the basis for a partial frequency comparison. An interrupt is generated only if the gate frequency is *greater than* the counter frequency. The cut-off frequency can be programmed because the counter reinitializes with the latch's stored value from the program initialization.

What effect does bit 2 have on the timer process? Recall from Sec. 14.4 that time-out for the dual 8-bit mode occurs once, at the end of each period. Thus, nothing is gained by using the dual 8-bit mode.

Case II.   Interrupt upon Time-Out

7	6	5	4	3	2	1	0
X	X	1	0	1	X	X	X

When bit 5 in the control register is set, an interrupt is generated upon counter time-out, but only if the signal transition occurs *after* time-out. Since the counter is always zero at the interrupt, no direct period measurements can be made. However, we do gain the opposing half of the frequency comparison. In Fig. 14.12(*a*) no interrupt is generated because time-out occurs after the gate signal transition. The counter is reinitialized by the transition, and the process continues. Once the gate signal frequency changes so that counter time-out is reached before the transition, Fig. 14.12(*b*), an interrupt signal is passed to the microprocessor. In this case it is a gate frequency which is *less than* the counter frequency which causes an interrupt.

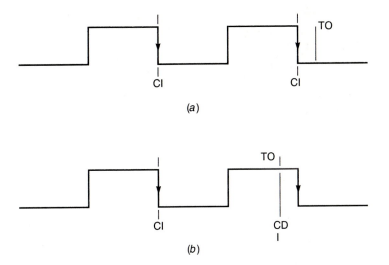

FIGURE 14.12
Frequency input mode: Bit 5 set. CI = counter initialized; CD = counter disabled; I = interrupt. (*a*) Transition before time-out (TO); (*b*) transition after time-out (TO).

Example 14.8. A circuit is built to measure the speed of a bullet. As sketched in Fig. 14.13, two LED-phototransistor pairs are located 1 ft apart. When the bullet passes an LED, a phototransistor is turned OFF, and the AND gate is driven LOW. The gate returns HIGH but is driven LOW when the bullet passes the second LED. The signal period gives the bullet time of passage. Set up counter #1 to receive the AND gate output. If the final counter state is FCBE, what is the bullet speed?

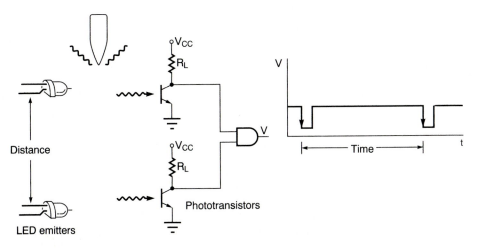

FIGURE 14.13
Measuring the speed of a bullet.

7	6	5	4	3	2	1	0
0	1	0	0	1	0	1	0

Bit 0:   Start all counters.
Bit 1:   Internal clock.
Bit 2:   16-bit mode.
Bits 3, 4:   Period measurement.
Bit 5:   Interrupt on gate transition.
Bit 6:   Enable interrupt.
Bit 7:   Disable output.

Program

```
LDA A   #4A        Counter initialization
STA A   5000
JSR     DELAY
LDX     5002
```

Counter latch does not require loading since it defaults to FFFF. PTM at address 5XXX. The speed is

$$FFFF - FCBE = 0341 = (833)_{10}$$
$$V = 1 \text{ ft}/833 \times 10^{-6} \text{ s} = 1200 \text{ ft/s}$$

14.7 INPUT MODE: PULSE MEASUREMENT

b_7	b_6	b_5	b_4	b_3	b_2	b_1	b_0
X	X	X	1	1	X	X	X

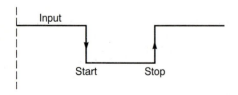

The pulse measurement mode is similar to the frequency measurement mode except that a LOW-to-HIGH signal transition disables the counter. The pulse is referred to as a negative pulse.

Case I.   Interrupt upon Gate Transition

7	6	5	4	3	2	1	0
X	X	0	1	1	X	X	X

With bit 5 equal to zero the interrupt is generated on a positive transition of the gate signal but only if the counter has not timed out. We see in Fig. 14.14(*a*) that the counter is initialized (and enabled) on the negative transition. When the positive transition occurs before counter time-out, the count difference gives us the time of the pulse. After the interrupt flag is cleared, the next negative transition initializes the counter again. If time-out occurs before the positive transition, there is no interrupt. See Fig. 14.14(*b*). The counter reinitializes on the next pulse start and the cycle is repeated until the pulse length is *shorter* than time-out. A gate pulse time to counter time comparison can be made similar to frequency comparison.

Case II.   Interrupt upon Time-Out

7	6	5	4	3	2	1	0
X	X	1	1	1	X	X	X

With bit 5 set counter time-out fixes the interrupt but only if time-out occurs before the signal positive transition. See Fig. 14.15. As with the period comparison counterpart, the pulse comparison continues until the pulse length is *greater* than time-out.

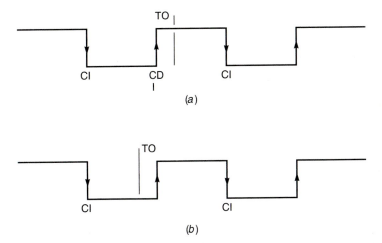

FIGURE 14.14
Pulse input: Bit 5 cleared. (*a*) Transition before time-out (TO); (*b*) transition after time-out (TO).

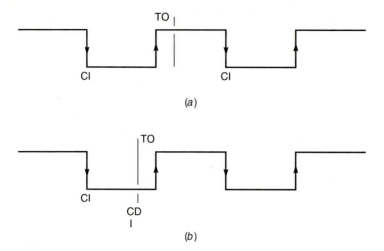

FIGURE 14.15
Pulse input: Bit 5 set. (*a*) Transition before time-out (TO); (*b*) transition after time-out (TO).

Example 14.9. Crankshaft position sensors are the most important of all engine control sensors. It provides signals essential for ignition and fuel injection timing. In addition, this sensor allows accurate measurement of crankshaft speed. There is a variety of position transducer concepts available to the design engineer, including reluctance sensors, Hall-effect sensors, and optical sensors. Figure 14.16 illustrates the concept. A pulse ring is bolted to the crankshaft at the engine front behind the belt drive pulley. Teeth on the ring are located at each piston top dead center (TDC) position for a four-cylinder or eight-cylinder engine. Each tooth passing the sensor produces a positive timing pulse whose width is proportional to the crankshaft speed. Determine speed S for pulse width W, tooth width $d(\frac{1}{2}$ in), and the tooth end radius R (2 in). Initialize the PTM and write the interrupt service routine.

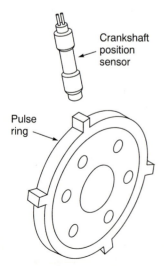

FIGURE 14.16
Determining crankshaft position and speed.

Solution No. 1. Calculate the speed based on the pulse width:

$$S \text{ (rpm)} = \frac{60d}{2\pi RW} = \frac{2.39}{W}$$

or

$$S \text{ (rpm)} = \frac{2.39 \times 10^6}{(65,535 - N)}$$

where N is the final count. Since the numerator exceeds 2-byte arithmetic, separate as

$$S \text{ (rpm)} = \frac{64,698}{(65,535 - N)} \times 32$$

Let us assume an allowable error of 1 percent. Then N *itself* must be held to a minimum count (one count in one hundred) of $(100)_{10}$, and the count *difference* must be held to a minimum of $(100)_{10}$. Therefore,

$$S_{min} = 36 \text{ rpm}$$

$$S_{max} = 23,983 \text{ rpm}$$

PTM timer #1 control register is configured as follows:

7	6	5	4	3	2	1	0
0	1	0	1	1	0	1	0

Bit 0:   Start timer.
Bit 1:   Internal clock.
Bit 2:   16-bit mode.
Bits 3, 4:   Pulse measurement.
Bit 5:   Interrupt on gate transition.
Bit 6:   Enable interrupt.
Bit 7:   Disable output.

Program

```
LDA A   #5A     Load CR1 at
STA A   5000    address 5000.
```

Interrupt routine

```
LDX     5002    Get N.
CPX     #0064   Compare N to (100)₁₀.
BMI     ERROR

STX     0000
LDA A   0000
SUB A   0002    Put difference (FFFF–N)
STA A   0002    at addresses 0002, 0003.
LDA A   0001
SBC A   0003
STA A   0003
```

LDX	0002	Compare difference
CPX	#0064	to $(100)_{10}$.
BPL	ERROR	
LDX	#FCBA	$(64,698)_{10}$
STX	0000	
LDA A	#20	$(32)_{10}$
STA A	0004	
JSR	SOLVE	$(64,698)_{10}$ at addresses 0000, 0001
		$(FFFF–N)$ at addresses 0002, 0003
		$(32)_{10}$ at address 0004

Solution No. 2. Calculate the speed by counting pulses. Each revolution produces four counts. If T_0 is the time period (s) for counting,

$$S \text{ (rpm)} = \frac{(\text{counts}/4)}{T_0} \times 60 \frac{\text{s}}{\text{min}}$$

By selecting a counting period of 15 s, a 2-byte memory location will display the speed directly in rpm. We use counter #3 as an external clock to drive counter #2. The output from counter #2 will not be used, but time-out at the one-half period point will generate the interrupt to halt the process:

$$(N_2 + 1)T_2 = (N_2 + 1)[2(N_1 + T)] = 15 \text{ s}$$
$$T = 10^{-6}$$
$$N_1 = (65,535)_{10} = \text{FFFF}$$
$$N_2 = (113)_{10} = 71$$

Counter #1 receives the pulsed signal. Its interrupt signals the interrupt subroutine to increase the index register count by one. We use the time-out interrupt mode in Fig. 14.15. Latch #1 is loaded with a small number $[N_1 = (20)_{10}$ or 12, e.g.]. Time-out occurs soon after the signal starts transition. The counter is reinitialized on each successive negative transition and the time-out interrupt follows each. Place the PTM at address 5XXX.

Control register #3

7	6	5	4	3	2	1	0
1	0	0	0	0	0	1	0

Bit 0:	Divide clock by 1.
Bit 1:	Internal clock.
Bit 2:	16-bit counting mode.
Bits 3, 4, 5:	Continuous output.
Bit 6:	Interrupt disabled.
Bit 7:	Output enabled.

Control register #2

7	6	5	4	3	2	1	0
0	1	0	0	0	0	0	1

Bit 0:   Select control register #1 for write.
Bit 1:   External clock.
Bit 2:   16-bit counting mode.
Bits 3, 4, 5:   Continuous output.
Bit 6:   Interrupt enabled.
Bit 7:   Output disabled.

Control register #1

7	6	5	4	3	2	1	0
0	1	1	1	1	0	1	0

Bit 0:   Start all counters.
Bit 1:   Internal clock.
Bit 2:   16-bit mode.
Bits 3, 4:   Pulse measurement.
Bit 5:   Interrupt on TO.
Bit 6:   Enable interrupt.
Bit 7:   Disable output.

Program

```
LDX     #0071
STX     5004
LDX     #0012
STX     5002
LDA A   #82
STA A   5000
LDA A   #41
STA A   5001
LDA A   #7A
STA A   5000
```

Interrupt routine

The old index register is saved at addresses 0000 (MSByte) and 0001 (LSByte). The PTM status register is read. If the interrupt was caused by timer #2, disable both interrupts and return to the main program. A read of bit 6 in (CR1 or CR2) at any time in the main program will establish when the counting is complete. If timer #1 caused the interrupt, the count or speed (rpm) is increased and saved at the 2-byte address 0002 (MSByte) and 0003 (LSByte), which is cleared originally. The old index register is retrieved, and control is returned to the main program:

```
STX     0000     Save old X.
LDA A   5001     Read status.
LSR A
BCS     OC       Branch ahead if #1 interrupt set, or
LDA A   #01      disable both interrupts.
STA A   5001
LDA A   #3A
STA A   5000
RTI

LDX     0002     Retrieve count.
INX
STX     0002     Store count.
LDA A   5004     Clear both interrupts.
LDA A   5002

LDX     0000     Retrieve old X.
RTI
```

14.8   THE 555 (FIXED) TIMER

In those cases where timing or clock signal requirements are fixed, the task can be performed better by the 555 timer. It is a cheaper IC that has no programming or bus interface. The 555 timer is capable of continuous outputs (astable vibrator) with a variable duty cycle and single-pulse outputs (monostable vibrator). Timing intervals are set with external capacitors and resistors.

Signetics Corporation first introduced this device in 1972. Other vendors have manufactured their own versions over the years, and it remains the most popular timer today.

Vendor	Number
Signetics	*SE555/NE555
Fairchild	NE555
National Semiconductor	LM555
Motorola	*MC1455/MC1555
Texas Instruments	*SN52555/SN72555

*Military type.

Pin-out and function remain the same regardless of the maker. Figure 14.17 shows the pin-out on the metal can and DIP packages.

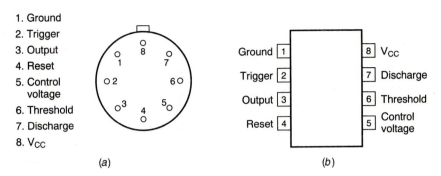

1. Ground
2. Trigger
3. Output
4. Reset
5. Control voltage
6. Threshold
7. Discharge
8. V_{CC}

(a)

(b)

FIGURE 14.17
Pin assignments for the 555 timer. (*a*) 8-pin T package; (*b*) 8-pin DIP package.

Monostable (Pulse) Circuit

A simplified equivalent circuit of the 555 timer is shown in Fig. 14.18. External resistors and capacitors are connected for the monostable mode which produces a single pulse on the output. Pulse width is a function of the connector values.

Internally the 555 timer has five basic components:

1. Lower comparator
2. Upper comparator
3. Flip-flop (FF)
4. (Capacitor) discharge transistor
5. (Inverter) output driver

An internal 5-kΩ resistor ladder network divides the supply voltage into thirds, $\frac{2}{3}$ V_{CC} to the plus input of the upper comparator and $\frac{1}{3}$ V_{CC} to the minus input of the lower comparator. V_{CC} can be any voltage from 4.5 to 18 V. An external control voltage (pin 5) is also tied into the upper comparator plus input. When a voltage is applied to pin 5, this voltage overrides the ladder voltage. Normally, pin 5 is not used and is optionally tied to ground through a 0.01 μF capacitor to improve noise immunity.

The lower comparator compares the trigger input with $\frac{1}{3}$ V_{CC}. If the voltage on pin 2 is less than $\frac{1}{3}$ V_{CC}, the comparator output is LOW, which CLEARS the FF. A flip-flop LOW produces a HIGH driver output and turns OFF the discharge transistor. The external capacitor will charge.

The upper comparator compares the threshold input, which is the capacitor voltage, with $\frac{2}{3}$ V_{CC}. If pin 6 is greater than $\frac{2}{3}$ V_{CC}, it SETS the FF. A flip-flop HIGH produces a LOW driver output and turns ON the discharge transistor, connecting the capacitor to ground.

In the quiescent state the FF is SET, the discharge transistor is ON, and the timer output is LOW. The voltage at pin 6 is effectively ($V_{CE})_{SAT} \approx 0.2$ V. An output pulse (one-shot) is generated by a HIGH-to-LOW-to-HIGH transition (negative pulse) on the trigger input. Once the input goes LOW the FF is CLEARED and the timer output goes HIGH.

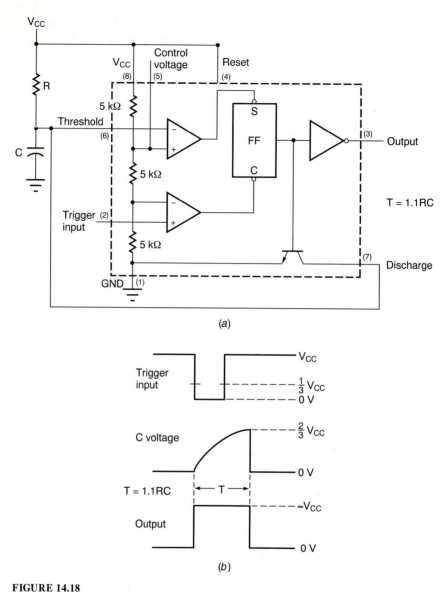

FIGURE 14.18
The 555 timer: Monostable mode. (*a*) Circuit; (*b*) waveform.

Since the transistor is OFF, the capacitor C begins charging through R. When the voltage on pin 6 reaches $\frac{2}{3} V_{CC}$, the upper comparator output switches to HIGH. This SETs the FF, which immediately returns the timer output to LOW and discharges the capacitor through the transistor. Pulse width T is found from

$$\tfrac{2}{3} V_{CC} = V_{CC}(1 - e^{-T/RC})$$

or

$$T = 1.1RC$$

1. The trigger input pulse *must* be shorter than the output pulse. Minimum input pulse width is approximately one clock cycle, typically 1 μs. While the timer output pulse is HIGH additional triggers have no effect on the timer or output.

2. The limits are 1 kΩ < R < 10 MΩ

   $$100 \text{ pF} < C < \text{any value up to leakage}$$

   A reasonable lower limit for R is 1 kΩ based simply on power dissipation of the resistor. The upper limit for R is governed by the high-impedance leakage current of the comparator. A practical minimum value for C is about 100 pF. Below this value stray capacitance becomes a factor. Maximum values are limited by capacitor leakage. Low leakage capacitors are available up to 10 μF.

3. Maximum pulse periods T are on the order of 100 s. Internal propagation delays limit T to about 10 μs.

4. RESET (pin 4) is normally connected to V_{CC}. If the pin temporarily goes LOW the discharge capacitor is turned ON, and the timer output goes LOW and remains LOW until the timer is retriggered.

Astable (Clock) Circuit

A variable duty cycle continuous output or clock generator can be achieved with the same 555 timer. Figure 14.19 shows that two external resistors are used; both comparator inputs (pins 2 and 6) are connected to the capacitor. The circuit operates as follows. When the discharge transistor is switched OFF (at the same instant timer output goes HIGH), the capacitor charges through R_A and R_B. When the capacitor voltage reaches $\frac{2}{3} V_{CC}$, the upper comparator SETs the FF. This turns ON the discharge transistor, shorting the capacitor to ground through R_B. The timer output goes LOW. Since the lower comparator input falls below $\frac{1}{3} V_{CC}$, the FF is CLEARED. The discharge transistor is turned OFF and the output goes HIGH. This process is repeated indefinitely. Relationships between component values and operating parameters are

$$\text{Charging time} = 0.685(R_A + R_B)C \qquad \text{output HIGH}$$

$$\text{Discharging time} = 0.685 R_B C \qquad \text{output LOW}$$

$$\text{Frequency} = \frac{1}{\text{period}} = \frac{1.46}{(R_A + 2R_B)C}$$

$$\text{Duty cycle} = \frac{\text{time HIGH}}{\text{period}} = \frac{1 + R_B/R_A}{1 + 2R_B/R_A}$$

1. The circuit is capable of periods from tens of microseconds to minutes. If leakage error is ignored, the range can be extended to hours with 1000-μF capacitors.

2. Duty cycle can be controlled by adjusting the resistance ratio. As R_B becomes larger with respect to R_A the duty cycle approaches 50% (square wave). Conversely, small ratios produce duty cycles approaching 100 percent.

3. The internal driver can supply a large output current (up to 200 mA with the Motorola unit), which can drive most loads directly.

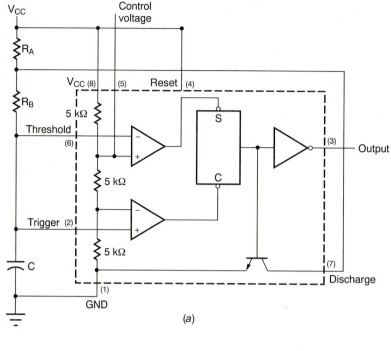

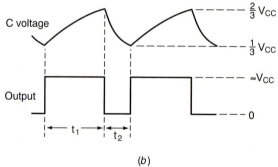

FIGURE 14.19
The 555 timer: Astable mode. (*a*) Circuit; (*b*) waveform.

Figure 14.20 offers two useful circuits. One circuit extends the low range of the output frequency duty cycle to zero. A second circuit converts thermistor resistance to frequency for counting.

Control Pin 5

Suppose a voltage V_5 is applied to pin 5. The upper comparator now uses V_5 instead of $\frac{2}{3} V_{CC}$ to SET the FF. Thus, in the monostable mode the 555 timer output pulse width is

$$T = -RC \ln\left(1 - \frac{V_5}{V_{CC}}\right)$$

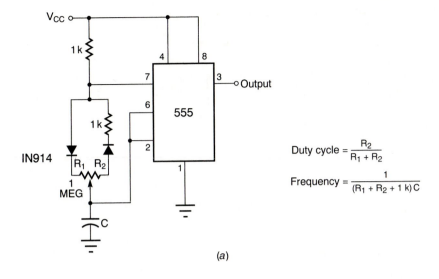

$$\text{Duty cycle} = \frac{R_2}{R_1 + R_2}$$

$$\text{Frequency} = \frac{1}{(R_1 + R_2 + 1\,k)\,C}$$

(a)

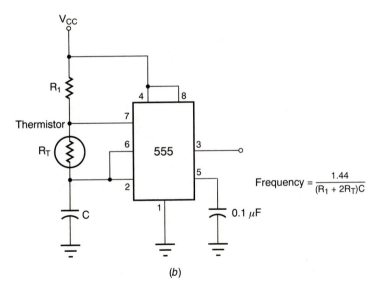

$$\text{Frequency} = \frac{1.44}{(R_1 + 2R_T)C}$$

(b)

FIGURE 14.20
Useful 555 timer circuits. (a) Full-range duty cycle; (b) temperature measurement.

In the astable clock mode the frequency is

$$f = \frac{1}{K(R_A + 2R_B)C}$$

where

$$K = \ln\left[\frac{V_{CC} - V_5}{V_{CC} - \frac{1}{3}\,V_{CC}}\right]$$

Both pulse width and frequency are logarithmic functions of the control voltage. For the innovative designer, control voltage has two aspects. First, timing parameters can be extended beyond the range limited by the external resistors and capacitor. Second, a cheap 555 timer can replace a more expensive A/D converter and free up I/O. It will require additional software to interpret timer output. These aspects have not been sufficiently utilized in the past.

Accuracy

The 555 timer is a simple device which performs accurately. In the monostable mode the initial error is typically within 1 percent. Drift with temperature is approximately 50 ppm/°C. For the astable mode the initial error is somewhat more at 2 percent. Astable drift with temperature is 150 ppm/°C. Timer errors due to supply voltage drift or ripple are negligible (0.1 percent per V).

Most of the initial error is caused by inaccurate external components. If timing is critical, a good quality multiturn pot can be used in series with the resistor. High-quality, low leakage capacitors should be used. Avoid ceramic disk capacitors, selecting polystyrene or mylar types.

For accuracy greater than 0.1 percent, crystals should be considered. While the 555 timer is adequate for most purposes, computers and precise instrumentation require more stable crystal clocks.

PROBLEMS

14.1. What does the term "time-out" mean?

14.2. Suppose timer control registers #1, #2, and #3 contain B3, 3B, and C9 respectively. How does each control the timer?

14.3. Example 14.3 uses timer #3 to generate a square wave output. If timer #2 is used instead, how does the program change? If the frequency is 0.1 Hz, how does the program change?

14.4. A batching operation uses two heaters. Heater #1 must be turned on for 1 min every 5 min. Heater #2 must be turned on 15 min every hour. Configure the programmable timer module (PTM) to produce the proper outputs. Use interrupts. Sketch the hardware interface.

14.5. An electronic fuel injection system delivers fuel at the rate of 200 pulses per second. A microprocessor updates the duty cycle for the pulse width modulated (PWM) signal from the PTM. Assume the processor is continually placing any required PTM data at addresses 0000–0004. Write the PTM program.

14.6. The PTM in Example 14.3 outputs (timer #3) a square wave with a frequency of 30 Hz. The PTM in Example 14.6 outputs (timer #2) a single pulse with a 1 ms width. Write the program for a single PTM at address 4XXX to output both signals.

14.7. Configure the 6840 PTM to output a single pulse with a 1-h duration.

14.8. If a 60-Hz clock signal deviates by ∓ 3 percent, an experiment is ruined. Devise a computer interrupt scheme using a PTM and two timers. Give the PTM control register contents and the PTM program initialization.

14.9. If the projectile in Example 14.8 is 0.4 in long, discuss the pros and cons when using a single LED-phototransistor to determine the projectile speed.

CHAPTER
15

INTERFACING
POWER
ELECTRONICS

Interfacing the data bus to the real world invariably leads to devices that demand current (power). At the lowest level TTL/CMOS gates drive similar logic. A modest boost in drive capability can be obtained with TTL/CMOS buffers. The 74S241 octal buffer was introduced earlier as a multiple gate driver and as an LED lamp driver. When the current from several segments of a 7-segment LED display was switched, a general-purpose (small signal) transistor was used. The highest level of control is achieved with power semiconductors. High-current drives are necessary for a variety of products, including motors, solenoids, magnets, heaters (dryers), lights, power supplies, and audio amplifiers. Current source capability of semiconductors is given in Table 15.1.

Power electronics center mainly on three terminal, gate-(base) controlled devices. These devices fall into two classes: (1) power transistors and (2) thyristors. In the low-to-medium DC power range *power transistors* are the choice. They include the bipolar transistor, the Darlington (or dual bipolar) transistor, and the field-effect transistor (MOSFET).

TABLE 15.1
Current source capability of several semiconductor groups

Group	Current
TTL Gate	1 mA
TTL buffer	15 mA
General-purpose transistor	<0.5 A
Power semiconductors (transistors, thyristors)	>1.0 A

Each remains ON only as long as the gate is active. MOSFET is preferred because the gate is voltage controlled rather than current controlled. Recent design advances have enabled MOSFETs to compete more effectively with Darlingtons in the medium power range. From the early 1960s until quite recently, the silicon-controlled rectifier (SCR) member of the *thyristor* family was universally used as the high-power semiconductor switch. Today the words thyristor and SCR are interchangeable. SCRs are *pn*-type devices that are turned ON by a simple voltage pulse. However, they latch ON until the load current goes to (almost) zero. Conduction is unidirectional. Recently, interest has been growing in the gate turn OFF thyristor (GTO), which can be turned ON by a positive voltage pulse and turned OFF by a negative voltage pulse. SCRs and GTOs control DC or rectified AC power. The TRIAC consists of opposing SCRs and can conduct in both directions. It is normally used to switch AC circuits in the low-to-medium power range.

Table 15.2 gives the present rating limits for power semiconductors. Voltage limits relate to the *pn*-junction breakdown in the OFF state. Current limits depend upon chip wire capacity. Power limits pertain to the load power being switched, and they are directly tied to the chip power dissipation limits which are a much lower number.

Power semiconductors are placed into an assortment of packages. These packages are designed to dissipate heat. All but the low-power ones require heat sink attachments to meet power specifications. Figure 15.1 shows the most common packages and their approximate current range.

Power electronics is based primarily on the switching of semiconductor devices rather than on their linear operation. There are several reasons: (1) the impact of microprocessors on control strategy, (2) fast-switching speeds, and (3) reduced semiconductor power dissipation, which allows larger loads to be controlled.

The ideal switch has infinite impedance in the OFF-state and no voltage drop (no power dissipation) in the ON-state. Semiconductors are not ideal, but they are reasonably close. Figure 15.2 illustrates the voltage-current of a switched device. In the OFF-state leakage current is on the order of microamps. In the ON-state the voltage drop across the device is on the order of 1 to 3 V. Switching times are on the order of microseconds. Power consumption is greatest as the device switches through the linear range.

The measure of acceptability of any power conversion is efficiency. This is a comparison of the load power against the combined load power and wasted power of the control unit. The most efficient mode is the fully ON (saturated mode) with efficiencies over 70%. In some cases it is more appropriate to operate in the linear range, i.e., audio amplifiers and low-power (analog) DC motors, where designs can be simplified if control is exercised with power ICs such as power op amps.

TABLE 15.2
Limits of present day power semiconductors

	Transistors		Thyristors	
	Darlington	**MOSFET**	**TRIAC**	**SCR**
I (A)	400	100	50 (rms)	2000
V (V)	1000	1000	1000	4000
P (W)	20 k	5 k	2 k	100 k

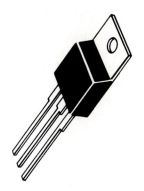

TO-39
metal can
low power
I = 1 to 2 A

TO-220
plastic
medium power
I = 2 to 25 A

TO- 3
metal can
medium power
I = 2 to 25 A

TO-59
stud
high power
I > 25 A

FIGURE 15.1
Power semiconductor packages.
(*Drawings copyright of Motorola,
Inc. Used by permission.*)

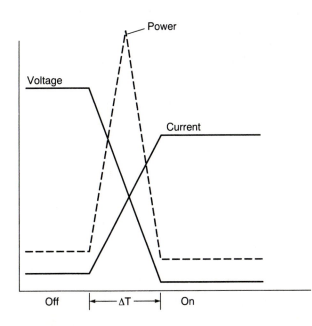
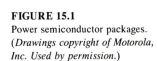

Power

Voltage

Current

Off   ← ΔT →   On

FIGURE 15.2
Voltage-current for a switched-ON
semiconductor.

15.1   BIPOLAR POWER TRANSISTORS

Parameters controlling the performance of bipolar transistors are essentially the same regardless of chip size and power handling capacity. The basic theory of operation was documented in Chap. 9. Let us review with Fig. 15.3. Bipolar transistors can be placed into three states: (1) cutoff, (2) active or linear, and (3) saturation. To turn the transistor ON, input voltage V_i must be increased above V_{BE}, the base to emitter voltage drop (0.4 V). Input voltages are generally 2 to 5 V. In the active (linear) state the collector current is

$$I_C = h_{FE} I_B$$

where h_{FE} is the characteristic current gain of the transistor and I_B is the base current. Collector current increases with I_B until the device enters the saturation region at $(V_{CE})_{SAT}$.

The collector current in saturation will be nearly constant and given by

$$I_{CS} = \frac{V_{CC} - (V_{CE})_{SAT}}{R_C} \tag{15.1}$$

To ensure that the transistor is driven into saturation, we must force a base current of at least

$$I_{BS} = \frac{I_{CS}}{h_{FE}} \tag{15.2}$$

Normally, one designs the circuit such that I_B exceeds I_{BS} by a factor of 2 to 10, called the overdrive factor (ODF):

$$ODF = \frac{I_B}{I_{BS}} \tag{15.3}$$

ODF covers up a considerable degree of uncertainty in the value of h_{FE}. A high value of overdrive factor will not affect the collector-emitter voltage significantly. However, the increased base current will increase transistor base power loss.

Performance Parameters

It is not possible to achieve in a few devices an all-encompassing performance representing low-cost, high-current gain, low "ON" resistance, high-operating voltage, and high-speed operation. This explains the sometimes bewildering selection on the market. Although a

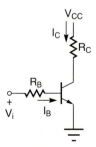

FIGURE 15.3
Basic circuit for bipolar transistor.

transistor can be found to optimize a given design with regard to a set of particular parameters, many transistors will perform satisfactorily. Figure 15.4 gives the performance parameters for a TIP73 power transistor as they are found in the Texas Instruments *Power Products Data Book*. Some data books, including Texas Instruments (TI), will give additional information in the form of characteristic curves. See Fig. 15.5.

MAXIMUM RATINGS. Transistor circuits must be designed so that operating conditions do not exceed their suggested maximum ratings. These limits include current (base and collector), voltage (terminal to terminal), and power dissipation. Of particular interest are the collector conditions. Together they form a safe operating area.

TIP73 npn Silicon Power Transistor

TO-220AB package

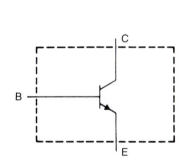
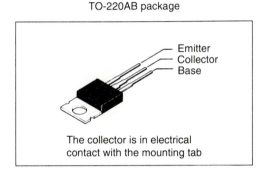

Maximum ratings at 25°C case temperature

$(V_{CE})_{BR}$ at $i_B = 0$: 40 V
V_{EB}: 5 V P_D at 25°C case: 80 W
I_C: 15 A P_D at 25°C free air: 2 W
I_B: 5 A T_J: −65°C to 150°C

Electrical characteristics at 25°C case temperature

I_C at $I_B = 0$: 50 μA(max) $(V_{CE})_{SAT}$ at $I_C = 5$ A: 1.3 V(max)
h_{FE} at $I_C = 5$ A: 20(min) $(V_{CE})_{SAT}$ at $I_C = 15$ A: 3.5 V(max)
                150(max)
h_{FE} at $I_C = 15$ A: 15(min) $(V_{BE})_{SAT}$ at $I_C = 5$ A: 1.3V(max)
                                      $(V_{BE})_{SAT}$ at $I_C = 15$ A: 3.5 V(max)

Thermal characteristics

$R_{\theta JC}$: 1.56°C/W
$R_{\theta JA}$: 62.5°C/W

Switching time

t_d: 20 ns t_s: 500 ns
t_r: 350 ns t_f: 400 ns

FIGURE 15.4
Bipolar power transistor performance characteristics. (*Drawings reprinted by permission of Texas Instruments.*)

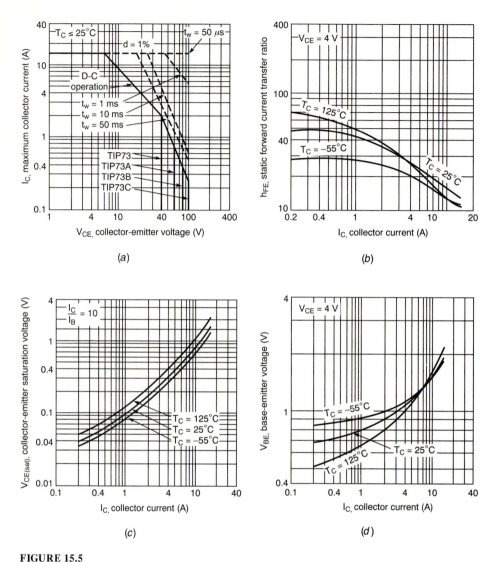

FIGURE 15.5

Typical characteristics of a TIP73 *npn*-bipolar power transistor. (*a*) Safe operating area; (*b*) current gain; (*c*) collector-emitter voltage; (*d*) base-emitter voltage. (*Reprinted by permission of Texas Instruments.*)

SAFE OPERATING AREA (SOA). All transistors and power ICs have safe operating areas similar to Fig. 15.5(*a*) for TIP73. The current limit (15 A) arises because of limitations in the chip or bonding wires. Wires may melt if forced to carry more than the rated collector current. All points on the power limit portion have constant power dissipation (80 W). Thermal resistance, junction temperature T_J, and the attachment package limit the dissipation. We will examine this limit later in Sec. 15.4. Finally, the voltage limit is the maximum-rated collector-to-emitter voltage. This voltage can be sustained from cutoff to about 2 A of collector current in the active state. The reader should observe that a saturated transistor operates in the extreme left region of an SOA curve.

CURRENT GAIN (h_{FE}). Figure 15.5(b) shows that current gain varies considerably with temperature and collector current. h_{FE} decreases from 70 to 10 over the collector current range and increases to a lesser extent over the temperature range. It is also a parameter over which the manufacturer has little control. Devices in the same batch, even after they are selected, can vary by at least 2:1 in gain. Manufacturer curves represent the worst case, and higher-current gains will be found in a random purchase. When designing a circuit, the engineer assumes the worst condition and the poorest device. The actual current gains are much smaller for power transistors than for general-purpose transistors. Gains can be improved by testing and making a selection from within a group of similar devices.

COLLECTOR-TO-EMITTER SATURATION VOLTAGE ($V_{CE})_{SAT}$. Low values of ($V_{CE})_{SAT}$ will reduce the ON-state transistor losses. For general-purpose transistors the collector-emitter voltage at saturation is approximately 0.3 V. However, power transistors have values which can range to several volts. As Fig. 15.5(c) shows, ($V_{CE})_{SAT}$ increases with collector current to 2 V for $h_{FE} = 10$. The reason is that load voltages are usually higher for power transistors. Thus, the collector breakdown voltage ($V_{CE})_{BRO}$ must be made correspondingly higher than general-purpose transistors. This cannot be accomplished without increasing the ON-state resistance.

BASE-TO-EMITTER SATURATION VOLTAGE ($V_{BE})_{SAT}$. The base-to-emitter saturation voltage also depends upon the collector current. With general-purpose transistors ($V_{BE})_{SAT}$ is approximately 0.7 V. However, values for power transistors can be much higher as shown in Fig. 15.5(d). Since current gains tend to be low for power transistors, base power loss $I_B \times (V_{BE})_{SAT}$ cannot be ignored. The loss is compounded with high overdrive factors.

Switching Times

Because of internal capacitive effects, transistors do not switch in zero time. Figure 15.6 illustrates the switch ON and switch OFF times. A finite delay time t_d elapses before the base is charged to ($V_{BE})_{SAT}$. After charging the collector current begins an exponential rise time t_r to $0.9 I_{CS}$. A more popular measure is the turn-ON time t_{ON} as indicated in the figure. When the transistor is switched OFF, there is a delay, called the storage time t_s, to remove the saturation charge from the base. Storage time is increased by ODF and constitutes the limiting factor in switching speed. Once the charge is removed the current decays exponentially in time t_f, the fall time.

> **Example 15.1.** A TIP73 power transistor switches a 2-Ω load from a 20-V source. The transistor operating temperature should not exceed 25°C. Determine the base resistor.

Solution. If $I_{CS} \cong 20$ V/2 Ω or 10 A, then ($V_{CE})_{SAT}$ at 10 A = 1.0 V:

$$I_{CS} = \frac{V_{CC} - (V_{CE})_{SAT}}{R_C} = \frac{20 - 1.0}{2} = 9.5 \text{ A}$$

From Fig. 15.4, h_{FE} (min) = 15:

$$I_{BS} = \frac{I_{CS}}{h_{FE}} = \frac{9.5}{15} = 0.63 \text{ A}$$

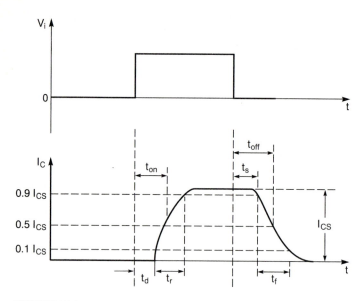

FIGURE 15.6
Bipolar power transistor switching time.

Using an overdrive factor of 2, $I_B = 1.2$ A. This high value is rather typical for bipolar power transistors. Driving R_B with the source, we obtain

$$R_B = \frac{V_{CC} - V_{BE}}{I_B} = \frac{20 - 1.0}{1.2} = 15 \ \Omega$$

Switching Inductance

All loads show some degree of inductance and resistance. Solenoids, stepping motors, and DC motors are highly inductive. When the transistor switches OFF, voltage buildup

$$V_{CE} = \frac{L \, dI}{dt} = \frac{L I_C}{t_f}$$

is sufficient to exceed the breakdown voltage. $(V_{CE})_{BR}$ should never be exceeded. To avoid transistor burn out, a diode is placed in parallel with the load (see Fig. 11.7). Better yet, a zener diode is placed in series with the regular diode to protect $(V_{CE})_{BR}$. The inductor energy is allowed to dissipate by looping the current back through the load resistance while clamping the voltage at $(V_{CE})_{BR}$. More details are covered in Chap. 11.

Example 15.2. A permanent magnet DC motor is connected in a T bridge for simple ON-OFF control. See Fig. 15.7. Motor current requirements are 2 A in the continuous mode and 10 A in a start/stall mode, both at motor voltage $V_m = 10$ V. The driver selected is the TIP73 bipolar power transistor rated at 15 A continuous collector current. Its case temperature is maintained at 25°C with a heat sink. Supply voltage is 12 V. Evaluate the design.

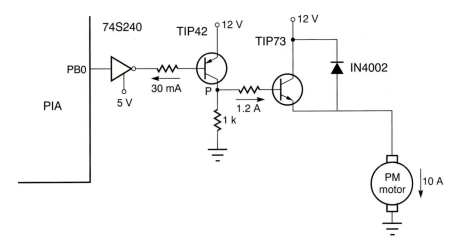

FIGURE 15.7
ON-OFF control of a DC motor with bipolar power transistor drive.

Power transistor only

Continuous mode

$I_C = 2$ A $(P_D)_{base} = 12(\frac{4}{35}) = 1.4$ W (includes R_b)

$(V_{CE})_{SAT} = 0.17$ V $(P_D)_{collector} = (0.17)(2) = 0.35$ W

$h_{FE} = 35$ $(P_D)_{total} = (12)(2) = 24$ W

$I_B = \frac{2}{35}$ A

$\quad = \frac{4}{35}$ (with overdrive) $\text{Eff} = \dfrac{24 - (1.4 + 0.35)}{24} = 93\%$

$V_m = (12 - 0.17) = 11.8$ V

Start/stall mode

$I_C = 10$ A $(P_D)_{base} = 12(\frac{20}{13}) = 18.5$ W

$(V_{CE})_{SAT} = 0.8$ V $(P_D)_{collector} = (0.8)(10) = 8$ W

$h_{FE} = 17$ $(P_D)_{total} = 12(10) = 120$ W

$I_B = \frac{10}{17}$ A

$\quad = \frac{20}{17}$ (with overdrive) $\text{Eff} = \dfrac{120 - (18.5 + 8)}{120} = 78\%$

$V_m = (12 - 0.8) = 11.2$ V

A major drawback for bipolar power transistors is the large base current drive. To meet start conditions, the base current must be supplied by another power transistor. The second transistor must be driven by a buffer (or general-purpose transistor) before interfacing a PIA.

The design is based on the "typical characteristics" plot of Fig. 15.7, where $(V_{CE})_{SAT}$ is 0.8 V. A more conservative design evaluates parameters with the worst possible transistor; i.e., use the electrical characteristics data in Fig. 15.4, where $(V_{CE})_{SAT} = 3.5$ V (max). Suppose we had selected a TIP33 with a current rating of 10 A. The data sheet has no typical characteristics plot for $(V_{CE})_{SAT}$. We would be forced to use the listed $(V_{CE})_{SAT} = 4$ V (max). Needless to say, the design efficiency would be severely reduced for the worst case.

15.2 DARLINGTON TRANSISTORS

Bipolar power transistors have low-current gains with high base currents. Its driver is usually another lower-rated power transistor. If both transistors are mounted on heat sinks, a considerable amount of space will be occupied by the two transistors. This problem is circumvented by encapsuling the two transistors on the same chip, called a Darlington transistor.

The Darlington circuit is shown in Fig. 15.8. Emitter current from Q1 is also the base current of Q2. The combination has three external connections (B, C, and E), which act as a single transistor whose gain equals Q1 gain × Q2 gain. Current gains of the two transistors are controlled during manufacture so that the overall gain varies linearly over a range of collector current when it is used as an amplifier. As a result, Darlingtons have high values of $(V_{CE})_{SAT}$, which cause larger switching losses than the separated design.

Performance Parameters

Specifications for the TIP663 *npn* Darlington transistor are listed in Fig. 15.9. Typical characteristic curves are also shown in Fig. 15.10. Although the transistors are bipolar, there are major differences from the single unit.

CURRENT GAIN (h_{FE}). Careful matching of the transistor pair in Darlingtons produces some rather startling current gains with h_{FE} up to 10,000 possible. Darlingtons often can be driven directly from the PIA or other TTL logic gates. We can see in Fig. 15.10(b) that h_{FE} varies significantly with collector current and case temperature. Optimum h_{FE} is approximately equal to 2000 (25°C case temperature) and occurs at a collector current nearly one-half the rated 20 A. To employ Darlingtons at their rated collector current is a mistake since h_{FE} is not much better than a single power transistor. Chip temperature presents a somewhat different problem. Unless the case temperature is known, h_{FE} is uncertain by a factor of 3. Therefore, a large overdrive factor will be necessary to be certain that saturation is reached.

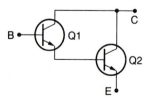

FIGURE 15.8
Darlington circuit.

TIP663 npn Darlington Power Transistor

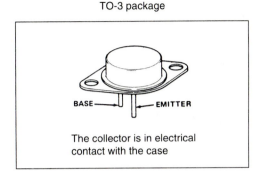

TO-3 package

The collector is in electrical
contact with the case

Maximum ratings at 25°C case temperature

$(V_{CE})_{BR}$ at $I_B = 0$: 300 V
V_{EB}: 8 V
I_C: 30 A
I_B: 5 A

P_D at 25 case: 150 W
P_D at 25 free air: 5.5 W
T_J: −65 C to 200 C

Electrical characteristics at 25°C case temperature

I_C at $I_B = 0$: 250 μA
h_{FE} at $I_C = 5$ A: 500(min)
           10,000(max)
h_{FE} at $I_C = 20$ A: 25(min)

$(V_{CE})_{SAT}$ at $I_C = 10$ A: 1.3 V(max)
$(V_{CE})_{SAT}$ at $I_C = 20$ A: 3 V(max)

$(V_{BE})_{SAT}$ at $I_C = 10$ A: 2.1 V(max)
$(V_{BE})_{SAT}$ at $I_C = 20$ A: 25 V(max)

Thermal characteristics

$R_{\theta JC}$: 0.67°C/W
$R_{\theta JA}$: 31.8°C/W

Switching time

t_d: 50 ns t_s: 6500 ns
t_r: 220 ns t_f: 1300 ns

FIGURE 15.9
Darlington power transistor performance characteristics. (*Drawings by permission of Texas Instruments.*)

COLLECTOR-TO-EMITTER SATURATION VOLTAGE. Figure 15.10(*c*) shows that $(V_{CE})_{SAT}$ is higher than bipolar power transistor values at all collector currents. The voltage drop is far worse at collector currents approaching the maximum-rated current. In spite of the higher-power losses in Darlingtons, they are preferred because the high h_{FE} will simplify driver hardware. Power losses are minimized by using Darlingtons in load circuits that have low currents and high voltages across the load.

SWITCHING TIME. Darlington switch-ON times are comparable to bipolar transistors. However, the switch-OFF times are an order of magnitude longer than bipolar power transistor t_{OFF}.

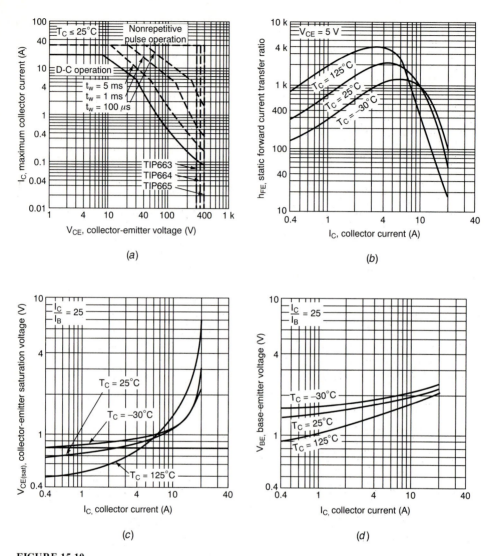

FIGURE 15.10
Typical characteristics of a TIP663 Darlington power transistor. (*a*) Safe operating area; (*b*) current gain;
(*c*) collector-emitter voltage; (*d*) base-emitter voltage. (*Reprinted by permission of Texas Instruments.*)

Free-Wheeling Diode

Texas Instruments' Darlington transistors have a built-in free-wheeling diode across the
collector-emitter terminals. In some configurations it will eliminate the need for an external diode to shunt the switch-OFF voltage buildup with load inductance. The key is placing the load between the transistor and ground. This point is illustrated in Fig. 15.11.
Only case (*a*) should have an external protecting diode.

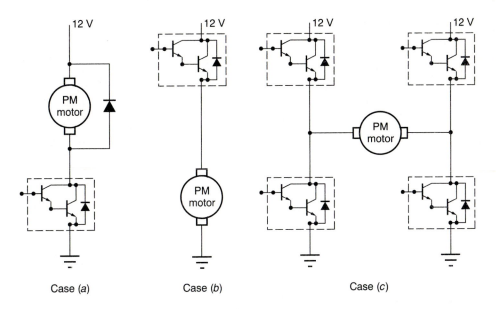

Case (*a*)   Case (*b*)   Case (*c*)

FIGURE 15.11
When to include external diodes if using power transistors with built-in free-wheeling diodes. Case (*a*) only.

Example 15.3. The T bridge motor design in Example 15.2 replaces the bipolar power transistor with the TIP663 Darlington rated 20-A continuous collector current. A driver circuit is given in Fig. 15.12. Compare the two designs at 25°C case temperature.

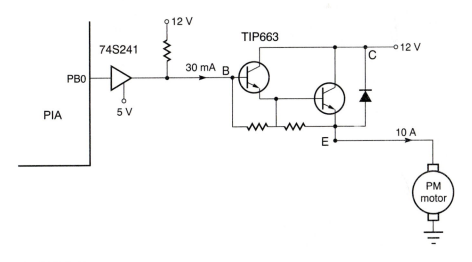

FIGURE 15.12
ON-OFF control of a DC motor with Darlington power transistor drive.

Continuous mode

$$I_C = 2 \text{ A} \qquad\qquad (P_D)_{\text{base}} = 12\left(\tfrac{4}{1500}\right) = 0.03 \text{ W}$$

$$(V_{CE})_{SAT} = 0.8 \text{ V} \qquad\qquad (P_D)_{\text{collector}} = (0.8)(2) = 1.6 \text{ W}$$

$$h_{FE} = 1500 \qquad\qquad (P_D)_{\text{total}} = 1.6 \text{ W}$$

$$I_B = 1.3 \text{ mA}$$

$$= 4 \text{ mA (with overdrive)} \qquad \text{Eff} = \frac{24 - 1.6}{24} = 93\%$$

$$V_m = (12 - 0.8) = 11.2 \text{ V}$$

Start/stall mode

$$I_C = 10 \text{ A} \qquad\qquad (P_D)_{\text{base}} = 12(0.01) = 0.1 \text{ W}$$

$$(V_{CE})_{SAT} = 1.2 \text{ V} \qquad\qquad (P_D)_{\text{collector}} = (1.2)(10) = 12 \text{ W}$$

$$h_{FE} = 1000 \qquad\qquad (P_D)_{\text{total}} = 12(10) = 120 \text{ W}$$

$$I_B = 10 \text{ mA}$$

$$= 30 \text{ mA (with overdrive)} \qquad \text{Eff} = \frac{120 - 12.1}{120} = 90\%$$

$$V_m = (12 - 1.2) = 10.8 \text{ V}$$

The drive requirements are substantially reduced from the bipolar power transistor design. Still, a 74S241 buffer/driver and pull-up resistor must be placed between the PIA and the Darlington. The Darlington dissipation is greater than the bipolar transistor, which is expected from its higher $(V_{CE})_{SAT}$. We should point out that using this transistor for a motor rated at 15-A start/stall is perfectly acceptable. Yet the collector voltage drop $(V_{CE})_{SAT} = 3.3$ V. If a heat sink design allows the case temperature to stabilize much above 100°C, $(V_{CE})_{SAT}$ could double this value. The motor drive voltage V_m is supply voltage minus the transistor voltage drop. In some cases, particularly H bridge drives which have two transistor voltage drops, there may be insufficient voltage to meet the design requirements (10 V). Darlingtons are susceptible to this problem.

15.3   POWER MOSFET TRANSISTORS

Power MOSFETs have made great strides since the early 1980s, and they are rapidly becoming the transistor of choice in the low-to-medium power range. At one time cost was a factor, but new designs and improved manufacturing have led to competitive prices. There are two reasons for selecting a MOSFET:

1. Bipolar transistors are current-driven devices, whereas MOSFETs are voltage-driven devices. The only current is leakage current on the order of microamperes. Drive circuitry is greatly simplified. Gates can be driven directly with TTL logic using pull-up resistors or by CMOS logic. MOSFET is fully switched ON with a gate voltage of 10 V.

2. MOSFET switching speeds depend upon the charging of an input capacitance. Thus, switching speeds are inherently faster than bipolar transistors. This makes them more suitable for PWM drives and chopper drives in motor applications.

There are two types of MOSFET: (1) depletion MOSFETs and (2) enhancement MOSFETs. Recall that depletion types have a p or n channel and that a given device is active with both gate voltage polarities. On the other hand, enhancement types have no p or n channel, depending upon induced voltages to enhance an "electron channel" from the substrate. It is activated by a single polarity. Practically all MOSFETs are the enhancement type. Figure 15.13 shows circuits for the two versions. An n-channel enhancement MOSFET is activated by a positive gate voltage, and drain (i.e., load) current is from drain-to-source. The distinguishing feature is an inward pointing "base" arrow. This type is analogous to npn-bipolar transistors. It is the more common of the two versions. A p-channel enhancement MOSFET is activated by a negative gate voltage, and the drain current is source-to-drain. It is analogous to pnp-bipolar transistors.

Operating Characteristics

The operating characteristics of a power MOSFET depend upon the following parameters:

$$V_{GS} = \text{gate-to-source voltage}$$

$$(V_{GS})_{TH} = \text{gate threshold voltage}$$

$$V_{DS} = \text{drain-to-source voltage}$$

$$I_D = \text{drain current}$$

There are three regions of operation:

1. A cutoff region where $V_{GS} \leqq (V_{GS})_{TH}$
2. An active or saturation region where $V_{DS} \geqq V_{GS} - (V_{GS})_{TH}$
3. A linear or constant resistance region where $V_{DS} \leqq V_{GS} - (V_{GS})_{TH}$

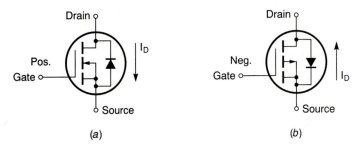

(a)   (b)

FIGURE 15.13
Types of power MOSFET transistors. (*a*) n channel; (*b*) p channel.

In the saturation region the drain current depends only upon the gate voltage. Transistor gain is defined as transconductance g_{FS}:

$$g_{FS} = \left.\frac{I_D}{V_{GS}}\right|_{V_{DS}=\text{constant}}$$

Transconductance denotes the gain of MOSFETs much as h_{FE} denotes the gain of bipolar transistors. In the linear region the drain current varies proportionally with V_{DS}. ON resistance is defined as

$$(R_{DS})_{ON} = \frac{\Delta V_{DS}}{\Delta I_D}$$

We are much more interested in this parameter since MOSFETs are mostly used as switching devices. When switching MOSFET from OFF to ON, the drain-to-source resistance changes from megaohms to $(R_{DS})_{ON} < 0.1\ \Omega$. It should be noted that the words "linear" and "saturation" have the exact opposite meaning for bipolar transistors.

Performance Parameters

Figure 15.14 gives the performance parameters for an MTM20N08 n-channel enhancement power MOSFET as found in the Motorola *Power MOSFET Transistor Data* book. Typical characteristic curves are shown in Fig. 15.15.

MAXIMUM RATINGS. The maximum drain-to-source voltage (in the switched OFF mode) is $V_{DSS} = 80$ V. MOSFETs can go as high as 1000 V, so this device is low voltage, and they typically have a lower ON resistance. The gate-to-source maximum voltage is $V_{GS} = 20$ V. MOSFET gates are susceptible to damage from static electricity. They are not so susceptible as CMOS gates because power devices have large input capacitances which absorb energy. However, the usual antistatic precautions should be taken. Some circuits are protected by placing a zener diode between the gate and the source.

OPERATING CHARACTERISTICS. Figure 15.15(a) shows that the cut-off region occurs for gate voltage $V_{GS} \approx 5$ V. The electrical characteristics list defines $(V_{GS})_{TH} = 2$ V (min), 4.5 V (max). The actual value varies with the junction temperature and is found in Fig. 15.15(b). The high drain-to-source voltage V_{DS} region to the right in Fig. 15.15(a) is the amplification region where transconductance $g_{FS} = 6$ mhos is defined. The region has little interest to us. The low V_{DS} region to the left of $V_{GS} = 10$ V is the linear switching region. The actual gate voltage to switch the transistor ON depends upon the drain current, and it is found in Fig. 15.15(c).

ON RESISTANCE $(R_{DS})_{ON}$. The maximum $(R_{DS})_{ON}$ is defined in the electrical characteristics list as $(R_{DS})_{ON} = 0.15\ \Omega$ (max) at $I_D = 10$ A. Resistance is quite independent of drain current up to a rated current of 20 A. For 50- to 100-V devices, MOSFET ON resistance approximately equals that of bipolars. But for higher voltages, bipolar ON resistance is ap-

MTM20N08 Power Field Effect Transistor (MOSFET)

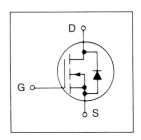

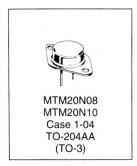

MTM20N08
MTM20N10
Case 1-04
TO-204AA
(TO-3)

MTP20N08
MTP20N10
Case 221A-02
TO-220AB

Maximum ratings

V_{DS} at $V_{GS} = 0$: 80 V P_D at 25° case: 100 W
V_{GS}: ±20 V P_D at 25° free air: not given
I_D: 20 A T_J: −65°C to 150°C

Electrical characteristics at 25°C case temperature

I_D at $V_{GS} = 0$: 0.2 mA(max) $(V_{DS})_{ON}$ at $I_D = 20$ A: 3.5 V(max)
$(V_{GS})_{TH}$ at $I_D = 1$ mA: 2 V(min)   at $I_D = 10$ A: 3 V(max)
    4.5 V(max)
$(R_{DS})_{ON}$ at $T_J = 100°C$: 0.15 Ω g_{FS} at $I_D = 10$ A: 6 mhos(min)
    at $T_J = 25°C$: 0.10 Ω

Thermal characteristics

$R_{\theta JC}$: 1.25°C/W
$R_{\theta JA}$: 30°C/W for TO-3
    62.5°C/W for TO-220

Switching time

$(t_D)_{ON}$: 50 ns $(t_D)_{OFF}$: 100 ns
t_R: 450 ns t_F: 200 ns

FIGURE 15.14
Power MOSFET performance characteristics. (*Drawings by permission of Motorola, Inc.*)

preciably lower. Resistance increases with junction temperature. The effect is thermally stabilizing since drain current decreases. Bipolar ON resistance decreases with junction temperature, and they sometimes fail from thermal runaway. Power dissipation in the MOSFET is $P_D = R_{DS}I_{DS}^2$.

SAFE OPERATING AREA (SOA). The SOA for MOSFETs is quite similar to bipolar transistors as shown in Fig. 15.15(*d*). There is a maximum limit on the value of I_D and a maximum limit on V_{DS}; the latter is the same as the drain-to-source breakdown voltage. For the amplification region with high V_{DS}, the maximum power dissipation line is operating. For the switched ON region with low V_{DS}, the SOA differs from bipolar plots since the transistor operates along an $(R_{DS})_{ON}$ line.

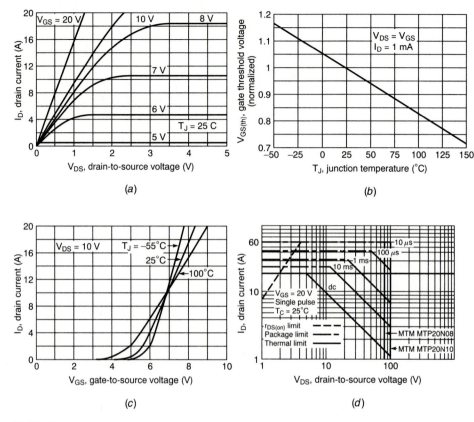

FIGURE 15.15
Typical characteristics of an MTM20N08 MOSFET power transistor. (*a*) ON-region characteristics; (*b*) gate threshold; (*c*) gate-to-source voltage; (*d*) safe operating area. (*Used by permission of Motorola, Inc.*)

Example 15.4. An MTM20N08 power MOSFET is to be operated as a switch for a 2-Ω load. The heat sink is chosen to give a junction temperature $T_J = 100°C$ and a case temperature $T_C = 25°C$ (see Sec. 15.4). The load supply voltage is $V_{DD} = 20$ V. Determine the (1) ON-state resistance R_{DS}, (2) current I_{DS}, (3) gate voltage if the overdrive factor ODF = 2, (4) MOSFET power dissipation, and (5) the peak current I_{DS} if the transistor is switched ON once for 1 ms:

1. Since $R_{DS} = 0.15$ Ω at $T_J = 100°C$

2. $V_{DD} = I_{DS} R_{total}$

   20 V $= I_{DS}(2 + 0.15)$, $\quad I_{DS} = 9.3$ A

   where $(I_{DS})_{max} = 20$ A

3. From Fig. 15.15(*c*)

   $(V_{GS})_{TH} \times$ ODF $= 13.6$ V

4. $P_D = V_{DS} I_{DS} = (9.3 \times 0.15)(9.3)$

   $= 13$ W

5. Assuming there is no change in the load resistance, R_L, V_{DS}, and I_{DS} remain the same as the preceding. If the value of the resistance is left open, Fig. 15.15(d) indicates $(I_{DS})_{max} = 32$ A at $V_{DS} \geqq 2.8$ V.

Example 15.5. The T bridge motor design in Example 15.2 replaces the bipolar power transistor with the MTM20N08 MOSFET transistor. A driver circuit is given in Fig. 15.16. Compare the two designs assuming the junction temperature is 25°C.

Continuous mode

$$I_{DS} = 2 \text{ A} \qquad\qquad (P_D)_{gate} \approx 0$$

$$R_{DS} = 0.10 \ \Omega \qquad\qquad (P_D)_{DS} = 0.40 \text{ W}$$

$$V_{DS} = 0.20 \text{ V} \qquad\qquad (P_D)_{total} = 12 \text{ V} (2 \text{ A}) = 24 \text{ W}$$

$$V_{GS} = 5.5 \text{ V}$$

$$= 12 \text{ V (with overdrive)} \qquad \text{Eff} = \frac{24 - 0.40}{24} = 98\%$$

Start/stall mode

$$I_{DS} = 10 \text{ A} \qquad\qquad (P_D)_{gate} = 0$$

$$R_{DS} = 0.10 \ \Omega \qquad\qquad (P_D)_{DS} = 10 \text{ W}$$

$$V_{DS} = 1.0 \text{ V} \qquad\qquad (P_D)_{total} = 120 \text{ W}$$

$$V_{GS} = 6.8 \text{ V}$$

$$= 12 \text{ V (with overdrive)} \qquad \text{Eff} = \frac{120 - 10}{120} \cong 92\%$$

The advantage of MOSFET is very LOW gate drive power. At times the gate voltage can exceed the power supply voltage, and a step-up transformer or voltage multiplier circuit

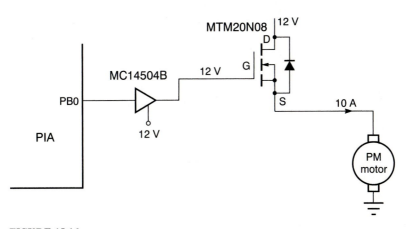

FIGURE 15.16
ON-OFF control of a DC motor with MOSFET power transistor drive.

TABLE 15.3
Motorola 80 V MOSFET specifications and power dissipation for a 10-A load

Part Number	R_{DS} (Ω)	Rated I_{DS} (A)	Rated power dissipation (W)	Voltage drop (V)	Actual power dissipation (W)
MTM10N08	0.37	10	75	3.7	37
MTM12N08	0.23	12	75	2.3	23
MTM20N08	0.15	20	100	1.5	15
MTM25N08	0.10	25	150	1.0	10
MTM55N08	0.03	55	250	0.3	3

would be necessary to power the Motorola MC 14504B level shifter. Thus, circuit simplicity, a MOSFET feature, can be lost. Here the shifter is powered directly from the 12-V supply. If the power supply voltage exceeds the maximum gate voltage, a zener diode/resistor circuit can step down the voltage.

MOSFET dissipates slightly less power than the Darlington but more than the bipolar transistor. These comparisons depend upon maintaining specified junction/case temperature(s) with a proper heat sink design, Sec. 15.4. Cooling requirements often can be simplified by picking an oversized MOSFET. A cost increase due to the larger size is offset by (1) eliminating the need for a fan, (2) decreasing the heat sink size, or (3) eliminating the heat sink altogether. Table 15.3 lists five MOSFETs to handle a 10-A load current if the junction temperature is allowed to rise to 100°C. Note that the larger MOSFETs have lower ON-resistance, and thus less power dissipation.

Parallel Operation

In some applications the most beneficial property of power MOSFET is its ability to be paralleled to meet load demands. As shown in Fig. 15.17, each of two (or more) MOSFETs share the load current equally. Production devices are mismatched in their individual voltage drops. If one transistor carries more current, its "resistance" decreases and its current increases further. MOSFETs have positive coefficients which will establish a more equal distribution of the total load current in the two branches.

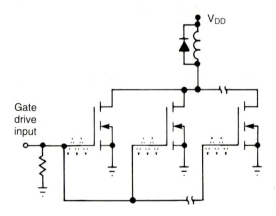

FIGURE 15.17
Increasing drive current by paralleling MOSFET. (*Used by permission of Motorola, Inc.*)

When gates are driven directly from a common node, a high-Q network is established which may cause the device to oscillate as it switches through the active region. This problem can be averted by inserting a 10- to 20-Ω resistor in the gate path.

15.4 THERMAL CONSIDERATIONS AND HEAT SINKS

By most frames of reference, semiconductor junctions are quite small. Discrete junctions must be encased in a comparatively larger structure so that electrical connections are convenient to the user. For general-purpose transistors where power dissipation is less than a watt, the loss of heat to the surroundings is adequate without any means of heat transfer enhancement. Therefore, the manufacturer quotes maximum dissipation ratings which are referenced to ambient conditions.

With a power transistor the physical size is much too small to allow sufficient heat transfer to the surroundings. The necessary enhancement is in the form of a "heat sink." The term is a misnomer because the main purpose of a heat sink is to improve the steady heat transfer from the transistor to ambient conditions by increasing the surface area. Heat sinks are usually made of finned, extruded aluminum with the appearance in Fig. 15.18. Sometimes the designer can "make do" by mounting the transistor to a simple vertical plate. They operate by natural convection of the ambient air although blown air is used where very large heat transfer rates are necessary. In most applications heat sinking is vital.

Heat transfer in electronics is calculated under steady-state conditions. When wave forms are periodic, graphical integration may be performed with an oscilloscope.

FIGURE 15.18
Typical heat sinks for power semiconductors.

Voltage-current products are added by increments, and the sum is divided by an appropriate time base. Heat transfer rate can be expressed by

$$q = \frac{\Delta T}{\Sigma R_\theta}$$

where q = rate of heat transfer or power dissipation
  R_θ = thermal resistance
  ΔT = temperature difference between regions of heat transfer

Heat transfer can be conductive or convective and is highly dependent upon how the heat sink is mounted. Figure 15.19 shows the recommended method. The current (not heat) insulator between the transistor case and heat sink is necessary because the case is commonly at collector voltage.

There are three important thermal resistances in series. The equivalent electrical circuit gives

$$P_D = \frac{T_J - T_A}{R_{\theta JC} + R_{\theta CS} + R_{\theta SA}}$$

where P_D = power dissipation (Watts)
  T_J = npn-junction temperature (°C)
  $R_{\theta JC}$ = junction to transistor case thermal resistance (°C/W)
  $R_{\theta CS}$ = insulator interface thermal resistance (°C/W)
  $R_{\theta SA}$ = heat sink to ambient thermal resistance (°C/W)
  T_A = ambient temperature (°C)

Remember that ambient temperature pertains to the surroundings of the heat sink. This may not be room temperature when electronics are enclosed, which is why good air circulation through the enclosure is needed.

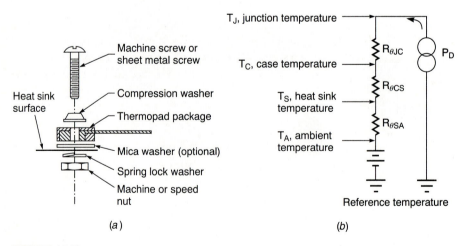

(a)   (b)

FIGURE 15.19
Assembly of power semiconductor and heat sink with thermal/resistance model. (a) Assembly; (b) model. (*Used with permission of Motorola, Inc.*)

Suppose you wish to mount the transistor without a heat sink. What is the maximum power dissipation capability of a particular semiconductor package? In this case the heat transfer path is not to the heat sink but directly to ambient. Thus,

$$P_D = \frac{T_J - T_A}{R_{\theta JA}} \quad \text{(no heat sink)}$$

where $R_{\theta JA}$ is the junction to ambient thermal resistance.

Power product specifications are readily made available by the manufacturer. A typical transistor data book would give the following information necessary to size the heat sink.

75W MOSFET IRF730

Maximum ratings

Total power dissipation at $T_C = 25°C$:	75 W
Derate above 25°C:	0.6 W/°C
Junction temperature:	−55 to 150°C

Thermal characteristics

Thermal resistance $T_{\theta JC}$:	1.67°C/W
Thermal resistance $R_{\theta JA}$:	62.5°C/W

Such maximum ratings refer to the temperature of the case, *assuming it is mounted to an appropriate heat sink*. This bit of news is never made explicit. If these devices were operated without a heat sink, the power rating could be two orders of magnitude less. In this case

$$P_D = \frac{T_J - T_A}{R_{\theta JA}} = \frac{150°C - 20°C}{62.5°C/W} \cong 2 \text{ W}$$

Transistor collectors are common to the case. Ideally, it is best to isolate the heat sink/transistor assembly from ground rather than use insulators between the heat sink and transistor. Sometimes the chassis itself is used as a heat sink. If the chassis is a printed circuit board, no insulator is necessary. When an insulator is necessary, it can have a significant impact on the heat sink selection. Since mating materials have a degree of unevenness, a thermal grease is generally used. Table 15.4 gives the differences between insulators with and without grease. Other useful information includes

$$(T_J)_{max} \text{ metal package} = 200°C$$

$$(T_J)_{max} \text{ plastic package} = 150°C$$

$$R_{\theta SA} \text{ (typical heat sink)} = \frac{200}{\mathcal{A}}$$

where $\mathcal{A}$ is the area in cm^2.

$$R_{\theta SA} \text{ (vertical plate)} = \frac{0.5}{L} °C/W$$

TABLE 15.4
Interface thermal resistance $R_{\theta CS}$ (°C/W)

	Case			
	TO-220		TO-3	
Interface	Grease	No grease	Grease	No grease
Mica (0.003″)	1.8	3.7	0.35	1.3
Thermalfilm (0.002″)	2.3	4.5	0.65	1.5
Silicone (0.012″)	—	—	0.70	1.0
BrO$_3$ (0.062)	—	—	0.20	0.6
No insulator	1.0	1.3	0.15	0.5

where L is the length in meters.

$$R_{\theta SA} \text{ (forced convection)} = \frac{R_{\theta SA}}{\sqrt{3 \mathcal{V}/2}}$$

where $\mathcal{V}$ is the velocity in m/s.

> **Example 15.6.** The MOSFET IRF730 power transistor is to be used in a design where the power dissipation will be 50 W. It will be mounted to the chassis for a heat sink. The chassis temperature has been known to reach 60°C. Determine if this approach is feasible. Transistor case is TO-220.
>
> Let us try a mica/grease interface:
>
> $$P_D = \frac{T_J - T_{CH}}{R_{\theta JC} + R_{\theta CS}}$$
>
> $$50 \text{ W} = \frac{T_J - 60°C}{1.67 + 1.8}$$

where $T_J = 235°C$, clearly too high. In fact, with no insulation the junction temperature $T_J = 143°C$. Mount the transistor on a heat sink isolated from the chassis.

15.5 THYRISTORS: SCRs

The silicon-controlled rectifier (SCR) or thyristor is a *pnpn* structure similar to diodes and bipolar transistors. It has three transistorlike terminals: anode, cathode, and gate or trigger. Figure 15.20 shows the symbol and a sectional view. The properties of an SCR are similar to a (power) diode. An actual current-voltage characteristic curve is shown in Fig. 15.21.

Consider the case with the gate current OFF. When the anode is positive with respect to the cathode, the device is forward biased. The thyristor is said to be in the forward-blocking state with a small leakage current (microamperes). Likewise, when the cathode is positive with respect to the anode, the device is reverse biased or in a reverse-blocking state. If the anode-to-cathode voltage is increased to a sufficiently large value, the junction breaks down and the thyristor reverts to a conducting state with large currents under negligible resistance. Unlike a diode the forward and reverse breakdown voltages (hundreds of volts) are the same.

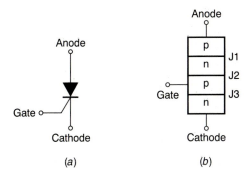

FIGURE 15.20
Tyristor: Silicon-controlled rectifier (SCR).
(*a*) Symbol; (*b*) structure.

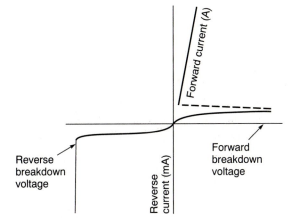

FIGURE 15.21
SCR current-voltage characteristic.

With the gate current ON the SCR conducts with no forward-blocking traits. Reverse-blocking traits remain. Load current is limited by the external load impedance or resistance, and there is a small voltage drop, typically 1 V, across the SCR. In this respect it behaves much like a saturated bipolar transistor. With tens of milliamps to the gate, hundreds of load amps can be switched ON with very little power loss in the SCR. However, the anode (or load) current must be more than the latching current I_L (tens of milliamps) or the SCR will revert back to its blocking state.

Once the thyristor conducts, there is no control over the device. It continues to conduct even though the gate current is OFF. Thus, the SCR can be turned ON with a DC gate current or a single-pulse gate current. The device can be turned OFF only by switching OFF the anode current. Actually, the SCR will revert to the blocking state if the forward current is reduced below a level known as the holding current I_H. The holding current is less than the latching current.

SCRs can be considered as two independent interconnected transistors, Fig. 15.22. In fact, a model of an SCR can be assembled from *pnp* and *npn* transistors. In the initial state the gate is OFF and there is no anode-to-cathode current. When the gate is activated, it turns ON the *npn* transistor Q2. Thus, the base of the *pnp* transistor goes LOW, turning ON the *pnp* transistor. The anode current flows from the emitter to the base of Q1 and to the cathode. At the same time current flows from the anode to the collector of Q1, which

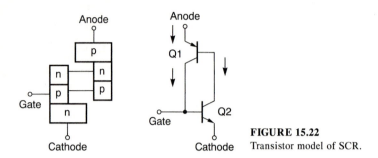

FIGURE 15.22
Transistor model of SCR.

is connected to the gate. Once the gate is pulsed ON, the SCR maintains its own gate current to lock the anode current.

Control of SCRs

There are two aspects to SCR control: (1) The anode current must be (near) zero before the SCR will switch OFF. This is referred to as commutation. (2) The gate current should be switched in some manner to control the average power delivered to the load, thus regulating the power to a heater or the speed of a motor.

With an AC power source the line voltage naturally commutates through zero. If the gate is OFF, the device will switch OFF. In Fig. 15.23(*a*) an AC source across an SCR with an active gate produces a half-rectified anode current for a DC load because the thyristor reverse-blocking effect prevents any current during the negative voltage cycle. If we employ a diode bridge rectifier, the full-rectified wave becomes available for the DC load. See Fig. 15.23(*b*). For a constant voltage DC source, there must be some special means to commutate (turn-OFF) the anode current. In other words, we must have forced commutation. This generally involves additional circuits which incorporate capacitors, inductors, and switching devices. See Fig. 15.23(*c*). The circuit repeatedly drives the anode voltage to zero momentarily in constant anticipation of gate switch OFF.

The most common method of non-microprocessor-based gate control is called phase control, shown in Fig. 15.24. It takes advantage of the fact that the gate requires a minimum voltage ($\sim$ 1 V) to fire. The circuit uses a variable resistor in parallel with the SCR power source to drive the gate. When the source voltage reaches some value at phase angle θ determined by the resistor setting, the gate is triggered. Load power can be controlled from maximum to one-half the maximum power. Microprocessor-based gate control uses PWM techniques. We will discuss interfacing methods in the section on TRIACs. TRIACs are a type of SCR for controlling AC loads.

GTOs

Gate turn-OFF (GTO) thyristors are three terminal *pnpn* silicon devices similar to the SCR. As the name implies, the GTO can be turned OFF by a negative gate signal. GTOs have advantages over SCRs: (1) The turn-OFF characteristic eliminates the need for forced commutation circuits with DC power supplies. (2) GTOs have faster switch OFF, permitting higher switching frequencies than SCRs: 100 k vs. 10 kHz. There are several disad-

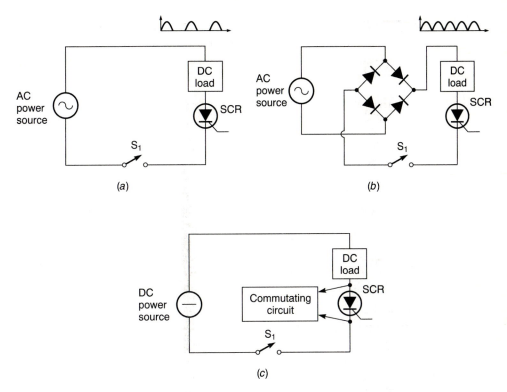

FIGURE 15.23
Driving DC loads with SCRs. (*a*) Direct AC source; (*b*) rectified AC source; (*c*) DC source.

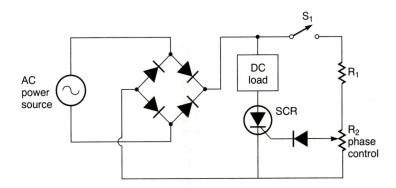

FIGURE 15.24
Phase control of load power.

vantages: (1) Gate interfacing circuits are more complicated, requiring both positive and negative pulses to control the gate. (2) Its ON-state voltage drop is typically 3.5 V, higher than SCRs. (3) GTOs cannot control the same range of load power as SCRs, which places them in competition with power transistors.

Thyristor Protection Circuits

There are several simple components that can protect a thyristor against signal stress. See Fig. 15.25. Although most of them are not necessary in a typical application, their inclusion will assure trouble-free operation:

Gate G. Most gate drive sources have high impedance and can interface the gate directly. However, a series resistor R_S should be inserted for low-impedance gate sources to prevent gate burnout. A parallel resistor R_G (1 k typical) to cathode drains off current leaking from anode to gate. This leakage can be sufficient to turn ON the thyristor. It is highly recommended for sensitive gates, i.e., trigger currents on the order of 10 mA.

Cathode K. Thyristors can block very high voltages between the anode and cathode. But they are susceptible to cathode voltages exceeding gate voltages by more than 10 V (typical). This may happen in noisy currents. The parallel resistor R_G offers some protection, but a diode will limit the voltage difference to (about) 1 V.

Anode A. The anode is susceptible to transients when the load power is switched ON. Each thyristor carries a *dI/dt* and a *dV/dt* rating. If the anode current rises too fast, the current becomes localized rather than spread out over the junction and excessive temperatures occur. In practice, *dI/dt* is limited with series inductance. Motors have enough inductance, but resistance heaters sometimes need a series inductor:

$$\left(\frac{dI}{dt}\right)_{max} = \frac{(V_A)_{max}}{L_L}$$

The *dI/dt* rating can be improved approximately 25% if the gate is overdriven. Increase the gate trigger magnitude to several times the minimum, and apply it with a fast rise time (1 μs).

The *dV/dt* effect. When a source voltage is suddenly applied to the anode with the gate OFF, the thyristor may switch from an OFF-state to an ON-state. As long as the voltage source is AC, the false start will persist for only a one-half cycle. However, if the source is a DC voltage, the transient may trip the thyristor to a conducting state in which turn

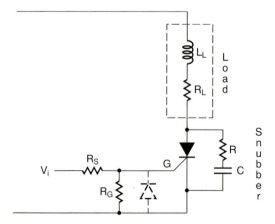

FIGURE 15.25
Protection elements in an SCR circuit.

OFF can be achieved only by a circuit interruption. An *RC* network, called a *snubber,* is used to keep the thyristor static *dV/dt* within limits. The snubber/load corresponds to a series *RCL* circuit where the snubber capacitance is found from

$$C = \frac{(V_A)^2_{max}}{L_L(dV/dt)^2_{max}}$$

If the resistance is selected for a critically damped system,

$$(R + R_L) = 2\sqrt{\frac{L_L}{C}}$$

Example 15.7. A thyristor and load are driven by a 230-V, 60-Hz rectified supply. The DC motor load is specified at 5 Ω and 1.0 mH. If the thyristor is rated at *dV/dt* = 50 V/μs find (*a*) the snubber *RC* values, and (*b*) the power rating of the snubber resistor.

Solution

(a) From the preceding equations, capacitance and resistance can be calculated as

$$C = \frac{[230\sqrt{2}]^2}{10^{-3} \times (50 \times 10^6)^2} = 0.04 \ \mu F$$

$$R + 5 \ \Omega = 2\left[\frac{10^{-3}}{C}\right]^{1/2} = 316 \ \Omega$$

(b) Assume all the energy stored in the capacitance is dissipated in *R* only:

$$P = \tfrac{1}{2}CV_A^2 f$$

$$= \tfrac{1}{2} \times 0.04 \times 10^{-6} \times (230)^2 \times 60 = 0.06 \ W$$

15.6 THYRISTORS: TRIACS

Loads that need AC sources can be controlled by two antiparallel SCRs with a common gate connection as shown in Fig. 15.26(*a*). If this concept is reduced to a single *pnpn* junction, the three terminal thyristor is called a TRIode AC semiconductor. The TRIAC symbol is given in Fig. 15.26(*b*). Since the device is bidirectional, its terminals cannot be designated as anode or cathode.

TRIACs are a simple, economical means for controlling AC power. They can be triggered by positive or negative DC, AC, rectified AC, or pulses. Since TRIACs conduct current in both directions through a single junction, there is only a brief interval during which the sine wave current passes through zero and must revert to its blocking state. For this reason TRIACs are limited to low/medium power applications with currents less than 100 A and to line frequencies 60 Hz or lower.

Performance Parameters

Thyristors are characterized by certain performance parameters which manufacturers specify in data sheets. Although there are various types of thyristors, the parameters are the

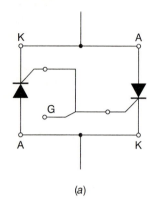

(a)

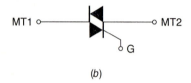

(b)

FIGURE 15.26

Full wave control of AC loads with thyristors. (*a*) Two SCRs for high-power loads; (*b*) TRIAC (combined SCRs) for medium power loads.

same, except for different gate characteristics. Figure 15.27 shows the data sheet for the Motorola MAC212 Series TRIAC:

OFF-state voltage, V_{DRM}. The maximum allowed value of repetitive OFF-state voltage which may be applied and not switch ON the TRIAC.

ON-state RMS current, $I_{T(RMS)}$. The maximum value of ON-state RMS current the device may conduct. This limit is due to junction heating effects.

Peak forward surge current, I_{TSM}. The maximum allowable nonrepetitive surge current the device will withstand.

Circuit fusing, I^2t. The maximum forward nonrepetitive overcurrent capability. Usually specified for one-half the cycle at 60 Hz.

Forward average gate power, $P_{G(AVE)}$. The maximum allowable average gate power that may be dissipated between the gate and the cathode: $I_{GT} \times V_{GT}$.

Forward peak gate current, I_{GM}. The maximum gate current which may be applied to the device to cause conduction.

Peak-forward blocking current, I_{DRM}. The maximum value of current which will flow at V_{DRM} for the specified temperature.

Peak ON-state voltage, V_{TM}. The maximum voltage drop across terminals at stated conditions. Typical voltage drop V_T vs. instantaneous current I_T is given in Fig. 15.28(*a*).

Gate trigger current, I_{GT}. The maximum expected value of current required to switch the device from the OFF-state to the ON-state. TRIACs are switched ON by both positive and negative signals. The level of I_{GT} varies with both the sign of MT2 and gate. Typical values are less than the maximum values. Typical (normalized) I_{GT} vs. case temperature is shown in Fig. 15.28(*b*).

Triacs
Silicon Bidirectional Thyristors

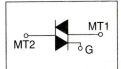

**TRIACs
12 A RMS
200 thru 800 V**

. . . designed primarily for full-wave AC control applications, such as light dimmers, motor controls, heating controls and power supplies; or wherever full-wave silicon gate-controlled solid-state devices are needed. Triac-type thyristors switch from a blocking to a conducting state for either polarity of applied anode voltage with positive or negative gate triggering.

- Blocking voltage to 800 V
- All diffused and glass passivated junctions for greater parameter uniformity and stability
- Small, rugged, thermowatt construction for low thermal resistance, high heat dissipation, and durability
- Gate triggering guaranteed in three modes (MAC212 Series) or four modes (MAC212A Series)

Case 221A-02
TO-220AB

Maximum Ratings

Rating	Symbol	Value	Unit
Repetitive peak OFF-state voltage[(1)] ($T_J = -40$ to $+125°C$) $\frac{1}{2}$ sine wave 50 to 60 Hz, gate open	V_{DRM}		V
MAC212-4, A4		200	
MAC212-6, A6		400	
MAC212-8, A8		600	
MAC212-10, A10		800	
ON-state current RMS ($T_C = +85°C$) Full cycle sine wave 50 to 60 Hz	$I_{T(RMS)}$	12	A
Peak nonrepetitive surge current (one full cycle, 60 Hz, $T_C = +85°C$) preceded and followed by rated current	I_{TSM}	100	A
Circuit fusing considerations ($T_C = +85°C$, t = 1 to 8.3 ms)	I^2t	35	A^2s
Peak gate power ($T_C = +85°C$, pulse width = $10\,\mu s$)	P_{GM}	20	W
Average gate power ($T_C = +85°C$, t = 8.3 ms)	$P_{G(AV)}$	0.35	W
Peak gate current ($T_C = +85°C$, pulse width = $10\,\mu s$)	I_{GM}	2	A
Operating junction temperature range	T_J	-40 to $+125$	°C
Storage temperature range	T_{stg}	-40 to $+150$	°C

(1) Ratings apply for open gate conditions. Thyristor devices shall not be tested with a constant current source for blocking capability such that the voltage applied exceeds the rated blocking voltage.

Thermowatt is a trademark of Motorola Inc.

FIGURE 15.27
The MAC212 Series TRIAC data sheet. (*Used by permission of Motorola, Inc.*)

MAC212,A Series

Thermal Characteristics

Characteristic	Symbol	Max	Unit
Thermal resistance, junction to case	$R_{\theta JC}$	2.1	°C W

Electrical Characteristics (T_C = +25°C unless otherwise noted)

Characteristic	Symbol	Min	Typ	Max	Unit
Peak blocking current (either direction) rated V_{DRM}, gate open T_J = +125°C	I_{DRM}	—	—	0.1 2	mA
Peak ON-state voltage (either direction) I_{TM} = 17 A peak; pulse width = 1 to 2 ms. duty cycle ≤ 2%	V_{TM}	—	1.3	1.75	V
Gate trigger current, continuous DC Main terminal voltage = 12 V DC, R_L = 100 Ω Minimum gate pulse width = 2 μs MT2(+), G(+) all types MT2(+), G(−) all types MT2(−), G(−) all types MT2(−), G(+) "A" suffix only	I_{GT}	— — — —	12 12 20 35	50 50 50 75	mA
Gate trigger voltage, continuous DC Main terminal voltage = 12 V DC, R_L = 100 Ω Minimum gate pulse width = 2 μs MT2(+), G(+) all types MT2(+), G(−) all types MT2(−), G(−) all types MT2(−), G(+) "A" suffix only Main terminal voltage = rated V_{DRM}, R_L = 10 kΩ; T_J = +125°C MT2(+), G(+); MT2(−), G(−); MT2(+), G(−) all types; MT2(−), G(+) "A" suffix only	V_{GT}	— — — — 0.2 0.2	0.9 0.9 1.1 1.4 — —	2 2 2 2.5 — —	V
Holding current (either direction) Main terminal voltage = 12 V DC, gate open, Initiating current = 500 mA, T_C = +25°C	I_H	—	6	50	mA
Turn-on time Rated V_{DRM}, I_{TM} = 17 A, I_{GT} = 120 mA. Rise time = 0.1 μs, pulse width = 2 μs	t_{gt}	—	1.5	2	μs
Critical rate of rise of commutation voltage Rated V_{DRM}, I_{TM} = 17A, commutating di/dt = 4.3 A/ms Gate unenergized, T_C = +85°C	$dv/dt_{(c)}$	5	—	—	V/μs
Critical rate of rise of OFF-state voltage (V_D = V_{DROM}, exponential voltage rise, gate open, T_C = +85°C)	dv/dt	100	—	—	V/μs

FIGURE 15.27 (continued)

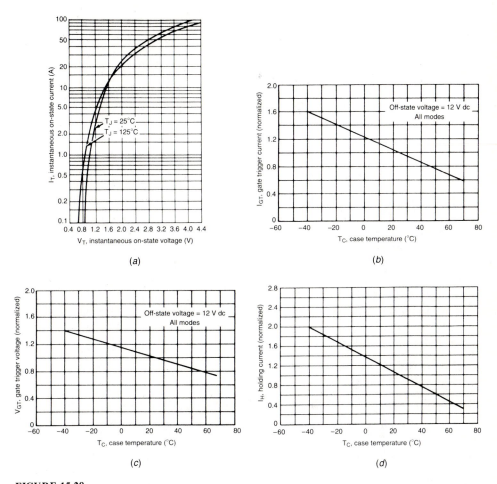

FIGURE 15.28
Typical characteristics of an MAC212 Series TRIAC. (*a*) ON-state voltage drop; (*b*) normalized gate trigger current; (*c*) normalized gate trigger voltage; (*d*) normalized holding current. (*Used by permission of Motorola, Inc.*)

Gate trigger voltage, V_{GT}. The gate DC voltage required to produce the maximum gate trigger current. Typical (normalized) V_{GT} is shown as a function of case temperature in Fig. 15.28(*c*).

Holding current, I_H. The minimum forward current which allows the device to remain in conduction. Below this value the device will return to a forward-blocking state at prescribed gate conditions. Typical (normalized) I_H is shown as a function of case temperature in Fig. 15.28(*d*).

Example 15.8. A Motorola MAC 212 TRIAC is used to drive a 230-V AC motor. A positive PWM gate signal controls the speed. What series, gate voltage, and current are recommended?

The TRIAC should have a blocking voltage in the gate-OFF state of at least $230\sqrt{2}$ V, the peak AC line voltage. Select the MAC 212.6 with a 400 V blocking voltage. Gate drive requirements vary from device to device. The worst case (maximum) requirement is

$$\text{MT2}(+), G(+): \quad I_{GT} = 50 \text{ mA} \quad V_{GT} = 2 \text{ V}$$

$$\text{MT2}(-), G(+): \quad I_{GT} = 75 \text{ mA} \quad V_{GT} = 2.5 \text{ V}$$

Use 3 V and 110 mA (75 mA × overdrive factor of 1.5 or 110/35 mA ≅ 3 overdrive factor for typical device). 3 V × 110 mA < 0.35 W, maximum allowable gate power.

Snubbing TRIACs

Static $(dV/dt)_s$ occurs in thyristors when the source voltage changes rapidly with the gate OFF. Commutating $(dV/dt)_c$ is another type of transient unique to TRIACs with the gate ON because they conduct in both directions. The problem is more severe with inductive loads than resistive loads and is caused by a phase lag between voltage and current. Figure 15.29 shows the waveform. When the forward (+) current reaches zero (actually the holding current), the forward half of the TRIAC ceases to conduct. The reverse half begins to conduct upon the reverse (−) current. Since the voltage lags, there is a suddenly applied voltage to the now conducting half. The result can be a loss of control with the device remaining ON in the absence of a gate signal. The rate of rise of the opposite polarity voltage the TRIAC can handle without remaining ON when the gate is turned off is the commutating $(dV/dt)_c$. It is more critical than the static $(dV/dt)_s$. Snubber RC values for TRIACs are calculated the same as for SCRs.

Interfacing Microprocessors

Microprocessors control thyristor loads with ON/OFF bursts or with more defined PWM signals. Commonly, the MPU drives a PIA which drives the next stage. The PIA output port is TTL compatible but has slightly less current source and sink capability than standard TTL. Thyristors require gate currents on the order of 50 to 100 mA (low-power, sensitive gate thyristors require about 10 mA). Thus, additional interfacing units are necessary. The preferred interfacing units are optocouplers, discussed in the following section. However, if logic circuit/power circuit isolation is not a problem, transistors may be used.

When the *npn* transistor is activated by a logic HIGH, its base must receive sufficient current to drive the transistor hard into saturation. There are two methods to source current greater than the TTL HIGH. (1) Use a 74LS244 octal buffer (15 mA at 3.4 V). (2) Use a pull-up resistor; R (minimum) set by the logic-LOW current to the PIA. An al-

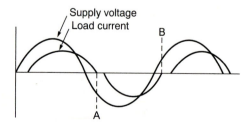

FIGURE 15.29
Commutating voltage in TRIACS; $(dV/dt)_c$.

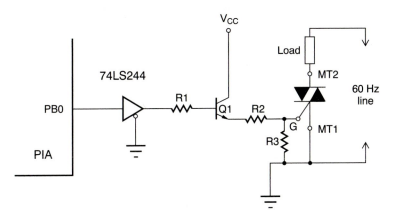

FIGURE 15.30
Interfacing TRIACs to the PIA.

ternative method uses a *pnp* transistor, sinking the active-LOW base directly to the PIA. Figure 15.30 shows the buffer method.

> **Example 15.9.** The selected TRIAC in Fig. 15.30 has a positive gate requirement of 1 V and 150 mA. The interfacing transistor is an *npn* 2N4401. It has a specified minimum h_{FE} of 100 at a collector current of 150 mA. If we use a saturation overdrive factor of 2.5, the forced h_{FE} is 40. The 74244 octal buffer supplies 15 mA at 3.4 V. Determine resistors R1 and R2.
> The transistor base current $I_B = I_G/h_{FE}$ or 150 mA/40 = 3.75 mA. Then

$$R1 = \frac{V_{BUF} - V_{BE} - V_G}{I_B}$$

$$= \frac{3.4 - 0.7 - 1.0}{3.75}$$

$$= 450 \ \Omega \ \text{(maximum)}$$

Leakage resistance R3 is set at 1 kΩ as designated previously. Its current drain will be negligible compared to the gate current. So,

$$R2 = \frac{V_{CC} - V_{CE(SAT)} - V_G}{I_G}$$

$$= \frac{5 - 1 - 1}{150}$$

$$= 20 \ \Omega$$

15.7   OPTOCOUPLERS

Optocouplers, also known as an optoisolators, combine an LED source in the same package with some type of solid-state photo transistor/thyristor. Because information is passed optically across an insulating air gap, the detector side of the circuit cannot affect the input side. This is important because the LED is driven by low-voltage circuits utilizing an

MPU or logic gate. The photo semiconductor side is often part of a high-voltage DC or AC circuit. Optical isolation prevents interaction or even damage to the input circuit caused by transients or noise spikes on the high-voltage side. Optocouplers offer other advantages in addition to electrical isolation. People can be isolated from high voltages. Ground loop problems are eliminated. They are particularly troublesome for circuits having separate power supplies for logic and load.

Figure 15.31 shows typical optocoupler packages. The most common package has an internal light transmission path and is used for isolation. The external reflective and slotted air gap packages detect proximity or motion in the environment as well.

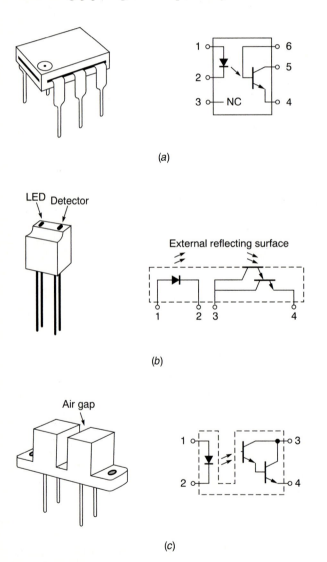

FIGURE 15.31
Typical optocoupler packages. (*a*) Internal light path; (*b*) reflective light path; (*c*) external gap light path.

There is a wide selection of optocouplers based on transistor/thyristor detectors. Figure 15.32 gives the main configurations. An important operating parameter is optical efficiency. It defines the amount of input current that is required to obtain a desired output. With the transistor and Darlington isolators the efficiency is given by the *current transfer ratio* or CTR. This is the guaranteed output current divided by the input current, expressed as a percent. Do not confuse CTR with transistor DC current gain h_{FE}, which is referenced to the base current generated in the detector by the LED. Thyristor efficiency is given by the amount of input current to trigger the output.

Figure 15.33(*a*) gives the collector (i.e., load) current vs. collector-emitter voltage for various inputs to the Motorola H112AV transistor optocoupler. At a typical input current $I_F = 10$ mA the output current $I_C = 5$ mA just before the curve breaks (flat). Then

$$\text{CTR} = \frac{5}{10} = 0.5 \text{ or } 50\%$$

The minimum guaranteed CTR is 20 percent. The maximum input current that can be applied is 60 mA for an output of 25 mA. Other transistor configurations have CTRs from

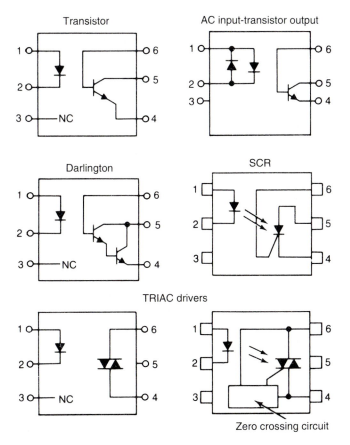

FIGURE 15.32
Optocoupler detector types. (*Used by permission of Motorola, Inc.*)

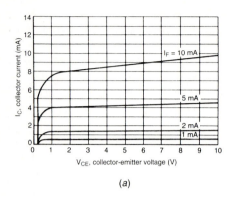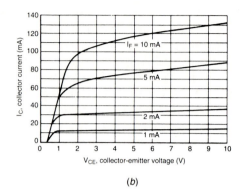

(a)   (b)

FIGURE 15.33
Comparing input current I_F to output current I_C for several optocouplers. (*a*) Transistor detector; (*b*) Darlington detector. (*Copyright of Motorola, Inc. Used by permission.*)

2 percent to 200 percent. Figure 15.33(*b*) for the H11B shows that much higher collector currents are possible with Darlington optocouplers. Current transfer ratios up to 1000 percent can be found. The limiting factor is the detector output maximum power dissipation of 150 mW. Most optocouplers fall in the range 100 to 400 mW.

Optocoupler outputs can be used to switch low-power load circuits directly although their main purpose is to drive (or trigger) medium to high-power circuits. Figure 15.34 illustrates this point. A very small DC motor can be powered directly from AC by using a

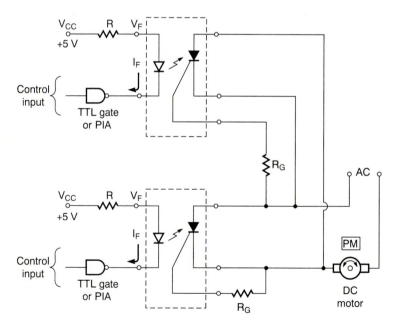

FIGURE 15.34
Bidirectional drive of low-power DC motor with dual SCR optocouplers and AC supply.

pair of SCR optocouplers. When the top optocoupler is active, the half-rectified current goes clockwise through the motor. When the bottom optocoupler is active, the current is counterclockwise, reversing the motor.

One of the more popular optocouplers is the Motorola MOC3011 TRIAC driver shown in Fig. 15.35(*a*). (The MOC3031 includes a zero crossing circuit, which turns ON the TRIAC only when the voltage goes through zero. This reduces turn ON transients and prevents electromagnetic interference.) Important design parameters are

LED trigger current I_F:	10 mA (max)
	5 mA (typical)
LED forward voltage V_F:	1.5 V (max)
Peak repetitive surge current I_{TSM}:	1 A (max)
Total power dissipation P_D:	300 mW
ON-state current I_{TM}:	See Fig. 15.35(*b*)
ON-state voltage drop V_{TM}:	See Fig. 15.35(*b*)

Optocouplers are driven directly from the PIA by sinking LED currents with an active LOW. Figure 15.36 shows the standard circuit for TRIAC-controlled AC loads. A

MOC3011

CASE 730A-02
PLASTIC

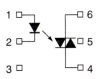

1. Anode
2. Cathode
3. NC
4. Main terminal
5. Substrate
   Do not connect
6. Main terminal

(*a*)

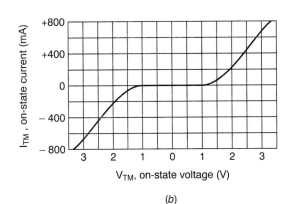

(*b*)

FIGURE 15.35
The MOC3011 TRIAC optocoupler. (*a*) Optocoupler and pin-out; (*b*) ON-state voltage drop. (*Used by permission of Motorola, Inc.*)

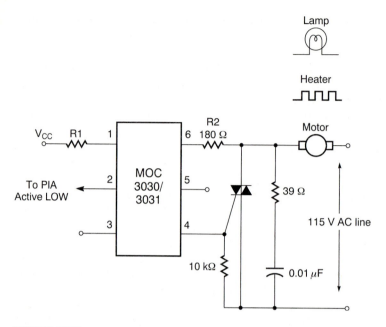

FIGURE 15.36
Interfacing medium power AC loads to the PIA with an optocoupler and external TRIAC. (*Used by permission of Motorola, Inc.*)

snubber should be used with inductive loads such as motors. R1 is set by the maximum trigger current.

$$V_{CC} = V_{LED} + R1 \times I_F + V_{OL}$$

$$5\text{ V} = 1.5\text{ V} + R1 \times 10\text{ mA} + 0.4\text{ V}$$

$$R1 = 310\ \Omega\ (\text{max})$$

$$R1 = 155\ \Omega\ (\text{min})$$

Optocoupler output resistance R2 is based on the worse case ($R_{LOAD} = 0$).

$$R2\ (\text{min}) = \frac{V_{in}\ (\text{peak})}{I_{TSM}} = \frac{115V_{ac}\sqrt{2}}{1\text{ A}} = 163\ \Omega$$

In practice, this would be a 180-Ω resistor. The power TRIAC is not reversed at zero voltage crossover but is reversed at

$$V_{in} = \pm(R2\ I_{GT} + V_{TM} + V_{GT})$$

15.8   POWER ICs

There is an increasing trend by electronic chip manufacturers to include more and more functions on the same chip. While their cost can be higher, they greatly simplify the designer's task. When low-to-medium power transistors are combined with logic and protection circuits, they are called power-integrated circuits or drivers (because they supply high-current levels). Power ICs are designed to interface both CMOS and TTL levels. Output ratings up to 20 A and 400 V are feasible. The L298 stepper motor driver and the HCTL-1100 motion controller for DC motors are two power ICs that were conveniently introduced in earlier chapters. Allegro MicroSystems Inc. (formerly Sprague Semiconductor Group) is a dedicated manufacturer of power ICs. We will review several of their devices.

The ULN-2003A Darlington array in Fig. 15.37 is composed of seven *npn*-Darlington pairs on a common substrate. They are low-power ICs with sink current levels to 500 mA each. Internal diodes allow for inductive load transient suppression. The transistor base has a series resistor which limits the output current to the specified value. The arrays are ideal for driving tungsten filament lamps. Filaments have a cold resistance many times less than the warm resistance. Other drives require a current limiting resistor in series with the lamp to prevent burnout at start-up. The recent UGQ-5140K lamp/solenoid driver (300 mA sink) has a Hall switch for magnet activation.

In Chapter 10, seven-segment and dot-matrix LED displays were multiplexed with discrete transistors. A simple compact solution can be achieved with the UDN-2982A (350-mA source for each outlet) segment driver and the UDN-2068B (1.5-A sink for each outlet) digit selector. The common cathode arrangement is given in Fig. 15.38.

The UDN-2878W quad driver in Fig. 15.39 is designed to interface a variety of peripheral power devices such as solenoids, motors, incandescent displays, heaters, and similar loads up to 320 W per channel. Its Darlington transistors with flyback diodes can sink load currents to 4 A. All drivers rated over 1 A will need heat sinks. The UDN-2549B (600 mA sink) quad driver adds over current and thermal protection for each driver.

A driver for a four-phase *L/R* stepper motor is shown in Fig. 15.40. The UCN-5804B provides complete control with continuous current ratings to 1.25 A per phase and 35 V. There are three drive formats: one-phase (wave drive), two-phase (step drive), and half step. The wave drive consumes the least power, whereas the step drive offers the best torque-speed product. Free-wheeling diodes are provided as well as a thermal protection circuit to disable the output when temperatures are excessive.

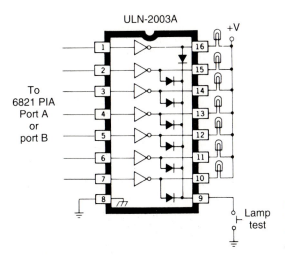

FIGURE 15.37
Sprague ULN-2003A lamp driver (to 500 mA each). (*Courtesy of Sprague Semiconductor Group.*)

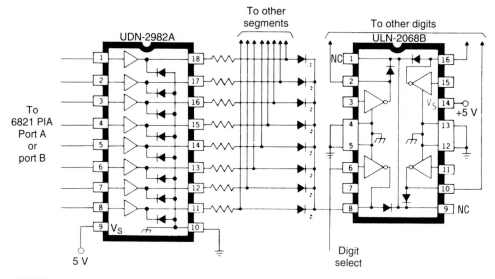

FIGURE 15.38
Sprague UDN-2982A common cathode display driver shown with ULN-2068B for multiplexing. (*Courtesy of Sprague Semiconductor Group.*)

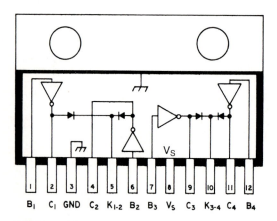

FIGURE 15.39
Sprague UDN-2878W quad driver (to 4 A each). (*Courtesy of Sprague Semiconductor Group.*)

The UDN-2952B power IC in Fig. 15.41 provides bidirectional control of DC motors or solenoids with a full bridge drive. Darlington outputs supply a continuous 2 A with peak start-up currents to 3.5 A. Thermal shutdown networks disable the motor drive if the power dissipation ratings are exceeded. Internal diode transient suppression is available on-chip. Output current can be limited with an external resistor $R_S = 0.6/I_{max}$. Copper tabs (SUB) enable easy attachment of a heat sink while fitting a standard IC socket.

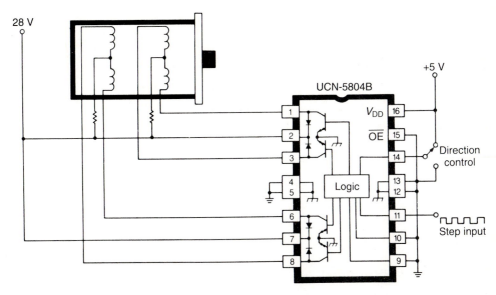

FIGURE 15.40
Sprague UDN-5804B *L/R* stepper motor driver. (*Courtesy of Sprague Semiconductor Group.*)

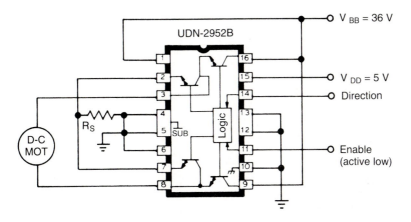

FIGURE 15.41
Sprague UDN-2952B full bridge motor driver. (*Courtesy of Sprague Semiconductor Group.*)

Figure 15.42 shows a three-phase brushless dc motor controller. The UDN-2936W combines commutation logic, PWM current control, free-wheeling diodes, and thermal shutdown protection. The commutating logic is programmed for 60° electrical separation and is compatible with digital or linear Hall-effect sensors. Outputs are rated at 45 V and 3 A (4-A peak). The external sense resistor $R_S = 0.3/I_{max}$ prevents overloading. The new 8901/02/03 three-phase brushless motor controller uses back EMF sensors, eliminating external Hall-effect position devices. It can be interfaced directly to the MPU bus. A serial port allows programming of motor speed.

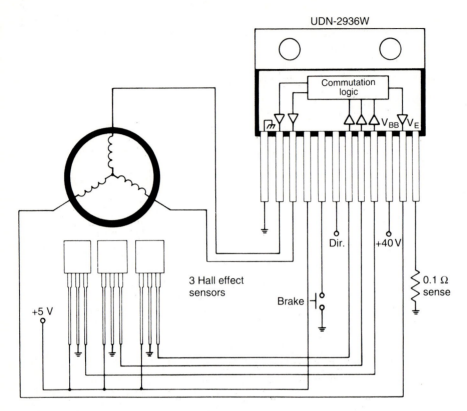

FIGURE 15.42
Sprague UDN-2936W brushless motor driver. (*Courtesy of Sprague Semiconductor Group.*)

15.9   POWER OP AMPS

In Chap. 12 we discussed the option of using linear amplifiers to control DC motors and audio equipment. These amplifiers are constructed of multiple components, including transistors operated in the linear range. Today these discrete circuits are increasingly being replaced by power op amps (POPs). While POPs cost more than components for the discrete circuits, they offer many advantages.

1. Easy to use; no circuit analysis necessary.
2. Obey all the rules of op amps; high-input impedance, LOW output impedance, and high gain.
3. Matched components give a more linear output.
4. Device protection circuits built in.
5. Produce less electrical noise.

   Power op amps are constructed differently from their low-power counterparts. The majority of op amps are single ICs. Their output is limited to 25 mA (100-mA in special designs). POPs essentially combine a power transistor with an op amp at the front end. Supply voltages (thus, output voltages) above 300 V are available. The power champ is the PA03 with an internal dissipation of 500 W and an output current up to ±30 A. It is manufactured by APEX, a leader in the power op-amp field. All POPs should have a heat sink.

A photograph and pin-out of the PA01 are shown in Fig. 15.43. Pins 2 and 8 enable the user to limit the output current. This protects the POP from an overload if the output is short-circuited or if a driven motor stalls, assuming the POP is not oversized to accommodate stall. The matched external resistors are calculated from

$$R_{CL} = \frac{0.65 \text{ V}}{I_{LIM} - 0.54 \text{ A}}$$

$$\text{Power (resistor)} = 0.65 \times I_{LIM}$$

where I_{LIM} is the designed current limit of the output. Some APEX pin-outs use pin 7 to balance the input offset voltage, whereas others have an internal circuit, keyed to junction temperature, to limit the output current. All POPs have a built-in free-wheeling diode for turning OFF inductive loads.

Several analog feedback circuits were covered in Chap. 12. Figure 15.44 illustrates two DC motor control circuits with power op amps. Each input is a DAC output. In Fig. 15.44(*a*) a unidirectional speed control system has an optical pulse generator sensing speed, a frequency-to-voltage converter supplying the required analog voltage feedback. The PA01 power op amp compares the feedback voltage with the command voltage and integrates the difference (note the op amp capacitor) prior to driving the motor. The equivalent block diagram is left as an exercise for the reader. In Fig. 15.44(*b*) the amplifier/motor system controls the torque applied by a robot gripper. Command voltage from the DAC represents the desired torque. POP output current to the DC motor passes through a small, high-wattage resistor, called the sense resistor. Torque is proportional to motor current; therefore, the feedback voltage is proportional to the torque. Since the output signal from the POP is negative, the closed-loop transfer function for torque is

$$T = \frac{K_S \, R2 V_R}{R1 \, R5} \left[\frac{1}{\tau s + 1} \right]$$

where $\tau = (1 + R_m/R5)R2C1$ and R_m is the motor winding resistance.

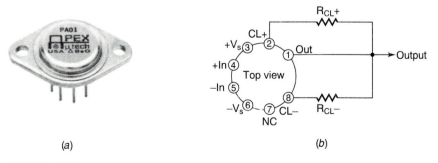

(a)   *(b)*

FIGURE 15.43
Apex PA01 power op amp (to 30 A). (*a*) Package; (*b*) pin-out. (*Courtesy of APEX Microtechnology Corp.*)

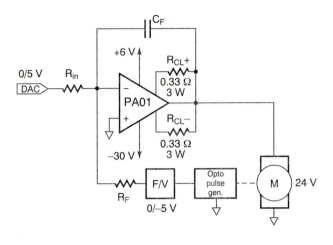

(a)

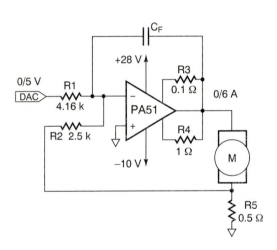

(b)

FIGURE 15.44

Power op-amp feedback control circuits. (*a*) Speed control; (*b*) torque control. (*Courtesy of APEX Microtechnology Corp.*)

PROBLEMS

15.1. What advantage does the TO-3 package have over the TO-220 package?

15.2. Compare the current ratio h_{FE} found on the data sheet in Fig. 15.4 with the plot given in Fig. 15.5. What is the difference in terms of a design with the TIP73 power transistor?

15.3. In Fig. 15.7 two bipolar transistors are used to drive the DC motor. Assume that the TIP42 characteristics are nearly the same as the TIP73. For the currents shown, find values for each base resistor.

15.4. In Example 15.3 a TIP663 Darlington power transistor drives a DC motor. The problem assumes the case is at 25°C. If the case is actually at 125°C, how does the solution change?

15.5. The speed of a DC motor is to be controlled similar to the diagram in Fig. 15.12. The small motor has a stall current of 2 A and a continuous current of 0.3 A. An oversized power transistor in the form of a TIP663 Darlington is used to drive the motor.
(a) Evaluate the design.
(b) Are there any suggested changes for the diagram?

15.6. At a case temperature of 125°C would you recommend a heat sink for the power transistor in Prob. 15.5?

15.7. The MTM20N08 MOSFET power transistor is to be used in an H bridge circuit for bidirectional ON-OFF DC motor control. The motor draws 4 A at continuous running speed. A 12-V supply is available.
(a) Draw a circuit which can be interfaced to the PIA.
(b) Evaluate the design for continuous running.

15.8. The last three MOSFETs in Table 15.3 all have a junction-to-ambient resistance $R_{\theta JA} = 30°C/W$. Would any of these power transistors be adequate in Example 15.5 without a heat sink?

15.9. Repeat Example 15.6 with a heat sink. Estimate the heat sink size. Could a heat sink be avoided if air is forced over the circuit at 5 m/s. $R_{\theta JA} = 62.5°C/W$.

15.10. Draw a circuit for microprocessor-based control of a high-power AC load using a pair of SCRs.

15.11. Complete the interface to a PIA for the SCR network in Fig. 15.25. Determine R_S and R_G for your interface. Assume $I_G = 100$ mA.

15.12. TRIACs switch AC loads. Can power transistors in the circuits presented here do the same? Why?

15.13. The MAC212 TRIAC is to be interfaced to the PIA similar to Fig. 15.30. Select a general-purpose drive transistor from Chap. 9. Size resistors R1, R2, and R3. Attempt to replace the buffer with a pull-up resistor.

15.14. Repeat Prob. 15.13 if the selected transistor has a minimum $h_{FE} = 150$.

15.15. A common problem with sheet paper conveying systems in copying machines is the inadvertent transport of two sheets of paper instead of one. A gap optocoupler can detect copy numbers by paper opacity. Devise a circuit to operate the 200 V, 6-A AC drive motor, using the optocoupler as a switch to activate the motor only if a single sheet of paper is present. An AC power supply is available.

15.16. Evaluate the possibility of using either transistor-optocouplers or Darlington optocouplers to replace the buffer for motor drive circuit in Fig. 15.7.

15.17. The Sprague motor driver in Fig. 15.41 is described as a full bridge drive. What does this mean? Can the motor speed be controlled? Is it an open-loop or a closed-loop system?

15.18. Figure 15.44(b) shows a power op amp in a torque control circuit for a DC motor. Develop the feedback block diagram for the system represented by this circuit.

CHAPTER
16

MICROPROCESSOR SYSTEM ON A BREADBOARD

A complete microprocessor-based control system has three ingredients:

1. Digital (microprocessor, memory, and ports)
2. Outside world (motors, displays, etc.)
3. Interface (analog, power electronics, etc.)

Each ingredient was discussed separately in previous chapters. Interface wiring diagrams between the PIA and the outside world were explained in detail. Memory and PIA lines to the microprocessor were given pin-by-pin together with typical software to address peripherals. Yet our understanding is not complete until we see how these elements can be combined into a working unit. This chapter illustrates the designing and building of a complete 6802 microprocessor system.

Alternatives to a 6802 system should be considered by the designer. The 6802 operates at the rate of several microseconds per instruction. Shorter program execution time or higher data transfer rates may require alternative MPUs with faster clocks. In addition, chip count and cost is reduced if the task can be performed with microcontrollers. In some instances a system of gates, flip-flops, and programmable logic arrays (PLAs) only can handle a control problem, but a microprocessor system is likely to be easier and cheaper to implement than the former design.

A systematic design methodology consists of four steps:

Problem statement. The design begins with a complete description of what the system should accomplish. Questions must be resolved with the customer. What are the performance requirements? What are the system inputs and outputs? How does a human interact with the system? What type of errors can occur? Are there any constraints? What are the alternatives?

Hardware design. Hardware design is dominated by memory and I/O. EPROMs (or ROMs) are needed for program storage. Is the 6802 on-board RAM necessary? Data collection schemes will probably require additional RAM. How much memory is necessary? Assignment of memory space should be carefully considered. What location is assigned to the stack? Where is the I/O located? The EPROM must be assigned at high addresses for reset and interrupts. Memory allocation is crucial to address decoding chip number.

The number of I/O lines will dictate the number of PIAs. Can PIA control bits be used for I/O? Do peripheral chips have on-board latches to bypass a PIA? Are interrupts used?

Always strive for the minimum number of chips. It is more cost-effective to replace hardware with software.

Software design. Software is written after the hardware is designed. Memory space allocation must be known as well as PIA pin function. Real-world hardware interface dictates how the programming is structured.

The software designer should use a top-down approach to programming. The total system function is partitioned into less complex subfunctions which perform specific tasks. Subfunctions are further subdivided. At the lowest level the programmer can rely on a library of tested subroutines for look-up tables, multiplication, number conversion, etc.

When a complete system is undertaken, program length can make hand assembly a tedious process. It is easier to write assembly language on personal computers (PCs) with cross-assembler software which will edit and generate the machine code. Most EPROM programmers will accept code directly from the PC.

Testing the design. A system that works perfectly the first time will be unusual. Errors should be located one at a time and the system should be retested after each correction, a process called *debugging*. The only reasonable strategy is bottom-up debugging.

Testing first divides the problem into software or hardware. An EPROM with only enough instructions to cause HIGH/LOWs on a PIA port is used to replace the system EPROM. Failure of the port to respond signifies a hardware problem. ICs represent modules that can be debugged individually for wiring mistakes.

In bottom-up software testing subroutines are debugged first. Then groups of related subroutines are tested until the entire software operates as a unit. The student should resist any temptation to "burn in" a new EPROM after finding a mistake. Software errors are usually multiple. Software testing can be done with simulation packages for the PC or with microprocessor trainers in a single-step mode. Alternatively, the EPROM can be programmed in a series of test routines.

16.1 PROBLEM STATEMENT

The project undertaken in this chapter is a digital readout bathroom scale similar to the unit described in Chap. 1. The prototype system will be constructed on a breadboard for testing and evaluation. The following requirements are to be met.

Input/output

1. Weight is sensed by a strain gauge transducer. The transducer analog signal should be conditioned for an analog-to-digital converter (ADC).
2. Weight is viewed on three 7-segment LED displays.
3. The scale is activated with a pushbutton on the front. It stays ON for 10 s, then shuts OFF.

Software

1. After the unit is activated, the right display shows zero and the two left displays are blank.
2. When the scale is stepped on, the message "ICU" appears for 3 s. The delay allows for the mechanical system to settle.
3. The unit then averages a series of readings and displays the result every second until time-out.
4. Maximum range of the scale is 250 lbf. If this weight is exceeded, the message "2HI" appears on the displays.

The designer should minimize the number of ICs to reduce the selling cost when the unit is put into production. No additional features are warranted for future expansion. The unit is powered with a 9-V battery.

16.2   HARDWARE DESIGN

The hardware design is relatively straightforward. The strain gauge analog output is amplified with an op amp and is converted to digital with an ADC. Three 7-segment displays (common anode) are driven with a PIA in the multiplex mode. Three general-purpose transistors control individual displays. Thus, two ports are used instead of three. A look-up table is used in place of decoder/driver ICs.

The single PIA must be dedicated to the displays. In order to save a second PIA, we choose an ADC with an on-board latch which can interface directly to the data bus. No interrupts will be employed. However, some scratch pad RAM is usually necessary. Memory will include the 6802 on-board RAM in addition to an EPROM for the program. Address decoding is handled by a 3- to 8-line decoder and support gates.

Battery supply and ON timer hardware are presented as a separate item. The breadboard system $\overline{\text{RST}}$ is activated by the simple method shown in Fig. 8.4. Several $\overline{\text{RST}}$ switch debounce schemes are offered later.

Table 16.1 gives a summary of the components in the breadboard system. We now discuss each of the IC subsystems and their pin assignments.

Power Supply

120-V AC LINE. TTL compatible ICs operate on a 5-V DC supply. To make a DC voltage source from an AC line voltage, several components are arranged as shown in Fig. 16.1(*a*).

TABLE 16.1
Breadboard system components

Component	Type	Quantity	Remarks
MPU	MC 6802	1	On-board RAM
PIA	MC 6821	1	Two ports
EPROM	2716	1	2048-byte storage
ADC	NS ADC0809	1	8-bit with latch
Decoder	74LS138	1	3- to 8-line
NORgate	7404	2	2-input, 4 to a package
Inverter	7402	1	6 to a package
Transistors	general-purpose	3	To carry 100 mA
Displays	5082–7651	3	Common anode
Resistors	10 k	2	Pull-up
Resistors	120 Ω	7	Current limiting LEDs
Resistors	220 Ω	3	Current limiting transistor drive
Op amp	779	1	Single supply voltage
Resistor	1 k	1	Op amp
Resistor	576 k	1	Op amp
Potentiometer	10 Ω	1	Op amp
Capacitor	27 pF	2	XTAL
Capacitor	0.1 μF	1	Reset
Crystal	4 MHz	1	

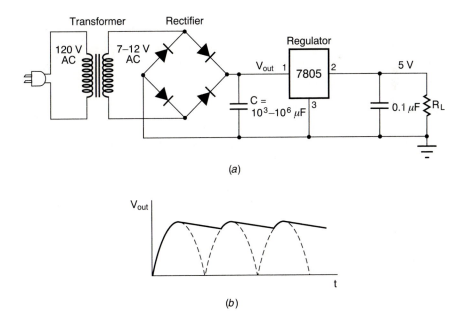

(a)

(b)

FIGURE 16.1
5-V power supply from 120-VAC line. (a) Components; (b) waveform after rectifier.

Line voltage is first stepped down from 120 V AC to a voltage level close to the desired 5 V with a transformer. At this point the voltage is still AC. The supply is then rectified to DC with a full wave bridge rectifier.

The waveform from the rectifier appears as the series of positive half cycles in Fig. 16.1(*b*). Output ripple from the rectifier is too high for most electronic circuits although small motors and lamps are not adversely affected. Ripple is further reduced with a large capacitor C in parallel; typical size is several thousand microfarads. Capacitance stores charge and supplies current between peaks of the rectified voltage. If V_r is the ripple voltage,

$$\frac{V_r}{V_{DC}} \cong 4CR_L f \text{ (no regulator)}$$

where R_L is the load resistance and f is the line frequency.

Most power supplies are regulated because the supply voltage will change if the load resistance or AC line voltage changes. Also, regulation eliminates ripple. Here we use a 7805 voltage regulator which will maintain a 5-V output over a wide range of input voltages. The three-pin (input, output, ground) device handles up to 1.5 A if it is properly heat sunk. A thermal shutdown circuit on-board will turn off the regulator if the heat sink is too small. The 0.1-μF capacitor helps trap voltage spikes that bother logic circuits.

Harris Semiconductor offers a single chip power supply to convert 120 V AC to 5 V DC, the HV-1205. Output current is limited to 50 mA. Many companies make integrated monoliths with much higher currents.

BATTERIES. Common dry cell batteries come in three grades: general-purpose (ZnC), heavy duty (ZnCl$_2$), and alkaline (ZnMnO$_2$). Alkaline batteries cost more, but they have almost twice the energy per unit volume (2.5 vs. 4.5 Wh/in^3). Table 16.2 gives the milliamp-hours of energy at rated voltage. Batteries can be wired in parallel for more current or in series for more voltage.

How much current does the proposed system draw? Table 16.3 gives typical supply current requirements for each IC in the system. In addition to the ICs, each display segment will draw an average 7 mA (21 mA one-third of the time) by design. The most segments that can be ON at any given weight is 16 (the number 250), assuming preceding zeros are blanked. Therefore, an average current demand for all possible numbers is approximately 56 mA. Total current demand for the system is 406 mA. For a 9-V transistor battery (500 mAh), a 10-s reading will result in approximately 420 weighings.

TABLE 16.2
Current ratings of common alkaline batteries

Type	Voltage (V)	Size	Current × time (mAh)
Penlight AA	1.5	2″ × 0.5″ diameter	900
Standard C	1.5	2″ × 1″ diameter	3,000
Standard D	1.5	2.3″ × 1.3″ diameter	10,000
Transistor	9	1.75″ × 1″ × 0.625″	500
Lantern	6	2.5″ × 2.5″ × 3.75″	9,000

TABLE 16.3
Current demand of system components

Component	Current	Component	Current
6802 MPU	150 mA	LS gates	< 3 mA
6821 PIA	110	ADC	25
2716 EPROM	57	Three displays	56
74LS138 Decoder	6	Total	406 mA

TIMED POWER ON CIRCUIT. A circuit for timed power ON is illustrated in Fig. 16.2. The battery is wired to the regulator input through a transistor switch. A 555 CMOS timer controls the transistor ON time. The timer is wired in the monostable or one-shot mode. Each time the START button is pushed, a 10-s positive pulse turns ON the transistor.

The 6802 MPU

The 6802/6808 are 8-bit microprocessors which have the same registers and instruction set as the 6800 except the external clock for the 6800 has been moved on-board. Both microprocessors require an external crystal to drive the clock. The 6802 differs from the 6808 by having 128 bytes of RAM on-board. RAM is essential for any software that uses the stack or temporary storage:

1. JSR instructions
2. Interrupts
3. PSH, PUL instructions
4. Memory write instructions

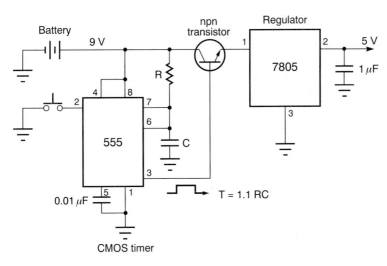

FIGURE 16.2
Circuit to shut off system 10 s after push button is activated.

Pin assignments for the 6802 were given in Fig. 3.1. Many of these pins have been discussed. Several new pins need to be explained. All inputs and outputs are TTL compatible with one standard TTL drive capability.

Revisited pins

Name	Pin No.	Function
A0–A15	9–20, 22–25	Address
D0–D7	33–26	Data
V_{CC}	8	Power (5 V)
V_{SS}	1, 21	Ground
E	37	$\phi2$ clock control
VMA	5	Valid memory address control
R/$\overline{W}$	34	Read/write control
$\overline{IRQ}$	4	Maskable interrupt
$\overline{NMI}$	6	Nonmaskable interrupt
$\overline{RST}$	40	Start/reset

New pins

$\overline{HALT}$ (2)   When this input is LOW, all MPU activity stops. VMA will be LOW, and the address bus will display the address of the next instruction. Any interrupt occurring in a halt state will be latched and serviced as soon as the MPU is taken out of the halt mode. Normally, the line is simply tied HIGH to 5 V. The input must be pulled LOW (use a pull-up resistor) by external circuitry when an idle state is desired.

MR (3)   Memory ready (MR) controls the stretching of clock E. In rare cases it may be desirable to slow the $\phi2$ control signal, thus allowing an interface to slow memory. Special circuitry for the procedure is given in the data sheet on the 6802. Normally, the line is tied HIGH.

BA (7)   Bus available (BA) is an output line. It is LOW when the MPU is using the buses. In the intervals between fetching and decoding instructions this line is HIGH. The purpose is to signal external peripherals when the buses are free for memory access (called cycle stealing) while the MPU is operating. Advanced techniques must be mastered for this application. Normally, the line is not connected (NC).

V_{CC} (35) (Standby)   This pin supplies voltage to the first 32 bytes of RAM, drawing only 8 mA. Thus, memory can be retained in this portion of RAM during power-down if desired. Normally, the line is tied directly to V_{CC}.

RE (36)   RAM enable (RE) controls the on-board RAM. When placed in the HIGH state, the on-board RAM at addresses 0000–007F is enabled. If the RAM is not used, the pin should be tied to ground.

XTAL (38) EXTAL (39)   These inputs are for the external quartz crystal, a $1/2'' \times 1/2'' \times 1/4''$ silver box with two leads. It times the 6802 internal oscillator. The oscillator incorporates a $\div4$ frequency divider; thus, the internal clock frequency is one-quarter the indicated crystal frequency. Allowable crystal frequencies are 0.4 to 4 MHz inclusive (up to 8 MHz for the 68B02 MPU).

Connections. Connecting the 6802 MPU to the remaining system is relatively straight-forward. The on-board RAM will be found useful, if not necessary, for most projects. There are no interrupts. Besides the RAM enable and crystal inputs, the new pins are rarely needed. Unused *input* lines should be tied to power or ground as appropriate.

Figure 16.3 demonstrates the 6802 connections. The EPROM calls for only 11 address lines: A0–A10. The top three address lines are reserved for decoding. Unused output lines A11, A12, and BA are not connected.

A 4-MHz crystal is placed across pins 38 and 39 in no preferred orientation. Two 27-pF capacitors between pins and ground are specified by the 6802 data sheet.

The $\overline{\text{RESET}}$ or $\overline{\text{RST}}$ pin must be brought LOW and held for three clock cycles. When the $\overline{\text{RST}}$ is released to a HIGH state, the MPU starts execution at the $\overline{\text{RST}}$ vector address and continues to run as long as the pin remains HIGH. A technique for implementing the procedure simply uses a 10 k pull-up resistor. Inserting a pin-connecting wire into a ground socket inactivates the MPU with a $\overline{\text{RST}}$ LOW. Pulling the wire from the socket promotes a $\overline{\text{RST}}$ HIGH through the pull-up resistor.

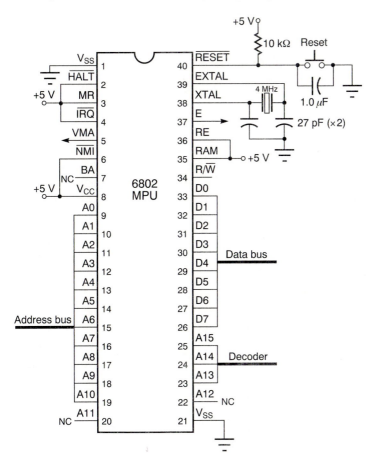

FIGURE 16.3
Pin connections for system microprocessor.

RST Methods

The preceding method for start/reset has certain drawbacks. First, switches (or pulled wires) bounce. If a bounce OFF period is less than the required three clock cycles, the MPU operation may be improper. However, a bounce this short is extremely unlikely. More important, a switch bounces 4 to 10 times over a period of approximately 20 ms before contact is permanent. Will a burst of multiple starts be detrimental to system behavior? Not in the present project, but start switch bounce should be weighed in any system design. Methods to eliminate false starts have included latches, flip-flops, timers, and simple capacitors. Figure 16.4 shows a 555 timer circuit for debouncing the start/reset switch. Figure 10.7 illustrated the use of a 7474 flip-flop.

A second possible RST feature is power-ON start/reset. Figure 16.5 shows a circuit which will start the system operating upon power-ON. The resistor-capacitor charges as a first-order circuit. Therefore, RST goes from logic LOW to HIGH in approximately one time constant RC. The problem is that many power systems themselves power-up as a first-order system, and the MPU power pin V_{CC} must be at least 4.75 V before a RST commences. Line voltage-to-DC systems can take as long as 100 ms (ripple capacitance × system resistance) or more to reach this level. Thus, RC should have a comparable value.

Once the system is operating, it can be manually reset by the push-button switch. Pressing the switch discharges the capacitor to 0 V and starts another charge process to logic HIGH. Note the switch is automatically debounced by any time constant over 20 ms.

2716 EPROM

Pin connections for the 2716 EPROM are given in Fig. 16.6. The 25-V line V_{PP} for programming purposes should be tied to the 5-V source. Programming pulse pin PGM should be grounded. All address and data lines should be tied to their respective buses.

Inexperience with EPROM software is a common reason that systems fail to work. *Do not attempt to store data into the EPROM address area.* It is a read-only memory

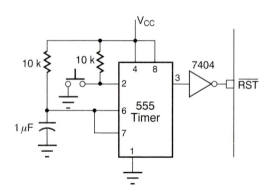

FIGURE 16.4
555 timer to debounce start/reset switch.

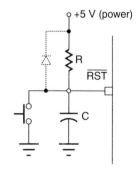

FIGURE 16.5
Power-ON start/reset circuit.

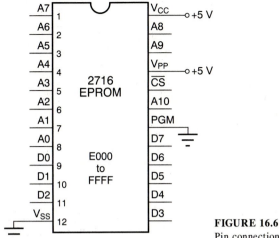

FIGURE 16.6
Pin connections for system EPROM.

(ROM). This mistake is difficult to detect because program testing will likely be done on a RAM. Thus,

STA (port) is valid.

STA (RAM) is valid.

STA (EPROM) is *not* valid.

As an example, the look-up table routine in Sec. 4.7 includes the line, "STA A OFFSET" where address OFFSET is within the routine. If the routine was thoughtlessly placed into the EPROM, it would fail.

Look-up table for EPROM (with 6802 RAM)

```
LDX     #TABLE
STX     RAM
LDA A   INPUT
STA A   (RAM + 1)
LDX     RAM
LDA A   0,X
STA A   OUTPUT
```

6821 PIA and Displays

Three displays are driven in multiplex mode by a 6821 PIA. Each common anode display is switched ON by its own transistor. Port B with PB0, PB1, and PB2 controls the hundreds, tens, and units display, respectively. Figure 16.7 shows the hundred display pin connections. The transistor switches power to three pins which route power to seven LED anodes. Cathodes are wired to port A through 120-Ω resistors, which limit the PIA sink

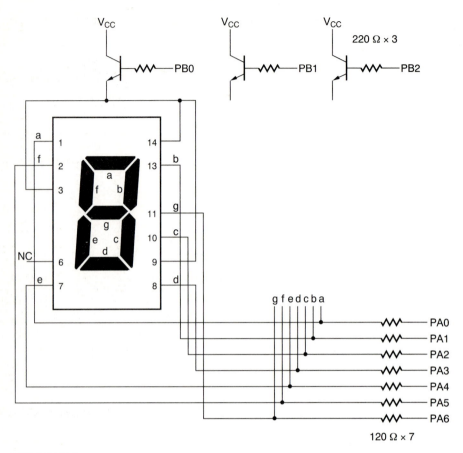

FIGURE 16.7
Pin connections for system displays.

currents to 20 mA at 33% duty cycle. Common anode display code will light the correct digit when segment designations *a* to *f* are tied to PA0 to PA6, respectively. The tens and units displays are connected similarly with exception of the transistor drive source.

Figure 16.8 gives pin connections for the PIA. Port A is common to all displays. Resistors (220 Ω) between port B and the transistors limit base current, and thus transistor power dissipation. Only one chip select $\overline{CS2}$ is used for decoding. Interrupt and program control pins, as well as the remaining port B pins, are not connected (NC).

0809 ADC and Op Amp

A latch version of an ADC is recommended to save an additional PIA. We are familiar with the 0809 ADC. Pin connections and local decoding for the project are given in Fig. 16.9. The 74LS138 decoder sends a LOW when the ADC is addressed. A pair of NOR gates combines this LOW with the MPU read/write signal to operate the ADC.

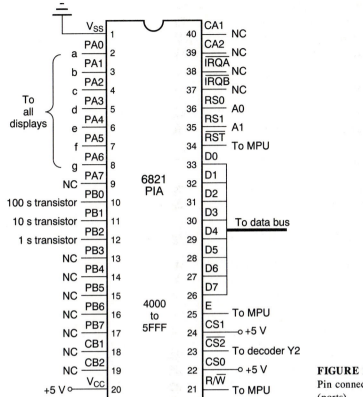

FIGURE 16.8
Pin connections for system PIA
(ports).

During a write—STA (address)—the second NOR-gate output produces a positive pulse to both START and ALE. Analog signal IN0 is latched into the ADC, and conversion starts. To select line IN0, a binary 000 must be present on pins 23 to 25. Ground these pins permanently because we do not want to multiplex inputs.

During a read—LDA (address)—the first NOR gate produces a positive pulse. Thus, the output buffers are enabled, and the binary conversion is placed on the data bus for the MPU. The end of conversion pin EOC is not connected because reads will be spaced much farther apart than the $100\text{-}\mu s$ conversion time.

Input IN0 is from a 799 op amp which supplies gain for the strain gauge voltage. The 799 is a general-purpose, single voltage supply amplifier. The strain gauge output is 0 to 10 mV for a weight of 0 to 300 lbf. The ADC output is FF for a full-scale input of 4.98 V. If weight corresponds to the base 10 representation of the ADC output, then the op-amp gain is given by

$$\text{Gain} = \frac{4.98 \text{ V}}{10 \text{ mV}} \times \frac{300}{255} = 1 + \frac{R_f}{R_1} \qquad \text{(noninverter)}$$

or $R_f/R_1 = 585$. Laboratory system tests can run without a strain gauge by applying voltage directly to IN0.

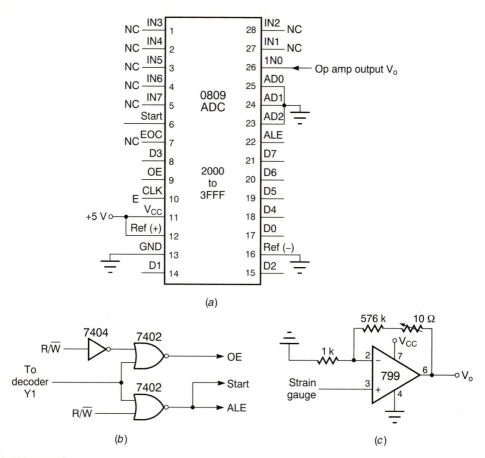

FIGURE 16.9

Pin connections for system analog-to-digital converter. (*a*) ADC; (*b*) decoding; (*c*) op amp for sensor conditioning.

74LS138 Decoder

Decoding circuits earlier in the textbook used gates, particularly NAND-NOT combinations, to decode space for individual ICs. For a complete system, the 74LS138 decoder offers the simplest method to assign address space. Figure 16.10 shows the pin connections.

Three inputs ABC are decoded into eight outputs Y0 to Y7. Only one output is LOW at any given time. That output number corresponds to the binary number on the inputs. Thus, output Y5 is LOW when ABC = 101. By connecting address lines A15–A13 to ABC, respectively, the address space is divided into eight regions. See Table 16.4.

The EPROM must be assigned to the uppermost space because it contains the reset vector FFFE–FFFF. Space 0000–1FFF cannot be assigned to any ICs because the MPU's on-board RAM uses 0000–007F. ADC and PIA assignments are entirely arbitrary.

There are three enables or chip selects for the 74LS138. VMA enables the decoder through enable G1. LOW enables $\overline{G2A}$ and $\overline{G2B}$ are tied to ground.

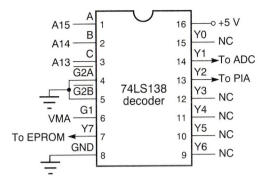

FIGURE 16.10
Pin connections for system decoder.

TABLE 16.4
74LS138 address space decoding

A15	A14	A13	Output (LOW only)	Address space	IC
1	1	1	Y7	E000–FFFF	EPROM
1	1	0	Y6	C000–DFFF	
1	0	1	Y5	A000–BFFF	
1	0	0	Y4	8000–9FFF	
0	1	1	Y3	6000–7FFF	
0	1	0	Y2	4000–5FFF	PIA
0	0	1	Y1	2000–3FFF	ADC
0	0	0	Y0	0000–1FFF	

16.3   SCHEMATIC TO PRINTED CIRCUIT

Now that we have examined the pin connections for individual ICs, the components must be joined together into an overall wiring diagram for the system. In the trade the diagram is called a *schematic.* In lieu of a simple hand sketch it is common practice to draw the diagram with any number of software packages made for PCs. These packages rely on an extensive library of IC pin-outs that can be retrieved and located on the drawing with the aid of a mouse. Figure 16.11 offers a schematic of the bathroom scale project drawn with typical software. Pin assignments are not necessarily in numerical order but are arranged to show the clearest interconnections between ICs. Most software provides a parts list from the drawing.

Since the 1940s all electronic circuits have been assembled on a printed circuit board. With a *printed circuit,* connections between mounted components are formed by strips of copper foil bonded to an insulating board. The circuit connections are created by a printing and etching process which virtually eliminates wiring errors in the manufacturing.

A fiberglass board is first clad on one side with a thin sheet (25 to 50μm) of copper. The copper is then coated with a layer of "negative" acid-resistant photographic emulsion, called photoresist. A photograph of the desired circuit connections, called a mask, is superimposed on the foil. In the next step the mask is exposed to ultraviolet light, causing the image to be transferred to the photoresist. The board is immersed in a copper etchant ($FeCl_4$), removing all underlying copper that is not protected by the imaged photoresist. After remaining photoresist and etchant are removed, a copper pattern of the mask only is

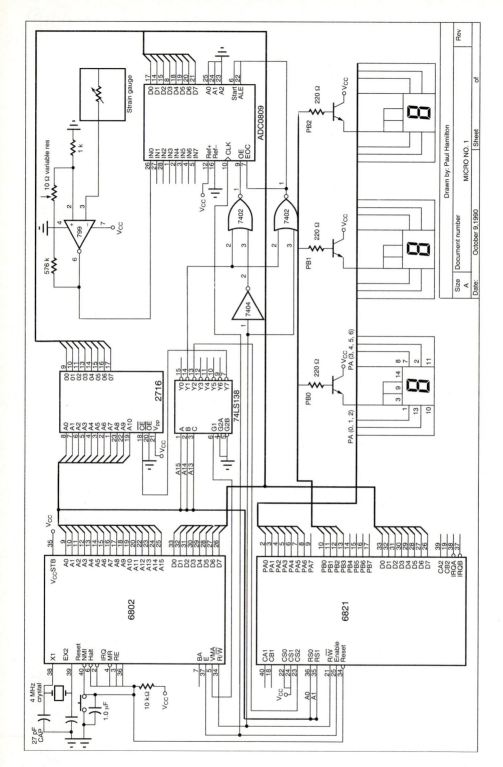

FIGURE 16.11
Schematic of bathroom scale system using the 6802 microprocessor.

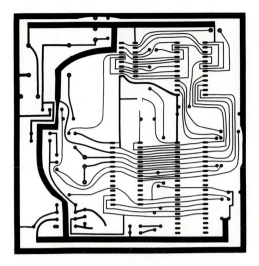

FIGURE 16.12
Circuit board art for a microcontroller programming module. (*Courtesy of Motorola, Inc.*)

left on the board. In the last step holes are drilled into the board to mount the components. Component pins are soldered to the copper mask in a single-step operation where the foil side is dipped into a bath of molten solder.

The intermediate step between schematic and printed circuit board is the "artwork" for the mask. Figure 16.12 shows an example of artwork. In the past artwork was done by the designer in draftsmanlike fashion. Today, software can be found for PCs which simplify the task. A widely used product is smARTWORK by Wintek Corporation, Lafayette, IN; smARTWORK outlines the components, numbers the pins, and automatically connects selected pins. Proper trace widths, spacing, and pin footprints are drawn for you. The final production design can be generated with dot-matrix or laser printers. HiWIRE-Plus by Wintek combines schematic and artwork software in a single package.

Printed circuit boards should not be a concern for the designer. Small vendors, which specialize in the manufacture, can be found in most cities. Vendors can also be found to perform the artwork from a customer schematic.

16.4 BREADBOARD PROTOTYPE

Before the printed circuit board is manufactured, a prototype of the system must be built to validate both hardware and software designs. It would not be unusual to find that major modifications are necessary. The ideal prototype vehicle is a patchboard or *breadboard* where circuitry can be easily constructed and altered. Generally, the ICs are placed on the breadboard in the same configuration as the schematic. Figure 16.13 illustrates a breadboard layout of the bathroom scale project.

A breadboard consists of rows of solderless sockets (holes) spaced (0.1 in pitch) to accommodate all IC pins. Spring clips are embedded in the plastic carrier to grip each pin or connecting wire. The actual connectors should be #22-AWG equipment wire with the ends stripped approximately $\frac{3}{8}$ in. Components such as capacitors, resistors, transistors, etc. are inserted directly into the spring-clip sockets.

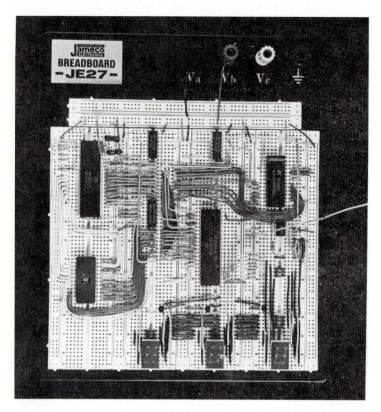

FIGURE 16.13
Breadboard layout of bathroom scale system.

Each breadboard is assembled from several of two basic carriers. The main carrier has a deep groove or well running down the middle lengthwise. JameCo's J-27 breadboard in the figure has four main carriers. ICs are inserted by straddling the insulated groove. Radiating outward from the groove are rows of five sockets with a common contact. Once the IC is in place anywhere from two to four common sockets will be exposed (depending upon IC width) for connector wires to each pin. A second plastic carrier has two rows of sockets. *Each row* from end to end is common, i.e., electrically connected. They are typically used for 5-V power and ground. The J-27 board has two horizontal and five vertical small carriers.

Wiring Tips

1. *Start with a schematic which shows all IC pin numbers and connections between ICs.* Group similarly on the breadboard.
2. *Make sure the IC pins are straight.* Carefully but firmly push the IC into the sockets until it is solid against the carrier. One pin bending underneath the IC and *out* of its socket is not unusual.

3. *Color-coded wiring is helpful.* Keep the length just long enough to route *around* the IC but stay flat on the board. You may want to test individual pins or even remove an IC. The first troubleshooting a student wants to do is to replace an IC. IC failure is rare. When ICs are removed, rock them out alternately from each end. Any attempt to remove in one thrust invariably rotates the IC at right angles to the board, leaving bent pins.

4. *Wire all V_{CC} and ground pins first.* Red and black connectors are commonly used. A 0.1-μF capacitor between V_{CC} and ground is optional in prototype or finished designs to despike the supply voltage. They are recommended for input lines where environs are noisy, i.e., automobile control modules.

5. *Wire the data and address lines next.* Check each connection several times as you proceed. A typical wiring error is the reversal of adjacent wires at the terminal IC, particularly data lines. If blue is the color code for data lines, alternate blue and violet. Similarly, alternate yellow and orange for address lines. Of course, most wire supplies may not permit this versatility.

6. *To prevent ground loops* (several grounds at different potential), *wire each IC ground to a ground carrier, not to another IC ground.* Digital and analog grounds should be separate. See Fig. 16.14. Each component ground should connect directly to the digital or analog ground (single-point grounding). Since both systems share a common power supply, the two grounds must be common at the power supply.

The ideal method to separate the microprocessor system from high-power motor systems controlled by transistors is optocouplers. Separate power sources and ground returns

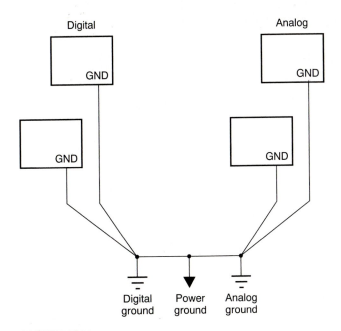

FIGURE 16.14
Grounding of digital and analog circuits in one system.

are possible because the only link between the digital system with the LED and motor circuit with the phototransistor is a light beam.

16.5   SOFTWARE DESIGN

A systematic software design strategy first identifies major functional modules where tasks are clearly defined. Each module is then subdivided into more easily managed submodules. The subdividing process is continued until a level is reached where the algorithm is straightforward, e.g., a timing loop.

There are as many software solutions as there are designers. The only meaningful difference between designs is execution time. Few control applications are so critical that a 10 percent reduction in program length will make a difference. Therefore, don't waste time (and development costs) by trying to improve a *workable* design.

Begin with a review of the software requirements in Sec. 16.1. Besides the routines for time delays and message displays, three fundamental algorithms should be considered. First, input from the ADC is binary. But the three 7-segment displays give the weight in decimal, i.e., three BCD numbers. Therefore, a binary-to-BCD subroutine will be offered to the reader. The subroutine will be entered with the binary number in accumulator A. The three BCD numbers will be stored in the on-board RAM. Locations 0000 to 0002 are assigned to the BCD outputs. Second, averages are accomplished by sum and divide. The fastest method to divide is register shifting, i.e., divide by 2, 4, 8, 16, etc. For example, add two numbers and shift the result once to the right. Add four numbers and shift the result twice to the right. If we add an 8-bit number to a 16-bit storage location 256 times, the rounded-off average appears in the HIGH-byte—or in the LOW-byte if we shift the result to the right eight times. Locations 0003 to 0004 are assigned to input additions with the average appearing in the HIGH-byte address. Third, a look-up table subroutine will be necessary, namely, the routine in Sec. 16.2. RAM locations 0005 to 0006 are assigned to look-up table scratch pad duties for the code vector address. After each BCD number is converted to code, the code is successively stored at locations 0007 to 0009.

Software Functions

Functions

1. Initialization
2. Get weight
3. Display

Subfunctions I

1. Initialization
   1.1 Save memory.
   1.2 Initialize registers, ports, etc.
   1.3 Load data.

2. Get weight

    2.1 Is it within resolution?

    2.2 Is it over 250?

3. Display

    3.1 Display zero.

    3.2 Display ICU.

    3.3 Display 2HI.

    3.4 Display weight.

Subfunctions II

1. Initialization

    1.1 Save memory space.

        1.1.1 Save space for BCD.

        1.1.2 Save space for output code.

        1.1.3 Save space for 16-bit average.

        1.1.4 Save space for look-up table scratch pad.

    1.2 Initialize registers and ports.

        1.2.1 Initialize SP.

        1.2.2 Initialize PIA.

    1.3 Load data.

        1.3.1 Load reset vector.

        1.3.2 Load look-up table.

2. Get weight

    2.1 Is it within resolution?

        2.1.1 No, display zero.

        2.1.2 Yes, display ICU for 3 s.

    2.2 Is it over 250?

        2.2.1 Yes, display 2HI.

        2.2.2 No, display average every one second.

3. Display

    3.1 Display zero.

        3.1.1 Load code for zero.

        3.1.2 Multiplex display.

    3.2 Display ICU.

        3.2.1 Load code for ICU.

        3.2.2 Multiplex display.

        3.2.3 Delay 3 s.

    3.3 Display 2HI.

        3.3.1 Load code for 2HI.

        3.3.2 Multiplex display.

    3.4 Display weight.

        3.4.1 Convert A/D to BCD.

        3.4.2 Convert BCD to code via look-up table.

        3.4.3 Multiplex display.

        3.4.4 Average A/D for one second as multiplexing.

Program

	ORG	F000	Program origin
HDRDS	EQU	0000	BCD weight
TENS	EQU	0001	
UNITS	EQU	0002	
AVE	EQU	0003	16-bit averaging
AVE+1	EQU	0004	
RAM	EQU	0005	Look-up table vector address
RAM+1	EQU	0006	
CODEH	EQU	0007	BCD code for display
CODET	EQU	0008	
CODEU	EQU	0009	
BINARY	EQU	000A	Scratch pad for BCD routine
INPUT	EQU	2000	Base address for A/D
PORTA	EQU	4000	Base addresses for PIA
CRA	EQU	4001	
PORTB	EQU	4002	
CRB	EQU	4003	
RESET	EQU	FFFE	Address for loading reset vector F000

Program starts at address F000

	LDS	#007F	Load stack pointer.
	LDA A	#FF	Initialize PIA; both
	STA A	PORTA	ports are outputs.
	STA A	PORTB	
	LDA A	#04	
	STA A	CRA	
	STA A	CRB	
ZERO	LDA A	INPUT	Is someone on the scale?
	CMP A	#01	
	BHI	ICU	Yes, go to ICU routine.
	LDA A	#FF	No, place code to display
	STA A	CODEH	two blanks and a zero.
	STA A	CODET	
	LDA A	#CO	
	STA A	CODEU	
	JSR	DISPLAY	Displays code for three characters for 1.3 ms each.
	JMP	ZERO	Repeat to multiplex.
ICU	LDA A	#F9	Place code to display ICU.
	STA A	CODEH	
	LDA A	#C6	
	STA A	CODET	
	LDA A	#C1	

	STA A	CODEU	
	LDX	#0300	Load count of 3 × 256.
DELAY3	JSR	DISPLAY	Display each character for 1.3 ms
	DEX		
	BNE	DELAY3	Count down for 3 s.
2HI	LDA A	INPUT	Is weight over 250?
	CMP A	#FA	
	BLS	WEI	No, go to weight display.
	LDA A	#A4	Yes, place code to display 2HI.
	STA A	CODEH	
	LDA A	#89	
	STA A	CODET	
	LDA A	#F9	
	STA A	CODEU	
	JSR	DISPLAY	
	JMP	2HI	Repeat to multiplex.
WEI	LDA B	#FF	Load counter for one second average.
	CLR	AVE	
	CLR	AVE+1	
	STA A	BINARY	
	JSR	BCD	Convert input to BCD.
	JSR	LOOKUP	Convert BCD to output code.
SUM	JSR	DISPLAY	Display characters for 1.3 ms each.
	LDA A	INPUT	Get input and add to 16-bit address.
	CLC		
	ADD A	AVE	
	STA A	AVE	
	LDA A	#00	
	ADC A	AVE+1	
	STA A	AVE+1	
	DEC B		
	BNE	SUM	Continue multiplexing for one second.
	LDA A	AVE+1	Get new weight.
	JMP	WEI	

Subroutine 1: Convert binary to BCD

	CLR	HDRDS	
BCD	CLR	TENS	
	CLR	UNITS	
LOOP1	LDA A	BINARY	
	SUB A	#64	Subtract one hundred.
	BCS	LOOP2	Did borrow occur?
	STA A	BINARY	No, save result as new binary.
	LDA A	HDRDS	Increment hundreds digit until a
	INC A		borrow occurs.

```
              STA A    HDRDS
              BRA      LOOP1
LOOP2         LDA A    BINARY      Binary now less than one hundred.
              SUB A    #0A         Subtract ten.
              BCS      LOOP3       Repeat process for tens.
              STA A    BINARY
              LDA A    TENS
              INC A
              STA A    TENS
              BRA      LOOP2
LOOP3         LDA A    BINARY      Only units left.
              STA A    UNITS
              RTS
```

Subroutine 2: Look-up table routine

```
LOOKUP        LDX      #TABLE      Convert BCD hundreds to code.
              STX      RAM
              LDA A    HDRDS
              STA A    RAM+1
              LDX      RAM
              LDA A    0,X
              STA A    CODEH
              LDX      #TABLE      Convert BCD tens to code.
              STX      RAM
              LDA A    TENS
              STA A    RAM+1
              LDX      RAM
              LDA A    0,X
              STA A    CODET
              LDX      #TABLE      Convert BCD units to code.
              STX      RAM
              LDA A    UNITS
              STA A    RAM+1
              LDX      RAM
              LDA A    0,X
              STA A    CODEU
              RTS
```

Subroutine 3: Multiplex code to display weight

```
DISPLAY       LDA A    CODEH       Latch hundreds code.
              STA A    PORTA
              LDA A    #01         Turn on hundreds display.
              STA A    PORTB
              JSR      DELAY       Delay 1.3 ms.
              LDA A    CODET       Latch tens code.
              STA A    PORTA
              LDA A    #02         Turn on tens display.
```

```
        STA A    PORTB
        JSR      DELAY      Delay 1.3 ms.
        LDA A    CODEU      Latch units code.
        STA A    PORTA
        LDA A    #04        Turn on units display.
        STA A    PORTB
        JSR      DELAY      Delay 1.3 ms.
        RTS
```

Subroutine 4: delay for 1.3 ms

```
DELAY   LDA A    #D8
        DEC A
        BNE      DELAY
        RTS
```

Load look-up table code

```
TABLE   FCB      C0, F9, A4, B0, 99, 92, 82, F8, 80, 98
```

Load reset vector address

```
RESET   FCB      F0,00
        END
```

16.6   TESTING AND DEBUGGING

Now that the microprocessor prototype system has been designed, built, and programmed, a certain suspense surrounds tripping the reset to activate the system. Even a small system will have hundreds of connections and lines of programming. Success is measured in how much discipline was used in applying what you know. Expect the system to function perhaps 20 percent of the time on the first try. Therefore, testing and debugging are the natural order of events in system design. The idea is to divide the testing problem into small hardware and software modules. The method is as personal and varied as the original design. We will offer a few suggestions.

Equipment

A wide variety of equipment and software is available for testing and debugging your breadboard. Logic analyzers and in-circuit emulators are expensive and take a large investment in learning time. Other tools are advocated for small projects. Every laboratory should have three items of test equipment: (1) a logic probe, (2) a multimeter, and (3) an oscilloscope.

Logic probes are handheld devices that have two LED lamp displays. If an individual line has a logic-LOW, its lamp will light. A logic-HIGH will light the other lamp. If the line switches between HIGH and LOW (pulses), the lamp that is predominantly HIGH or LOW will pulse, i.e., VMA will pulse HIGH. Some logic probes will detect invalid logic levels. Oscilloscopes can also check logic levels but not so conveniently.

A handheld multimeter can test circuit voltage, current, or resistance with the twist of a dial. Some units check transistors by inserting the three leads into openings. Analog voltages (particularly V_{CC} and ground) and unknown resistors or potentiometer settings are prime targets. Multimeters are especially useful when tracing short circuits and transposed wires.

The oscilloscope should be a high-quality dual trace, triggered-sweep, DC-coupled, analog device with a bandwidth of at *least* five times the microprocessor clock frequency (5-MHz scope for a 1-μs clock). More than two channels to view multiple lines is very desirable. MPU output lines are TTL compatible and should pull LOW to about 0.1 V and HIGH to about 4.0 V. Readings outside this range are probably shorted to ground or V_{CC}. Adjacent lines are shorted if they show the same waveform. If opposite ends of the presumed same line don't show the same waveform, they are transposed or open. A functioning MPU will have a repetitive clock output E of 1 μs. Decoders have pulsing address lines as inputs but the CE (or $\overline{CE}$) line will show a valid HIGH (or LOW) only when the decoded chip is addressed.

Software or Hardware

A strategy to correct a malfunctioning system should begin with several questions. Have good wiring practices been followed? Are ICs pushed tightly into the board? Have ground and power lines been run to each IC? Is V_{CC} at 5 V? It is easy to short the system by connecting a ground pin to the power bus, or vice versa. Are there any unused pins? A missing or transposed wire will be apparent if all nonconnected pins on each IC are verified. Have software modules on the microprocessor trainer unit been debugged?

Now a division process will reveal if the problem is hardware or software. First, disconnect the PIA port lines. A simple reliable routine is loaded into a separate EPROM. The PIA is programmed for output ports, latching 55 into port A and 01 into port B. If the digital side of the hardware is working, port A pins will be alternately HIGH and LOW while pin PB0 will be HIGH. The problem will be software, provided the I/O circuits are operating. Take disconnected port B lines which drive a display control transistor and connect them to PB0 one at a time. Each seven-segment display should show a meaningless pair of parallel segments. The isolated op amp and ADC pair can be tested separately by permanently enabling the ADC latch and comparing analog inputs to digital outputs.

The following program sends 55 to port A and 01 to port B:

```
          ORG      F000
          LDA A    #FF
          STA A    PORTA
          STA A    PORTB
          LDA A    #04
          STA A    CRA
          STA A    CRB
          LDA A    #55
          STA A    PORTA
          LDA A    #01
          STA A    PORTB
          WAI
RESET     FCB      F0,00
```

The 6802 MPU

If the preceding test program fails to operate the system, a hardware problem exists. The procedure is to isolate the troublespot(s): MPU, EPROM, decoder, or PIA. Start with the MPU.

A simple test for a functioning MPU is the following. Disconnect the address and data lines to prevent possible line shorts from interfering with the test. Connect ground to data pins D7 to D1. At the same time connect V_{CC} to data pin D0 through a 1 k resistor. The assembly language code for NOP or 01 is now permanently wired to the data bus.

Keep the reset switch closed. Even though the MPU is not fetching instructions, the MPU is powered and the clock is active. The following measurements can be observed with a logic probe:

Address bus:	FFFE
VMA:	LOW
R/W:	HIGH
E(Clock):	pulses

Open the reset switch. The MPU successively places the reset addresses FFFE and FFFF on the address bus, but only fetches 0101 for the starting address. The effective program being executed is NOP, NOP, NOP, ... Every two cycles, the program counter is incremented and placed onto the address bus. The starting address is irrelevant, but the bus behaves like a 16-bit counter. Address bit A0 pulses every two microseconds. Higher address bits slow their pulsing by a factor of 2. Line A15 pulses every $2 \times 65,536$ microseconds or 0.13 seconds. The probe will show higher bits pulsing but lower bits will show HIGH because the eye cannot detect rapid change.

The 6802 MPU and 2716 EPROM

Your MPU is now working properly. Address and data lines to the MPU should be reconnected. The next step troubleshoots the MPU/EPROM system. To test the system, any program will suffice providing it executes a predictable pattern on the address and data lines. Good diagnostic technique requires a short program that is easily observed with the oscilloscope or logic probe, preferably the latter.

The following program activates all address and data lines except A8 and D0.

Address	Content	Mnemonics
FEF7	CE	LDX #0000
FEF8	00	
FEF9	00	
FEFA	08	INX
FEFB	7E	JMP FEFA
FEFC	FE	
FEFD	FA	
FFFE	FE	Reset
FFFF	F7	Vector

In the first instance the reset vector is loaded and instruction (LDX) is executed. Thereafter, only addresses FEFA–FEFD appear on the line as the MPU continually executes a short loop. If these addresses are decoded to binary, the pattern will show all the address bits are active except line A8. Only A8 will register LOW with a logic probe. Remaining address lines will be HIGH. Similarly, data lines will have active bits on all lines except line D0. Only data line D0 will register LOW.

If the preceding pattern does not develop, the hardware problem is either the EPROM or the decoder. 74LS138 decoders are very easy to check. Isolate it and impose binary input groups using supply and ground. The decimal equivalent output line should go LOW. A satisfactory decoder will pinpoint the EPROM as a problem area. If the address/data pattern does develop, only the PIA remains a candidate.

PROBLEMS

16.1. Revise the bathroom scale software/hardware to turn OFF when an individual steps from the scale.

16.2. Revise the bathroom scale software/hardware for push-button ON and a repeat push for OFF.

16.3. A bathroom scale is to be designed with a lbf/kg switch. Implement the design with (*a*) software, and (*b*) hardware.

16.4. Modify the bathroom scale design to accommodate a 300-lbf limit.

16.5. Design an alternate system for the bathroom scale which uses the 74LS138 decoder to assign address space to each transistor. What latch options are available?

16.6. Devise a $\overline{\text{RST}}$ start circuit using (*a*) 555 timer, and (*b*) 7476 flip-flop with SPDT switch.

16.7. Write a look-up table routine for the EPROM which does not use the 6802 on-board RAM.

16.8. The bathroom scale design has low current in the display LEDs. The level is satisfactory for dim to moderate lighting but may be too low for some surroundings. How would you revise the design for 20-mA average current levels? How does higher current affect battery life?

16.9. Suppose we want to simplify the bathroom scale project for laboratory demonstration. Change the software to only continually display the input.

16.10. The EPROM is enabled by the decoder for EXXX to FXXX space. Yet the EPROM programmer enables the EPROM starting at address 0000. How do you explain the difference? Where is the reset vector loaded?

16.11. If you test the display multiplex routine on the trainer, how do you avoid the delay routine trap of endless cycles?

16.12. A test for the multiplex hardware was not specifically mentioned in the text. Establish one.

16.13. Design a microprocessor-based system for one of the following:
(*a*) A thermometer that saves the high and low temperatures over a 24-h period
(*b*) A bicycle computer to display speed and total mileage
(*c*) A microwave oven
(*d*) A three-axis robot driven by DC motors that takes position from a master computer latch
(*e*) An idea of your own

CHAPTER 17

MOTOROLA 6805
MICROCONTROLLER

Early 8-bit microprocessors, such as the 6800, have evolved in two directions: systems applications and dedicated control applications. Systems use the microprocessor for computation and data processing where microprocessors have grown in speed, power, and complexity to 32-bit wonders. Although older 8-bit microprocessors are still employed in control applications, the market is now dominated by microcomputers-on-a-chip, more commonly called microcontroller units (MCUs). Dedicated applications include appliances, automobiles, games, instrumentation, industrial controls, etc. Microprocessor power is not important, but cost reduction through integrated functionality is the driving factor. Microcontrollers combine a microprocessor, memory, I/O, and decoding on the same chip. Other on-chip functions include A/D conversion and timers.

Many companies successfully market microcontrollers today, including Motorola, Intel, Texas Instruments, and National Semiconductor. Motorola manufactures several microcontrollers which are upgraded versions of the 6802: the 6801 and the more recent 68HC11. Instruction sets are compatible with the 6802, and several new instructions have been added, including an unsigned multiply. Both can function as a stand-alone microcontroller or as a microprocessor in which several ports can be programmed to act as address and data buses.

The most popular microcontroller over the years is the Motorola 6805. It has a *downgraded* 6802 MPU and as few as 28 pins. These aspects translate into a cheaper controller, yet one powerful enough for most control applications.

6805 MCU vs. 6802 MPU

The 6805 microcontroller contains a simplified 6802 microprocessor unit. Briefly, their differences are summarized as follows:

1. *Accumulator B* is removed to free up valuable OP-code space. All instructions and addressing modes pertaining to the B register are eliminated.
2. The *V-flag bit* is removed from the conditional code register because small controllers generally do not need signed arithmetic operations. However, unsigned arithmetic operations are still available.
3. Because the controller addresses only small memories, the *program counter* and *stack pointer* are reduced in bit size.
4. The *index register* is reduced to 8 bits. This change in size requires new addressing modes so that all memory can be accessed through the X register. It also doubles as a second accumulator.
5. A comparison of the 6805 and 6802 instruction sets shows many additions and deletions, but the majority of the 6802 mnemonics remain intact. Instruction hexadecimal codes are not compatible.

17.1 THE 6805 FAMILY

The 6805 is not a single microcontroller but a family with different functions built into the chip. All have at least 64 bytes of RAM, a 1-k-byte ROM (or EPROM), a timer, and 20 bits of parallel I/O. Table 17.1 shows other versions of the 6805 series. Each version

TABLE 17.1
M6805 family selector guide. *(Reprinted with permission of Motorola, Inc.)*

Device	Technology	Pins	RAM (bytes)	ROM (bytes)	EPROM (bytes)	I/O	Timer bit	A/D	SPI	Packaging
6805P2	HMOS	28	64	1110	—	20	8	—	—	P, S, FN
6805P6	HMOS	28	64	1804	—	20	8	—	—	P, S, FN
68705P3	HMOS	28	112	—	1804	20	8	—	—	S
68705P5	HMOS	28	112	—	1804	20	8	—	—	S
6805R2	HMOS	40/44	64	2048	—	32	8	Yes	—	P, S, FN
6805R3	HMOS	40/44	112	3776	—	32	8	Yes	—	P, S, FN
68705R3	HMOS	40	112	—	3776	32	8	Yes	—	S
68705R5	HMOS	40	112	—	3776	32	8	Yes	—	S
6805S2	HMOS	28	64	1480	—	21	8	Yes	Yes	P, S, FN
6805S3	HMOS	28	104	2720	—	21	8	Yes	Yes	P, S, FN
68705S3	HMOS	28	104	—	3752	21	8	Yes	Yes	S
6805U2	HMOS	40/44	64	2048	—	32	8	—	—	P, S, FN
6805U3	HMOS	40/44	112	3776	—	32	8	—	—	P, S, FN
68705U3	HMOS	40	112	—	3776	32	8	—	—	S
68705U5	HMOS	40	112	—	3776	32	8	—	—	S

has unique features which are especially suited to different control applications. A suffix distinguishes one from another. MCUs with a *P* designation are the minimum unit. An *R* designation denotes an added ADC, whereas an *S* signifies a serial interface. EPROM versions have a "7" inserted in the number: 68705.

Motorola's first MCUs, the 6805 series, were produced in HMOS technology (high-density *N*-channel metal oxide on silicon). It is a highly efficient fabrication process which produces a low-cost, high-speed chip. The 146805 series followed using CMOS. This technology has very low-power consumption, and it is important for battery-powered applications. CMOS also operates over a wide range of supply voltages. However, since CMOS requires a larger silicon area than HMOS, it is more expensive and slower. In 1983 the introduction of the HCMOS series 68HC05 offered the best of both worlds—high-density and low-power consumption. Table 17.2 lists the CMOS and HCMOS series that are available. The HCMOS series, which is replacing the CMOS, has greatly expanded functions. It also has more pins and higher cost over the HMOS series. Two serial interfaces are offered. The serial communications interface (SCI) is used for long-range communications as in data transfer from an MCU to a terminal or modem. The serial

TABLE 17.2
M6805 HCMOS/CMOS family selector guide. *(Reprinted with permission of Motorola, Inc.)*

Device	Technology	Pins	RAM (bytes)	ROM (bytes)	EPROM (bytes)	EEPROM (bytes)	I/O	Timer bit	SPI	SCI	A/D	Packaging
68HC05A6	HCMOS	40/44	176	4160	—	2056	32	16	Yes	Yes	—	P, FN
68HC05B4	HCMOS	48/52	176	4160	—	—	32	16	—	Yes	Yes	P, FN
68HC05B6	HCMOS	40/52	176	5952	—	256	32	16	—	Yes	Yes	P, FN
68HC05C2	HCMOS	40	176	2096	—	—	32	16	—	—	—	P
68HC05C3	HCMOS	40	176	2096	—	—	32	16	Yes	Yes	—	P
68HC05C4	HCMOS	40/44	176	4160	—	—	32	16	Yes	Yes	—	P, FN
68HC05C8	HCMOS	40/44	176	7700	—	—	32	16	Yes	Yes	—	P, FN
68HC05L6	HCMOS	68	176	6208	—	—	32	16	Yes	—	—	FN
68HC05M4	HCMOS	52	128	4K	—	—	32	8/16	—	—	Yes	FN
68HCL05C4	HCMOS	40/44	176	4160	—	—	32	16	Yes	Yes	—	P, FN
68HCL05C8	HCMOS	40/44	176	8K	—	—	32	16	Yes	Yes	—	P, FN
68HSC05C4	HCMOS	40/44	176	4160	—	—	32	16	Yes	Yes	—	P, FN
68HSC05C8	HCMOS	40/44	176	8K	—	—	32	16	Yes	Yes	—	P, FN
68HC705C8	HCMOS	40/44	304	—	—	—	32	16	Yes	Yes	—	P, FN
68HC805B6	HCMOS	48/52	176	—	8K	6208	32	16	—	Yes	—	P, FN
68HC805C4	HCMOS	40/44	176	—	—	4160	32	16	Yes	Yes	—	P, FN
146805E2	CMOS	40	112	0	—	—	16	8	—	—	—	P, S, FN
146805F2	CMOS	28	64	1089	—	—	20	8	—	—	—	P, S, FN
146805G2	CMOS	40	112	2106	—	—	32	8	—	—	—	P, S, FN

Definitions:

P = plastic
S = cerdip
FN = plastic leaded chip carrier
I/O = input/output
A/D = analog/digital converter
SCI = serial communications interface

SPI = serial peripheral interface
RAM = random access memory
ROM = read only memory
EPROM = eraseable programmable ROM
EEPROM = electrical eraseable ROM

peripheral interface (SPI) is used primarily for serial communication with chips on the same printed circuit board.

MCU versions are being continually added to the HCMOS series. With all that variety, designers may not know every available version for their needs or the most cost-effective solution. In 1989 the microcontroller group of Motorola Inc., Austin, Texas, developed a disk-based questionnaire that runs on IBM PCs or Apple Macintosh computers. Designers can specify their ideal microcontroller. If a standard chip cannot be used, a customer-specific chip can be created with the company's design automation system.

In this chapter we concentrate on the 68705R3 MPU unit, which is an EPROM version with an ADC. Microcontrollers with EPROMs are primarily used for system development before ROM masks are set or for small production runs. Except for on-board functions, most of the chapter material on the 68705R3 pertains to all 6805 family members.

17.2   THE 68705R3 MICROCONTROLLER

Pin assignments and block diagram for the 68705R3 microcontroller are shown in Fig. 17.1. The 40-pin device has the following features:

> 112 bytes of RAM
> 3776 bytes of EPROM
> 32 I/O pins
> 8-bit timer
> Four channels of A/D
> Two interrupt pins

The 32 I/O lines are arranged into four 8-bit ports (A, B, C, and D). Ports A, B, and C are programmable as either inputs or outputs similar to the PIA. Port D is a fixed input port. It can be read directly as binary inputs, or it can double as four channels of an ADC and a second interrupt.

The timer pin controls the interval timer/counter circuitry. The timer includes a 7-bit prescaler that allows the timing interval to be extended by multiples of 2 up to 2^7.

The 68705R3 emulates the 6805R3 ROM version exactly. In the 68705R3 programming pin V_{PP} is tied to V_{CC} during normal operations, whereas in the 6805R3 this pin is not connected.

Memory Map

Like the 6802, the 6805 family uses a single address map with memory-mapped I/O. Since ports, timer, and ADC are on the chip, these peripheral blocks have been decoded for addressable locations by the manufacturer. Figure 17.2 shows the memory map for the 68705R3 EPROM version.

The 128 bytes of RAM are located at the beginning of the 12-bit address space 000–07F. The first 16 bytes are reserved for the port, timer, and A/D registers. Each of the ports requires a data register, although only the three programmable ports have data direction registers. Both the timer and A/D converter are given two addresses.

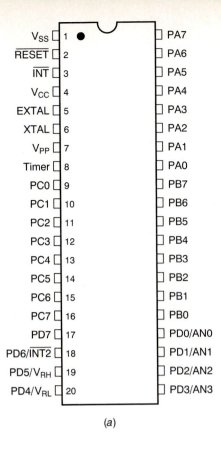

(a)

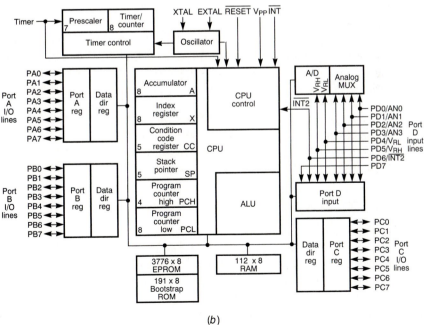

(b)

FIGURE 17.1

68705R3 EPROM microcontroller. (*a*) Pin assignments; (*b*) block diagram. (*Reprinted with permission of Motorola, Inc.*)

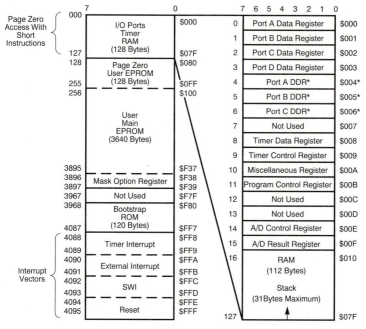

Caution: Data direction registers (DDRs) are write-only; they read as FF.

FIGURE 17.2
68705R3 microcontroller memory map. (*Reprinted with permission of Motorola, Inc.*)

The miscellaneous register at 00A is associated with interrupt INT2. Programming the EPROM temporarily stores information in the program control register at 00B, so the user does not need to be concerned with it. The remaining RAM is available for scratch pad, but the top of RAM is assigned to the stack.

EPROM locations start at address 080 and go to address F37. Many of the 6805 instructions support only the direct mode; thus, page zero takes on special significance for the programmer. The mask option register is utilized for timer operations. At the top of the memory map FF8–FFF various interrupt vectors are located as expected. All registers and vectors will be discussed later.

17.3 68705R3 MCU REGISTERS

The 6805 MCU register structure is shown in Fig. 17.3. There is one accumulator, an index register, a stack pointer, program counter, and a conditional code register. It is similar to the 6802, but aside from the accumulator, the registers are somewhat shortened. The internal address bus is sized according to the on-board memory space. Thus, the stack pointer and program counter registers can be 11-bits (2K memory) to 13-bits (8K memory) long.

Accumulator (A). Accumulator A is a general-purpose 8-bit register which performs the same functions as its counterpart in the 6802. The B accumulator is missing.

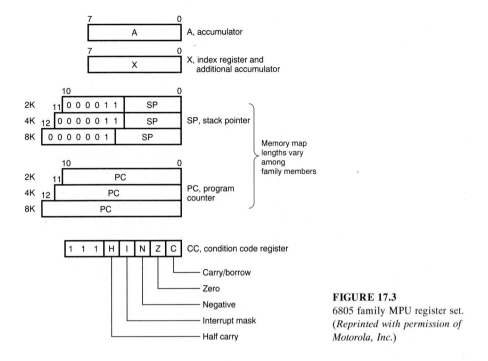

FIGURE 17.3
6805 family MPU register set.
(*Reprinted with permission of Motorola, Inc.*)

Index register (X). The index register has been reduced to 8-bits. Obviously, we would like to point anywhere in memory similar to what the 6802's 16-bit index register does. The problem is solved by adding an indexed offset mode of addressing to form an effective 16-bit address. An advantage of an 8-bit index register is that it can double as a second accumulator. A reduced index register and a single accumulator are used in the Apple's 6502 MPU.

Stack pointer (SP). The 11- to 13-bit stack pointers of the 6805 series point to the address of the next free location in RAM for temporary storage. A hardware reset or the reset stack pointer instruction (RSP) sets the stack pointer at location 07F, the top of RAM. The stack pointer is then decremented as data is pushed onto the stack and incremented as data is pulled from the stack. Unlike the 6802, the only stack instruction is a new instruction: reset stack pointer (RSP). Only subroutines and interrupts make use of the stack. They may be nested down to location 061 (31 bytes). The 7 MSBs of the stack pointer are permanently set at 0000011.

Program Counter (PC). The program counter holds the address of the next instruction to be executed. The address is placed onto the internal bus and the PC is incremented. The program starting address is obtained from the reset vector upon a hardware reset. Instructions are executed in order unless the PC is altered by a jump or branch instruction.

Conditional Code Register (CCR). The conditional code register on the 6805 is 5 bits long compared to 6 bits on the 6802. The V flag is not included in the 6805 family CCR because the MCU is intended for simple control applications, not number-crunching applications. Branch instructions for signed arithmetic have been dropped.

17.4   68705R3   INSTRUCTION SET

The 6805 family, including the 68705R3, has 59 instructions compared to 72 instructions for the 6802 MPU. Basic differences in the instruction sets are as follows:

1. Removal of accumulator B from the 6802 MPU registers has eliminated instructions ABA, CBA, TAB, TBA, and SBA.
2. Omission of the V-flag from the conditional code register has eliminated instructions BGE, BGT, BLE, BLT, BVC, BVS, CLV, and SEV.
3. Although the stack reset instruction RSP has been added to the 6805 instruction set, all other stack-altering instructions have been deleted: DES, INS, LDS, PSH, STS, TSX, TXS, and WAI.
4. Since the index register is now 8 bits, register transfer instructions are added: TXA, TAX.
5. Several instructions are deleted for no apparent reason: DAA, TAP, TPA. For programmers who need to add BCD numbers, the *Motorola M6805 HMOS M146805 CMOS Family Users' Manual* offers a software routine for DAA.
6. One instruction mnemonic is slightly altered: ASL to LSL. DEX and INX are no longer available, but their function is embedded in existing instructions (index mode): DEC X and INC X. Most assemblers recognize either form.

The instruction set can be divided into five different groups: register/memory, read-modify-write, branch, bit manipulation, and control. Figure 17.4 gives the mnemonic and its function for each group. A more detailed description is found in Appendix B.

Register/Memory Instructions

Most register/memory instructions have one operand in either the accumulator or the index register and a second operand in memory. Exceptions are the JMP and JSR instructions, which have no register operand.

Read-Modify-Write Instructions

These instructions read memory (or registers), modify its contents, and write the modified value back to memory. The TST instruction is an exception since it does not modify the value.

Branch Instructions

A number of new branch instructions are added. Branching on the interrupt mask flag and the half carry (H) flag in the conditional code register is now available. Branch Never (BRN), Branch if Higher or Same (BHS), and Branch if Lower (BLO) are added to pro-

Register/Memory Instructions

Function	Mnemonic
Load A from memory	LDA
Load X from memory	LDX
Store A in memory	STA
Store X in memory	STX
Add memory to A	ADD
Add memory and carry to A	ADC
Subtract memory	SUB
Subtract memory from A with borrow	SBC
AND memory to A	AND
OR memory with A	ORA
Exclusive-OR memory with A	EOR
Arithmetic compare A with memory	CMP
Arithmetic compare X with memory	CPX
Bit test memory with A (logical compare)	BIT
Jump unconditional	JMP
Jump to subroutine	JSR

Branch Instructions

Function	Mnemonic
Branch always	BRA
Branch never	BRN
Branch if higher	BHI
Branch if lower or same	BLS
Branch if carry clear	BCC
(Branch if higher or same)	(BHS)
Branch if carry set	BCS
(Branch if lower)	(BLO)
Branch if not equal	BNE
Branch if equal	BEQ
Branch if half carry clear	BHCC
Branch if half carry set	BHCS
Branch if plus	BPL
Branch if minus	BMI
Branch if interrupt mask bit is clear	BMC
Branch if interrupt mask bit is set	BMS
Branch if interrupt line is low	BIL
Branch if interrupt line is high	BIH
Branch to subroutine	BSR

Read-Modify-Write Instructions

Function	Mnemonic
Increment	INC
Decrement	DEC
Clear	CLR
Complement	COM
Negate (2s Complement)	NEG
Rotate left through carry	ROL
Rotate right through carry	ROR
Logical shift left	LSL
Logical shift right	LSR
Arithmetic shift right	ASR
Test for negative or zero	TST

Bit Manipulation Instructions

Function	Mnemonic
Branch if bit n is set	BRSET n ($n = 0 \ldots 7$)
Branch if bit n is clear	BRCLR n ($n = 0 \ldots 7$)
Set bit n	BSET n ($n = 0 \ldots 7$)
Clear bit n	BCLR n ($n = 0 \ldots 7$)

Control Instructions

Function	Mnemonic
Transfer A to X	TAX
Transfer X to A	TXA
Set carry bit	SEC
Clear carry bit	CLC
Set interrupt mask bit	SEI
Clear interrupt mask bit	CLI
Software interrupt	SWI
Return from subroutine	RTS
Return from interrupt	RTI
Reset stack pointer	RSP
No-operation	NOP

FIGURE 17.4

6805 family instruction set. (*Reprinted with permission of Motorola, Inc.*)

vide an inverse symmetry to the set. BRN may be used as a double NOP instruction (four clock cycles).

Bit Manipulation Instructions

These instructions allow the setting or clearing of any bit which resides in the first 256 bytes of RAM memory space including ports and timers. An additional feature allows branching on the state of virtually any bit within the 256 locations. The value of the bit tested is also placed in the carry bit of the conditional code register.

Control Instructions

Control instructions refer to the microprocessor registers and are used to control processor operation during program execution.

17.5   68705R3 ADDRESSING MODES

To overcome the scaled back microprocessor registers in the 6805, Motorola has added several new addressing modes which make the 6805 more powerful than the 6802 for control applications. There are 7 to 10 addressing modes altogether depending upon how they are grouped. We are already familiar with the immediate, direct, extended, inherent, and relative modes of the 6802. However, the 6805 index register is only 8 bits long and cannot hold an address. The index mode is now given three submodes to handle addressing through the offset. Bit set/clear mode and bit test and branch mode are confined to the corresponding new instructions much as the relative mode is confined to branching instructions.

Indexed, No Offset

This mode is a single byte instruction where the content of the X register is the effective address (EA). In this mode the EA can only be in the lowest 256 bytes of the address space. But this includes the I/O, timer, A/D, RAM, and part of EPROM. It may be used to point to table values, point to a frequently used address such as I/O, or hold the address of data that is calculated by a program.

Example 17.1. Suppose we want to obtain data from port A at many different lines in a program. The indexed, no offset mode will handle the task more efficiently than the 6802.

	Mnemonics	Comments
AE,00	LDX #00	Port address into X once.
F6	LDA, X	Get data in one byte for each request.

In contrast, the 6802 would be programmed:

CE,40,00	LDX	4000	Port address at 4000.
A6,00	LDA A	00,X	Data in two bytes for each request.

Example 17.2. The A/D converter continually converts analog data and places the binary result into address 00F. Fill up the first 100 bytes of RAM (010 to 073) with data read every second.

```
        LDX    #10      Load X register with RAM start.
AGAIN   LDA    0F       Get A/D conversion.
        JSR    DELAY    Delay one second.
        STA    ,X       Store conversion in RAM.
        INC X           Next location.
        CPX    #074     Last one?
        BNE    AGAIN    Repeat if more.
```

Indexed, 8-Bit Offset

This mode is a 2-byte instruction. The second byte offset is a positive 8-bit number that is added to the contents of the index register when the instruction is executed. The sum is an address located anywhere in the first two pages of the address space: 0000 to 01FE. The address 01FF is not accessible because the largest number that can be placed in the X register is FF and the largest 8-bit offest is FF. Thus,

$$FF + FF = 01FE$$

Efficient programming encourages the inclusion of tables in page 0 and page 1 of memory.

Example 17.3. Interpret the following instructions which use the indexed, 8-bit offset mode:

```
AE,60   LDX   #60
E6,70   LDA   70,X
```

The first instruction loads the X register with 60. The second instruction loads the accumulator with data using the indexed, 8-bit offset mode. The data is found at address 060+070 or 0D0.

Example 17.4. A look-up table for a 7-segment display starts at address 0080. The binary input for the display is received from port B, and the 7-segment code is then placed on port C. Write a look-up table routine for the 68705R3 based on the same routine for the 6802 found in Chap. 4, Example 4.12.

```
BE,01   LDX   PORTB      Get data at port B:direct.
E6,80   LDA   TABLE,X    Load A with code at address TABLE+X.
B7,02   STA   PORTC      Send code to LED.
```

Indexed, 16-Bit Offset

This mode is a 3-byte instruction. The offset is a 16-bit positive number that is added to the contents of the index register when the instruction is executed. The sum is an address located anywhere in the memory space 000–FFF. The procedure is the reverse of the indexed mode for the 6802. There, the X register contains a 16-bit number that is added to an 8-bit offset, forming the address of the needed data.

Example 17.5. Interpret the following instructions which use the indexed, 16-bit offset mode:

```
AE,60     LDX   #60
D6,03,40  LDA   0340,X
```

The first instruction loads the X register with 60. The second instruction loads the accumulator with data using the indexed, 16-bit offset mode. The data is found at address 60+0340 or 03A0.

Example 17.6. Write a program to move a block of data starting at address TABLEA in the EPROM and locate the block in RAM starting at address TABLEB. The index register contains the block length:

```
               TABLEA  EQU   0A00
               TABLEB  EQU   0060

AE,20                  LDX   #20
D6,0A,00  AGAIN        LDA   TABLEA,X   Load A with last data at
                                        TABLEA + 20.
E7,60                  STA   TABLEB,X   Send to RAM at TABLEB + 20.
5A                     DECX             Next lower location.
26                     BNE   AGAIN      Repeat if more.
```

Bit Set/Clear

The bit set/clear mode is used to set or clear individual bits in memory. The byte following the Op code specifies the address of the data to which the bit belongs. Since the byte can be only 8 bits, direct addressing is automatically implied. Any read/write bit, i.e., RAM bit, in the first 256 memory locations can be selectively set or cleared.

The reader is cautioned that the data direction registers (DDR) on some 6805 family HMOS devices are write only. They read as FF. Therefore, bit set/clear instructions (and read-modify-write instructions) should not be used to manipulate the DDR.

Bit set/clear Op codes define which bit is to be set/cleared. Thus, there are a total of 16-bit set/clear instructions:

Bit set: Op code $= 10 + 2n$
Bit clear: Op code $= 11 + 2n$

where n is the bit to be manipulated.

Example 17.7. Interpret the following instruction:

```
BSET  3,02
```

BSET is mnemonic for Bit Set with instruction Op code equal to 16. The bit to be set is bit 3 of port C, which is located at address 002. It is a 2-byte instruction: 16,02.

Example 17.8. Port B is programmed as an output port. It interfaces eight devices which are to be turned ON in sequence with a one-second delay between each. A device is activated by setting its corresponding bit. Write the program. Can the program be entered in the EPROM?

	Mnemonics		Comments
	LDX	#10	Load X with bit set Op code base.
LOOP	BSET	n,01	Set bit zero, Op code changed to set other bits.
	JSR	DELAY	Delay one second.
	INC X		Change Op code to $10 + 2n$
	INC X		
	STX	LOOP	Store new Op code to BSET.
	CPX	#1E	Have we changed last bit Op code?
	BNE	LOOP	Repeat if not.

The program must be entered in the RAM because of the STX instruction.

Bit Test and Branch

The bit test and branch mode are used to test and branch on individual bits in memory just as the relative mode (branch instructions) is used to test and branch on conditional code register flags. The particular bit (0 to 7) to be tested is specified by the Op code in a manner similar to bit set/clear:

Branch set: Op code $= 2n$
Branch clear: Op code $= 01 + 2n$

where n is the bit. Two bytes follow the Op code. The direct 8-bit address of the byte containing the tested bit immediately follows the Op code. A signed relative 8-bit offset is the third byte. Thus, any readable bit in the first 256 locations can be tested, and the branching span is -128 to $+127$ from the next instruction.

Example 17.9. Interpret the following instruction:

    BRSET   3,01,FA

BRSET is mnemonic for BRanch if SET with an Op code equal to 06. The bit to be set is bit 3 of port B, which is located at address 001. Offset FA branches the program counter back six spaces. It is a 3-byte instruction: 06,01,FA.

Example 17.10. Port D is used to monitor sensors. If sensor bit 4 is set, device bit 2 on output port B should be turned OFF. Write the routine:

                ...
    09,03,02   BRCLR   4,PORTD,02
    15,01   BCLR   2,PORTB
                ...

BRCLR tests bit 4 of port D at address 03 direct mode. If bit 4 is set (sensor HIGH), no branch occurs and the instruction BCLR is executed. It clears bit 2 of port B, which turns the device OFF. If the tested bit is clear (sensor LOW), the branch instruction bridges BCLR and the device stays ON.

17.6 MULTIPLY ROUTINE

The following multiply subroutine is adapted from the *M6805 HMOS M146805 CMOS Family Users' Manual*. A divide subroutine is also available. Either subroutine operates with 16-bit numbers.

		MHIGH	EQU	0064	Multiplicand MSByte
		MLOW	EQU	0065	Multiplicand LSByte
		RESA	EQU	0066	Result MSByte
		RESB	EQU	0067	Result second SByte
		XHIGH	EQU	0068	Multiplier MSByte, result third SByte
		XLOW	EQU	0069	Multiplier LSByte, result LSByte
			ORG	080	
AE, 16	MULT	LDX	#16		
3F, 66		CLR	RESA		
3F, 67		CLR	RESB		
36, 68		ROR	XHIGH		
36, 69		ROR	XLOW		
24, OC	NXT	BCC	ROTAT		
B6,67		LDA	RESB		
BB, 65		ADD	MLOW		
B7, 67		STA	RESB		
B6, 66		LDA	RESA		
B9, 64		ADC	MHIGH		
B7, 66		STA	RESA		
36, 66	ROTAT	ROR	RESA		
36, 67		ROR	RESB		
36, 68		ROR	XHIGH		
36, 69		ROR	XLOW		
5A		DEC X			
26, E7		BNE	NXT		
81		RTS			

17.7 INPUT/OUTPUT (I/O) PORTS

The 68705R3 MCU has 32 input/output (I/O) lines arranged into four parallel 8-bit ports (A, B, C, and D). Ports A, B, and C are programmable as either inputs or outputs depending upon the contents of the data direction register (DDR). If an individual DDR bit is cleared, the associated port line is an input line. Similarly, if the data direction bit is set, the associated port line is an output line. Port D is a permanent input port and has no DDR.

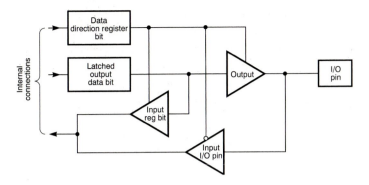

FIGURE 17.5
6805 Family I/O port circuitry for individual port bits. (*Reprinted with permission of Motorola, Inc.*)

Typical programmable port circuitry for each bit is shown in Fig. 17.5. A data register bit is latched and buffered from the I/O pin. The data direction register bit controls both input and output buffers (tristates). Upon hardware reset, all the DDRs are initialized to logic zero, placing the ports in the *input state*. The data registers are not initialized upon reset and will contain random values. Data may be written to a data register bit (latch), but the data will not appear on the I/O pin because the output buffer is inactive. A read of the data register when the DDR bit is zero gives the I/O pin data. A read of the DDR gives only logic HIGHs (FF).

When the I/O pin is programmed for the *output state* (DDR bit is HIGH), output and input register buffers are turned ON. The data register bit logic state is driven onto the I/O pin. A read of the data register gives its latched value via the input register buffer. Software intialization of the ports begins by writing output data to the data registers where they are latched. A write to the corresponding DDRs follows. Note that the procedure is exactly opposite from PIA port initialization with the 6802 MPU.

Data direction registers in the 6805 HMOS family are write-only registers and should not be used with any read-modify-write instructions or bit manipulation instructions. The 146805 CMOS family DDRs are read/write registers and may be used with these instructions.

Port A (pins 33–40)

Port A lines, programmed as output, are capable of sinking 1.6 mA at a maximum voltage of 0.4 V (logic-LOW). This is equivalent to one standard TTL load. When the output logic is HIGH, the lines can source 100 μA at 2.4 V or 10 μA at 4.0 V. The former is equivalent to two standard TTL loads. Port A has an internal pull-up resistor to V_{CC}, which enables the port to drive CMOS as well as TTL logic. Programmed as input lines, port A represents approximately one standard TTL load.

Port B (pins 25–32)

Port B lines, programmed as output, are capable of sinking 3.2 mA at a maximum voltage of 0.4 V (logic-LOW). This is equivalent to two standard TTL loads. A sinking current of

10 mA is available at 1 V, suitable for driving LEDs. When the output logic is HIGH, the lines can source a minimum of 1 mA (maximum of 10 mA) at a voltage of 1.5 V. This current is suitable for driving transistors. Output voltages are not CMOS compatible. Programmed as input lines, port B can be driven by either one standard TTL or CMOS logic.

Port C (pins 9–16)

Port C has characteristics equivalent to port A except there are no internal pull-up resistors. External pull-up resistors must be supplied to drive CMOS logic.

Port D (pins 17–24)

Port D lines are input lines only. Current requirements are typically under 1 μA. Thus, logic levels are compatible with both TTL and CMOS.

17.8   SYSTEM CLOCK

All 6805 versions have an on-chip clock oscillator circuit. The external frequency for this curcuit to complete the system clock can be provided in four ways. Each option is illustrated in Fig. 17.6.

Crystal

A crystal is necessary when using software timing loops or $\phi2$ to clock the internal timer. Recommended connections are shown in Fig. 17.6(a). Crystal frequency f should lie between 0.4 to 4.2 MHz. The resulting clock $\phi2$ cycle time is $4/f$. A 1-μs clock uses a 4-MHz crystal and a 27-pF external capacitor. For other crystals the total capacitance on *each* pin should be scaled as the inverse of the frequency ratio. There is already an internal 25-pF capacitor on the XTAL pin. For example, a 2-MHz crystal would use approximately 50 pF on EXTAL and 25 pF on XTAL (25 pF internal plus the 25 pF external).

External Resistor

A cheaper option does not use a crystal but simply connects a resistor from XTAL to V_{CC}. See Fig. 17.6(b). EXTAL is not connected. Select an 18-k resistor for a 4 MHz oscillator, 51-k resistor for a 1-MHz oscillator. Accuracy is approximately 10 to 25 percent excluding resistor tolerance.

External Jumper

The lowest-cost option connects a jumper wire and nothing else between XTAL and EXTAL. See Fig. 17.6(c). Typical clock cycle time is 1.25 μs with approximately 25 to 50 percent accuracy.

External Clock

An external clock may be applied directly to EXTAL while grounding XTAL to V_{SS}. See Fig. 17.6(d).

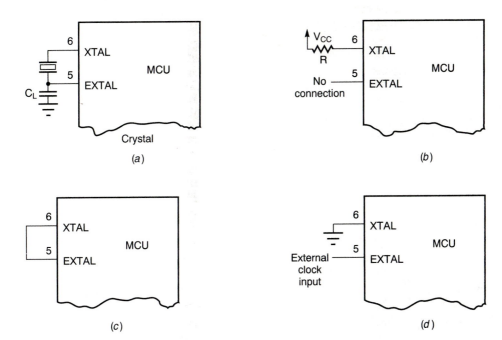

FIGURE 17.6
6805 Family system clock options. (*a*) Crystal; (*b*) external resistor—10% to 25% accuracy; (*c*) external jumper—25% to 50% accuracy; (*d*) external clock.

Clock options must be selected by the appropriate logic in bit 7 of the mask option register (MOR). This register is located at address F38 in the EPROM, and bit 7 is fixed during EPROM programming. MOR is primarily used for timer selections and will be discussed in a subsequent section.

17.9 RESETS

The 6805 family MCUs can be reset in two ways: initial power-up or external reset input pin ($\overline{\text{RESET}}$). Power-up reset occurs when a positive transition is detected on V_{CC}. Internal circuitry includes the equivalent of an internal pull-up resistor; therefore, only a capacitor is required externally, Fig. 17.7. A delay of 100 ms is necessary before allowing the $\overline{\text{RESET}}$ pin to go HIGH. A 1.0-μF capacitor typically provides sufficient delay.

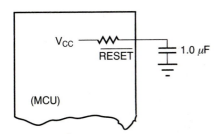

FIGURE 17.7
6805 Family power-up reset delay circuit.

The MPU can be externally reset by applying logic zero to the $\overline{\text{RESET}}$ pin for a period longer than one machine cycle. A switch between the pin and ground will do the trick. Debounce of the contact is automatically handled by the capacitor and internal pull-up resistor.

Any reset causes the following order of events:

1. All interrupt requests are cleared.
2. All interrupt masks are set.
3. All data direction registers are cleared.
4. The stack pointer is reset to 7F (top of RAM).
5. The reset vector at addresses FFE and FFF is fetched and placed in the program counter. The reset vector is the address of the program start.

17.10   INTERRUPTS

The 6805 family program execution can be interrupted four different ways:

1. External interrupt pin $\overline{\text{INT}}$
2. External Port D pin $\overline{\text{INT2}}$
3. Internal on-chip timer
4. Software interrupt instruction SWI

Upon receiving an interrupt request, the current instruction is first completed. Then the processor checks the I bit of the conditional code register and, if it is clear, proceeds with the following sequence:

1. Mask all interrupts (set I bit).
2. Place all MPU registers onto the stack.
3. Load the program counter with the appropriate interrupt vector address.
4. Execute the service routine starting at the vector address.

MPU registers are saved onto the stack in the order presented in Fig. 17.8. Service routines must end with the RTI instruction to recover register contents in the reverse order.

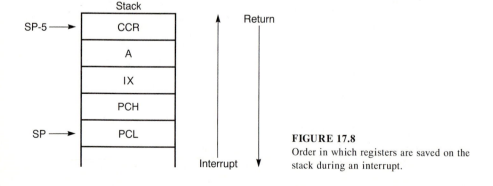

FIGURE 17.8
Order in which registers are saved on the stack during an interrupt.

All interrupts are maskable (except SWI). To use an interrupt, the instruction CLI must be placed in the beginning of the main program, clearing the interrupt mask bit set by $\overline{\text{RESET}}$.

External Interrupts

External interrupts include $\overline{\text{INT}}$ and $\overline{\text{INT2}}$. Both pins have internal edge-triggered flip-flops. See Fig. 17.9. An external interrupt request is triggered by a falling edge signal, provided the LOW duration exceeds one $\phi 2$ clock period plus 250 ns. If the interrupt is masked (I-bit set), the flip-flop holds the external interrupt request until the I bit is cleared. However, both a $\overline{\text{RESET}}$ and an external interrupt in the process of being serviced clear the flip-flop. Thus, an interrupt request from either $\overline{\text{INT}}$ or $\overline{\text{INT2}}$ arriving before the end of the service routine will be lost. A Schmitt trigger between input and flip-flop allows sinusoidal AC input signals to generate repetitive interrupts for a clock or AC power control applications.

The $\overline{\text{INT2}}$ has a second layer of control by way of the miscellaneous register (MR) at address 00A in the RAM. See Fig. 17.10. The $\overline{\text{INT2}}$ interrupt is inhibited when the MR

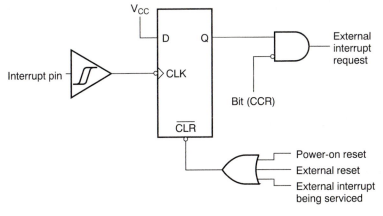

FIGURE 17.9
Interrupt request circuitry for external interrupt $\overline{\text{INT}}$ or $\overline{\text{INT2}}$.

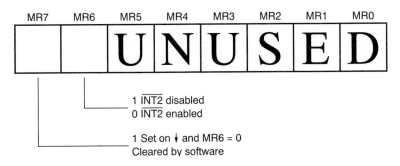

FIGURE 17.10
Miscellaneous register at address 00A.

bit 6 is set. Thus, the $\overline{\text{INT2}}$ always reads as a digital input on port D. An $\overline{\text{INT2}}$ interrupt is permitted only if the I bit and MR bit 6 are both clear. An interrupt signal on $\overline{\text{INT2}}$ will set MR bit 7, provided the MR bit 6 is clear. But the MR bit 7 is not cleared automatically upon RTI. It must be cleared by writing a logic zero to the register during the interrupt service routine. Upon RESET, MR bits 6 and 7 are cleared, disabling $\overline{\text{INT2}}$.

In systems *not* using interrupts input line $\overline{\text{INT}}$ can function as an extra I/O input line. Branch instructions BIH and BIL do *not* test the I bit in the conditional code register. They only test whether the $\overline{\text{INT}}$ line is HIGH or LOW.

Timer Interrupt

All 6805 family members have an internal timer than can be programmed to interrupt the processor when the timer counter decrements to zero. The procedure is similar to $\overline{\text{INT2}}$, although the timer interrupt is generated internally. The timer control register (TCR) at address 009 is used like the miscellaneous register (MR) for $\overline{\text{INT2}}$. More will be said about TCR in Sec. 17.12. Processor interrupts are generated only if the TCR bit 6 is clear in addition to the I bit in the conditional code register. A valid interrupt sets the TCR bit 7 flag, and the flag must be cleared during the service routine or the interrupt will repeat. RESET clears the TCR and disables the timer interrupt.

Interrupt vector addresses (HIGH-byte first) are as follows:

Reset	FFE/FFF
SWI	FFC/FFD
INT	FFA/FFB
Timer	FF8/FF9
INT2	FF8/FF9

External interrupt $\overline{\text{INT2}}$ and timer interrupt share the same vector address. Programming distinguishes between the two by reading the MR and TCR bit 7. Service routines for both may be located at FF8/FF9 by using the bit test and branch instructions at the beginning of the routine.

17.11 ANALOG-TO-DIGITAL CONVERTER

Both the 68705R3 MCU and the 6805R2/R3 ROM versions have an 8-bit ADC integrated on the chip. A block diagram is shown in Fig. 17.11. The converter uses a successive approximation technique described in Chap. 13. Conversion time is 30 machine cycles. With a 4-MHz crystal, the time is 30 μs. Up to four external analog inputs may be connected to the ADC through the multiplexer. These analog lines are shared simultaneously with port D digital lines PD0 to PD3. Port D lines that are not connected to analog inputs can be used as binary inputs. Analog inputs merely translate into logic-HIGH or LOW when port D is read via the data register. The analog channel input current is less than 1 μA. Thus, no additional buffering of most external transducer signals is required.

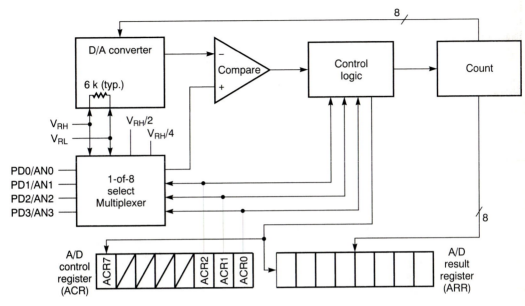

FIGURE 17.11
The 68705R3 A/D converter block diagram.

Registers

The multiplexer selection is controlled by the A/D control register (ACR) at address 00E.

AC2	AC1	AC0	Channel
0	0	0	AN0 (PD0)
0	0	1	AN1 (PD1)
0	1	0	AN2 (PD2)
0	1	1	AN3 (PD3)
1	X	X	Internal calibration

The selected input is converted continuously and the result is placed in the result register (ARR) at address 00F. Both registers are cleared upon RESET. The ADC immediately begins to convert data on AN0, updating the ARR every 30 machine cycles. At the end of the first conversion, bit 7 of the ACR goes HIGH to flag the user when the conversion is completed. However, bit 7 remains HIGH until cleared by programming. Clearing bit 7 during a write operation to the ACR aborts the conversion in progress and starts a new conversion. Writing a channel number to the ACR also starts a new conversion, but of the new channel.

Reference Voltages

The voltage reference high V_{RH} (PD5) and voltage reference low V_{RL} (PD6) pins establish the voltage range recognized by the converter. V_{RL} is typically ground ($V_{RL} = 0$ V). V_{RH}

may be V_{CC} but only if the power supply voltage is regulated and stable. Otherwise a voltage reference IC should be used.

The ADC channel input voltage should not exceed V_{RH} nor be less than V_{RL}. A voltage greater than V_{RH} converts to FF, whereas a voltage less than V_{RL} but greater than V_{SS} converts to 00. To maintain full accuracy, $V_{RH} - V_{RL}$ should be equal to or greater than 4 V.

The 8-bit ADC implemented in the 6805 family uses a unipolar code. Resolution is $(V_{RH} - V_{RL})/256$. Thus, a hexadecimal value of 00 represents an analog input voltage of 0.00 to 0.02 and a value of FF represents an analog input voltage of 4.98 to 5.00 given $V_{RH} = 5.00$ V and $V_{RL} = 0.00$ V. However, the quantizing error is ± 0.5 LSB. Thus, the first step occurs at 0.5 LSB or 0.01 V, and the transfer point between FE and FF is 4.97 V.

17.12   TIMER

The 6805 HMOS family MCUs contain an 8-bit timer, whereas the more recent 6805 HCMOS family MCUs have a 16-bit timer. The 68705 timer consists of an 8-bit down counter, which is driven by a 7-bit count scaler, Fig. 17.12. There are two timer registers that are directly accessible by software:

1. The timer data register at address 008, which is the 8-bit counter
2. The timer control register (TCR) at address 009

Timer Input Modes

The prescalar can be driven by the internal $\phi 2$ clock or by an external clock through TIMER pin 8. TIMER signals must be TTL compatible. Four different clock modes are available, depending upon the value written to TCR control bits 4 and 5.

Mode 1. When TCR5 = 0 and TCR4 = 0, the time input is from the internal $\phi 2$ clock, and the TIMER pin is disabled.

Mode 2. When TCR5 = 0 and TCR4 = 1, the internal $\phi 2$ clock is ANDed with the external TIMER clock to form the timer input. This mode can be used to measure the width of an external pulse (clock).

Mode 3. When TCR5 = 1 and TCR4 = 0, neither clock is connected to the prescaler, and the timer is disabled.

Mode 4. When TCR5 = 1 and TCR4 = 1, the timer input is from the external clock at pin TIMER. This mode can be used to count external events or generate periodic interrupts. Input frequency must be less than crystal frequency $\div$ 8.

Prescaler

The clock frequency can be selectively divided by powers of 2 up to 128 before the 8-bit counter is decremented. The 7-bit prescaler is programmed indirectly through the timer control register. Bits TCR2, TCR1, and TCR0 set the divisor at 2^n, where n is the number corresponding to the least three TCR bits. When divisor bits in the TCR are changed in

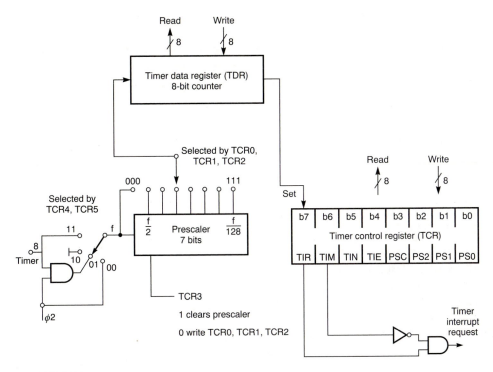

FIGURE 17.12
Schematic of the 6805 8-bit timer.

software, the prescaler should be cleared first by setting TCR3. The prescaler contents will remain zero (divisor one) as long as TCR3 = 1. TCR3 is then cleared with bits TCR2, TCR1, and TCR0 at the desired binary value. Otherwise prescaler truncation error may occur.

The timer counter is only 8 bits. Thus, the countdown time to interrupt is a multiple of the programmed count (1 to 256) and the counter-prescaled clock period. A 1-MHz $\phi2$ clock from a 4-MHz crystal gives a countdown time from 1 μs to 256 μs at 1-μs resolution for a prescaler divisor equal to one and a countdown time from 128 to 32,768 μs at 128-μs resolution for a prescaler divisor equal to 128. Slower crystals lengthen this period. However, significantly longer countdown periods are possible only with software. Each time the counter reaches zero, the interrupt service routine can decrement a preloaded memory location for a pseudo 16-bit timer.

Timer Counter

The 8-bit timer counter is programmable (STA 008) with any value between 00 and FF. It will decrement at the prescaled clock rate and can be read at any time (LDA 008). The elapsed count is found by 2s complementing the counter state. When the count reaches zero, the timer interrupt request bit (TCR7) is set. If the timer interrupt mask bit (TCR6) is clear, an interrupt request is sent to the processor. Providing the CCR I bit is clear, the processor jumps to the timer interrupt vector address. Bit TCR7 must be cleared in the interrupt service routine.

The timer counter will continue to decrement after it reaches zero, going from 00 to FF. The countdown continues at this point to the next zero. Therefore, the original count must be reloaded if it is to be repeated at the start of each cycle. In this respect the 68705 timer differs from the 6840 programmable timer.

Timer Control Register

Bits 2, 1, 0. Prescaler divisor (PS2, PS1, PS0)

PS2	PS1	PS0	Divide by
0	0	0	1
0	0	1	2
0	1	0	4
0	1	1	8
1	0	0	16
1	0	1	32
1	1	0	64
1	1	1	128

Bit 3. Prescaler clear (PSC)

Binary state	*Remarks*
0	Prescaler accumulates clock pulses.
1	Prescaler contents set to zero as long as the bit is set.

Bit 4. TIMER external enable (TIE)

0	Disables TIMER pin.
1	Enables TIMER pin.

Bit 5. Clock select (TIN)

0	Selects internal clock for prescaler.
1	Selects external clock for prescaler.

Bit 6. Timer interrupt mask (TIM)

0	Inhibits timer interrupt.
1	Enables timer interrupt.

Bit 7. Timer interrupt request (TIR)

0	Cleared by reset or programming.
1	Set when timer counter reaches zero.

Masked Option Register (MOR)

Since the 68705R3 EPROM emulates a 6805R2/R3 ROM, it is often used to test system hardware and software where the ROM version is the final design. However, the 6805R2 clock input mode and prescaler divisor is set by the manufacturing mask, not the timer control register. Thus, the 6805R2 TCR looks like Fig. 17.13(a). TCR bits 5, 4, 2, 1, 0 are not used. To make the 68705R3 EPROM truly emulate the 6805R2 ROM, a RAM space F38 is retained for the mask option register (MOR). If bit MOR6 is reset (MOR6 = 0), the 68705R3 timer control register will be under program control as described in the previous section. If bit MOR6 is set (MOR6 = 1), the 68705R3 timer control register will behave like a 6805R2/R3. See Fig. 17.13(b). The masked prescale divisor is "faked" by programming MOR bits 2, 1, 0 with the desired binary number.

Bits 2, 1, 0. Prescaler divisor (PS2, PS1, PS0)

PS2	PS1	PS0	Divided by
0	0	0	1
0	0	1	2
0	1	0	4
0	1	1	8
1	0	0	16
1	0	1	32
1	1	0	64
1	1	1	128

Bit 3. Not used.

Bit 4. Not used.

Bit 5. Clock select.

Binary state	*Remarks*
0	External TIMER pin
1	Internal clock ANDed with TIMER

7	6	5	4	3	2	1	0
TIR	TIM	1	1	PSC	1	1	1

(a)

7	6	5	4	3	2	1	0
TIR	TIM	TIN	TIE	PSC	PS2	PS1	PS0

(b)

FIGURE 17.13
Comparison of (a) 6805R2 and (b) 6805R3/68705R3 timer control registers. The MOR register enables the 68705R3 to emulate the 6805R2 timer.

Bit 6. Timer option.

0	All TCR bits programmable
1	Emulates 6805R2/R3.

Bit 7. Clock.

0	Crystal.
1	Resistor capacitor *RC* or jumper

17.13   PROJECT: HARDWARE

The 68705R3 MCU project undertaken in this chapter is the digital readout bathroom scale designed in Chap. 16 with the 6802 MPU. Design requirements in Sec. 16.1 are followed. For the hardware, the task is simply to read a strain gauge voltage representing weight and to display the results on three 7-segment LED displays. A requirement to turn OFF the system after 10 s is omitted for clarity, but the option is easily implemented with the hardware to generate supply voltage V_{CC} in Fig. 16.2. A project comparison between the 68705R3 MCU and 6802 MPU will give the reader a real appreciation for microcontrollers.

A schematic diagram for the system hardware is shown in Fig. 17.14. Of the 40 pins on the 68705R3 MCU, 32 pins are I/O, leaving only 6 pins in addition to power V_{CC} and ground V_{SS}. Programming pin V_{PP} is always tied HIGH when the MCU is placed in the system. Interrupt pins $\overline{INT}$ and $\overline{INT2}$ are typically tied HIGH when the design doesn't have interrupts. This action is optional since no interrupts can take place without the CLI instruction. No external timer is connected to the MCU. However, the 6805R2/R3, which the 68705R3 emulates, ANDs TIMER pin 8 input, and the internal clock $\phi2$ as the internal clock mask option. Thus, it is good practice to tie pin 8 HIGH when the on-chip timer is not used rather than leave it not connected. The external resistor option rather than a standard crystal drives the internal clock. A 20-k resistor should produce a $\phi2$-clock period approximately equal to 1 μs $\pm$ 25%. Clock accuracy is not important since there are no software timing loops which are critical. The $\overline{RESET}$ circuit is standard for power-up starts with optional reset button.

In stark contrast to the 6802 system, there are no address bus pins and data bus pins. Ports, memory, and A/D conversion are on-chip, so there are no decoding/timing problems—a major design hurdle. We are left with a greatly reduced chip count and only the strain gauge input and display output to interface the 68705R3 MCU.

MCU Input

The on-chip ADC requires reference voltages V_{RH} and V_{RL} (ground) at pins 19 and 20. V_{RH} must be greater than 4.0 V but equal to or less than V_{CC}. A 7805 regulator cannot guarantee better accuracy than $\pm$5% V over the input range, although somewhat better accuracy would be expected for a 9-V battery. Although the voltage source could be used to drive the ADC, it is not recommended. A more accurate source is an external voltage reference IC. A cheaper alternative is the 1N5521D 400 mW zener diode, which will hold V_{RH} at 4.3 V $\pm$ 1%. A standard 750-Ω resistor (or thereabouts) will maintain V_{CC} and V_{RH} at specified values, yet limit the current drain through the diode to less than 1 mA.

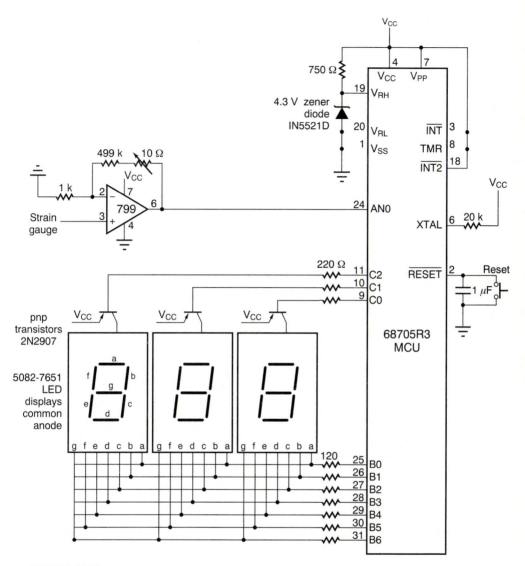

FIGURE 17.14

Schematic of bathroom scale system using the 68705R3 microcontroller.

Analog input AN0 is from a 799 op amp, which provides gain for the strain gauge voltage. The 799 amplifier has a single voltage supply compatible with the system supply. Strain gauge output is 0 to 10 mV for a weight 0 to 300 lbf. The ADC output is FF for a full-scale input of 4.3 V. If weight corresponds to the base 10 representation of the ADC output, then the op-amp gain is given by

$$\text{Gain} = \frac{4.30 \text{ V}}{10 \text{ mV}} \times \frac{300}{255} = 1 + \frac{R_f}{R_1} \qquad \text{(noninverter)}$$

or $R_f/R_1 = 505$. Use a standard 499-k metal film resistor ($\pm 1\%$) in series with a 10-Ω potentiometer for R_f and a 1-k resistor for R_1.

MCU Output

Three LED displays are driven in multiplex fashion by the MCU software. Each common anode display is switched ON by a transistor. Since only one display is active at a given time, the displays share current-limiting resistors and port B. An active-LOW output on any port B bit will light the corresponding segment. The 20-mA sink current will raise port B bit to 0.5 V but average $\frac{20}{3}$ mA over the switching cycle.

General-purpose switching transistors for the displays are the *pnp* type. Only port B can supply the milliamp current at logic-HIGH to switch *npn*-type transistors; *pnp* transistors switch ON to a logic-LOW and are inactive with a logic-HIGH at the base. Port C and other TTL ports can sink more than enough current in the LOW state to switch general-purpose *pnp* transistors. Resistors (220 Ω or thereabouts) limit the current to port C bits.

17.14   PROJECT: SOFTWARE

Software design follows the methodology outlined in Sec. 16.5. We duplicate the 6802 project programming as closely as possible for the present 68705R3 MCU project. The reader should consider several necessary software changes to conform with 6805 software requirements:

1. RAM assignments must be moved since beginning RAM on the 68705 is occupied by registers.
2. The index register in the ICU routine cannot be used as a 16-bit counter.
3. Accumulator B is no longer available for the WEI routine.
4. The LOOK-UP routine must be modified for an 8-bit index register.

Program

The program displays a single zero until the bathroom scale detects weight. It then displays the message ICU for 3 s. If the weight is over 250 lb, the message 2HI appears; otherwise the weight appears. Weight is found by converting INPUT to BCD prior to looking up the tabled code for each numeral:

```
          ORG   080        Program origin
HDRDS     EQU   010        BCD weight
TENS      EQU   011
UNITS     EQU   012
AVE       EQU   013        16-bit averaging
AVE+1     EQU   014
RAM       EQU   015        Look-up table vector address
RAM+1     EQU   016
CODEH     EQU   017        BCD code for display
CODET     EQU   018
CODEU     EQU   019
BINARY    EQU   01A        Scratch pad for BCD routine
COUNT1    EQU   01B
COUNT2    EQU   01C
```

INPUT	EQU	00F	Base address for A/D
PORTB	EQU	001	Base addresses for ports
DDRB	EQU	005	
PORTC	EQU	002	
DDRC	EQU	006	
MOR	EQU	F38	
RESET	EQU	FFE/	Address for loading reset vector 0080
		FFF	

Program starts at address 080

INIT	LDA	#FF	Latch data to turn OFF displays
	STA	PORTB	
	STA	PORTC	
	STA	DDRB	Program ports B and C as outputs
	STA	DDRC	
	LDA	#40	Disable timer
	STA	009	
	LDA	#7F	Disable INT2
	STA	00A	
	LDA	#00	
	STA	00E	Pick AN0.
ZERO	LDA	INPUT	Is someone on the scale?
	CMP	#01	
	BHI	ICU	Yes, go to ICU routine.
	LDA	#FF	No, place code to display
	STA	CODEH	two blanks and a zero.
	STA	CODET	
	LDA	#C0	
	STA	CODEU	
	JSR	DISPLAY	Displays code for three characters for 1.3 ms each.
	JMP	ZERO	Repeat to multiplex.
ICU	LDA	#F9	Place code to display ICU.
	STA	CODEH	
	LDA	#C6	
	STA	CODET	
	LDA	#C1	
	STA	CODEU	
	LDA	#03	Load count for 3 × 256.
	STA	COUNT1	
BACK	LDX	#FF	
DELAY3	JSR	DISPLAY	Display each character for 1.3 ms.
	DEX		
	BNE	DELAY3	Is count 1 × 256?
	DEC	COUNT1	Yes, get next 256.
	BNE	BACK	

```
2HI     LDA     INPUT       Is weight over 250?
        CMP     #FA
        BLS     WEI         No, go to weight display.
        LDA     #A4         Yes, place code to display 2HI.
        STA     CODEH
        LDA     #89
        STA     CODET
        LDA     #F9
        STA     CODEU
        JSR     DISPLAY
        JMP     2HI         Repeat to multiplex.

WEI     CLR     AVE
        CLR     AVE+1
        STA     BINARY
        LDA     #FF
        STA     COUNT2
        JSR     BCD         Convert input to BCD.
        JSR     LOOKUP      Convert BCD to output code.
SUM     JSR     DISPLAY     Display characters for 1.3 ms each.
        LDA     INPUT       Get input and add to 16-bit address.
        CLC
        ADD     AVE
        LDA     #00
        ADC     AVE+1
        DEC     COUNT2
        BNE     SUM         Continue multiplexing for one second.
        LDA     AVE+1       Get new weight.
        JMP     WEI
```

Subroutine 1: Convert binary to BCD

```
BCD     CLR     HDRDS
        CLR     TENS
        CLR     UNITS
LOOP1   LDA     BINARY
        SUB     #64         Subtract one hundred.
        BCS     LOOP2       Did borrow occur?
        STA     BINARY      No, save result as new binary.
        LDA     HDRDS       Increment hundreds digit until a
        INC                 borrow occurs.
        STA     HDRDS
        BRA     LOOP1
LOOP2   LDA     BINARY      Binary now less than one hundred.
        SUB     #0A         Subtract ten.
        BCS     LOOP3       Repeat process for tens.
        STA     BINARY
```

```
          LDA   TENS
          INC
          STA   TENS
          BRA   LOOP2
LOOP3     LDA   BINARY       Only units left.
          STA   UNITS
          RTS
```

Subroutine 2: Look-up table routine

```
LOOKUP    LDX   HDRDS        Convert BCD hundreds to code.
          LDA   TABLE,X
          STA   CODEH
          LDX   TENS         Convert BCD tens to code.
          LDA   TABLE,X
          STA   CODET
          LDX   UNITS        Convert BCD units to code.
          LDA   TABLE,X
          STA   CODEU
          RTS
```

Subroutine 3: Multiplex code to display weight

```
DISPLAY   LDA   CODEH        Latch hundreds code.
          STA   PORTB
          LDA   #FE          Turn on hundreds display.
          STA   PORTC
          JSR   DELAY        Delay 1.3 ms.
          LDA   CODET        Latch tens code.
          STA   PORTB
          LDA   #FD          Turn on tens display.
          STA   PORTC
          JSR   DELAY        Delay 1.3 ms.
          LDA   CODEU        Latch units code.
          STA   PORTB
          LDA   #FB          Turn on units display.
          STA   PORTC
          JSR   DELAY        Delay 1.3 ms.
          RTS
```

Subroutine 4: Delay for 1.3 ms

```
DELAY     LDA   #D8
          DEC
          BNE   DELAY
          RTS
```

Load look-up table code

```
TABLE     FCB   C0,F9,A4,B0,99,92,82,F8,80,98
```

Select RC clock

```
MOR              FCB   80
```

∗Load reset vector address∗

```
RESET            FCB   00,80

                 END
```

17.15   DEVELOPMENT SYSTEM

A development system for the Motorola 68705 single-chip microcontroller is available from Engineers Collaborative, Inc., West Glover, VT. It is a set of software and hardware tools used with an IBM PC/XT/AT personal computer. The system consists of a cross-assembler program, a simulator/debugger program, and a programming circuit board with driver program.

The cross-assembler supports the standard 6805 instruction set mnemonics and allows the user to format output, include other source files in the assembly, and control the output generated. Descriptive error messages prevent errors from becoming bugs.

The simulator/debugger allows the user to simulate and verify program logic before the program is burned into the microcontroller EPROM. It provides a full screen display of memory, registers, and I/O ports. The user can single step the program and view results as each instruction is executed. Programs can be modified at any point.

The programming board holds the microcontroller and connects to the PC via an RS-232 port. A menu driver downloads the files to the microcontroller. Board memory can be examined or printed, and self-tests of the board can be performed.

PROBLEMS

17.1. Rewrite the look-up table for EPROM routine in Chap. 16 using 6805 instructions.

17.2. Rewrite the message program in Sec. 4.7 using 6805 instructions.

17.3. Construct a timing loop to give a 20-ms delay.

17.4. A program reads port bit PD6. If the bit is HIGH, the program blinks an LED lamp at half-second intervals. The lamp is driven by a LOW on PA7. Write the program branch and subsequent blink subroutine.

17.5. Review the microcontroller program in Sec. 17.14 for opportunities to use bit set/clear and bit branch instructions and make any worthwhile changes in the program.

17.6. Ports A and B are output ports and ports C and D are input ports. All port A bits should be HIGH at start-up. Write the initialization routine.

17.7. What software step must be taken to use an external crystal to drive the clock?

17.8. A signal interrupts the 68705R3 regularly via $\overline{\text{INT}}$. Upon each interrupt the MCU reads port C and stores the reading in succession starting at address 010 until the RAM is filled. At this point the interrupts are ignored. Write the interrupt service routine.

17.9. The on-board timer is used together wth the ADC for data collection. Every 20 ms the MCU subtracts the analog signal on channel AN3 from the analog signal on channel AN4 and stores the result at address 020. Write the service routine. What initialization procedures are necessary, if any, for the timer and ADC? List any assumptions.

17.10. Explain how you would measure the width of a pulse with the 68705R3 MCU. Outline the software.

17.11. Repeat Prob. 16.13 but with a 68705R3 MCU.

6802
INSTRUCTION
SET

Nomenclature

The following nomenclature is used in the subsequent definitions.

(a) *Operators*

( )	=	contents of
←	=	is transferred to
↑	=	"is pulled from stack"
↓	=	"is pushed into stack"
·	=	Boolean AND
⊙	=	Boolean (Inclusive) OR
⊕	=	Exclusive OR
≈	=	Boolean NOT

(b) *Registers in the MPU*

ACCA	=	Accumulator A
ACCB	=	Accumulator B
ACCX	=	Accumulator ACCA or ACCB
CC	=	Condition codes register
IX	=	Index register, 16 bits
IXH	=	Index register, higher order 8 bits
IXL	=	Index register, lower order 8 bits
PC	=	Program counter, 16 bits
PCH	=	Program counter, higher order 8 bits
PCL	=	Program counter, lower order 8 bits
SP	=	Stack pointer
SPH	=	Stack pointer high
SPL	=	Stack pointer low

(c) *Memory and Addressing*

M	=	A memory location (one byte)
M +1	=	The byte of memory at 0001 plus the address of the memory location indicated by "M."
Rel	=	Relative address (i.e. the two's complement number stored in the second byte of machine code corresponding to a branch instruction).

(d) *Bits 0 thru 5 of the Condition Codes Register*

C	=	Carry — borrow	bit — 0
V	=	Two's complement overflow indicator	bit — 1
Z	=	Zero indicator	bit — 2
N	=	Negative indicator	bit — 3
I	=	Interrupt mask	bit — 4
H	=	Half carry	bit — 5

(e) *Status of Individual Bits BEFORE Execution of an Instruction*

An	=	Bit n of ACCA (n=7,6,5,...,0)
Bn	=	Bit n of ACCB (n=7,6,5,...,0)
IXHn	=	Bit n of IXH (n=7,6,5,...,0)

IXLn = Bit n of IXL (n=7,6,5,...,0)
Mn   = Bit n of M (n=7,6,5,...,0)
SPHn = Bit n of SPH (n=7,6,5,...,0)
SPLn = Bit n of SPL (n=7,6,5,...,0)
Xn   = Bit n of ACCX (n=7,6,5,...,0)

(f) *Status of Individual Bits of the RESULT of Execution of an Instruction*

   (i) For 8-bit Results

       Rn = Bit n of the result (n =7,6,5,...,0)

           This applies to instructions which provide a result contained in a single byte of memory or in an 8-bit register.

   (ii) For 16-bit Results

       RHn = Bit n of the more significant byte of the result
              (n =7,6,5,...,0)

       RLn = Bit n of the less significant byte of the result
              (n =7,6,5,...,0)

           This applies to instructions which provide a result contained in two consecutive bytes of memory or in a 16-bit register.

Executable Instructions (definition of)

Detailed definitions of the 72 executable instructions of the source language are provided on the following pages.

ABA

ABA Add Accumulator B to Accumulator A **ABA**

Operation: ACCA ← (ACCA) + (ACCB)

Description: Adds the contents of ACCB to the contents of ACCA and places the result in ACCA.

Condition Codes:
H: Set if there was a carry from bit 3; cleared otherwise.
I: Not affected.
N: Set if most significant bit of the result is set; cleared otherwise.
Z: Set if all bits of the result are cleared; cleared otherwise.
V: Set if there was two's complement overflow as a result of the operation; cleared otherwise.
C: Set if there was a carry from the most significant bit of the result; cleared otherwise.

Addressing Modes	Execution Time (No. of cycles)	Number of bytes of machine code	Coding of First (or only) byte of machine code		
			HEX.	OCT.	DEC.
Inherent	2	1	1B	033	027

ADC

ADC Add with Carry **ADC**

Operation: ACCX ← (ACCX) + (M) + (C)

Description: Adds the contents of the C bit to the sum of the contents of ACCX and M, and places the result in ACCX.

Condition Codes:
H   Set if there was a carry from bit 3; cleared otherwise.
I: Not affected.
N: Set if most significant bit of the result is set; cleared otherwise.
Z: Set if all bits of the result are cleared; cleared otherwise.
V: Set if there was two's complement overflow as a result of the operation; cleared otherwise.
C: Set if there was a carry from the most significant bit of the result; cleared otherwise.

Addressing Modes		Execution Time (No. of cycles)	Number of bytes of machine code	Coding of First (or only) byte of machine code		
				HEX.	OCT.	DEC.
A	IMM	2	2	89	211	137
A	DIR	3	2	99	231	153
A	EXT	4	3	B9	271	185
A	IND	5	2	A9	251	169
B	IMM	2	2	C9	311	201
B	DIR	3	2	D9	331	217
B	EXT	4	3	F9	371	249
B	IND	5	2	E9	351	233

ADD

ADD

Operation: ACCX ← (ACCX) + (M)

Description: Adds the contents of ACCX and the contents of M and places the result in ACCX.

Condition Codes:
- H: Set if there was a carry from bit 3; cleared otherwise.
- I: Not affected.
- N: Set if most significant bit of the result is set; cleared otherwise.
- Z: Set if all bits of the result are cleared; cleared otherwise.
- V: Set if there was two's complement overflow as a result of the operation; cleared otherwise.
- C: Set if there was a carry from the most significant bit of the result; cleared otherwise.

Addressing Modes	Execution Time (No. of cycles)	Number of bytes of machine code	Coding of First (or only) byte of machine code		
			HEX.	OCT.	DEC.
A IMM	2	2	8B	213	139
A DIR	3	2	9B	233	155
A EXT	4	3	BB	273	187
A IND	5	2	AB	253	171
B IMM	2	2	CB	313	203
B DIR	3	2	DB	333	219
B EXT	4	3	FB	373	251
B IND	5	2	EB	353	235

AND

AND

Operation: ACCX ← (ACCX) · (M)

Description: Performs logical "AND" between the contents of ACCX and the contents of M and places the result in ACCX. (Each bit of ACCX after the operation will be the logical "AND" of the corresponding bits of M and of ACCX before the operation.)

Condition Codes:
- H: Not affected.
- I: Not affected.
- N: Set if most significant bit of the result is set; cleared otherwise.
- Z: Set if all bits of the result are cleared; cleared otherwise.
- V: Cleared.
- C: Not affected.

Addressing Modes	Execution Time (No. of cycles)	Number of bytes of machine code	Coding of First (or only) byte of machine code		
			HEX.	OCT.	DEC.
A IMM	2	2	84	204	132
A DIR	3	2	94	224	148
A EXT	4	3	B4	264	180
A IND	5	2	A4	244	164
B IMM	2	2	C4	304	196
B DIR	3	2	D4	324	212
B EXT	4	3	F4	364	244
B IND	5	2	E4	344	228

ASL

Arithmetic Shift Left

ASL

Operation:

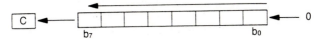

Description: Shifts all bits of the ACCX or M one place to the left. Bit 0 is loaded with a zero. The C bit is loaded from the most significant bit of ACCX or M.

Condition Codes:
H: Not affected.
I: Not affected.
N: Set if most significant bit of the result is set; cleared otherwise.
Z: Set if all bits of the result are cleared; cleared otherwise.
V: Set if, after the completion of the shift operation, EITHER (N is set and C is cleared) OR (N is cleared and C is set); cleared otherwise.
C: Set if, before the operation, the most significant bit of the ACCX or M was set; cleared otherwise.

Addressing Modes	Execution Time (No. of cycles)	Number of bytes of machine code	Coding of First (or only) byte of machine code		
			HEX.	OCT.	DEC.
A	2	1	48	110	072
B	2	1	58	130	088
EXT	6	3	78	170	120
IND	7	2	68	150	104

ASR

Arithmetic Shift Right

ASR

Operation:

Description: Shifts all bits of ACCX or M one place to the right. Bit 7 is held constant. Bit 0 is loaded into the C bit.

Condition Codes:
H: Not affected.
I: Not affected.
N: Set if the most significant bit of the result is set; cleared otherwise.
Z: Set if all bits of the result are cleared; cleared otherwise.
V: Set if, after the completion of the shift operation, EITHER (N is set and C is cleared) OR (N is cleared and C is set); cleared otherwise.
C: Set if, before the operation, the least significant bit of the ACCX or M was set; cleared otherwise.

Addressing Modes	Execution Time (No. of cycles)	Number of bytes of machine code	Coding of First (or only) byte of machine code		
			HEX.	OCT.	DEC.
A	2	1	47	107	071
B	2	1	57	127	087
EXT	6	3	77	167	119
IND	7	2	67	147	103

BCC

Branch if Carry Clear

BCC

Operation: PC ← (PC) + 0002 + Rel if (C)=0

Description: Tests the state of the C bit and causes a branch if C is clear.

See BRA instruction for further details of the execution of the branch.

Condition Codes: Not affected.

Addressing Modes	Execution Time (No. of cycles)	Number of bytes of machine code	Coding of First (or only) byte of machine code		
			HEX.	OCT.	DEC.
REL	4	2	24	044	036

BCS

Branch if Carry Set

BCS

Operation: PC ← (PC) + 0002 + Rel if (C)=1

Description: Tests the state of the C bit and causes a branch if C is set.

See BRA instruction for further details of the execution of the branch.

Condition Codes: Not affected.

Addressing Modes	Execution Time (No. of cycles)	Number of bytes of machine code	Coding of First (or only) byte of machine code		
			HEX.	OCT.	DEC.
REL	4	2	25	045	037

BEQ

Branch if Equal

BEQ

Operation: PC ← (PC) + 0002 + Rel if (Z)=1

Description: Tests the state of the Z bit and causes a branch if the Z bit is set.

See BRA instruction for further details of the execution of the branch.

Condition Codes: Not affected.

Addressing Modes	Execution Time (No. of cycles)	Number of bytes of machine code	Coding of First (or only) byte of machine code		
			HEX.	OCT.	DEC.
REL	4	2	27	047	039

BGE
Branch if Greater than or Equal to Zero
BGE

Operation: $PC \leftarrow (PC) + 0002 + Rel$ if $(N) \oplus (V) = 0$

i.e. if $(ACCX) \geqslant (M)$

(Two's complement numbers)

Description: Causes a branch if (N is set and V is set) OR (N is clear and V is clear).

If the BGE instruction is executed immediately after execution of any of the instructions CBA, CMP, SBA, or SUB, the branch will occur if and only if the two's complement number represented by the minuend (i.e. ACCX) was greater than or equal to the two's complement number represented by the subtrahend (i.e. M).

See BRA instruction for details of the branch.

Condition Codes: Not affected.

Addressing Modes	Execution Time (No. of cycles)	Number of bytes of machine code	Coding of First (or only) byte of machine code		
			HEX.	OCT.	DEC.
REL	4	2	2C	054	044

BGT
Branch if Greater than Zero
BGT

Operation: $PC \leftarrow (PC) + 0002 + Rel$ if $(Z) \odot [(N) \oplus (V)] = 0$

i.e. if $(ACCX) > (M)$

(two's complement numbers)

Description: Causes a branch if [Z is clear] AND [(N is set and V is set) OR (N is clear and V is clear)].

If the BGT instruction is executed immediately after execution of any of the instructions CBA, CMP, SBA, or SUB, the branch will occur if and only if the two's complement number represented by the minuend (i.e. ACCX) was greater than the two's complement number represented by the subtrahend (i.e. M).

See BRA instruction for details of the branch.

Condition Codes: Not affected.

Addressing Modes	Execution Time (No. of cycles)	Number of bytes of machine code	Coding of First (or only) byte of machine code		
			HEX.	OCT.	DEC.
REL	4	2	2E	056	046

BHI   Branch if Higher   BHI

Operation: $PC \leftarrow (PC) + 0002 + Rel$ if $(C) \cdot (Z)=0$

i.e. if (ACCX) > (M)
(unsigned binary numbers)

Description: Causes a branch if (C is clear) AND (Z is clear).

If the BHI instruction is executed immediately after execution of any of the
instructions CBA, CMP, SBA, or SUB, the branch will occur if and only if the
unsigned binary number represented by the minuend (i.e. ACCX) was greater
than the unsigned binary number represented by the subtrahend (i.e. M).

See BRA instruction for details of the execution of the branch.

Condition Codes: Not affected.

Addressing Modes	Execution Time (No. of cycles)	Number of bytes of machine code	Coding of First (or only) byte of machine code		
			HEX.	OCT.	DEC.
REL	4	2	22	042	034

BIT   Bit Test   BIT

Operation: $(ACCX) \cdot (M)$

Description: Performs the logical "AND" comparison of the contents of ACCX and the contents
of M and modifies condition codes accordingly. Neither the contents of ACCX or M
operands are affected. (Each bit of the result of the "AND" would be the logical
"AND" of the corresponding bits of M and ACCX.)

Condition Codes: H: Not affected.
I: Not affected.
N: Set if the most significant bit of the result of the "AND" would be set; cleared
otherwise.
Z: Set if all bits of the result of the "AND" would be cleared; cleared otherwise.
V: Cleared.
C: Not affected.

Addressing Modes	Execution Time (No. of cycles)	Number of bytes of machine code	Coding of First (or only) byte of machine code		
			HEX.	OCT.	DEC.
A IMM	2	2	85	205	133
A DIR	3	2	95	225	149
A EXT	4	3	B5	265	181
A IND	5	2	A5	245	165
B IMM	2	2	C5	305	197
B DIR	3	2	D5	325	213
B EXT	4	3	F5	365	245
B IND	5	2	E5	345	229

BLE

BLE   Branch if Less than or Equal to Zero   **BLE**

Operation: $PC \leftarrow (PC) + 0002 + Rel$ if $(Z)\odot[(N) \oplus (V)]=1$

         i.e. if $(ACCX) \leq (M)$

         (two's complement numbers)

Description: Causes a branch if [Z is set] OR [(N is set and V is clear) OR (N is clear and V is set)].

If the BLE instruction is executed immediately after execution of any of the instructions CBA, CMP, SBA, or SUB, the branch will occur if and only if the two's complement number represented by the minuend (i.e. ACCX) was less then or equal to the two's complement number represented by the subtrahend (i.e. M).

See BRA instruction for details of the branch.

Condition Codes: Not affected.

Addressing Modes	Execution Time (No. of cycles)	Number of bytes of machine code	Coding of First (or only) byte of machine code		
			HEX.	OCT.	DEC.
REL	4	2	2F	057	047

BLS

BLS   Branch if Lower or Same   **BLS**

Operation: $PC \leftarrow (PC) + 0002 + Rel$ if $(C)\odot(Z) = 1$

         i.e. if $(ACCX) \leq (M)$

         (unsigned binary numbers)

Description: Causes a branch if (C is set) OR (Z is set).

If the BLS instruction is executed immediately after execution of any of the instructions CBA, CMP, SBA, or SUB, the branch will occur if and only if the unsigned binary number represented by the minuend (i.e. ACCX) was less than or equal to the unsigned binary number represented by the subtrahend (i.e. M).

See BRA instruction for details of the execution of the branch.

Condition Codes: Not affected.

Addressing Modes	Execution Time (No. of cycles)	Number of bytes of machine code	Coding of First (or only) byte of machine code		
			HEX.	OCT.	DEC.
REL	4	2	23	043	035

BLT
Branch if Less than Zero
BLT

Operation: $PC \leftarrow (PC) + 0002 + Rel$ if $(N) \oplus (V) = 1$

i.e. if $(ACCX) < (M)$
(two's complement numbers)

Description: Causes a branch if (N is set and V is clear) OR (N is clear and V is set).

If the BLT instruction is executed immediately after execution of any of the instructions CBA, CMP, SBA, or SUB, the branch will occur if and only if the two's complement number represented by the minuend (i.e. ACCX) was less than the two's complement number represented by the subtrahend (i.e. M).

See BRA instruction for details of the branch.

Condition Codes: Not affected.

Addressing Modes	Execution Time (No. of cycles)	Number of bytes of machine code	Coding of First (or only) byte of machine code		
			HEX.	OCT.	DEC.
REL	4	2	2D	055	045

BMI
Branch if Minus
BMI

Operation: $PC \leftarrow (PC) + 0002 + Rel$ if $(N) = 1$

Description: Tests the state of the N bit and causes a branch if N is set.

See BRA instruction for details of the execution of the branch.

Condition Codes: Not affected.

Addressing Modes	Execution Time (No. of cycles)	Number of bytes of machine code	Coding of First (or only) byte of machine code		
			HEX.	OCT.	DEC.
REL	4	2	2B	053	043

BNE
Branch if Not Equal
BNE

Operation: $PC \leftarrow (PC) + 0002 + Rel$ if $(Z) = 0$

Description: Tests the state of the Z bit and causes a branch if the Z bit is clear.

See BRA instruction for details of the execution of the branch.

Condition Codes: Not affected.

Addressing Modes	Execution Time (No. of cycles)	Number of bytes of machine code	Coding of First (or only) byte of machine code		
			HEX.	OCT.	DEC.
REL	4	2	26	046	038

BPL   **Branch if Plus**   **BPL**

Operation: $PC \leftarrow (PC) + 0002 + Rel$ if $(N) = 0$

Description: Tests the state of the N bit and causes a branch if N is clear.

                   See BRA instruction for details of the execution of the branch.

Condition Codes: Not affected.

Addressing Modes	Execution Time (No. of cycles)	Number of bytes of machine code	Coding of First (or only) byte of machine code		
			HEX.	OCT.	DEC.
REL	4	2	2A	052	042

BRA   **Branch Always**   **BRA**

Operation: $PC \leftarrow (PC) + 0002 + Rel$

Description: Unconditional branch to the address given by the foregoing formula, in which R is the relative address stored as a two's complement number in the second byte of machine code corresponding to the branch instruction.

                   Note: The source program specifies the destination of any branch instruction by its absolute address, either as a numerical value or as a symbol or expression which can be numerically evaluated by the assembler. The assembler obtains the relative address R from the absolute address and the current value of the program counter PC.

Condition Codes: Not affected.

Addressing Modes	Execution Time (No. of cycles)	Number of bytes of machine code	Coding of First (or only) byte of machine code		
			HEX.	OCT.	DEC.
REL	4	2	20	040	032

BSR
Branch to Subroutine
BSR

Operation:

PC ← (PC) + 0002

↓ (PCL)

SP ← (SP) − 0001

↓ (PCH)

SP ← (SP) − 0001

PC ← (PC) + Rel

Description: The program counter is incremented by 2. The less significant byte of the contents of the program counter is pushed into the stack. The stack pointer is then decremented (by 1). The more significant byte of the contents of the program counter is then pushed into the stack. The stack pointer is again decremented (by 1). A branch then occurs to the location specified by the program.

See BRA instruction for details of the execution of the branch.

Condition Codes: Not affected.

Addressing Modes	Execution Time (No. of cycles)	Number of bytes of machine code	Coding of First (or only) byte of machine code		
			HEX.	OCT.	DEC.
REL	8	2	8D	215	141

BVC
Branch if Overflow Clear
BVC

Operation: PC ← (PC) + 0002 + Rel if (V) = 0

Description: Tests the state of the V bit and causes a branch if the V bit is clear.

See BRA instruction for details of the execution of the branch.

Condition Codes: Not affected.

Addressing Modes	Execution Time (No. of cycles)	Number of bytes of machine code	Coding of First (or only) byte of machine code		
			HEX.	OCT.	DEC.
REL	4	2	28	050	040

BVS
Branch if Overflow Set
BVS

Operation: PC ← (PC) + 0002 + Rel if (V) = 1

Description: Tests the state of the V bit and causes a branch if the V bit is set.

See BRA instruction for details of the execution of the branch.

Condition Codes: Not affected.

Addressing Modes	Execution Time (No. of cycles)	Number of bytes of machine code	Coding of First (or only) byte of machine code		
			HEX.	OCT.	DEC.
REL	4	2	29	051	041

CBA   Compare Accumulators   CBA

Operation: (ACCA) − (ACCB)

Description: Compares the contents of ACCA and the contents of ACCB and sets the condition codes, which may be used for arithmetic and logical conditional branches. Both operands are unaffected.

Condition Codes:
- H: Not affected.
- I: Not affected.
- N: Set if the most significant bit of the result of the subtraction would be set; cleared otherwise.
- Z: Set if all bits of the result of the subtraction would be cleared; cleared otherwise.
- V: Set if the subtraction would cause two's complement overflow; cleared otherwise.
- C: Set if the subtraction would require a borrow into the most significant bit of the result; clear otherwise.

Addressing Modes	Execution Time (No. of cycles)	Number of bytes of machine code	Coding of First (or only) byte of machine code		
			HEX.	OCT.	DEC.
INHERENT	2	1	11	021	017

CLC   Clear Carry   CLC

Operation: C bit ← 0

Description: Clears the carry bit in the processor condition codes register.

Condition Codes:
- H: Not affected.
- I: Not affected.
- N: Not affected.
- Z: Not affected.
- V: Not affected.
- C: Cleared

Addressing Modes	Execution Time (No. of cycles)	Number of bytes of machine code	Coding of First (or only) byte of machine code		
			HEX.	OCT.	DEC.
INHERENT	2	1	0C	014	012

CLI
Clear Interrupt Mask
CLI

Operation: I bit ← 0

Description: Clears the interrupt mask bit in the processor condition codes register. This enables the microprocessor to service an interrupt from a peripheral device if signalled by a high state of the "Interrupt Request" control input.

Condition Codes: H: Not affected.
                   I: Cleared.
                   N: Not affected.
                   Z: Not affected.
                   V: Not affected.
                   C: Not affected.

Addressing Modes	Execution Time (No. of cycles)	Number of bytes of machine code	Coding of First (or only) byte of machine code		
			HEX.	OCT.	DEC.
INHERENT	2	1	0E	016	014

CLR
Clear
CLR

Operation: ACCX ← 00
or: M ← 00

Description: The contents of ACCX or M are replaced with zeros.

Condition Codes: H: Not affected.
                   I: Not affected.
                   N: Cleared
                   Z: Set
                   V: Cleared
                   C: Cleared

Addressing Modes	Execution Time (No. of cycles)	Number of bytes of machine code	Coding of First (or only) byte of machine code		
			HEX.	OCT.	DEC.
A	2	1	4F	117	079
B	2	1	5F	137	095
EXT	6	3	7F	177	127
IND	7	2	6F	157	111

CLV

Clear Two's Complement Overflow Bit

CLV

Operation: V bit ← 0

Description: Clears the two's complement overflow bit in the processor condition codes register.

Condition Codes: H: Not affected.
I: Not affected.
N: Not affected.
Z: Not affected.
V: Cleared.
C: Not affected.

Addressing Modes	Execution Time (No. of cycles)	Number of bytes of machine code	Coding of First (or only) byte of machine code		
			HEX.	OCT.	DEC.
INHERENT	2	1	0A	012	010

CMP

Compare

CMP

Operation: (ACCX) – (M)

Description: Compares the contents of ACCX and the contents of M and determines the condition codes, which may be used subsequently for controlling conditional branching. Both operands are unaffected.

Condition Codes: H: Not affected.
I: Not affected.
N: Set if the most significant bit of the result of the subtraction would be set; cleared otherwise.
Z: Set if all bits of the result of the subtraction would be cleared; cleared otherwise.
V: Set if the subtraction would cause two's complement overflow; cleared otherwise.
C: Carry is set if the absolute value of the contents of memory is larger than the absolute value of the accumulator; reset otherwise.

Addressing Modes		Execution Time (No. of cycles)	Number of bytes of machine code	Coding of First (or only) byte of machine code		
				HEX.	OCT.	DEC.
A	IMM	2	2	81	201	129
A	DIR	3	2	91	221	145
A	EXT	4	3	B1	261	177
A	IND	5	2	A1	241	161
B	IMM	2	2	C1	301	193
B	DIR	3	2	D1	321	209
B	EXT	4	3	F1	361	241
B	IND	5	2	E1	341	225

COM

Complement

COM

Operation: ACCX ← ≈ (ACCX) = FF − (ACCX)

or: M ← ≈ (M) = FF − (M)

Description: Replaces the contents of ACCX or M with its one's complement. (Each bit of the contents of ACCX or M is replaced with the complement of that bit.)

Condition Codes:
H: Not affected.
I: Not affected.
N: Set if most significant bit of the result is set; cleared otherwise.
Z: Set if all bits of the result are cleared; cleared otherwise.
V: Cleared.
C: Set.

Addressing Modes	Execution Time (No. of cycles)	Number of bytes of machine code	Coding of First (or only) byte of machine code		
			HEX.	OCT.	DEC.
A	2	1	43	103	067
B	2	1	53	123	083
EXT	6	3	73	163	115
IND	7	2	63	143	099

CPX

Compare Index Register

CPX

Operation: (IXL) − (M+1)

(IXH) − (M)

Description: The more significant byte of the contents of the index register is compared with the contents of the byte of memory at the address specified by the program. The less significant byte of the contents of the index register is compared with the contents of the next byte of memory, at one plus the address specified by the program. The Z bit is set or reset according to the results of these comparisons, and may be used subsequently for conditional branching.

The N and V bits, though determined by this operation, are not intended for conditional branching.

The C bit is not affected by this operation.

Condition Codes:
H: Not affected.
I: Not affected.
N: Set if the most significant bit of the result of the subtraction from the more significant byte of the index register would be set; cleared otherwise.
Z: Set if all bits of the results of both subtractions would be cleared; cleared otherwise.
V: Set if the subtraction from the more significant byte of the index register would cause two's complement overflow; cleared otherwise.
C: Not affected.

Addressing Modes	Execution Time (No. of cycles)	Number of bytes of machine code	Coding of First (or only) byte of machine code		
			HEX.	OCT.	DEC.
IMM	3	3	8C	214	140
DIR	4	2	9C	234	156
EXT	5	3	BC	274	188
IND	6	2	AC	254	172

Decimal Adjust ACCA

Operation: Adds hexadecimal numbers 00, 06, 60, or 66 to ACCA, and may also set the carry bit, as indicated in the following table:

State of C-bit before DAA (Col. 1)	Upper Half-byte (bits 4-7) (Col. 2)	Initial Half-carry H-bit (Col.3)	Lower to ACCA (bits 0-3) (Col. 4)	Number Added after by DAA (Col. 5)	State of C-bit DAA (Col. 6)
0	0-9	0	0-9	00	0
0	0-8	0	A-F	06	0
0	0-9	1	0-3	06	0
0	A-F	0	0-9	60	1
0	9-F	0	A-F	66	1
0	A-F	1	0-3	66	1
1	0-2	0	0-9	60	1
1	0-2	0	A-F	66	1
1	0-3	1	0-3	66	1

Note: Columns (1) through (4) of the above table represent all possible cases which can result from any of the operations ABA, ADD, or ADC, with initial carry either set or clear, applied to two binary-coded-decimal operands. The table shows hexadecimal values.

Description: If the contents of ACCA and the state of the carry-borrow bit C and the half-carry bit H are all the result of applying any of the operations ABA, ADD, or ADC to binary-coded-decimal operands, with or without an initial carry, the DAA operation will function as follows.

Subject to the above condition, the DAA operation will adjust the contents of ACCA and the C bit to represent the correct binary-coded-decimal sum and the correct state of the carry.

Condition Codes: H: Not affected.

              I: Not affected.

              N: Set if most significant bit of the result is set; cleared otherwise.

              Z: Set if all bits of the result are cleared; cleared otherwise.

              V: Not defined.

              C: Set or reset according to the same rule as if the DAA and an immediately preceding ABA, ADD, or ADC were replaced by a hypothetical binary-coded-decimal addition.

Addressing Modes	Execution Time (No. of cycles)	Number of bytes of machine code	Coding of First (or only) byte of machine code		
			HEX.	OCT.	DEC.
INHERENT	2	1	19	031	025

DEC

Decrement

DEC

Operation: ACCX ← (ACCX) − 01

or: M ← (M) − 01

Description: Subtract one from the contents of ACCX or M.

The N, Z, and V condition codes are set or reset according to the results of this operation.

The C bit is not affected by the operation.

Condition Codes:
- H: Not affected.
- I: Not affected.
- N: Set if most significant bit of the result is set; cleared otherwise.
- Z: Set if all bits of the result are cleared; cleared otherwise.
- V: Set if there was two's complement overflow as a result of the operation; cleared otherwise. Two's complement overflow occurs if and only if (ACCX) or (M) was 80 before the operation.
- C: Not affected.

Addressing Modes	Execution Time (No. of cycles)	Number of bytes of machine code	Coding of First (or only) byte of machine code		
			HEX.	OCT.	DEC.
A	2	1	4A	112	074
B	2	1	5A	132	090
EXT	6	3	7A	172	122
IND	7	2	6A	152	106

DES

Decrement Stack Pointer

DES

Operation: SP ← (SP) − 0001

Description: Subtract one from the stack pointer.

Condition Codes: Not affected.

Addressing Modes	Execution Time (No. of cycles)	Number of bytes of machine code	Coding of First (or only) byte of machine code		
			HEX.	OCT.	DEC.
INHERENT	4	1	34	064	052

DEX Decrement Index Register **DEX**

Operation:	IX ← (IX) − 0001
Description:	Subtract one from the index register.
	Only the Z bit is set or reset according to the result of this operation.
Condition Codes:	H: Not affected.
	I: Not affected.
	N: Not affected.
	Z: Set if all bits of the result are cleared; cleared otherwise.
	V: Not affected.
	C: Not affected.

Addressing Modes	Execution Time (No. of cycles)	Number of bytes of machine code	Coding of First (or only) byte of machine code		
			HEX.	OCT.	DEC.
INHERENT	4	1	09	011	009

EOR Exclusive OR **EOR**

Operation:	ACCX ← (ACCX) $\oplus$ (M)
Description:	Perform logical "EXCLUSIVE OR" between the contents of ACCX and the contents of M, and place the result in ACCX. (Each bit of ACCX after the operation will be the logical "EXCLUSIVE OR" of the corresponding bit of M and ACCX before the operation.)
Condition Codes:	H: Not affected.
	I: Not affected.
	N: Set if most significant bit of the result is set; cleared otherwise.
	Z: Set if all bits of the result are cleared; cleared otherwise.
	V: Cleared
	C: Not affected.

Addressing Modes	Execution Time (No. of cycles)	Number of bytes of machine code	Coding of First (or only) byte of machine code		
			HEX.	OCT.	DEC.
A IMM	2	2	88	210	136
A DIR	3	2	98	230	152
A EXT	4	3	B8	270	184
A IND	5	2	A8	250	168
B IMM	2	2	C8	310	200
B DIR	3	2	D8	330	216
B EXT	4	3	F8	370	248
B IND	5	2	E8	350	232

INC

Increment

INC

Operation: ACCX ← (ACCX) + 01

or: M ← (M) + 01

Description: Add one to the contents of ACCX or M.

The N, Z, and V condition codes are set or reset according to the results of this operation.

The C bit is not affected by the operation.

Condition Codes:
- H: Not affected.
- I: Not affected.
- N: Set if most significant bit of the result is set; cleared otherwise.
- Z: Set if all bits of the result are cleared; cleared otherwise.
- V: Set if there was two's complement overflow as a result of the operation; cleared otherwise. Two's complement overflow will occur if and only if (ACCX) or (M) was 7F before the operation.
- C: Not affected.

Addressing Modes	Execution Time (No. of cycles)	Number of bytes of machine code	Coding of First (or only) byte of machine code		
			HEX.	OCT.	DEC.
A	2	1	4C	114	076
B	2	1	5C	134	092
EXT	6	3	7C	174	124
IND	7	2	6C	154	108

INS

Increment Stack Pointer

INS

Operation: SP ← (SP) + 0001

Description: Add one to the stack pointer.

Condition Codes: Not affected.

Addressing Modes	Execution Time (No. of cycles)	Number of bytes of machine code	Coding of First (or only) byte of machine code		
			HEX.	OCT.	DEC.
INHERENT	4	1	31	061	049

526

INX

Increment Index Register

INX

Operation: $IX \leftarrow (IX) + 0001$

Description: Add one to the index register.

Only the Z bit is set or reset according to the result of this operation.

Condition Codes: H: Not affected.
                   I: Not affected.
                   N: Not affected.
                   Z: Set if all 16 bits of the result are cleared; cleared otherwise.
                   V: Not affected.
                   C: Not affected.

Addressing Modes	Execution Time (No. of cycles)	Number of bytes of machine code	Coding of First (or only) byte of machine code		
			HEX.	OCT.	DEC.
INHERENT	4	1	08	010	008

JMP

Jump

JMP

Operation: $PC \leftarrow$ numerical address

Description: A jump occurs to the instruction stored at the numerical address. The numerical address is obtained according to the rules for EXTended or INDexed addressing.

Condition Codes: Not affected.

Addressing Modes	Execution Time (No. of cycles)	Number of bytes of machine code	Coding of First (or only) byte of machine code		
			HEX.	OCT.	DEC.
EXT	3	3	7E	176	126
IND	4	2	6E	156	110

JSR
Jump to Subroutine
JSR

Operation:

Either: $\quad$ PC ← (PC) + 0003 (for EXTended addressing)

or: $\quad$ PC ← (PC) + 0002 (for INDexed addressing)

Then: $\quad$ ↓ (PCL)

$\quad$ SP ← (SP) − 0001

$\quad$ ↓ (PCH)

$\quad$ SP ← (SP) − 0001

$\quad$ PC ← numerical address

Description: The program counter is incremented by 3 or by 2, depending on the addressing mode, and is then pushed onto the stack, eight bits at a time. The stack pointer points to the next empty location in the stack. A jump occurs to the instruction stored at the numerical address. The numerical address is obtained according to the rules for EXTended or INDexed addressing.

Condition Codes: Not affected.

Addressing Modes	Execution Time (No. of cycles)	Number of bytes of machine code	Coding of First (or only) byte of machine code		
			HEX.	OCT.	DEC.
EXT	9	3	BD	275	189
IND	8	2	AD	255	173

LDA
Load Accumulator
LDA

Operation: $\quad$ ACCX ← (M)

Description: Loads the contents of memory into the accumulator. The condition codes are set according to the data.

Condition Codes:
H: Not affected.
I: Not affected.
N: Set if most significant bit of the result is set; cleared otherwise.
Z: Set if all bits of the result are cleared; cleared otherwise.
V: Cleared.
C: Not affected.

Addressing Modes	Execution Time (No. of cycles)	Number of bytes of machine code	Coding of First (or only) byte of machine code		
			HEX.	OCT.	DEC.
A IMM	2	2	86	206	134
A DIR	3	2	96	226	150
A EXT	4	3	B6	266	182
A IND	5	2	A6	246	166
B IMM	2	2	C6	306	198
B DIR	3	2	D6	326	214
B EXT	4	3	F6	366	246
B IND	5	2	E6	346	230

LDS
Load Stack Pointer
LDS

Operation: SPH ← (M)
SPL ← (M+1)

Description: Loads the more significant byte of the stack pointer from the byte of memory at the address specified by the program, and loads the less significant byte of the stack pointer from the next byte of memory, at one plus the address specified by the program.

Condition Codes:
H: Not affected.
I: Not affected.
N: Set if the most significant bit of the stack pointer is set by the operation; cleared otherwise.
Z: Set if all bits of the stack pointer are cleared by the operation; cleared otherwise.
V: Cleared.
C: Not affected.

Addressing Modes	Execution Time (No. of cycles)	Number of bytes of machine code	Coding of First (or only) byte of machine code		
			HEX.	OCT.	DEC.
IMM	3	3	8E	216	142
DIR	4	2	9E	236	158
EXT	5	3	BE	276	190
IND	6	2	AE	256	174

LDX
Load Index Register
LDX

Operation: IXH ← (M)
IXL ← (M+1)

Description: Loads the more significant byte of the index register from the byte of memory at the address specified by the program, and loads the less significant byte of the index register from the next byte of memory, at one plus the address specified by the program.

Condition Codes:
H: Not affected.
I: Not affected.
N: Set if the most significant bit of the index register is set by the operation; cleared otherwise.
Z: Set if all bits of the index register are cleared by the operation; cleared otherwise.
V: Cleared.
C: Not affected.

Addressing Modes	Execution Time (No. of cycles)	Number of bytes of machine code	Coding of First (or only) byte of machine code		
			HEX.	OCT.	DEC.
IMM	3	3	CE	316	206
DIR	4	2	DE	336	222
EXT	5	3	FE	376	254
IND	6	2	EE	356	238

LSR
Logical Shift Right
LSR

Operation:

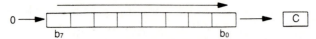

Description:
Shifts all bits of ACCX or M one place to the right. Bit 7 is loaded with a zero. The C bit is loaded from the least significant bit of ACCX or M.

Condition Codes:
H: Not affected.
I: Not affected.
N: Cleared.
Z: Set if all bits of the result are cleared; cleared otherwise.
V: Set if, after the completion of the shift operation, EITHER (N is set and C is cleared) OR (N is cleared and C is set); cleared otherwise.
C: Set if, before the operation, the least significant bit of the ACCX or M was set; cleared otherwise.

Addressing Modes	Execution Time (No. of cycles)	Number of bytes of machine code	Coding of First (or only) byte of machine code		
			HEX.	OCT.	DEC.
A	2	1	44	104	068
B	2	1	54	124	084
EXT	6	3	74	164	116
IND	7	2	64	144	100

NEG
Negate
NEG

Operation:
or:
ACCX ← − (ACCX) = 00 − (ACCX)
M ← − (M) = 00 − (M)

Description:
Replaces the contents of ACCX or M with its two's complement. Note that 80 is left unchanged.

Condition Codes:
H: Not affected.
I: Not affected.
N: Set if most significant bit of the result is set; cleared otherwise.
Z: Set if all bits of the result are cleared; cleared otherwise.
V: Set if there would be two's complement overflow as a result of the implied subtraction from zero; this will occur if and only if the contents of ACCX or M is 80.
C: Set if there would be a borrow in the implied subtraction from zero; the C bit will be set in all cases except when the contents of ACCX or M is 00.

Addressing Modes	Execution Time (No. of cycles)	Number of bytes of machine code	Coding of First (or only) byte of machine code		
			HEX.	OCT.	DEC.
A	2	1	40	100	064
B	2	1	50	120	080
EXT	6	3	70	160	112
IND	7	2	60	140	096

NOP

No Operation

NOP

Description: This is a single-word instruction which causes only the program counter to be incremented. No other registers are affected.

Condition Codes: Not affected.

Addressing Modes	Execution Time (No. of cycles)	Number of bytes of machine code	Coding of First (or only) byte of machine code		
			HEX.	OCT.	DEC.
INHERENT	2	1	01	001	001

ORA

Inclusive OR

ORA

Operation: $ACCX \leftarrow (ACCX) \odot (M)$

Description: Perform logical "OR" between the contents of ACCX and the contents of M and places the result in ACCX. (Each bit of ACCX after the operation will be the logical "OR" of the corresponding bits of M and of ACCX before the operation).

Condition Codes:
H: Not affected.
I: Not affected.
N: Set if most significant bit of the result is set; cleared otherwise.
Z: Set if all bits of the result are cleared; cleared otherwise.
V: Cleared.
C: Not affected.

Addressing Modes	Execution Time (No. of cycles)	Number of bytes of machine code	Coding of First (or only) byte of machine code		
			HEX.	OCT.	DEC.
A IMM	2	2	8A	212	138
A DIR	3	2	9A	232	154
A EXT	4	3	BA	272	186
A IND	5	2	AA	252	170
B IMM	2	2	CA	312	202
B DIR	3	2	DA	332	218
B EXT	4	3	FA	372	250
B IND	5	2	EA	352	234

PSH

Push Data Onto Stack

PSH

Operation: ↓ (ACCX)
SP ← (SP) − 0001

Description: The contents of ACCX is stored in the stack at the address contained in the stack pointer. The stack pointer is then decremented.

Condition Codes: Not affected.

Addressing Modes	Execution Time (No. of cycles)	Number of bytes of machine code	Coding of First (or only) byte of machine code		
			HEX.	OCT.	DEC.
A	4	1	36	066	054
B	4	1	37	067	055

PUL

Pull Data from Stack

PUL

Operation: SP ← (SP) + 0001
↑ ACCX

Description: The stack pointer is incremented. The ACCX is then loaded from the stack, from the address which is contained in the stack pointer.

Condition Codes: Not affected.

Addressing Modes	Execution Time (No. of cycles)	Number of bytes of machine code	Coding of First (or only) byte of machine code		
			HEX.	OCT.	DEC.
A	4	1	32	062	050
B	4	1	33	063	051

ROL
Rotate Left
ROL

Operation:

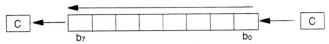

Description: Shifts all bits of ACCX or M one place to the left. Bit 0 is loaded from the C bit. The C bit is loaded from the most significant bit of ACCX or M.

Condition Codes:
H: Not affected.
I: Not affected.
N: Set if most significant bit of the result is set; cleared otherwise.
Z: Set if all bits of the result are cleared; cleared otherwise.
V: Set if, after the completion of the operation, EITHER (N is set and C is cleared) OR (N is cleared and C is set); cleared otherwise.
C: Set if, before the operation, the most significant bit of the ACCX or M was set; cleared otherwise.

Addressing Modes	Execution Time (No. of cycles)	Number of bytes of machine code	Coding of First (or only) byte of machine code		
			HEX.	OCT.	DEC.
A	2	1	49	111	073
B	2	1	59	131	089
EXT	6	3	79	171	121
IND	7	2	69	151	105

ROR
Rotate Right
ROR

Operation:

Description: Shifts all bits of ACCX or M one place to the right. Bit 7 is loaded from the C bit. The C bit is loaded from the least significant bit of ACCX or M.

Condition Codes:
H: Not affected.
I: Not affected.
N: Set if most significant bit of the result is set; cleared otherwise.
Z: Set if all bits of the result are cleared; cleared otherwise.
V: Set if, after the completion of the operation, EITHER (N is set and C is cleared) OR (N is cleared and C is set); cleared otherwise.
C: Set if, before the operation, the least significant bit of the ACCX or M was set; cleared otherwise.

Addressing Modes	Execution Time (No. of cycles)	Number of bytes of machine code	Coding of First (or only) byte of machine code		
			HEX.	OCT.	DEC.
A	2	1	46	106	070
B	2	1	56	126	086
EXT	6	3	76	166	118
IND	7	2	66	146	102

Operation: SP ← (SP) + 0001 , ↑CC
                 SP ← (SP) + 0001 , ↑ACCB
                 SP ← (SP) + 0001 , ↑ACCA
                 SP ← (SP) + 0001 , ↑IXH
                 SP ← (SP) + 0001 , ↑IXL
                 SP ← (SP) + 0001 , ↑PCH
                 SP ← (SP) + 0001 , ↑PCL

Description: The condition codes, accumulators B and A, the index register, and the program
                counter, will be restored to a state pulled from the stack. Note that the interrupt
                mask bit will be reset if and only if the corresponding bit stored in the stack is zero.

Condition Codes: Restored to the states pulled from the stack.

Addressing Modes	Execution Time (No. of cycles)	Number of bytes of machine code	Coding of First (or only) byte of machine code		
			HEX.	OCT.	DEC.
INHERENT	10	1	3B	073	059

Operation: SP ← (SP) + 0001
                 ↑ PCH
                 SP ← (SP) + 0001
                 ↑ PCL

Description: The stack pointer is incremented (by 1). The contents of the byte of memory, at the
                address now contained in the stack pointer, are loaded into the 8 bits of highest
                significance in the program counter. The stack pointer is again incremented (by
                1). The contents of the byte of memory, at the address now contained in the stack
                pointer, are loaded into the 8 bits of lowest significiance in the program counter.

Condition Codes: Not affected.

Addressing Modes	Execution Time (No. of cycles)	Number of bytes of machine code	Coding of First (or only) byte of machine code		
			HEX.	OCT.	DEC.
INHERENT	5	1	39	071	057

SBA   Subtract Accumulators   SBA

Operation: ACCA ← (ACCA) − (ACCB)

Description: Subtracts the contents of ACCB from the contents of ACCA and places the result in ACCA. The contents of ACCB are not affected.

Condition Codes: H: Not affected.

I: Not affected.

N: Set if most significant bit of the result is set; cleared otherwise.

Z: Set if all bits of the result are cleared; cleared otherwise.

V: Set if there was two's complement overflow as a result of the operation.

C: Carry is set if the absolute value of accumulator B plus previous carry is larger than the absolute value of accumulator A; reset otherwise.

Addressing Modes	Execution Time (No. of cycles)	Number of bytes of machine code	Coding of First (or only) byte of machine code		
			HEX.	OCT.	DEC.
INHERENT	2	1	10	020	016

SBC   Subtract with Carry   SBC

Operation: ACCX ← (ACCX) − (M) − (C)

Description: Subtracts the contents of M and C from the contents of ACCX and places the result in ACCX.

Condition Codes: H: Not affected.

I: Not affected.

N: Set if most significant bit of the result is set; cleared otherwise.

Z: Set if all bits of the result are cleared; cleared otherwise.

V: Set if there was two's complement overflow as a result of the operation; cleared otherwise.

C: Carry is set if the absolute value of the contents of memory plus previous carry is larger than the absolute value of the accumulator; reset otherwise.

Addressing Modes		Execution Time (No. of cycles)	Number of bytes of machine code	Coding of First (or only) byte of machine code		
				HEX.	OCT.	DEC.
A	IMM	2	2	82	202	130
A	DIR	3	2	92	222	146
A	EXT	4	3	B2	262	178
A	IND	5	2	A2	242	162
B	IMM	2	2	C2	302	194
B	DIR	3	2	D2	322	210
B	EXT	4	3	F2	362	242
B	IND	5	2	E2	342	226

SEC

Set Carry

SEC

Operation: C bit ← 1

Description: Sets the carry bit in the processor condition codes register.

Condition Codes: H: Not affected.
                 I: Not affected.
                 N: Not affected.
                 Z: Not affected.
                 V: Not affected.
                 C: Set.

Addressing Modes	Execution Time (No. of cycles)	Number of bytes of machine code	Coding of First (or only) byte of machine code		
			HEX.	OCT.	DEC.
INHERENT	2	1	0D	015	013

SEI

Set Interrupt Mask

SEI

Operation: I bit ← 1

Description: Sets the interrupt mask bit in the processor condition codes register. The microprocessor is inhibited from servicing an interrupt from a peripheral device, and will continue with execution of the instructions of the program, until the interrupt mask bit has been cleared.

Condition Codes: H: Not affected.
                 I: Set.
                 N: Not affected.
                 Z: Not affected.
                 V: Not affected.
                 C: Not affected.

Addressing Modes	Execution Time (No. of cycles)	Number of bytes of machine code	Coding of First (or only) byte of machine code		
			HEX.	OCT.	DEC.
INHERENT	2	1	0F	017	015

SEV

Set Two's Complement Overflow Bit

SEV

Operation: V bit ← 1

Description: Sets the two's complement overflow bit in the processor condition codes register.

Condition Codes:
- H: Not affected.
- I: Not affected.
- N: Not affected.
- Z: Not affected.
- V: Set.
- C: Not affected.

Addressing Modes	Execution Time (No. of cycles)	Number of bytes of machine code	Coding of First (or only) byte of machine code		
			HEX.	OCT.	DEC.
INHERENT	2	1	0B	013	011

STA

Store Accumulator

STA

Operation: M ← (ACCX)

Description: Stores the contents of ACCX in memory. The contents of ACCX remains unchanged.

Condition Codes:
- H: Not affected.
- I: Not affected.
- N: Set if the most significant bit of the contents of ACCX is set; cleared otherwise.
- Z: Set if all bits of the contents of ACCX are cleared; cleared otherwise.
- V: Cleared.
- C: Not affected.

Addressing Modes	Execution Time (No. of cycles)	Number of bytes of machine code	Coding of First (or only) byte of machine code		
			HEX.	OCT.	DEC.
A DIR	4	2	97	227	151
A EXT	5	3	B7	267	183
A IND	6	2	A7	247	167
B DIR	4	2	D7	327	215
B EXT	5	3	F7	367	247
B IND	6	2	E7	347	231

STS
Store Stack Pointer
STS

Operation: M ← (SPH)

                 M + 1 ← (SPL)

Description: Stores the more significant byte of the stack pointer in memory at the address specified by the program, and stores the less significant byte of the stack pointer at the next location in memory, at one plus the address specified by the program.

Condition Codes: H: Not affected.

                       I: Not affected.

                       N: Set if the most significant bit of the stack pointer is set; cleared otherwise.

                       Z: Set if all bits of the stack pointer are cleared; cleared otherwise.

                       V: Cleared.

                       C: Not affected.

Addressing Modes	Execution Time (No. of cycles)	Number of bytes of machine code	Coding of First (or only) byte of machine code		
			HEX.	OCT.	DEC.
DIR	5	2	9F	237	159
EXT	6	3	BF	277	191
IND	7	2	AF	257	175

STX
Store Index Register
STX

Operation: M ← (IXH)

                 M + 1 ← (IXL)

Description: Stores the more significant byte of the index register in memory at the address specified by the program, and stores the less significant byte of the index register at the next location in memory, at one plus the address specified by the program.

Condition Codes: H: Not affected.

                       I: Not affected.

                       N: Set if the most significant bite of the index register is set; cleared otherwise.

                       Z: Set if all bits of the index register are cleared; cleared otherwise.

                       V: Cleared.

                       C: Not affected.

Addressing Modes	Execution Time (No. of cycles)	Number of bytes of machine code	Coding of First (or only) byte of machine code		
			HEX.	OCT.	DEC.
DIR	5	2	DF	337	223
EXT	6	3	FF	377	255
IND	7	2	EF	357	239

Operation: ACCX ← (ACCX) − (M)

Description: Subtracts the contents of M from the contents of ACCX and places the result in ACCX.

Condition Codes:
- H: Not affected.
- I: Not affected.
- N: Set if most significant bit of the result is set; cleared otherwise.
- Z: Set if all bits of the result are cleared; cleared otherwise.
- V: Set if there was two's complement overflow as a result of the operation; cleared otherwise.
- C: Set if the absolute value of the contents of memory are larger than the absolute value of the accumulator; reset otherwise.

Addressing Modes	Execution Time (No. of cycles)	Number of bytes of machine code	Coding of First (or only) byte of machine code		
			HEX.	OCT.	DEC.
A IMM	2	2	80	200	128
A DIR	3	2	90	220	144
A EXT	4	3	B0	260	176
A IND	5	2	A0	240	160
B IMM	2	2	C0	300	192
B DIR	3	2	D0	320	208
B EXT	4	3	F0	360	240
B IND	5	2	E0	340	224

SWI

Operation: PC ← (PC) + 0001
                 ↓ (PCL) , SP ← (SP)-0001
                 ↓ (PCH) , SP ← (SP)-0001
                 ↓ (IXL) , SP ← (SP)-0001
                 ↓ (IXH) , SP ← (SP)-0001
                 ↓ (ACCA) , SP ← (SP)-0001
                 ↓ (ACCB) , SP ← (SP)-0001
                 ↓ (CC) , SP ← (SP)-0001
                 I ← 1
                 PCH ← (n-0005)
                 PCL ← (n-0004)

Description: The program counter is incremented (by 1). The program counter, index register, and accumulator A and B, are pushed into the stack. The condition codes register is then pushed into the stack, with condition codes H, I, N, Z, V, C going respectively into bit positions 5 thru 0, and the top two bits (in bit positions 7 and 6) are set (to the 1 state). The stack pointer is decremented (by 1) after each byte of data is stored in the stack.

                 The interrupt mask bit is then set. The program counter is then loaded with the address stored in the software interrupt pointer at memory locations (n-5) and (n-4), where n is the address corresponding to a high state on all lines of the address bus.

Condition Codes: H: Not affected.
                 I: Set.
                 N: Not affected.
                 Z: Not affected.
                 V: Not affected.
                 C: Not affected.

Addressing Modes	Execution Time (No. of cycles)	Number of bytes of machine code	Coding of First (or only) byte of machine code		
			HEX.	**OCT.**	**DEC.**
INHERENT	12	1	3F	077	063

TAB

Transfer from Accumulator A to Accumulator B

TAB

Operation: ACCB ← (ACCA)

Description: Moves the contents of ACCA to ACCB. The former contents of ACCB are lost. The contents of ACCA are not affected.

Condition Codes:
- H: Not affected.
- I: Not affected.
- N: Set if the most significant bit of the contents of the accumulator is set; cleared otherwise.
- Z: Set if all bits of the contents of the accumulator are cleared; cleared otherwise.
- V: Cleared.
- C: Not affected.

Addressing Modes	Execution Time (No. of cycles)	Number of bytes of machine code	Coding of First (or only) byte of machine code		
			HEX.	OCT.	DEC.
INHERENT	2	1	16	026	022

TAP

Transfer from Accumulator A to Processor Condition Codes Register

TAP

Operation: CC ← (ACCA)

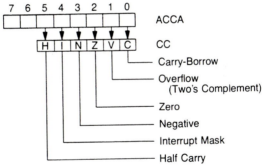

Bit Positions

Description: Transfers the contents of bit positions 0 thru 5 of accumulator A to the corresponding bit positions of the processor condition codes register. The contents of accumulator A remain unchanged.

Condition Codes: Set or reset according to the contents of the respective bits 0 thru 5 of accumulator

Addressing Modes	Execution Time (No. of cycles)	Number of bytes of machine code	Coding of First (or only) byte of machine code		
			HEX.	OCT.	DEC.
INHERENT	2	1	06	006	006

TBA — Transfer from Accumulator B to Accumulator A — TBA

Operation: ACCA ← (ACCB)

Description: Moves the contents of ACCB to ACCA. The former contents of ACCA are lost. The contents of ACCB are not affected.

Condition Codes:
- H: Not affected.
- I: Not affected.
- N: Set if the most significant accumulator bit is set; cleared otherwise.
- Z: Set if all accumulator bits are cleared; cleared otherwise.
- V: Cleared.
- C: Not affected.

Addressing Modes	Execution Time (No. of cycles)	Number of bytes of machine code	Coding of First (or only) byte of machine code		
			HEX.	OCT.	DEC.
INHERENT	2	1	17	027	023

TPA — Transfer from Processor Condition Codes Register to Accumulator A — TPA

Operation: ACCA ← (CC)

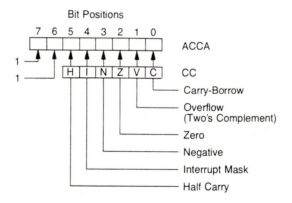

Description: Transfers the contents of the processor condition codes register to corresponding bit positions 0 thru 5 of accumulator A. Bit positions 6 and 7 of accumulator A are set (i.e. go to the "1" state). The processor condition codes register remains unchanged.

Condition Codes: Not affected.

Addressing Modes	Execution Time (No. of cycles)	Number of bytes of machine code	Coding of First (or only) byte of machine code		
			HEX.	OCT.	DEC.
INHERENT	2	1	07	007	007

TST   Test   TST

Operation: (ACCX) − 00
                  (M) − 00

Description: Set condition codes N and Z according to the contents of ACCX or M.

Condition Codes: H: Not affected.
                  I: Not affected.
                  N: Set if most significant bit of the contents of ACCX or M is set; cleared
                      otherwise.
                  Z: Set if all bits of the contents of ACCX or M are cleared; cleared otherwise.
                  V: Cleared.
                  C: Cleared.

Addressing Modes	Execution Time (No. of cycles)	Number of bytes of machine code	Coding of First (or only) byte of machine code		
			HEX.	OCT.	DEC.
A	2	1	4D	115	077
B	2	1	5D	135	093
EXT	6	3	7D	175	125
IND	7	2	6D	155	109

TSX   Transfer from Stack Pointer to Index Register   TSX

Operation: IX ← (SP) + 0001

Description: Loads the index register with one plus the contents of the stack pointer. The
                  contents of the stack pointer remain unchanged.

Condition Codes: Not affected.

Addressing Modes	Execution Time (No. of cycles)	Number of bytes of machine code	Coding of First (or only) byte of machine code		
			HEX.	OCT.	DEC.
INHERENT	4	1	30	060	048

TXS   Transfer From Index Register to Stack Pointer   TXS

Operation: SP ← (IX) − 0001

Description: Loads the stack pointer with the contents of the index register, minus one.
                  The contents of the index register remain unchanged.

Condition Codes: Not affected.

Addressing Modes	Execution Time (No. of cycles)	Number of bytes of machine code	Coding of First (or only) byte of machine code		
			HEX.	OCT.	DEC.
INHERENT	4	1	35	.065	053

Operation:
PC ← (PC) + 0001
↓ (PCL) , SP ← (SP)-0001
↓ (PCH) , SP ← (SP)-0001
↓ (IXL) , SP ← (SP)-0001
↓ (IXH) , SP ← (SP)-0001
↓ (ACCA) , SP ← (SP)-0001
↓ (ACCB) , SP ← (SP)-0001
↓ (CC) , SP ← (SP)-0001

Condition Codes: Not affected.

Description: The program counter is incremented (by 1). The program counter, index register, and accumulators A and B, are pushed into the stack. The condition codes register is then pushed into the stack, with condition codes H, I, N, Z, V, C going respectively into bit positions 5 thru 0, and the top two bits (in bit positions 7 and 6) are set (to the 1 state). The stack pointer is decremented (by 1) after each byte of data is stored in the stack.

Execution of the program is then suspended until an interrupt from a peripheral device is signalled, by the interrupt request control input going to a low state.

When an interrupt is signalled on the interrupt request line, and provided the I bit is clear, execution proceeds as follows. The interrupt mask bit is set. The program counter is then loaded with the address stored in the internal interrupt pointer at memory locations (n-7) and (n-6), where n is the address corresponding to a high state on all lines of the address bus.

Condition Codes:
H: Not affected.
I: Not affected until an interrupt request signal is detected on the interrupt request control line. When the interrupt request is received the I bit is set and further execution takes place, provided the I bit was initially clear.
N: Not affected.
Z: Not affected.
V: Not affected.
C: Not affected.

Addressing Modes	Execution Time (No. of cycles)	Number of bytes of machine code	Coding of First (or only) byte of machine code		
			HEX.	OCT.	DEC.
INHERENT	9	1	3E	076	062

6805
INSTRUCTION
SET

NOMENCLATURE

The following nomenclature is used in the executable instructions which follow this paragraph.

(a) Operators:

( )	indirection, i.e., (SP) means the value pointed to by SP
←	is loaded with (read: "gets")
•	boolean AND
v	boolean (inclusive) OR
⊕	boolean EXCLUSIVE OR
~	boolean NOT
−	negation (twos complement)

(b) Registers in the MPU:

ACCA	Accumulator (shown as A in Boolean formula for condition codes and source forms)
CC	Condition Code Register
X	Index Register
PC	Program Counter
PCH	Program Counter High Byte
PCL	Program Counter Low Byte
SP	Stack Pointer

(c) Memory and Addressing:

M	Contents of any memory location (one byte)
Rel	Relative address (i.e., the twos complement number stored in the second byte of machine code in a branch instruction)

(d) Bits in the Condition Code Register:

C	Carry/Borrow, Bit 0
Z	Zero Indicator, Bit 1
N	Negative Indicator, Bit 2
I	Interrupt Mask, Bit 3
H	Half Carry Indicator, Bit 4

(e) Status of Individual Bits BEFORE Execution of an Instruction

An	Bit n of ACCA (n = 7, 6, 5, 4, 3, 2, 1, 0)
Xn	Bit n of X (n = 7, 6, 5, 4, 3, 2, 1, 0)
Mn	Bit n of M (n = 7, 6, 5, 4, 3, 2, 1, 0). In read/modify/write instructions, Mn is used to represent bit n of M, A or X.

(f) Status of Individual Bits AFTER Execution of an Instruction:

Rn	Bit n of the result (n = 7, 6, 5, 4, 3, 2, 1, 0)

(g) Source Forms:

P	Operands with IMMediate, DIRect, EXTended and INDexed (0, 1, 2 byte offset) addressing modes
Q	Operands with DIRect, INDexed (0 and 1 byte offset) addressing modes
dd	Relative operands
DR	Operands with DIRect addressing mode only.

(h) iff   abbreviation for if-and-only-if.

ADC   Add with Carry   ADC

Operation: ACCA — ACCA + M + C

Description: Adds the contents of the C bit to the sum of the contents of ACCA and M, and places the result in ACCA.

**Condition
Codes:**
    H: Set if there was a carry from bit 3; cleared otherwise.
    I: Not affected.
    N: Set if the most significant bit of the result is set; cleared otherwise.
    Z: Set if all bits of the result are cleared; cleared otherwise.
    C: Set if there was a carry from the most significant bit of the result; cleared otherwise.

Addressing Mode	Cycles HMOS	CMOS	Bytes	Opcode
Inherent				
Relative				
Accumulator				
Index Register				
Immediate	2	2	2	A9
Direct	4	3	2	B9
Extended	5	4	3	C9
Indexed 0 Offset	4	3	1	F9
Indexed 1-Byte	5	4	2	E9
Indexed 2-Byte	6	5	3	D9

ADD   Add   ADD

Operation: ACCA — ACCA + M

Description: Adds the contents of ACCA and the contents of M and places the result in ACCA.

**Condition
Codes:**
    H: Set if there was a carry from bit 3; cleared otherwise.
    I: Not affected.
    N: Set if the most significant bit of the result is set; cleared otherwise.
    Z: Set if all bits of the result are cleared; cleared otherwise.
    C: Set if there was a carry from the most significant bit of the result; cleared otherwise.

Addressing Mode	Cycles HMOS	CMOS	Bytes	Opcode
Inherent				
Relative				
Accumulator				
Index Register				
Immediate	2	2	2	AB
Direct	4	3	2	BB
Extended	5	4	3	CB
Indexed 0 Offset	4	3	1	FB
Indexed 1-Byte	5	4	2	EB
Indexed 2-Byte	6	5	3	DB

548

AND

Logical AND

AND

Operation: ACCA — ACCA · M

Description: Performs logical AND between the contents of ACCA and the contents of M and places the result in ACCA. Each bit of ACCA after the operation will be the logical AND result of the corresponding bits of M and of ACCA before the operation.

Condition Codes:

H: Not affected.
I: Not affected.
N: Set if the most significant bit of the result is set; cleared otherwise.
Z: Set if all bits of the result are cleared; cleared otherwise.
C: Not affected.

Addressing Mode	Cycles HMOS	Cycles CMOS	Bytes	Opcode
Inherent				
Relative				
Accumulator				
Index Register				
Immediate	2	2	2	A4
Direct	4	3	2	B4
Extended	5	4	3	C4
Indexed 0 Offset	4	3	1	F4
Indexed 1-Byte	5	4	2	E4
Indexed 2-Byte	6	5	3	D4

ASL

Arithmetic Shift Left

ASL

Operation:

Description: Shifts all bits of ACCA, X or M one place to the left. Bit 0 is loaded with a zero. The C bit is loaded from the most significant bit of ACCA, X or M.

Condition Codes:

H: Not affected.
I: Not affected.
N: Set if the most significant bit of the result is set; cleared otherwise.
Z: Set if all bits of the result are cleared; cleared otherwise.
C: Set if, before the operation, the most significant bit of ACCA, X or M was set; cleared otherwise.

Comments: Same opcode as LSL

Addressing Mode	Cycles HMOS	Cycles CMOS	Bytes	Opcode
Inherent				
Relative				
Accumulator	4	3	1	48
Index Register	4	3	1	58
Immediate				
Direct	6	5	2	38
Extended				
Indexed 0 Offset	6	5	1	78
Indexed 1-Byte	7	6	2	68
Indexed 2-Byte				

549

ASR

Arithmetic Shift Right

ASR

Operation:

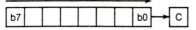

Description: Shifts all bits of ACCA, X or M one place to the right. Bit 7 is held constant. Bit 0 is loaded into the C bit.

Condition Codes:

- H: Not affected.
- I: Not affected.
- N: Set if the most significant bit of the result is set; cleared otherwise.
- Z: Set if all bits of the result are cleared; cleared otherwise.
- C: Set if, before the operation, the least significant bit of ACCA, X or M was set; cleared otherwise.

Addressing Mode	Cycles HMOS	CMOS	Bytes	Opcode
Inherent				
Relative				
Accumulator	4	3	1	47
Index Register	4	3	1	57
Immediate				
Direct	6	5	2	37
Extended				
Indexed 0 Offset	6	5	1	77
Indexed 1-Byte	7	6	2	67
Indexed 2-Byte				

BCC

Branch if Carry Clear

BCC

Operation: PC ← PC + 0002 + Rel iff C = 0

Description: Tests the state of the C bit and causes a branch iff C is clear. See BRA instruction for further details of the execution of the branch.

Condition Codes: Not affected.

Comments: Same opcode as BHS

Addressing Mode	Cycles HMOS	CMOS	Bytes	Opcode
Inherent				
Relative	4	3	2	24
Accumulator				
Index Register				
Immediate				
Direct				
Extended				
Indexed 0 Offset				
Indexed 1-Byte				
Indexed 2-Byte				

BCLR n

BCLR n   Clear Bit In Memory   **BCLR n**

Operation: Mn — 0

Description: Clear bit n (n = 0, 7) in location M. All other bits in M are unaffected.

**Condition
Codes:**   Not affected.

Addressing Mode	Cycles		Bytes	Opcode
	HMOS	CMOS		
Inherent				
Relative				
Accumulator				
Index Register				
Immediate				
Direct	7	5	2	11 + 2•n
Extended				
Indexed 0 Offset				
Indexed 1-Byte				
Indexed 2-Byte				

BCS

BCS   Branch if Carry Set   **BCS**

Operation: PC — PC + 0002 + Rel iff C = 1

Description: Tests the state of the C bit and causes a branch iff C is set. See BRA instruction for further details of the execution of the branch.

**Condition
Codes:**   Not affected.

Comments: Same opcode as BLO

Addressing Mode	Cycles		Bytes	Opcode
	HMOS	CMOS		
Inherent				
Relative	4	3	2	25
Accumulator				
Index Register				
Immediate				
Direct				
Extended				
Indexed 0 Offset				
Indexed 1-Byte				
Indexed 2-Byte				

BEQ

Branch if Equal

BEQ

Operation: PC ← PC + 0002 + Rel iff Z = 1

Description: Tests the state of the Z bit and causes a branch iff Z is set. Following a compare or subtract instruction BEQ will cause a branch if the arguments were equal. See BRA instruction for further details of the execution of the branch.

Condition Codes: Not affected.

Addressing Mode	Cycles HMOS	Cycles CMOS	Bytes	Opcode
Inherent				
Relative	4	3	2	27
Accumulator				
Index Register				
Immediate				
Direct				
Extended				
Indexed 0 Offset				
Indexed 1-Byte				
Indexed 2-Byte				

BHCC

Branch if Half Carry Clear

BHCC

Operation: PC ← PC + 0002 + Rel iff H = 0

Description: Tests the state of the H bit and causes a branch iff H is clear. See BRA instruction for further details of the execution of the branch.

Condition Codes: Not affected.

Addressing Mode	Cycles HMOS	Cycles CMOS	Bytes	Opcode
Inherent				
Relative	4	3	2	28
Accumulator				
Index Register				
Immediate				
Direct				
Extended				
Indexed 0 Offset				
Indexed 1-Byte				
Indexed 2-Byte				

552

BHCS

Branch if Half Carry Set

BHCS

Operation: PC ← PC + 0002 + Rel iff H = 1

Description: Tests the state of the H bit and causes a branch iff H is set. See BRA instruction for further details of the execution of the branch.

Condition Codes: Not affected.

Addressing Mode	Cycles HMOS	CMOS	Bytes	Opcode
Inherent				
Relative	4	3	2	29
Accumulator				
Index Register				
Immediate				
Direct				
Extended				
Indexed 0 Offset				
Indexed 1-Byte				
Indexed 2-Byte				

BHI

Branch if Higher

BHI

Operation: PC ← PC + 0002 + Rel iff (C v Z) = 0
i.e., if ACCA > M (unsigned binary numbers)

Description: Causes a branch iff both C and Z are zero. If the BHI instruction is executed immediately after execution of either of the CMP or SUB instructions, the branch will occur if and only if the unsigned binary number represented by the minuend (i.e., ACCA) was greater than the unsigned binary number represented by the subtrahend (i.e., M). See BRA instruction for further details of the execution of the branch.

Condition Codes: Not affected.

Addressing Mode	Cycles HMOS	CMOS	Bytes	Opcode
Inherent				
Relative	4	3	2	22
Accumulator				
Index Register				
Immediate				
Direct				
Extended				
Indexed 0 Offset				
Indexed 1-Byte				
Indexed 2-Byte				

BHS Branch if Higher or Same BHS

Operation: PC ← PC + 0002 + Rel iff C = 0

Description: Following an unsigned compare or subtract, BHS will cause a branch iff the register was higher than or the same as the location in memory. See BRA instruction for further details of the execution of the branch.

Condition Codes: Not affected.

Comments: Same opcode as BCC

Addressing Mode	Cycles HMOS	CMOS	Bytes	Opcode
Inherent				
Relative	4	3	2	24
Accumulator				
Index Register				
Immediate				
Direct				
Extended				
Indexed 0 Offset				
Indexed 1-Byte				
Indexed 2-Byte				

BIH Branch if Interrupt Line is High BIH

Operation: PC ← PC + 0002 + Rel iff $\overline{INT}$ = 1

Description: Tests the state of the external interrupt pin and branches iff it is high. See BRA instruction for further details of the execution of the branch.

Condition Codes: Not affected.

Comments: In systems not using interrupts, this instruction and BIL can be used to create an extra I/O input bit. This instruction does NOT test the state of the interrupt mask bit nor does it indicate whether an interrupt is pending. All it does is indicate whether the $\overline{INT}$ line is high.

Addressing Mode	Cycles HMOS	CMOS	Bytes	Opcode
Inherent				
Relative	4	3	2	2F
Accumulator				
Index Register				
Immediate				
Direct				
Extended				
Indexed 0 Offset				
Indexed 1-Byte				
Indexed 2-Byte				

BIL

Branch if Interrupt Line is Low

BIL

Operation: PC ← PC + 0002 + Rel iff $\overline{INT}$ = 0

Description: Tests the state of the external interrupt pin and branches iff it is low. See BRA instruction for further details of the execution of the branch.

Condition Codes: Not affected.

Comments: In systems not using interrupts, this instruction and BIH can be used to create an extra I/O input bit. This instruction does NOT test the state of the interrupt mask bit nor does it indicate whether an interrupt is pending. All it does is indicate whether the $\overline{INT}$ line is Low.

Addressing Mode	Cycles		Bytes	Opcode
	HMOS	CMOS		
Inherent				
Relative	4	3	2	2E
Accumulator				
Index Register				
Immediate				
Direct				
Extended				
Indexed 0 Offset				
Indexed 1-Byte				
Indexed 2-Byte				

BIT

Bit Test Memory with Accumulator

BIT

Operation: ACCA · M

Description: Performs the logical AND comparison of the contents of ACCA and the contents of M and modifies the condition codes accordingly. The contents of ACCA and M are unchanged.

Condition Codes:
H: Not affected.
I: Not affected.
N: Set if the most significant bit of the result of the AND is set; cleared otherwise.
Z: Set if all bits of the result of the AND are cleared; cleared otherwise.
C: Not affected.

Addressing Mode	Cycles		Bytes	Opcode
	HMOS	CMOS		
Inherent				
Relative				
Accumulator				
Index Register				
Immediate	2	2	2	A5
Direct	4	3	2	B5
Extended	5	4	3	C5
Indexed 0 Offset	4	3	1	F5
Indexed 1-Byte	5	4	2	E5
Indexed 2-Byte	6	5	3	D5

555

BLO

Branch if Lower

BLO

Operation: PC ← PC + 0002 + Rel iff C = 1

Description: Following a compare, BLO will branch iff the register was lower than the memory location. See BRA instruction for further details of the execution of the branch.

Condition Codes: Not affected.

Comments: Same opcode as BCS

Addressing Mode	Cycles HMOS	CMOS	Bytes	Opcode
Inherent				
Relative	4	3	2	25
Accumulator				
Index Register				
Immediate				
Direct				
Extended				
Indexed 0 Offset				
Indexed 1-Byte				
Indexed 2-Byte				

BLS

Branch if Lower or Same

BLS

Operation: PC ← PC + 0002 + Rel iff (C v Z) = 1
i.e., if ACCA ≤ M (unsigned binary numbers)

Description: Causes a branch if (C is set) OR (Z is set). If the BLS instruction is executed immediately after execution of either of the instructions CMP or SUB, the branch will occur if and only if the unsigned binary number represented by the minuend (i.e., ACCA) was less than or equal to the unsigned binary number represented by the subtrahend (i.e., M). See BRA instruction for further details of the execution of the branch.

Condition Codes: Not affected.

Addressing Mode	Cycles HMOS	CMOS	Bytes	Opcode
Inherent				
Relative	4	3	2	23
Accumulator				
Index Register				
Immediate				
Direct				
Extended				
Indexed 0 Offset				
Indexed 1-Byte				
Indexed 2-Byte				

BMC
Branch if Interrupt Mask is Clear
BMC

Operation: PC − PC + 0002 + Rel iff I = 0

Description: Tests the state of the I bit and causes a branch iff I is clear. See BRA instruction for further details of the execution of the branch.

Condition Codes: Not affected.

Comments: This instruction does NOT branch on the condition of the external interrupt line. The test is performed only on the interrupt mask bit.

Addressing Mode	Cycles		Bytes	Opcode
	HMOS	CMOS		
Inherent				
Relative	4	3	2	2C
Accumulator				
Index Register				
Immediate				
Direct				
Extended				
Indexed 0 Offset				
Indexed 1-Byte				
Indexed 2-Byte				

BMI
Branch if Minus
BMI

Operation: PC − PC + 0002 + Rel iff N = 1

Description: Tests the state of the N bit and causes a branch iff N is set. See BRA instruction for further details of the execution of the branch.

Condition Codes: Not affected.

Addressing Mode	Cycles		Bytes	Opcode
	HMOS	CMOS		
Inherent				
Relative	4	3	2	2B
Accumulator				
Index Register				
Immediate				
Direct				
Extended				
Indexed 0 Offset				
Indexed 1-Byte				
Indexed 2-Byte				

BMS

Branch if Interrupt Mask Bit is Set

BMS

Operation: PC ← PC + 0002 + Rel iff I = 1

Description: Tests the state of the I bit and causes a branch iff I is set. See BRA instruction for further details of the execution of the branch.

Condition Codes: Not affected.

Comments: This instruction does NOT branch on the condition of the external interrupt line. The test is performed only on the interrupt mask bit.

Addressing Mode	Cycles		Bytes	Opcode
	HMOS	CMOS		
Inherent				
Relative	4	3	2	2D
Accumulator				
Index Register				
Immediate				
Direct				
Extended				
Indexed 0 Offset				
Indexed 1-Byte				
Indexed 2-Byte				

BNE

Branch if Not Equal

BNE

Operation: PC ← PC + 0002 + Rel iff Z = 0

Description: Tests the state of the Z bit and causes a branch iff Z is clear. Following a compare or subtract instruction BNE will cause a branch if the arguments were different. See BRA instruction for further details of the execution of the branch.

Condition Codes: Not affected.

Addressing Mode	Cycles		Bytes	Opcode
	HMOS	CMOS		
Inherent				
Relative	4	3	2	26
Accumulator				
Index Register				
Immediate				
Direct				
Extended				
Indexed 0 Offset				
Indexed 1-Byte				
Indexed 2-Byte				

BPL
Branch if Plus
BPL

Operation: PC ← PC + 0002 + Rel iff N = 0

Description: Tests the state of the N bit and causes a branch iff N is clear. See BRA instruction for further details of the execution of the branch.

Condition Codes: Not affected.

Addressing Mode	Cycles HMOS	CMOS	Bytes	Opcode
Inherent				
Relative	4	3	2	2A
Accumulator				
Index Register				
Immediate				
Direct				
Extended				
Indexed 0 Offset				
Indexed 1-Byte				
Indexed 2-Byte				

BRA
Branch Always
BRA

Operation: PC ← PC + 0002 + Rel

Description: Unconditional branch to the address given by the foregoing formula, in which Rel is the relative address stored as a twos complement number in the second byte of machine code corresponding to the branch instruction.

NOTE: The source program specifies the destination of any branch instruction by its absolute address, either as a numerical value or as a symbol or expression which can be evaluated by the assembler. The assembler obtains the relative address Rel from the absolute address and the current value of the program counter.

Condition Codes: Not affected.

Addressing Mode	Cycles HMOS	CMOS	Bytes	Opcode
Inherent				
Relative	4	3	2	20
Accumulator				
Index Register				
Immediate				
Direct				
Extended				
Indexed 0 Offset				
Indexed 1-Byte				
Indexed 2-Byte				

BRCLR n Branch if Bit n is Clear BRCLR n

Operation: PC ← PC + 0003 + Rel iff bit n of M is zero

Description: Tests bit n (n = 0, 7) of location M and branches iff the bit is clear.

**Condition
Codes:**   H: Not affected.
           I: Not affected.
           N: Not affected.
           Z: Not affected.
           C: Set if Mn = 1; cleared otherwise.

Comments: The C bit is set to the state of the bit tested. Used with an appropriate rotate instruction, this instruction is an easy way to do serial to parallel conversions.

Addressing Mode	Cycles HMOS	CMOS	Bytes	Opcode
Inherent				
Relative	10	5	3	01 + 2•n
Accumulator				
Index Register				
Immediate				
Direct				
Extended				
Indexed 0 Offset				
Indexed 1-Byte				
Indexed 2-Byte				

BRN   Branch Never   BRN

Description: Never branches. Branch never is a 2 byte 4 cycle NOP.

**Condition
Codes:**   Not affected.

Comments: BRN is included here to demonstrate the nature of branches on the M6805 HMOS/M146805 CMOS Family. Each branch is matched with an inverse that varies only in the least significant bit of the opcode. BRN is the inverse of BRA. This instruction may have some use during program debugging.

Addressing Mode	Cycles HMOS	CMOS	Bytes	Opcode
Inherent				
Relative	4	3	2	21
Accumulator				
Index Register				
Immediate				
Direct				
Extended				
Indexed 0 Offset				
Indexed 1-Byte				
Indexed 2-Byte				

560

BRSET n   <small>Branch if Bit n is Set</small>   # BRSET n

Operation: $PC \leftarrow PC + 0003 + Rel$ iff Bit n of M is not zero

Description: Tests bit n (n = 0, 7) of location M and branches iff the bit is set.

**Condition
Codes:**
   H: Not affected.
   I: Not affected.
   N: Not affected.
   Z: Not affected.
   C: Set if Mn = 1; cleared otherwise.

Comments: The C bit is set to the state of the bit tested. Used with an appropriate rotate instruction, this instruction is an easy way to provide serial to parallel conversions.

Addressing Mode	Cycles HMOS	CMOS	Bytes	Opcode
Inherent				
Relative	10	5	3	2•n
Accumulator				
Index Register				
Immediate				
Direct				
Extended				
Indexed 0 Offset				
Indexed 1-Byte				
Indexed 2-Byte				

BSET n   <small>Set Bit in Memory</small>   # BSET n

Operation: $Mn \leftarrow 1$

Description: Set bit n (n = 0, 7) in location M. All other bits in M are unaffected.

**Condition
Codes:**   Not affected.

Addressing Mode	Cycles HMOS	CMOS	Bytes	Opcode
Inherent				
Relative				
Accumulator				
Index Register				
Immediate				
Direct	7	5	2	10 + 2•n
Extended				
Indexed 0 Offset				
Indexed 1-Byte				
Indexed 2-Byte				

BSR

Branch to Subroutine

BSR

Operation: PC ← PC + 0002
               (SP) ← PCL; SP ← SP − 0001
               (SP) ← PCH; SP ← SP − 0001
               PC ← PC + Rel

Description: The program counter is incremented by 2. The least (low) significant byte of the program counter contents is pushed onto the stack. The stack pointer is then decremented (by one). The most (high) significant byte of the program counter contents is then pushed onto the stack. Unused bits in the program counter high byte are stored as 1s on the stack. The stack pointer is again decremented (by one). A branch then occurs to the location specified by the relative offset. See the BRA instruction for details of the branch execution.

**Condition
Codes:**   Not affected.

Addressing Mode	Cycles		Bytes	Opcode
	HMOS	CMOS		
Inherent				
Relative	8	6	2	AD
Accumulator				
Index Register				
Immediate				
Direct				
Extended				
Indexed 0 Offset				
Indexed 1-Byte				
Indexed 2-Byte				

CLC

Clear Carry Bit

CLC

Operation: C bit ← 0

Description: Clears the carry bit in the processor condition code register.

**Condition
Codes:**   H: Not affected.
              I: Not affected.
              N: Not affected.
              Z: Not affected.
              C: Cleared.

Addressing Mode	Cycles		Bytes	Opcode
	HMOS	CMOS		
Inherent	2	2	1	98
Relative				
Accumulator				
Index Register				
Immediate				
Direct				
Extended				
Indexed 0 Offset				
Indexed 1-Byte				
Indexed 2-Byte				

562

CLI

Clear Interrupt Mask Bit

CLI

Operation: I bit — 0

Description: Clears the interrupt mask bit in the processor condition code register. This enables the microprocessor to service interrupts. Interrupts that were pending while the I bit was set will now begin to have effect.

Condition Codes:

H: Not affected.
I: Cleared
N: Not affected.
Z: Not affected.
C: Not affected.

Addressing Mode	Cycles HMOS	CMOS	Bytes	Opcode
Inherent	2	2	1	9A
Relative				
Accumulator				
Index Registers				
Immediate				
Direct				
Extended				
Indexed 0 Offset				
Indexed 1-Byte				
Indexed 2-Byte				

CLR

Clear

CLR

Operation:
X — 00 or,
ACCA — 00 or,
M — 00

Description: The contents of ACCA, X, or M are replaced with zeroes.

Condition Codes:

H: Not affected.
I: Not affected.
N: Cleared.
Z: Set.
C: Not affected.

Addressing Mode	Cycles HMOS	CMOS	Bytes	Opcode
Inherent				
Relative				
Accumulator	4	3	1	4F
Index Register	4	3	1	5F
Immediate				
Direct	6	5	2	3F
Extended				
Indexed 0 Offset	6	5	1	7F
Indexed 1-Byte	7	6	2	6F
Indexed 2-Byte				

CMP

Compare Accumulator with Memory

CMP

Operation: ACCA − M

Description: Compares the contents of ACCA and the contents of M and sets the condition codes, which may then be used for controlling the conditional branches. Both operands are unaffected.

Condition Codes:

H: Not affected.
I: Not affected.
N: Set if the most significant bit of the result of the subtraction is set; cleared otherwise.
Z: Set if all bits of the result of the subtraction are cleared; cleared otherwise.
C: Set if the absolute value of the contents of memory is larger than the absolute value of the accumulator; cleared otherwise.

Addressing Mode	Cycles		Bytes	Opcode
	HMOS	CMOS		
Inherent				
Relative				
Accumulator				
Index Register				
Immediate	2	2	2	A1
Direct	4	3	2	B1
Extended	5	4	3	C1
Indexed 0 Offset	4	3	1	F1
Indexed 1-Byte	5	4	2	E1
Indexed 2-Byte	6	5	3	D1

COM

Complement

COM

Operation:
X ← ~X = $FF − X or,
ACCA ← ~ACCA = $FF − ACCA or,
M ← ~M = $FF − M

Description: Replaces the contents of ACCA, X, or M with the ones complement. Each bit of the operand is replaced with the complement of that bit.

Condition Codes:

H: Not affected.
I: Not affected.
N: Set if the most significant bit of the result is set; cleared otherwise.
Z: Set if all bits of the result are cleared; cleared otherwise.
C: Set.

Addressing Mode	Cycles		Bytes	Opcode
	HMOS	CMOS		
Inherent				
Relative				
Accumulator	4	3	1	43
Index Register	4	3	1	53
Immediate				
Direct	6	5	2	33
Extended				
Indexed 0 Offset	6	5	1	73
Indexed 1-Byte	7	6	2	63
Indexed 2-Byte				

CPX

Compare Index Register with Memory

CPX

Operation: X − M

Description: Compares the contents of X to the contents of M and sets the condition codes, which may then be used for controlling the conditional branches. Both operands are unaffected.

Condition Codes:
- H: Not affected.
- I: Not affected.
- N: Set if the most significant bit of the result of the subtraction is set; cleared otherwise.
- Z: Set if all bits of the result of the subtraction are cleared; cleared otherwise.
- C: Set if the absolute value of the contents of memory is larger than the absolute value of the index register; cleared otherwise.

Addressing Mode	Cycles		Bytes	Opcode
	HMOS	CMOS		
Inherent				
Relative				
Accumulator				
Index Register				
Immediate	2	2	2	A3
Direct	4	3	2	B3
Extended	5	4	3	C3
Indexed 0 Offset	4	3	1	F3
Indexed 1-Byte	5	4	2	E3
Indexed 2-Byte	6	5	3	D3

DEC

Decrement

DEC

Operation:
X ← X − 01 or,
ACCA ← ACCA − 01 or,
M ← M − 01

Description: Subtract one from the contents of ACCA, X, or M. The N and Z bits are set or reset according to the result of this operation. The C bit is not affected by this operation.

Condition Codes:
- H: Not affected.
- I: Not affected.
- N: Set if the most significant bit of the result is set; cleared otherwise.
- Z: Set if all bits of the result are cleared; cleared otherwise.
- C: Not affected.

Addressing Mode	Cycles		Bytes	Opcode
	HMOS	CMOS		
Inherent				
Relative				
Accumulator	4	3	1	4A
Index Register	4	3	1	5A
Immediate				
Direct	6	5	2	3A
Extended				
Indexed 0 Offset	6	5	1	7A
Indexed 1-Byte	7	6	2	6A
Indexed 2-Byte				

EOR

EOR

Exclusive Or Memory with Accumulator

Operation: ACCA ← ACCA ⊕ M

Description: Performs the logical EXCLUSIVE OR between the contents of ACCA and the contents of M, and places the result in ACCA. Each bit of ACCA after the operation will be the logical EXCLUSIVE OR of the corresponding bit of M and ACCA before the operation.

Condition Codes:

H: Not affected.
I: Not affected.
N: Set if the most significant bit of the result is set; cleared otherwise.
Z: Set if all bits of the result are cleared; cleared otherwise.
C: Not affected.

Addressing Mode	Cycles HMOS	CMOS	Bytes	Opcode
Inherent				
Relative				
Accumulator				
Index Register				
Immediate	2	2	2	A8
Direct	4	3	2	B8
Extended	5	4	3	C8
Indexed 0 Offset	4	3	1	F8
Indexed 1-Byte	5	4	2	E8
Indexed 2-Byte	6	5	3	D8

INC

INC

Increment

Operation:
X ← X + 01 or,
ACCA ← ACCA + 01 or,
M ← M + 01

Description: Add one to the contents of ACCA, X, or M. The N and Z bits are set or reset according to the result of this operation. The C bit is not affected by this operation.

Condition Codes:

H: Not affected.
I: Not affected.
N: Set if the most significant bit of the result is set; cleared otherwise.
Z: Set if all bits of the result are cleared; cleared otherwise.
C: Not affected.

Addressing Mode	Cycles HMOS	CMOS	Bytes	Opcode
Inherent				
Relative				
Accumulator	4	3	1	4C
Index Register	4	3	1	5C
Immediate				
Direct	6	5	2	3C
Extended				
Indexed 0 Offset	6	5	1	7C
Indexed 1-Byte	7	6	2	6C
Indexed 2-Byte				

JMP

Jump

JMP

Operation: PC ← effective address

Description: A jump occurs to the instruction stored at the effective address. The effective address is obtained according to the rules for EXTended, DIRect or INDexed addressing.

Condition Codes: Not affected.

Addressing Mode	Cycles HMOS	Cycles CMOS	Bytes	Opcode
Inherent				
Relative				
Accumulator				
Index Register				
Immediate				
Direct	3	2	2	BC
Extended	4	3	3	CC
Indexed 0 Offset	3	2	1	FC
Indexed 1-Byte	4	3	2	EC
Indexed 2-Byte	5	4	3	DC

JSR

Jump to Subroutine

JSR

Operation:
PC ← PC + N
(SP) ← PCL; SP ← SP − 0001
(SP) ← PCH ; SP ← SP − 0001
PC ← effective address

Description: The program counter is incremented by N (N = 1, 2, or 3 depending on the addressing mode), and is then pushed onto the stack (least significant byte first). Unused bits in the program counter high byte are stored as 1s on the stack. The stack pointer points to the next empty location on the stack. A jump occurs to the instruction stored at the effective address. The effective address is obtained according to the rules for EXTended, DIRect, or INDexed addressing.

Condition Codes: Not affected.

Addressing Mode	Cycles HMOS	Cycles CMOS	Bytes	Opcode
Inherent				
Relative				
Accumulator				
Index Register				
Immediate				
Direct	7	5	2	BD
Extended	8	6	3	CD
Indexed 0 Offset	7	5	1	FD
Indexed 1-Byte	8	6	2	ED
Indexed 2-Byte	9	7	3	DD

567

LDA   Load Accumulator from Memory   LDA

Operation: ACCA ← M

Description: Loads the contents of memory into the accumulator. The condition code: are set according to the data.

Condition Codes:

H: Not affected.
I: Not affected.
N: Set if the most significant bit of the accumulator is set; cleared otherwise.
Z: Set if all bits of the accumulator are cleared; cleared otherwise.
C: Not affected.

Addressing Mode	Cycles		Bytes	Opcode
	HMOS	CMOS		
Inherent				
Relative				
Accumulator				
Index Register				
Immediate	2	2	2	A6
Direct	4	3	2	B6
Extended	5	4	3	C6
Indexed 0 Offset	4	3	1	F6
Indexed 1-Byte	5	4	2	E6
Indexed 2-Byte	6	5	3	D6

LDX   Load Index Register from Memory   LDX

Operation: X ← M

Description: Loads the contents of memory into the index register. The condition codes are set according to the data.

Condition Codes:

H: Not affected.
I: Not affected.
N: Set if the most significant bit of the index register is set; cleared otherwise.
Z: Set if all bits of the index register are cleared; cleared otherwise.
C: Not affected.

Addressing Mode	Cycles		Bytes	Opcode
	HMOS	CMOS		
Inherent				
Relative				
Accumulator				
Index Register				
Immediate	2	2	2	AE
Direct	4	3	2	BE
Extended	5	4	3	CE
Indexed 0 Offset	4	3	1	FE
Indexed 1-Byte	5	4	2	EE
Indexed 2-Byte	6	5	3	DE

LSL

Logical Shift Left

LSL

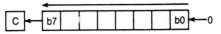

Operation:

Description: Shifts all bits of the ACCA, X or M one place to the left. Bit 0 is loaded with a zero. The C bit is loaded from the most significant bit of ACCA, X or M.

Condition Codes:

H: Not affected.
I: Not affected.
N: Set if the most significant bit of the result is set; cleared otherwise.
Z: Set if all bits of the result are cleared; cleared otherwise.
C: Set if, before the operation, the most significant bit of ACCA, X or M was set; cleared otherwise.

Comments: Same as ASL

Addressing Mode	Cycles HMOS	CMOS	Bytes	Opcode
Inherent				
Relative				
Accumulator	4	3	1	48
Index Register	4	3	1	58
Immediate				
Direct	6	5	2	38
Extended				
Indexed 0 Offset	6	5	1	78
Indexed 1-Byte	7	6	2	68
Indexed 2-Byte				

LSR

Logical Shift Right

LSR

 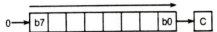

Operation:

Description: Shifts all bits of ACCA, X or M one place to the right. Bit 7 is loaded with a zero. Bit 0 is loaded into the C bit.

Condition Codes:

H: Not affected.
I: Not affected.
N: Cleared.
Z: Set if all bits of the result are cleared; cleared otherwise.
C: Set if, before the operation, the least significant bit of ACCA, X or M was set; cleared otherwise.

Addressing Mode	Cycles HMOS	CMOS	Bytes	Opcode
Inherent				
Relative				
Accumulator	4	3	1	44
Index Register	4	3	1	54
Immediate				
Direct	6	5	2	34
Extended				
Indexed 0 Offset	6	5	1	74
Indexed 1-Byte	7	6	2	64
Indexed 2-Byte				

NEG

Negate

NEG

Operation: X ← − X (i.e., 00 − X) or,
ACCA ← − ACCA (i.e., 00 − ACCA) or,
M ← − M (i.e., 00 − M)

Description: Replaces the contents of ACCA, X or M with its twos complement. Note that $80 is left unchanged.

Condition Codes:

H: Not affected.
I: Not affected.
N: Set if the most significant bit of the result is set; cleared otherwise.
Z: Set if all bits of the result are cleared; cleared otherwise.
C: Set if there would be a borrow in the implied subtraction from zero; the C bit will be set in all cases except when the contents of ACCA, X or M before the NEG is 00.

Addressing Mode	Cycles HMOS	CMOS	Bytes	Opcode
Inherent				
Relative				
Accumulator	4	3	1	40
Index Register	4	3	1	50
Immediate				
Direct	6	5	2	30
Extended				
Indexed 0 Offset	6	5	1	70
Indexed 1-Byte	7	6	2	60
Indexed 2-Byte				

NOP

No Operation

NOP

Description: This is a single-byte instruction which causes only the program counter to be incremented. No other registers are changed.

Condition Codes: Not affected.

Addressing Mode	Cycles HMOS	CMOS	Bytes	Opcode
Inherent	2	2	1	9D
Relative				
Accumulator				
Index Register				
Immediate				
Direct				
Extended				
Indexed 0 Offset				
Indexed 1-Byte				
Indexed 2-Byte				

ORA ORA

Inclusive OR

Operation: ACCA — ACCA v M

Description: Performs logical OR between the contents of ACCA and the contents of M and places the result in ACCA. Each bit of ACCA after the operation will be the logical (inclusive) OR result of the corresponding bits of M and ACCA before the operation.

Condition Codes:

H: Not affected.
I: Not affected.
N: Set if the most significant bit of the result is set; cleared otherwise.
Z: Set if all bits of the result are cleared; cleared otherwise.
C: Not affected.

Addressing Mode	Cycles HMOS	CMOS	Bytes	Opcode
Inherent				
Relative				
Accumulator				
Index Register				
Immediate	2	2	2	AA
Direct	4	3	2	BA
Extended	5	4	3	CA
Indexed 0 Offset	4	3	1	FA
Indexed 1-Byte	5	4	2	EA
Indexed 2-Byte	6	5	3	DA

ROL ROL

Rotate Left thru Carry

Operation: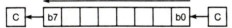

Description: Shifts all bits of the ACCA, X, or M one place to the left. Bit 0 is loaded from the C bit. The C bit is loaded from the most significant bit of ACCA, X, or M.

Condition Codes:

H: Not affected.
I: Not affected.
N: Set if the most significant bit of the result is set; cleared otherwise.
Z: Set if all bits of the result are cleared; cleared otherwise.
C: Set if, before the operation, the most significant bit of ACCA, X or M was set; cleared otherwise.

Addressing Mode	Cycles HMOS	CMOS	Bytes	Opcode
Inherent				
Relative				
Accumulator	4	3	1	49
Index Register	4	3	1	59
Immediate				
Direct	6	5	2	39
Extended				
Indexed 0 Offset	6	5	1	79
Indexed 1-Byte	7	6	2	69
Indexed 2-Byte				

ROR

Rotate Right Thru Carry

ROR

Operation:

```
          ┌─────────────────────────────────┐
          │                                 ↓
 ┌───┐   ┌────┬───┬───┬───┬───┬───┬────┐   ┌───┐
 │ C │──▶│ b7 │   │   │   │   │   │ b0 │──▶│ C │
 └───┘   └────┴───┴───┴───┴───┴───┴────┘   └───┘
```

Description: Shifts all bits of ACCA, X, or M one place to the right. Bit 7 is loaded from the C bit. Bit 0 is loaded into the C bit.

Condition Codes:

H: Not affected.
I: Not affected.
N: Set if the most significant bit of the result is set; cleared otherwise.
Z: Set if all bits of the result are cleared; cleared otherwise.
C: Set if, before the operation, the least significant bit of ACCA, X or M was set; cleared otherwise.

Addressing Mode	Cycles		Bytes	Opcode
	HMOS	CMOS		
Inherent				
Relative				
Accumulator	4	3	1	46
Index Register	4	3	1	56
Immediate				
Direct	6	5	2	36
Extended				
Indexed 0 Offset	6	5	1	76
Indexed 1-Byte	7	6	2	66
Indexed 2-Byte				

RSP

Reset Stack Pointer

RSP

Operation: SP ← $7F

Description: Resets the stack pointer to the top of the stack.

Condition Codes: Not affected.

Addressing Mode	Cycles		Bytes	Opcode
	HMOS	CMOS		
Inherent	2	2	1	9C
Relative				
Accumulator				
Index Register				
Immediate				
Direct				
Extended				
Indexed 0 Offset				
Indexed 1-Byte				
Indexed 2-Byte				

RTI
Return from Interrupt
RTI

Operation:
SP ← SP + 0001 ; CC ← (SP)
SP ← SP + 0001 ; ACCA ← (SP)
SP ← SP + 0001 ; X ← (SP)
SP ← SP + 0001 ; PCH ← (SP)
SP ← SP + 0001 ; PCL ← (SP)

Description: The condition codes, accumulator, index register, and the program counter are restored according to the state previously saved on the stack. Note that the interrupt mask bit (I bit) will be reset if and only if the corresponding bit stored on the stack is zero.

Condition Codes: Set or cleared according to the first byte pulled from the stack.

Addressing Mode	Cycles HMOS	CMOS	Bytes	Opcode
Inherent	9	9	1	80
Relative				
Accumulator				
Index Register				
Immediate				
Direct				
Extended				
Indexed 0 Offset				
Indexed 1-Byte				
Indexed 2-Byte				

RTS
Return from Subroutine
RTS

Operation:
SP ← SP + 0001 ; PCH ← (SP)
SP ← SP + 0001 ; PCL ← (SP)

Description: The stack pointer is incremented (by one). The contents of the byte of memory, pointed to by the stack pointer, are loaded into the high byte of the program counter. The stack pointer is again incremented (by one). The byte pointed to by the stack pointer is loaded into the low byte of the program counter.

Condition Codes: Not affected.

Addressing Mode	Cycles HMOS	CMOS	Bytes	Opcode
Inherent	6	6	1	81
Relative				
Accumulator				
Index Register				
Immediate				
Direct				
Extended				
Indexed 0 Offset				
Indexed 1-Byte				
Indexed 2-Byte				

573

SBC

SBC   Subtract with Carry   **SBC**

Operation: ACCA — ACCA − M − C

Description: Subtracts the contents of M and C from the contents of ACCA, and places the result in ACCA.

Condition Codes:

H: Not affected.
I: Not affected.
N: Set if the most significant bit of the result is set; cleared otherwise.
Z: Set if all bits of the result are cleared; cleared otherwise.
C: Set if the absolute value of the contents of memory plus the previous carry is larger than the absolute value of the accumulator; cleared otherwise.

Addressing Mode	Cycles HMOS	CMOS	Bytes	Opcode
Inherent				
Relative				
Accumulator				
Index Register				
Immediate	2	2	2	A2
Direct	4	3	2	B2
Extended	5	4	3	C2
Indexed 0 Offset	4	3	1	F2
Indexed 1-Byte	5	4	2	E2
Indexed 2-Byte	6	5	3	D2

SEC

SEC   Set Carry Bit   **SEC**

Operation: C bit ← 1

Description: Sets the carry bit in the processor condition code register.

Condition Codes:

H: Not affected.
I: Not affected.
N: Not affected.
Z: Not affected.
C: Set.

Addressing Mode	Cycles HMOS	CMOS	Bytes	Opcode
Inherent	2	2	1	99
Relative				
Accumulator				
Index Register				
Immediate				
Direct				
Extended				
Indexed 0 Offset				
Indexed 1-Byte				
Indexed 2-Byte				

SEI
Set Interrupt Mask Bit
SEI
SEI

Operation: I bit — 1

Description: Sets the interrupt mask bit in the processor condition code register. The microprocessor is inhibited from servicing interrupts, and will continue with execution of the instructions of the program until the interrupt mask bit is cleared.

Condition Codes:

H: Not affected.
I: Set
N: Not affected.
Z: Not affected.
C: Not affected.

Addressing Mode	Cycles HMOS	CMOS	Bytes	Opcode
Inherent	2	2	1	9B
Relative				
Accumulator				
Index Register				
Immediate				
Direct				
Extended				
Indexed 0 Offset				
Indexed 1-Byte				
Indexed 2-Byte				

STA
Store Accumulator in Memory
STA
STA

Operation: M — ACCA

Description: Stores the contents of ACCA in memory. The contents of ACCA remain the same.

Condition Codes:

H: Not affected.
I: Not affected.
N: Set if the most significant bit of the accumulator is set; cleared otherwise.
Z: Set if all bits of the accumulator are clear; cleared otherwise.
C: Not affected.

Addressing Mode	Cycles HMOS	CMOS	Bytes	Opcode
Inherent				
Relative				
Accumulator				
Index Register				
Immediate				
Direct	5	4	2	B7
Extended	6	5	3	C7
Indexed 0 Offset	5	4	1	F7
Indexed 1-Byte	6	5	2	E7
Indexed 2-Byte	7	6	3	D7

STX

Store Index Register in Memory

STX

Operation: M — X

Description: Stores the contents of X in memory. The contents of X remain the same.

Condition Codes:

H: Not affected.
I: Not affected.
N: Set if the most significant bit of the index register is set; cleared otherwise.
Z: Set if all bits of the index register are clear; cleared otherwise.
C: Not affected.

Addressing Mode	Cycles HMOS	CMOS	Bytes	Opcode
Inherent				
Relative				
Accumulator				
Index Register				
Immediate				
Direct	5	4	2	BF
Extended	6	5	3	CF
Indexed 0 Offset	5	4	1	FF
Indexed 1-Byte	6	5	2	EF
Indexed 2-Byte	7	6	3	DF

SUB

Subtract

SUB

Operation: ACCA — ACCA — M

Description: Subtracts the contents of M from the contents of ACCA and places the result in ACCA.

Condition Codes:

H: Not affected.
I: Not affected.
N: Set if the most significant bit of the result is set; cleared otherwise.
Z: Set if all bits of the results are cleared; cleared otherwise.
C: Set if the absolute value of the contents of memory are larger than the absolute value of the accumulator; cleared otherwise.

Addressing Mode	Cycles HMOS	CMOS	Bytes	Opcode
Inherent				
Relative				
Accumulator				
Index Register				
Immediate	2	2	2	A0
Direct	4	3	2	B0
Extended	5	4	3	C0
Indexed 0 Offset	4	3	1	F0
Indexed 1-Byte	5	4	2	E0
Indexed 2-Byte	6	5	3	D0

SWI

Software Interrupt

SWI

Operation: PC $\leftarrow$ PC + 0001
(SP) $\leftarrow$ PCL ; SP $\leftarrow$ SP $-$ 0001
(SP) $\leftarrow$ PCH ; SP $\leftarrow$ SP $-$ 0001
(SP) $\leftarrow$ X ; SP $\leftarrow$ SP $-$ 0001
(SP) $\leftarrow$ ACCA ; SP $\leftarrow$ SP $-$ 0001
(SP) $\leftarrow$ CC ; SP $\leftarrow$ SP $-$ 0001
I bit $\leftarrow$ 1
PCH $\leftarrow$ n $-$ 0003
PCL $\leftarrow$ n $-$ 0002

Description: The program counter is incremented (by one). The program counter, index register and accumulator are pushed onto the stack. The condition code register bits are then pushed onto the stack with bits H, I, N, Z, and C going into bit positions 4 through 0 with the top three bits (7, 6 and 5) containing ones. The stack pointer is decremented by one after each byte is stored on the stack.

The interrupt mask bit is then set. The program counter is then loaded with the address stored in the software interrupt vector located at memory locations n $-$ 0002 and n $-$ 0003, where n is the address corresponding to a high state on all lines of the address bus.

Condition Codes:

H: Not affected.
I: Set.
N: Not affected.
Z: Not affected.
C: Not affected.

Addressing Mode	Cycles		Bytes	Opcode
	HMOS	CMOS		
Inherent	11	10	1	83
Relative				
Accumulator				
Index Register				
Immediate				
Direct				
Extended				
Indexed 0 Offset				
Indexed 1-Byte				
Indexed 2-Byte				

TAX

Transfer Accumulator to Index Register

TAX

Operation: X — ACCA

Description: Loads the index register with the contents of the accumulator. The contents of the accumulator are unchanged.

Condition Codes: Not affected.

Addressing Mode	Cycles HMOS	CMOS	Bytes	Opcode
Inherent	2	2	1	97
Relative				
Accumulator				
Index Register				
Immediate				
Direct				
Extended				
Indexed 0 Offset				
Indexed 1-Byte				
Indexed 2-Byte				

TST

Test for Negative or Zero

TST

Operation:
X − 00 or,
ACCA − 00 or,
M − 0

Description: Sets the condition codes N and Z according to the contents of ACCA, X, or M.

Condition Codes:

H: Not affected.
I: Not affected.
N: Set if the most significant bit of the contents of ACCA, X, or M is set; cleared otherwise.
Z: Set if all bits of ACCA, X, or M are clear; cleared otherwise.
C: Not affected.

Addressing Mode	Cycles HMOS	CMOS	Bytes	Opcode
Inherent				
Relative				
Accumulator	4	3	1	4D
Index Register	4	3	1	5D
Immediate				
Direct	6	4	2	3D
Extended				
Indexed 0 Offset	6	4	1	7D
Indexed 1-Byte	7	5	2	6D
Indexed 2-Byte				

578

TXA

Transfer Index Register to Accumulator

TXA

Operation: ACCA ← X

Description: Loads the accumulator with the contents of the index register. The contents of the index register are unchanged.

Condition
Codes: Not affected.

Addressing Mode	Cycles		Bytes	Opcode
	HMOS	CMOS		
Inherent	2	2	1	9F
Relative				
Accumulator				
Index Register				
Immediate				
Direct				
Extended				
Indexed 0 Offset				
Indexed 1-Byte				
Indexed 2-Byte				

INDEX

INDEX OF IC NUMBERS